Next Generation Point-of-care Biomedical Sensors Technologies for Cancer Diagnosis

Pranjal Chandra • Yen Nee Tan
Surinder P. Singh
Editors

Next Generation Point-of-care Biomedical Sensors Technologies for Cancer Diagnosis

Springer

Editors
Pranjal Chandra
Department of Biosciences and Bioengineering
Indian Institute of Technology Guwahati
Guwahati
Assam
India

Yen Nee Tan
Institute of Materials Research and Engineering
A*STAR 2 Fusionopolis Way
Singapore
Singapore

Surinder P. Singh
Electron and Ion Microscopy Group
CSIR- National Physical Laboratory
New Delhi
India

ISBN 978-981-10-4725-1 ISBN 978-981-10-4726-8 (eBook)
DOI 10.1007/978-981-10-4726-8

Library of Congress Control Number: 2017959568

Printed on acid-free paper

This Springer imprint is published by Springer Nature
The registered company is Springer Nature Singapore Pte Ltd.
The registered company address is: 152 Beach Road, #21-01/04 Gateway East, Singapore 189721, Singapore

The book is dedicated in honour of
Prof. Yoon Bo Shim
Distinguished Professor and Director
Department of Chemistry and Institute of Biophysio Sensor Technology
Pusan National University, South Korea
on his
65th Birth Anniversary

Preface

Cancer today is one of the leading causes of mortality across the globe that can affect humans in more than 100 diverse forms. Conventional diagnostic approaches, biopsy for instance, have been used for several decades, but their limitations of being invasive and less sensitive have rendered them to be anachronistic. Seeing the upsurge in the incidence of cancer cases, its early diagnosis with high sensitivity in onsite mode has become extremely important.

This book contains 16 chapters that exclusively focus on different tactics of cancer diagnosis and prognosis. It provides a comprehensive fundamental understanding of different tools for cancer detection based on different tumour biomarkers and cancer cells. A detailed account of state-of-the-art cancer diagnostic approaches starting from labelled biosensors, label-free biosensors, implantable biosensors, integrated microfluidics systems, lateral flow devices, and biosensors based on application of various nano-biomaterials has been well stated in this book. Furthermore, development procedures of these diagnostic approaches along with their benefits, shortcomings, and future prospects are described in detail. This volume encompasses several illustrations and writing style is pedagogical to enable better understanding. The book can be used not only in formal courses at senior graduate level but also for self-study as the writing is very simple, interesting, and informative. The approach of this book is to generate a meticulous outlook of available cancer biosensors with an insight of new prospects.

Necessary compromises have been made between depth and breadth of different topics to give away a book of reasonable size. However, no compromises have been made in terms of delivering relevant information so that readers get full advantage of being enlightened.

The authors have received help from their colleagues and friends in country and overseas throughout the process of editing this book. We are especially grateful to our laboratory students and post-doctoral fellows for their diligent assistance with myriad details of preparation and production. We also thank our families for giving us the time, space, and freedom required to undertake and accomplish such an enormous project. Yet again, we owe thanks to many others who have directly or indirectly contributed to this book.

Guwahati, Assam, India — Pranjal Chandra
Singapore, Singapore — Yen Nee Tan
New Delhi, Delhi, India — Surinder P. Singh

Contents

About the Editors

Pranjal Chandra is currently employed as Assistant Professor and principal investigator at the Department of Biosciences and Bioengineering, Indian Institute of Technology Guwahati, Assam, India. He earned his Ph.D. from Pusan National University, South Korea, and did post-doctoral training at Technion-Israel Institute of Technology, Israel. He has published over 50 research articles in reputed journals and 1 book (IET, London). He is also a visiting scientist at IBST, South Korea. Pranjal's research contributions are highly interdisciplinary, spanning a wide range in nanobiotechnology, nanobiosensors, lab-on-chip systems for biomedical diagnostics, and nanomedicine. His work has been highlighted in the World news of the Royal Society of Chemistry, Cambridge, as "A new system for cancer detection" and also featured as a key scientific article in the Global Medical Discovery news, Canada. He is recipient of many prestigious awards and fellowships such as Ramanujan fellowship (Government of India), BK-21 and NRF fellowship of South Korea, Technion post-doctoral fellowship, Israel, University of Montreal Post-doc fellowship, Canada, and NMS Young scientist Award (2016). He is also an editorial board member of a dozen international journals including *World Journal of Methodology*, USA; *Frontiers in Bioscience*, USA; and *Journal of Biosensors and Bioelectronics*, USA.

Yen Nee Tan is the Principal Investigator and Program Manager of the Biomimetic and Biomedical Materials at the Institute of Materials Research and Engineering, under the Agency of Science, Technology and Research (A*STAR), Singapore. She is also the Adjunct Assistant Professor of Chemistry at the National University of Singapore. Her research focuses on the design of biogenic smart materials inspired by Nature and development of ultrasensitive technologies for biomedical diagnostics, food sensing, and environmental monitoring. She holds ten patents and two licenses on nanobiosensors. She is an editorial board member and guest editor of several scientific journals, such as *Austin Journal of Biosensors and Bioelectronics*, USA. Dr. Tan is the recipient of more than ten scientific awards, including the Young Giants of NanoScience Early Career Award 2016 (Hong Kong), Best Presenter Award at Asian Conference on Chemical Sensors 2015 (Malaysia), Finalist of L'Oreal for Women in Science National Fellowship 2013 (Singapore), Tan Kah Kee Young Inventors' Award 2012 (Singapore), Biomedical Engineering Society Best Technical Presentation Award 2011 (Singapore), and AsiaNANO Young Researcher Award 2010 (Japan).

Surinder P. Singh obtained his M.Sc. (1992) and Ph.D. (1998) degrees from G.B. Pant University of Agriculture & Tech., Pantnagar, India, in Physics. Currently he is working as a Sr. Scientist at National Physical Laboratory (CSIR), New Delhi, India. He served as Assistant Professor in Engineering Science and Materials Department at the University of Puerto Rico, Mayaguez, USA (2008–2011). His efforts at CSIR-NPL have initiated a national activity on biomedical instrument standardization. Dr. Singh is the project director of an Indo-US Joint centre on nanomedicine for Head and Neck Cancer in collaboration with Northeastern University and DFCI, Harvard medical school. He has 91 publications in his record. He has attended several national and international conferences and has delivered several invited talks. His research group is working on nanomaterials, optoelectronics, nanomedicine, nanobiointerface, bio-implants, sensors, and biosensors for various biomedical and environmental applications.

Cancer Biomarkers: Important Tools for Cancer Diagnosis and Prognosis

1

Ganesan Padmavathi, Devivasha Bordoloi, Kishore Banik, and Ajaikumar B. Kunnumakkara

1.1 Introduction

Cancer is one of the deadliest forms of diseases causing more than eight million deaths annually worldwide. Cancer can affect any organ and part of our body. Developed countries were reported to have the highest incidence rate of all cancers and the incidence rate of lung, colorectal, breast, and prostate cancers is high in developing and undeveloped nations as well (Torre et al. 2016). Moreover, in recent years, cancer incidence is increasing exponentially due to the drastic changes in lifestyle. Supporting this fact, approx. 90–95% of all cancers are known to be caused by environmental and lifestyle factors. For instance, cigarette smoking and smokeless tobacco, poor diet, consumption of alcohol, sustained exposure to environmental pollutants, infections, stress, obesity, and physical inactivity are some of the well-known risk factors for several commonly occurring cancers (Anand et al. 2008).

Depending on the cells, tissues, and organs involved, more than 200 types of cancers have been identified so far; among them cancers of lung, breast, colorectal, stomach, liver, cervical, head and neck, blood cells, etc. are the most commonly occurring cancers. Classification of cancer is extremely complex due to the occurrence of a wide variety of cancers arising in various tissues of the human body (Table 1.1). They can be categorized based on their primary site of origin or depending upon their tissue types (Guruvayoorappan et al. 2015). Notably, different body tissue types result in different tumors which can be either benign or malignant. Histologically, different types of cancers are grouped into six main classes:

G. Padmavathi • D. Bordoloi • K. Banik • A.B. Kunnumakkara, Ph.D. (✉)
Cancer Biology Laboratory, Department of Biosciences and Bioengineering, Indian Institute of Technology Guwahati, Guwahati, Assam 781039, India
e-mail: kunnumakkara@iitg.ernet.in, ajai78@gmail.com

P. Chandra et al. (eds.), *Next Generation Point-of-care Biomedical Sensors Technologies for Cancer Diagnosis*, DOI 10.1007/978-981-10-4726-8_1

Table 1.1 Different types of cancers

Cancers	Types
Adrenal cancer	–
Anal cancer	–
Astrocytomas	–
Bile duct cancer	Cholangiocarcinoma
Bladder cancer	–
Bone cancer	Ewing sarcoma, osteosarcoma, and malignant fibrous histiocytoma
Brain cancer	Atypical teratoid/rhabdoid tumor, embryonal tumors, germ cell tumor, craniopharyngioma and ependymoma
Breast cancer	Ductal carcinoma in situ (DCIS)
Cancer of unknown primary	–
Cardiac (heart) tumors	–
Cervical cancer	–
Endometrial cancer	–
Esophageal cancer	–
Eye cancer	Intraocular melanoma and retinoblastoma
Gallbladder cancer	–
Gastrointestinal cancer	Colorectal cancer, carcinoid tumor, and stomach cancer
Head and neck cancer	Hypopharyngeal cancer, laryngeal cancer, lip and oral cavity cancer, mouth cancer, nasal cavity and paranasal sinus cancer, nasopharyngeal cancer, and salivary gland cancer
Kidney cancer	Renal cell cancer, Wilms tumor, and transitional cell cancer
Leukemia	Acute lymphoblastic leukemia (ALL), acute myeloid leukemia (AML), chronic lymphocytic leukemia (CLL), chronic myelogenous leukemia (CML), chronic myelomonocytic leukemia (CMML), and hairy cell leukemia
Liver cancer	–
Lung cancer	Non-small-cell and small-cell lung cancers
Lymphoma	Cutaneous T-cell lymphoma, Hodgkin lymphoma, non-Hodgkin lymphoma
Malignant mesothelioma	–
Multiple myeloma	–
Neuroblastoma	–
NUT midline carcinoma	–
Ovarian cancer	–
Pancreatic cancer	–
Parathyroid cancer	–
Penile cancer	–
Pituitary tumor	–
Prostate cancer	–
Skin cancer	Melanoma, Merkel cell carcinoma, basal and squamous cell skin cancer

Table 1.1 (continued)

Cancers	Types
Soft-tissue tumors	Kaposi sarcoma, gastrointestinal stromal tumors (GIST), rhabdomyosarcoma, and vascular tumors
Testicular cancer	–
Thyroid cancer	–
Uterine cancer	–
Vaginal cancer	–
Vulvar cancer	–

1.1.1 Carcinoma

Carcinoma, which refers to the malignancies of epithelial tissue, constitutes around 80–90% of all cancer cases. They are mainly of two types such as adenocarcinoma and squamous cell carcinoma. The organs or glands affected by carcinoma are often capable of secretion including breast, lung, colon, prostate, and bladder (Guruvayoorappan et al. 2015; Cancer Types—National Cancer Institute).

1.1.2 Sarcoma

Sarcoma originates in supportive and connective tissues including bones, tendons, cartilage, muscle, and fat, for example osteosarcoma, chondrosarcoma, and leiomyosarcoma (Guruvayoorappan et al. 2015; Cancer Types—National Cancer Institute).

1.1.3 Myeloma

Myeloma is another cancer type which originates in the bone marrow plasma cells and is mostly associated with the overproduction of immature white blood cells, for example myelogenous or granulocytic leukemia, or lymphatic, lymphocytic, or lymphoblastic leukemia (Guruvayoorappan et al. 2015; Cancer Types—National Cancer Institute).

1.1.4 Lymphoma

Lymphomas, the "solid cancers," develop in the nodes or glands of the lymphatic system, particularly in spleen, tonsils, and thymus which are involved in the purification of bodily fluids and production of lymphocytes (Guruvayoorappan et al. 2015; Cancer Types—National Cancer Institute).

1.1.5 Leukemia

Leukemia begins in the blood forming tissue of the bone marrow and do not form solid tumors. The blood and the bone marrow develops abnormal white blood cells.

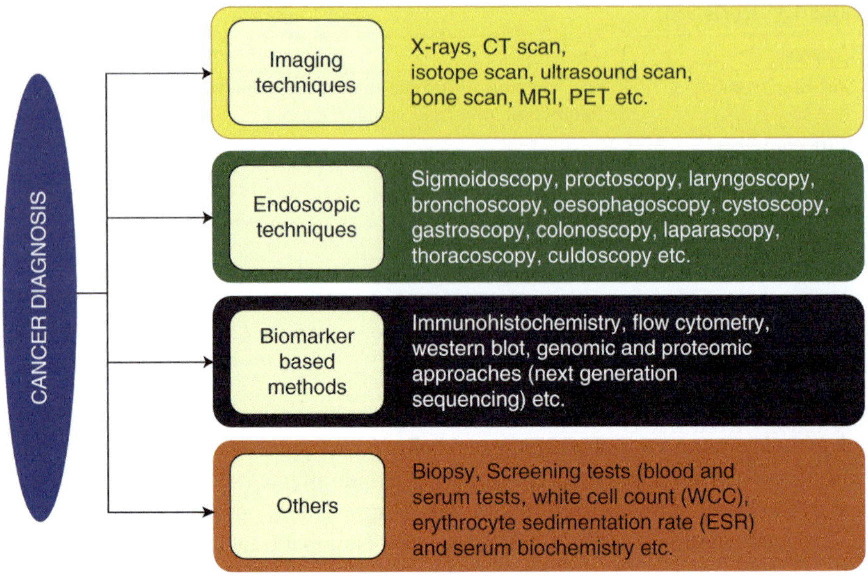

Fig. 1.1 Tools available for cancer diagnosis

It is mainly classified into two types lymphoblastic leukemia and myeloid leukemia (Guruvayoorappan et al. 2015; Cancer Types—National Cancer Institute).

1.1.6 Mixed Types

Adenosquamous carcinoma, mixed mesodermal tumor, carcinosarcoma, and teratocarcinoma are the mixed types of cancers (Guruvayoorappan et al. 2015; Cancer Types—National Cancer Institute).

1.2 Cancer Diagnosis

Cancer diagnosis has a great impact on the treatment outcome as late diagnosis is the major cause for treatment failure in most of the cancers (Richards 2009; Tørring et al. 2013). Therefore, numerous methods have been developed for the diagnosis of this disease including both invasive and noninvasive methods. Screening tests and imaging tests are the commonly employed techniques for cancer diagnosis where screening tests assist in predicting the risk of developing cancer in near future in the absence of any symptoms and imaging tests are useful in patients with symptoms of cancer (Fig. 1.1).

1.3 Imaging Techniques

Various imaging techniques used for cancer diagnosis include X-rays, CT scans, ultrasound scans, isotope scans, MRI scans, and PET scans. These techniques allow us to know the site and size of the tumor.

1.3.1 X-Rays

X-rays are used to detect tumors from lung, bowel, and bones. Detection of tumor is based on the difference in the permeability to X-rays. For instance, in case of lung imaging, normal lungs produce a relatively dark shadow on the X-ray film as X-rays are easily permeable through air. However, this technique is not suitable for detecting tumors from muscle and fat as they do not have any significant difference in permeability from that of normal muscle or fat tissues. Barium and iodine are used as contrasting materials that help in the X-ray imaging. Myelogram, radiographic screening, mammography, and angiography are some of the modified and sophisticated X-ray imaging methods used in cancer diagnosis (Stephens and Aigner 2009).

1.3.2 Computerized Tomography (CT) Scan

In CT scan, a series of cross-sectional X-ray images are taken in different angles and these images are combined using computer to produce a more detailed three-dimensional image which will clearly show position, size, and shape of the organ and presence of any abnormal tumor with more accuracy than that of plain X-ray images. This helps in assessment of tumors from head, abdomen, chest, limbs, and major blood vessels. However, it requires highly specialized equipment and skilled personnel and therefore it is expensive (Stephens and Aigner 2009).

1.3.3 Isotope Scans

Similar to X-ray imaging, isotope scans also record the shadows of a radioactive source on a film plate. However, the radioactive isotopes are injected into peripheral veins and allowed to spread through bloodstream. This isotope scan reveals the position, size, shape, and abnormal cellular activity in a particular organ which could be the result of cancer. The radioisotope is selected depending on the type of organ being examined. For example, radioactive iodine is used for thyroid gland, technetium is used for bone, and radioactive gallium is used for lymph nodes (Stephens and Aigner 2009).

1.3.4 Ultrasound Scans

Other imaging techniques including X-ray, CT scan, and isotope scan use harmful radiations for imaging purpose. Though the technical advancements in recent years tremendously reduced the dosage of radiation to be used, a better alternative would be more helpful to avoid any unwanted complications and side effects. Therefore, ultrasound waves are used as a safe alternative for these methods. Unlike the other methods, ultrasound waves can be used for imaging pregnant uterus or active ovaries as well. It produces images of tissues and organs comparable to that of CT scan; however it does not give more detailed information like CT scan. It relies on the change in reflection of sound waves by tissues of different nature or qualities. It is more suitable for detecting cyst type of lesions (Stephens and Aigner 2009).

1.3.5 Magnetic Resonance Imaging (MRI)

Similar to CT scan, MRI also produces three-dimensional cross-sectional images of the tissue being analyzed; however it uses a different principle. Computer-aided analysis of absorption and penetration of high-frequency radio waves by water molecules in a powerful magnetic field is the basis of MRI. It is considered to be better than CT scan for bone, muscle, brain, and spinal cord imaging. Moreover, it is preferred over the CT scan as it did not include the use of any damaging radiations such as X-rays. However, combination of both CT and MRI scans is used in cancer diagnosis as it would give more detailed information about the location, size, shape, consistency, and extent of spread of a tumor (Stephens and Aigner 2009).

1.3.6 Positron Emission Tomography (PET) Scan

Positron emission tomography (PET) scan is a more sophisticated noninvasive diagnostics where a radioactive drug (tracer) injected, swallowed, or inhaled is used to differentiate the cells with differential chemical activity. Cancer cells are found to be more active than the normal cells. For instance, increased uptake of glucose is associated with cancer cells than the normal cells. Such differential activities of cancer cells are used in PET scan to differentiate them from the normal cells unlike the other imaging techniques. It gives information about the extent of disease and metastatic status and also helps in knowing the response after therapy. Despite being the safest diagnostic technique, the use of PET is limited as it is highly expensive and less explored (Stephens and Aigner 2009).

1.4 Endoscopic Techniques

Endoscopes are used to analyze the internal organs such as esophagus, larynx, bladder, colon, lower bowel, rectum, anus, uterus, prostate, and peritoneal and pleural cavities. They help to visualize the organs, lesions, polyps, and tumors and also they are used for surgery and biopsy of tissues under study. Depending on the requirement, several types of endoscopies including rigid and flexible scopes such as sigmoidoscopy, proctoscopy, vaginal speculum, laryngoscopy, bronchoscopy, esophagoscopy, cystoscopy, gastroscopy, colonoscopy, laparoscopy, thoracoscopy, and culdoscopy are in practice. Most of these endoscopes contain a series of lenses, mirrors, and a light source which are aiding in visualization (Stephens and Aigner 2009).

1.4.1 Biopsy

Procedure of excising a small piece of tissue from tumor for analysis purpose is termed as biopsy and microscopic analysis of biopsy tissue is the ultimate test for diagnosing and characterizing cancer. It is also done for staging, grading, and classifying the type of cancers. In certain cases, biopsy is also used to decide the fate of treatment. Needle aspiration or "punch-out" biopsy, aspiration cytology, bone marrow biopsy, standard paraffin section biopsy, and frozen section biopsy are few of the methods used for biopsy (Stephens and Aigner 2009).

1.5 Screening Tests

Aforementioned, screening tests assist in predicting the risk of developing cancer in near future in the absence of any symptoms. They can be direct or indirect evidence of cancer. In general, cancer patients are known to have anemia due to reduced hemoglobin and RBC counts, change in WBC count, increased erythrocyte sedimentation rate (ESR), and altered serum components. Therefore, methods that determine these properties would give preliminary information about the disease. Such tests include blood and serum tests, white cell count (WCC), erythrocyte sedimentation rate (ESR), and serum biochemistry. However, most of these parameters would be more evident only in advanced stages of cancer, thus requiring more promising methods for the diagnosis and prognosis of cancer (Stephens and Aigner 2009).

Moreover, all the above mentioned techniques have their own disadvantages like labor intensive, time consumption, lack of specificity, side effects, cost, and complexity. Therefore, simple, cost-effective, and point-of-care methods are required for the early detection and treatment of cancer (Tothill 2009).

1.6 Importance of Tumor Biomarkers in the Management of Cancer

The common flaw in all the aforementioned diagnostic methods is the lack of accuracy. Therefore, for the past several decades researchers are involved in the identification of biomarkers that are exclusive for cancer cells thus helping in differentiating the tumor cell from the normal one ultimately facilitating the diagnosis of any particular cancer with great accuracy and specificity. As a result, several hundreds of novel biomarkers have been identified to be involved in the development of various cancers. However, only very few have been used in clinic for the diagnostic and prognostic purposes. In general, biomarkers are the cellular components present in tumor cells, blood, urine, or other body fluids that are overexpressed due to the commencement of disease (Rhea and Molinaro 2011).

Development of cancer is a multistage process which generally occurs in three defined stages namely initiation, promotion, and progression and these stages are associated with several molecular alterations in genomic, proteomic, and metabolomics levels (Fig. 1.2) (Pitot 1993; Hanahan and Weinberg 2011). Initiation occurs due to a defect in the cellular genome triggered by a genotoxic agent, whereas promotion involves an epigenetic change in genome expression and cell division (Tubiana 1989; Pitot 1993). Congregate evidences suggest that human cancers possess multiple genetic alterations caused by point mutations, recombinations, amplifications, deletions, etc. affecting both oncogenes and tumor-suppressor genes. Furthermore, diverse regulatory elements are also found to be altered during the process of multistage tumor progression, which primarily include the control of proliferation, balance between cell survival and apoptosis, interaction with nearby cells and the extracellular matrix, induction of angiogenesis, and ultimately cancer cell migration, invasion, and metastasis (Compagni and Christofori 2000). For example, genetic and epigenetic modification of transcription factors such as NF-κB, STAT, Notch, and PPAR-γ; growth factors including EGF, FGF, VEGF, TGF-β, and TF; protein kinases such as JAK, PKA, PKB, PKC, EGFR, ERK, MAPK, IKK, and PTK; apoptotic proteins viz. caspases, Bax, Bid, and PARP; survival and proliferative proteins such as Bcl-2, Bcl-xL, c-myc, survivin, inflammatory cytokines including interleukins, interferons, TNF, COX-2, and prostaglandins; adhesion molecules such as ICAMs, VCAMs, and MMPs; cell cycle regulators including cyclins and cyclin-dependent kinases; tumor-suppressor genes such as p53, BRCA1, BRCA2, and pRB; and oncogenes such as Ras, Raf, and Src are involved in the development of various cancers. Such molecular alterations are used as early diagnostic and treatment biomarkers for many cancers.

In addition, the molecules are regarded as diagnostic biomarkers if it explains the presence of a cancer and its type and the risk of developing cancer in future; prognostic biomarkers if it predicts the response of patient without treatment; predictive biomarkers if it predicts the outcome of treatment and tumor recurrence; and therapeutic biomarkers which are used as target for cancer therapy. Next part of this chapter discusses the important tumor biomarkers and their impact on the management of cancer.

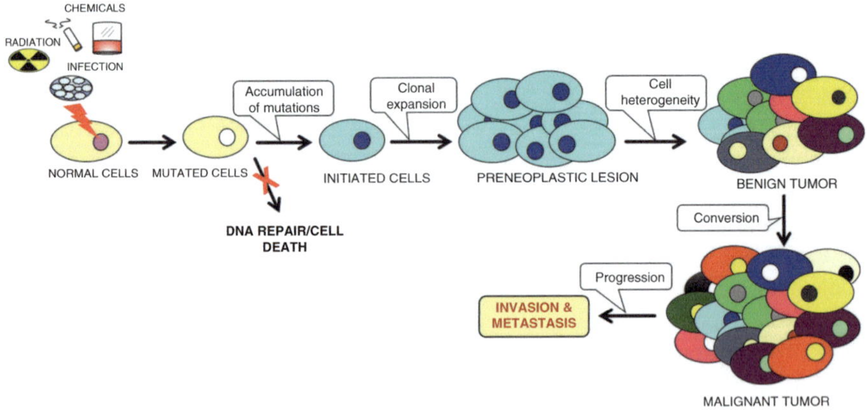

Fig. 1.2 Different stages of cancer progression

1.7 Transcription Factors

Proper control of gene expression is essential for maintaining the integrity of a cell and for the better maintenance of body physiology. This control is achieved by extremely complicated and interlinked molecular pathways. Ultimately, expression of any particular gene is controlled by complex of transcription factor proteins. Therefore, deregulation of transcription factors would critically affect the gene expression leaving it unregulated which leads to several developmental abnormalities and diseases including cancer. Deregulation of transcription factors was found to be constantly associated with the development of cancer. For example, NF-κB, a key transcription factor, is well known to be involved in the tumorigenesis of numerous malignancies. Similarly, other transcription factor families such as AP-1, ATF, FOS, HOX, JUN, MAF, and STAT also play a vital role in the initiation and progression of bladder cancer, breast cancer, colon cancer, endometrial cancer, esophageal cancer, gastric cancer, glioblastoma, leukemias, liver cancer, lung cancer, lymphomas, melanoma, mesothelioma, multiple myeloma, nasopharyngeal cancer, neuroblastoma, ovarian cancer, pancreatic cancer, prostate cancer, sarcomas, and thyroid cancer (Kontos et al. 2013; Ozanne et al. 2007; Vaiopoulos et al. 2010; Milde-Langosch 2005; Greene et al. 2009; Pradhan et al. 2012; Kannan et al. 2012; Eychène et al. 2008; Hurt et al. 2004; Rayet and Gélinas 1999; Subramaniam et al. 2013; Slattery et al. 2013; Gadducci et al. 2013; Karamouzis et al. 2007; Alharbi et al. 2013; Piérard and Piérard-Franchimont 2012; Kelly et al. 2011; Cantile et al. 2011). All these transcription factors were known to greatly impact the biology of a tumor, thus proving to be potential biomarkers for either diagnosis or treatment. Supporting this fact, thyroid transcription factor 1 (TTF-1) is used as a biomarker in the diagnosis of thyroid and lung carcinomas as it is mostly specific to these organs. However, expression of TTF-1 was also reported in ovarian, endometrial, colon, and breast cancers signifying the requirement of additional confirming factors for the diagnosis purpose (Ordóñez 2012). Distal-less homeobox 2 (DLX2) is a homeodomain transcription factor protein that was shown to be associated with progression of gastric cancer marking it to be a potential biomarker for gastric cancer (Zhang et al. 2016). EN2, another homeobox-containing transcription factor, is used to diagnose prostate cancer as it is specifically detected in the urine of patients with prostate cancer (McGrath et al. 2013). Likewise, specificity protein (Sp) family of transcription factors was found to be potential therapeutic targets for pancreatic cancer (Sankpal et al. 2012). Therefore, it is apparent that differential expression of transcription factors would be an effective biomarker for the diagnosis and prognosis of various cancers if precise detection is achieved through simple and cost-effective techniques.

1.8 Growth Factors and Their Tyrosine Kinase Receptors

By definition, cancer is the uncontrolled growth of abnormal cells which is ultimately governed by deregulation of growth factor signaling networks. Colony-stimulating factors (CSF), epidermal growth factor (EGF), fibroblast growth factor (FGF), hepatocyte growth factor (HGF), hepatoma-derived growth factor (HDGF), insulin, insulin-like growth factors (IGFs), interleukins, platelet-derived growth factor (PDGF), transforming growth factors (TGFs), tumor necrosis factor-alpha (TNF-α), and vascular endothelial growth factor (VEGF) are some of the growth factors that control proliferation, cell cycle progression, cellular differentiation, and apoptosis when interact with their respective receptors. Receptors of growth factors are often tyrosine kinase proteins which get activated upon binding with specific ligands through auto-phosphorylation. Activated tyrosine kinase receptors further initiate the phosphorylation-mediated activation of several downstream pathways including Raf/MEK/ERK and PI3k/Akt which in turn activates transcription factors ultimately leading to increased expression of proteins essential for survival, proliferation, and differentiation. Increased activation of these receptors in the absence of ligand was found to be the underlying mechanism for the development of cancer. In line with this, overexpression and/or mutation of epidermal growth factor receptor (EGFR), vascular endothelial growth factor receptor (VEGFR), platelet-derived growth factor (PDGFR), and insulin-like growth factor 1 receptor (IGF-1R) were associated with initiation and progression of numerous cancers (Wallace et al. 2007; Blume-Jensen and Hunter 2001; Porter and Vaillancourt 1998; Robertson et al. 2000). Moreover, some of the receptors are exclusive for certain cancers marking them as potential biomarkers for diagnosis and treatment. For instance, the chimeric tyrosine kinase receptors Bcr-Abl and PML-RARa are highly specific for chronic myeloid leukemia and acute promyelocytic leukemia (APL) respectively and are therefore used as diagnostic and therapeutic markers for these diseases (Sawyers 1999; De Braekeleer et al. 2014). Likewise, EGFR, HER2/neu, and mutated ALK are also used as tumor biomarkers for non-small-cell lung cancer, breast cancer, and anaplastic large-cell lymphoma (Allred 2010; Zhao et al. 2013; Mossé 2016). All these findings suggest the immense potential of receptor tyrosine kinases (RTKs) in the diagnosis and treatment of various cancers. However, out of the numerous RTKs found to be involved in tumorigenesis, only very few are progressed into tumor biomarkers in clinic implicating the essentialness of more advanced research which would produce added simple point-of-care assays for cancer diagnosis.

1.9 Inflammatory Cytokines

The link between inflammation and cancer development has become apparent in recent years. Supporting this, bacterial and viral infection-induced inflammation was evidenced to increase the risk of cancer development (de Martel and Franceschi 2009; Grivennikov et al. 2010; Grivennikov and Karin 2011). Likewise, other risk factors of cancer such as tobacco smoke and obesity are also known to induce cancer through inflammation (Takahashi et al. 2010; Park et al. 2010; Khasawneh et al. 2009; Grivennikov et al. 2010). Also, the tumor microenvironment was proved to be filled with several immune cells including dendritic cells, macrophages, mast cells, natural killer cells, neutrophils, and T and B lymphocytes which play a key role in cancer initiation and progression (Grivennikov et al. 2010; Grivennikov and Karin 2011; de Visser et al. 2006; Maletzki and Emmrich 2010). Moreover, the impact of inflammation on cancer development was found to be mediated through proinflammatory cytokines produced by the immune cells in the tumor microenvironment (Grivennikov and Karin 2011). Interestingly, it was also established that the inflammatory cytokines are actually involved in tumorigenesis rather than the immune cells. Inflammatory cytokines could be both anti-tumorigenic and pro-tumorigenic; however in most cases pro-tumorigenic cytokines dominate. For example, IL-1β, IL-6, IL-12, IL-17, IL-23, TRAIL, FasL, TNF-α, VEGF, EGFR ligands, and TGF-β are few of the cytokines that aid in tumor growth (Grivennikov et al. 2010; de Visser et al. 2006; Maletzki and Emmrich 2010). Therefore, increased levels of these cytokines indicate high risk of cancer initiation and progression which would be helpful in the diagnosis and in predicting the prognosis of cancer. Accordingly, increased levels of IL-6 and TNF-α in the serum and tissue samples signify the risk of colorectal adenomas and associated poor prognosis. They are also proved to be a potent therapeutic target for colorectal adenomas. Similar impact was observed in liver cancer as well (Grivennikov and Karin 2011; Kim et al. 2008; Wong et al. 2009). In addition to the inflammatory cytokines, inflammatory cells and costimulatory molecules in tumor cells are also being used as biomarkers for cancer diagnosis, prognosis, and treatment outcome. All these biomarkers are collectively known as inflammatory biomarkers. Further evidences reported that inflammatory cells including CD3 (cytotoxic T lymphocyte), CD8 (cytotoxic T lymphocyte), FOXP3 (regulatory T lymphocyte), CD68 (macrophages), and CD83 (dendritic cells); costimulatory molecules like PDL1, HLA, and HSP; and serum cytokines such as C-reactive protein (CRP), IL-6, and TNF-α are potential biomarkers for bladder cancer (Masson-Lecomte et al. 2014). Despite the identification of numerous potential inflammatory biomarkers, they failed to achieve successful clinical application due to unavailability of effective methods and assays, technical challenges, and improper validation of the results (Masson-Lecomte et al. 2014). Taken together, there is an urgent need for the development of effective methods for the detection of these inflammatory biomarkers which would have a great impact on cancer diagnostics.

1.10 FDA-Approved Tumor Biomarkers and Their Implications in Cancer Management

Aforementioned, tumor biomarkers can be of proteins, antibodies, nucleic acids, peptides, carbohydrates, or circulating tumor cells that can be either noninvasively assessed through circulatory body fluids, excretions, and secretions or assessed through biopsy or imaging of tissue (Henry and Hayes 2012; Scatena 2015). Among the several hundreds of biomarkers identified so far only very few have been approved by the Food and Drug Administration (FDA) and used in clinic for the diagnosis of specific cancers (Rhea and Molinaro 2011; Diamandis 2014). Some of the cancer biomarkers currently being used are listed below (Table 1.2) (Tumor Markers—National Cancer Institute):

- α-Fetoprotein (AFP) for non-seminomatous testicular cancer and hepatocellular carcinoma (Debruyne and Delanghe 2008; El-Bahrawy 2010; Wong et al. 2015)

Table 1.2 The important biomarkers used for the diagnosis of cancers

Biomarker	Cancer	Sample	Normal level	Reference
PSA	Prostate cancer	Serum	<20 ng/mL	Kamalov et al. (2012)
CA15.3	Metastatic breast cancer	Serum	<45 U/mL	O'Brien et al. (1994), Nicolini et al. (2015)
ER/PR/HER2	Breast cancer	Tumor tissue	<1% positive stain	Yip and Rhodes (2014)
CEA	CRC and lung cancer	Serum	<5 ng/mL (the cutoff ranges from 2.5 ng/mL to 40 ng/mL for diff. studies)	Swiderska et al. (2014), Grunnet and Sorensen (2012)
CA 19–9	CRC	Serum	<23.9 U/mL (the cutoff varied between 37 U/mL and 1000 U/mL for diff. studies)	Zhang et al. (2015), Grunnet and Mau-Sørensen (2014)
CA 125	Ovarian cancer	Serum	<35 U/mL	Bottoni and Scatena 2015
HE4	Ovarian cancer	Serum	<150 pM	Bottoni and Scatena (2015)
Tg	Thyroid cancer	Serum	<2 ng/mL (post-thyroidectomy)	Grebe (2009)
AFP	HCC	Serum	5–10 μg/L	Debruyne and Delanghe (2008)
Calcitonin	MTC	Blood	<10 pg/mL	Bae et al. (2015)
CgA	NET	Serum/ plasma	<34.7 U/L (10.5 U/L-avg. value)	Bajetta et al. (1999), Singh and Law (2012)

Abbreviations: *AFP* Alpha-fetoprotein, *CA 125* Cancer antigen 125, *CA 15–3* Cancer antigen 15–3, *CA 19–9* Cancer antigen 19–9, *CEA* Carcinoembryonic antigen, *CgA* Chromogranin A, *CRC* Colorectal cancer, *ER* Estrogen receptor, *HCC* Hepatocellular carcinoma, *HE4* Human epididymis protein 4, *HER2* Human epidermal growth factor receptor 2, *MTC* Medullary thyroid cancer, *NET* Neuroendocrine tumors, *PR* Progesterone receptor, *PSA* Prostate-specific antigen, *Tg* Thyroglobulin

- ALK gene for non-small-cell lung cancer and anaplastic large-cell lymphoma (Zhao et al. 2015; Mossé 2016)
- BCR-ABL fusion gene for chronic myeloid leukemia, acute lymphoblastic leukemia, and acute myelogenous leukemia (Sawyers 1999; Granatowicz et al. 2015; Voncken et al. 1995)
- Beta-2-microglobulin (B2M) for multiple myeloma, chronic lymphocytic leukemia, and some lymphomas (Bethea and Forman 1990)
- Beta-human chorionic gonadotropin (Beta-hCG) for choriocarcinoma and germ cell tumors (Sisinni and Landriscina 2015)
- BRAF V600 mutations for cutaneous melanoma and colorectal cancer (Lasota et al. 2014; Eklöf et al. 2013; Curry et al. 2012)
- BRCA1 and BRCA2 genes for ovarian cancer (Easton et al. 1995; Peshkin et al. 2001; Liu et al. 2012)
- CA15.3/CA27.29, estrogen receptor/progesterone receptor for breast cancer (Harris et al. 2007; EBCTCG 2011; Bast et al. 2001)

- CA-125 and human epididymis protein 4 (HE4) for ovarian cancer (Bottoni and Scatena 2015)
- CA19–9 for pancreatic cancer, gallbladder cancer, bladder cancer, bile duct cancer, and gastric cancer (Swiderska et al. 2014; Zhang et al. 2015; Shukla et al. 2006; Grunnet and Mau-Sørensen 2014; Wang et al. 2015)
- Calcitonin for medullary thyroid cancer (Bae et al. 2015; Brutsaert et al. 2015)
- Carcinoembryonic antigen (CEA) for breast, colorectal, gastrointestinal, lung, and pancreatic cancer (Harris et al. 2007; Duffy 2001; Grunnet and Sorensen 2012; Ballesta et al. 1995; Yasue et al. 1994)
- CD20 and CD3 for non-Hodgkin lymphomas (Kakinoki et al. 2015)
- Chromogranin A (CgA) for neuroendocrine tumors (Singh and Law 2012; Bajetta et al. 1999)
- C-kit/CD117 for gastrointestinal stromal tumor (Yamaguchi et al. 2004)
- Cytokeratin fragment 21–1 for lung cancer (Wieskopf et al. 1995)
- EGFR gene for non-small-cell lung cancer (Zhao et al. 2013)
- Fibrin/fibrinogen degradation product (FDP) for bladder cancer (Burchardt et al. 2000; Schmetter et al. 1997)
- HER2/neu for breast cancer, gastric cancer, and gastroesophageal junction adenocarcinoma (Allred 2010; Vakiani 2015; Park et al. 2015)
- Immunoglobulins or M protein for multiple myeloma (Rajkumar and Kumar 2016)
- KRAS for colorectal cancer and non-small-cell lung cancer (Allegra et al. 2009; Ying et al. 2015)
- Lactate dehydrogenase for germ cell tumors, lymphoma, melanoma, and neuroblastoma (Miao et al. 2013; Petrelli et al. 2015)
- Neuron-specific enolase (NSE) for small-cell lung cancer and neuroblastoma (Isgrò et al. 2015)
- Nuclear matrix protein 22 for bladder cancer (Burchardt et al. 2000)
- Programmed death ligand 1 (PD-L1) for non-small-cell lung cancer (Kerr et al. 2015)
- Prostate-specific antigen (PSA) for prostate cancer (Lin et al. 2008)
- Thyroglobulin (Tg) for thyroid cancer (Trimboli et al. 2015; Whitley and Ain 2004; Grebe 2009)
- Urokinase plasminogen activator (uPA) and plasminogen activator inhibitor (PAI-1) for breast cancer (Harris et al. 2007; Duffy et al. 2014)

Among these, few are used as diagnostic markers, few as prognostic, few as therapeutic, and few as predictive biomarkers for different cancers and assessed through either body fluids or biopsy. The following section briefs about few of the FDA-approved biomarkers and their use in cancer management.

1.10.1 Alpha-Fetoprotein (AFP)

Alpha-fetoprotein (AFP) is a plasma protein produced by yolk sac and fetal liver, highly expressed in maternal circulation during pregnancy and in human fetus. However, a tremendous reduction in the plasma levels of AFP is observed after birth to attain the normal range. Elevated levels of AFP are observed in adults with hepatocellular carcinoma (HCC) and non-seminomatous germ cell tumors. Therefore, it is used as tumor marker for diagnosis and follow-up of such cancers (Debruyne and Delanghe 2008; El-Bahrawy 2010). Evaluation of serum AFP level in high-risk patients is one of the conventional methods used for diagnosis and screening of HCC and considered as golden standard. Levels of AFP are measured by different immunoassays including radioimmunoassay (RIA), immunoradiometric assay (IRMA), ELISA, microparticle capture enzyme immunoassay (MEIA), nephelometry, and electrochemiluminescence (Debruyne and Delanghe 2008; Ruoslahti and Seppälä 1971; Chayvialle and Ganguli 1973; Suzuki 1988; Mancal et al. 1988; Fiore et al. 1988; Bernard et al. 1996; Blackburn et al. 1991). However, its diagnostic values are very limited due to lack of highly sensitive and specific assays for the detection of AFP. To overcome this issue, researchers introduced various AFP-related parameters such as AFP mRNA and AFP glycoforms. Interestingly, the AFP mRNA was found to have immense prognostic potential and AFP glycoforms to have diagnostic potential levels of which are measured by PCR-based techniques, isoelectric focusing, and lectin affinity electrophoretic methods (Debruyne and Delanghe 2008). Moreover, the FDA approved combination of lens culinaris agglutinin-reactive fraction of alpha-fetoprotein (AFP-L3, an isoform of AFP) with des-gamma-carboxy prothrombin (DCP) and AFP for the surveillance of HCC (Wong et al. 2015).

1.10.2 ALK Gene Abnormalities

Anaplastic lymphoma kinase (ALK) is an important oncogenic receptor tyrosine kinase, mutations and chromosomal rearrangements of which are found to be frequently involved in tumorigenesis of several malignancies including anaplastic large-cell lymphoma (ALCL), non-small-cell lung cancer (NSCLC), and neuroblastoma (Zhao et al. 2015; Mossé 2016). Expression of wild-type ALK is often found in developing neuronal tissues implicating its importance in the development of central and peripheral nervous system during fetal development and remarkably other normal tissues express no ALK. Therefore, abnormal expression and activation of ALK through point mutations or translocations are unique to the cancer cells and are thus being used as diagnostic, prognostic biomarkers and therapeutic targets for treating such neoplasms (Mossé 2016). Identification of t(2;5)(p23;q35) translocation resulting in the fusion of ALK with nucleophosmin (NPM) in non-Hodgkin's lymphoma led to the discovery of several other gene fusions and mutations involving ALK (Morris et al. 1994; Mossé 2016). Moreover, the NPM-ALK gene fusion was found to be the signature translocation of CD30+ ALCL and thus possesses a great diagnostic significance for ALCL (Shiota et al. 1995). Likewise, presence of another ALK fusion gene EML4-ALK in patients with congenital pulmonary airway malformation (CPAM) indicates the probability of developing lung adenocarcinoma in future, thus signifying its diagnostic potential for lung cancer (Tetsumoto et al. 2013). Expression of these fusion ALK genes is determined by IHC and molecular and cytogenetic techniques such as FISH (Mossé 2016). Moreover, due to their specificity, small-molecule ALK inhibitors exert significant anticancer properties and are used in the successful management of ALK-positive cancers implicating the role of ALK as therapeutic target as well (Mossé 2016). In addition, mutations in the ALK gene have been found to be the biomarker for poor prognosis in neuroblastoma patients (Mossé 2016; Bresler et al. 2014; Janoueix-Lerosey et al. 2008). R1060 between transmembrane and kinase domains of ALK, F1174, F1245, and R1275 mutations in kinase domain are the few examples of germline and somatic mutations observed in ALK gene of neuroblastoma patients (Bresler et al. 2014).

1.10.3 BCR-ABL

BCR-ABL is a chimeric tyrosine kinase receptor constitutively activated in chronic myeloid leukemia (CML). It is produced as a result of t(9;22)(q34;q11) translocation involving the long arms of chromosomes 9 and 22. Notably, the t(9;22) is the first identified chromosomal rearrangement which was known to produce the Philadelphia chromosome that was then considered as the driving factor for CML. Later, it was revealed that it is the fusion product of t(9;22) BCR-ABL that is responsible for the development of CML not the Philadelphia chromosome (Sawyers 1999; Granatowicz et al. 2015). Consequently, this chimeric protein is regarded as the characteristic feature of CML as approx. 95% of CML patients are found to carry the BCR-ABL oncogene. Therefore, ultimately the chimeric tyrosine kinase is used as a diagnostic biomarker for the detection of CML (Sawyers 1999; Granatowicz et al. 2015). In addition to serving as diagnostic biomarker, the BCR-ABL fusion kinase also aids in targeted therapy. For instance, imatinib (INN), a first-generation tyrosine kinase inhibitor targeted against BCR-ABL fusion kinase in fusion-positive CML patients, resulted in significant reduction of cancer burden (Al-Hadiya et al. 2014; Waller 2014). Moreover, it also serves as a prognostic biomarker for CML, as increased expression of BCR-ABL is often associated with poor survival (de França Azevedo et al. 2014). Apart from CML, BCR-ABL is also found to be associated with acute lymphoid leukemia (ALL) (Voncken et al. 1995).

1.10.4 BRAF V600

BRAF is a non-receptor serine/threonine-protein kinase, abnormally activated by gain-of-function mutations resulting in the activation of RAS/RAF/MAPK pathway. BRAF V600E is the most commonly found mutation of BRAF gene in a variety of cancers. Moreover, this BRAF V600 mutation is used as a prognostic and treatment target for colon cancer and melanoma (Lasota et al. 2014; Eklöf et al. 2013; Curry et al. 2012).

1.10.5 BRCA1 and BRCA2

BRCA1 and BRCA2 are tumor-suppressor genes involved in DNA repair, mutations in which are found to be associated with the development of breast and ovarian cancers and are discovered as germline mutations to be carried to the progenies (Mazoyer 2005; Sopik et al. 2015; Campeau et al. 2008). Further, it has been confirmed that women with mutations in BRCA genes are highly prone to these cancers. For instance, women carrying BRCA mutations are 80–90% prone to these cancers (Sopik et al. 2015; Ferla et al. 2007). Moreover, around 30–50% of hereditary breast and ovarian cancers are shown to have germline mutations in these tumor-suppressor genes. BRCA mutations accounts for 20–25% of hereditary breast cancers and 15% of ovarian cancers (Ferla et al. 2007; Szabo and King 1995; Pal et al. 2005; BRCA1 and BRCA2: Cancer Risk and Genetic Testing-National cancer institute). Therefore, analyzing the BRCA1 and BRCA2 status gives the essential information about the risk of developing breast and ovarian cancers implicating the use of BRCA gene mutations as effective biomarkers for screening, diagnosis, and prognosis of such cancers (Peshkin et al. 2001; Liu et al. 2012).

1.10.6 CA15.3/CA27.29

Carbohydrate antigen 15.3 (CA15.3) and CA27.29 are mucin-like tumor-associated glycoproteins secreted into serum and are widely used for the diagnosis and prognosis of breast cancer (Nicolini et al. 2015; Bast et al. 2001; Harris et al. 2007). The level of CA15.3 antigen was found to be increased with tumor progression and more than 45 U/mL of serum CA15.3 is regarded as an indicator of metastatic breast cancer (Nicolini et al. 2015; O'Brien et al. 1994). Besides breast cancer, CA15.3 was found to be overexpressed in ovarian, endometrial, and non-small-cell lung cancers as well (Molina et al. 2008; Nicolini et al. 2015). Likewise, another MUC-1-associated protein CA27.29 is also explored for its diagnostic potential. Interestingly, the diagnostic values of CA27.29 were found to be similar to those of CA15.3. However, unlike the CA15.3, it fails to differentiate between different stages of breast cancer (Gion et al. 2001; Gion et al. 1999; Nicolini et al. 2015; Hou et al. 1999). Moreover, these antigens are also used to predict the treatment outcome and to detect recurrence (Gion et al. 2001; Molina et al. 1995). In the past few decades, several assays including Truquant BR radioimmunoassay (RIA), AxSYM Abbott, and ACS:180 BR were developed and used for the detection of these tumor antigens (Gion et al. 2001). However, more sophisticated and easy assays are being designed every year for better precision and sensitivity.

1.10.7 CA19.9

Similar to CA15.3 and CA27.29, another carbohydrate antigen secreted into blood known as CA19.9 is also approved by the FDA as tumor marker for pancreatic, gallbladder, bile duct, and gastric cancers (Swiderska et al. 2014; Zhang et al. 2015; Shukla et al. 2006; Grunnet and Mau-Sørensen 2014). In addition, the level of this antigen was found to be higher in colorectal carcinoma (CRC) than the benign colorectal diseases suggesting its pivotal role in the diagnosis of CRC in combination with carcinoembryonic antigen (CEA). Moreover, combined analysis of the levels of CA 19.9 and CEA is used as a prognostic biomarker for CRC (Zhang et al.

2015; Swiderska et al. 2014). Further, it is also used as a prognostic biomarker for bladder cancer (Wang et al. 2015).

1.10.8 CA125 and HE4

CA125, a membrane mucin glycoprotein also recognized as mucin 16 or MUC16, is a widely used FDA-approved tumor marker for monitoring the treatment response, follow-up, and diagnosing recurrence of ovarian cancer. Serum level of CA125 is considered to be elevated if higher than 35 U/mL as the level in healthy individuals was found to be <35 U/mL (Bottoni and Scatena 2015). Other than ovarian cancer, increased levels of this antigen were observed in cancers of biliary tract, breast, cervical, colon, endometrial, fallopian tube, liver, lung, pancreatic, stomach, and uterine, thus limiting its specificity as ovarian cancer biomarker (Bottoni and Scatena 2015). Moreover, diagnosis of ovarian cancer in the asymptomatic early stages often fails with CA125 marker as the expression of the antigen is very less in contrast to the more than 90% increase observed in stage II, III, and IV diseases (Bottoni and Scatena 2015). Therefore, to overcome the lack of specificity and sensitivity of CA125, it is often used with human epididymis protein 4 (HE4), a small secretory protein recently approved for clinical use as biomarker for ovarian cancer. Around 80% of ovarian cancer patients were found to have elevated levels of HE4 (>150 pM) and unlike CA125, HE4 helps in differentiating the benign tumors from cancers. However, despite the dominating specificity of HE4 over CA125, estimation of both CA125 and HE4 serum levels is used in practice for accurate prediction of cancer (Bottoni and Scatena 2015).

1.10.9 Calcitonin

Calcitonin is an effective biomarker used for the diagnosis and follow-up of medullary thyroid cancer (MTC) (Bae et al. 2015; Brutsaert et al. 2015). Calcitonin is a hormone secreted only by the C-cells of thyroid gland, thus making it unique for MTC diagnosis (Bae et al. 2015). Serum concentration of <10 ng/L is considered to be the normal basal calcitonin level and >100 ng/L is regarded as an indicative factor for early MTC. In addition, for the follow-up after treatment, serum calcitonin level <10 ng/L indicates no residual tumor tissue; levels ranging from 10 to 150 ng/L indicate possible local disease in the neck and >150 ng/L indicate the possible distant metastases (Costante et al. 2009).

1.10.10 Carcinoembryonic Antigen (CEA)

CEA is a glycoprotein produced during fetal development which mediates cell adhesion. However, it has been found to be absent or very less in healthy individuals as the production of CEA stops before birth. Interestingly, increased serum levels of CEA have been found in certain cancers, thus distinguishing the cancer patients from healthy individuals. Therefore, CEA is widely used as a tumor marker for various cancers including gastrointestinal cancers, breast cancer, lung cancer, and pancreatic cancer (Harris et al. 2007; Duffy 2001; Locker et al. 2006; Grunnet and Sorensen 2012; Ballesta et al. 1995; Yasue et al. 1994). Due to lack of sensitivity and specificity, use of the CEA assay is limited for cancer diagnosis. However, it is used for prognosis, early diagnosis of recurrence, and follow-up of cancer patients. Especially, it is most commonly used for predicting the tumor recurrence and prognosis of colorectal cancers (Ballesta et al. 1995; Duffy et al. 2003).

1.10.11 CD20 and CD3

World Health Organization (WHO) classifies hematological malignancies based on the clinical manifestation, morphology, immunophenotype, and molecular genetics of the disease (Kakinoki et al. 2015). For example, CD20, a glycosylated phosphoprotein expressed by the B-cells, thus known as B-lymphocyte antigen CD20, is used as a unique immunomarker for classifying B-cell lymphomas (Kakinoki et al. 2015). Likewise, CD3 was found to be unique for T-cell lymphoma. Therefore, both CD20 and CD3 immunomarkers are used for the diagnosis of B- and T-cell lymphomas, respectively (Kakinoki et al. 2015).

1.11 Estrogen Receptor, Progesterone Receptor, and Human Epidermal Growth Factor Receptor 2

The estrogen, progesterone, and HER2/neu receptors are the most commonly used predictive and prognostic biomarkers of breast cancer. Immunohistochemistry (IHC) and fluorescent in situ hybridization (FISH) techniques are used to detect the expression levels of these receptor proteins (Allred 2010). HER2 receptor has been found to be highly overexpressed in approx. 25% of breast cancer cases and such cases were treated by HER2-targeted therapies like trastuzumab. Therefore, knowing the HER2 status of the tumor is helpful in deciding the therapeutic regimen (Dean-Colomb and Esteva 2008; Allred 2010). In addition to breast cancer, analyzing the HER2 status was found helpful in improving the treatment of gastric and gastroesophageal junction (GEJ) adenocarcinomas as well (Vakiani 2015; Park et al. 2015). Likewise, estrogen receptor is a nuclear receptor involved in the regulation of cell growth and differentiation and approx. 75% of all breast tumors are found to be ER positive (ER+). The estrogen hormone activates ERα which results in the growth of ER+ breast cancers. Further, it has been found to be associated with better prognosis marking ERα to be a prognostic factor for breast cancer, expression level of which is analyzed by radiolabeled biochemical ligand (i.e., estrogen)-binding assays (LBAs) and IHC. Moreover, it is also used as a predictive biomarker and therapeutic target for hormonal therapies such as tamoxifen and aromatase inhibitors (Keen and Davidson 2003; Clark 2000; Allred 2010). In addition to ERα, another ER known as ERβ is also used as a prognostic marker especially for

ERα-negative breast tumors (Tan et al. 2016). Similar to ER, progesterone receptor (PR) is also a nuclear receptor used as a predictive biomarker for treatment outcome of hormone therapy of breast cancer expression which is also detected by LBAs and IHC. Interestingly, around 50% of ER+ breast tumors are found to be PR+ as well and ~75% of ER/PR+ cancers respond positively to hormonal therapy (Keen and Davidson 2003; Elledge and Fuqua 2000; Allred 2010; Yip and Rhodes 2014). Further studies reported expression of PR to be an essential factor for better outcome of endocrine therapy (Keen and Davidson 2003; Elledge and Fuqua 2000). Taken together, it is apparent that it is indispensable to know the ER/PR/HER2 status of breast cancer which would serve as prognostic and predictive biomarkers and as therapeutic targets for endocrine and targeted therapies.

1.11.1 Prostate-Specific Antigen (PSA)

Prostate-specific antigen (PSA) or human kallikrein 3 (hK3) is a secreted serum protein encoded by *KLK3* gene which is used as diagnostic and prognostic biomarker for prostate cancer (PCa). It was first discovered in 1970 and approved by the FDA as a biomarker for PCa in 1986 (Filella and Foj 2016). Increased level of PSA is associated with increased possibility of prostate cancer. In general, the PSA level in prostate cancer patients will be >20 ng/mL (Kamalov et al. 2012). However, it varies with age of the patient and size of prostate giving false positives and overdiagnosis (Filella and Foj 2016). To overcome these issues, novel techniques are being used for the screening and detection of PCa where the PSA levels are used in addition to other factors. Prostate Health Index (PHI) was approved by the FDA for screening of PCa in men aged more than 50 with 4–10 μg/L of PSA level and a non-suspicious digital rectal examination (DRE). Calculating a 4Kscore using kallikrein panel is another method used for PCa screening. The panel is used to measure the levels of four kallikreins namely PSA, free PSA, intact PSA, and kallikrein-related peptidase 2 (hK2). The given 4Kscore with the patient age, DRE, and biopsy history is used to evaluate the risk of PCa; however this technique is yet to be approved by the FDA (Gupta et al. 2010; Vickers et al. 2008, 2010a, b, 2011; Filella and Foj 2016).

1.11.2 Thyroglobulin (Tg)

Thyroglobulin is a prohormone for liothyronine (T3) and thyroxine (T4) produced by thyroid follicular cells. Interestingly, normal and well-differentiated malignant thyroid cells are known to be the only source of Tg signifying it to be a highly specific marker for thyroid-related abnormalities such as cancers (Trimboli et al. 2015; Whitley and Ain 2004). Expression of this protein is assessed by immunohistochemical analysis of tissue specimen as it is mostly confined to the thyroid follicle cells. However, depending on the thyroid size, a small fraction of Tg enters circulatory system as well, concentration of which is measured through various immunoassays (Whitley and Ain 2004; Grebe 2009). Further, the concentration of circulatory Tg was shown to be increased during thyroid inflammation, hemorrhage, Graves' disease, and follicular cell-derived cancers (Grebe 2009). The serum level of Tg serves as an effective biomarker for advanced thyroid carcinoma of follicular origin and is highly tumor specific in patients undergoing thyroidectomy where Tg is used for follow-up and diagnosing residual or recurrent disease (Grebe 2009). Additionally, it is used to predict the treatment outcome. For example, in practice, postoperation, TSH-stimulated serum Tg level is measured to predict the possibility of cure. TSH-stimulated Tg level <1 ng/mL between 6 and 12 months following the initial treatment signifies the probable cure. However, despite the predictive and prognostic potentials, diagnostic values of Tg are highly limited as the preoperative serum concentration in small thyroid carcinoma patients is comparable with that of healthy individuals (Grebe 2009).

1.11.3 Assays for the Detection of Tumor Markers

Different types of cancers have been identified with various tumor markers like modified DNA, RNA, proteins, hormones, antigens, antibodies, oncogenes, and tumor-suppressor genes. Several assays were designed for detecting these molecular markers in patient samples with a range of sensitivity and specificity. This section of the chapter discusses various assays used for the detection of tumor biomarkers.

Biomarkers can be intracellular or extracellular. Intracellular markers are needed to be released from the cells and enriched (to increase the concentration) before analysis (Tothill 2009). In practice, immunoassays like ELISA and immunohistochemistry and enzyme activity assays are mostly used for the detection of cancer markers. However, immunoassays are very expensive and time consuming. Besides, it is less sensitive during early stages of disease due to low concentration of markers (Tothill 2009). Therefore, several new methods like DNA probes and biosensor-based and aptamer-based techniques have been developed for more sensitive, specific, simple, cost-effective, and point-of-care detection of tumor biomarkers (Zhang et al. 2016).

1.11.4 Development of Novel Biomarkers

Technical advancements in science, developed numerous latest technologies and genomics and proteomics approaches that are used for the identification of novel potential tumor markers (Fig. 1.3). Better understanding of the biology of tumorigenesis is essential for the identification of biomarkers. Advanced sequencing techniques such as high-throughput DNA sequencing, RNA-seq, transcriptome sequencing, microarray technologies such as DNA microarrays, tissue microarrays, gene expression arrays, antibody microarrays, and mass spectroscopy are of great

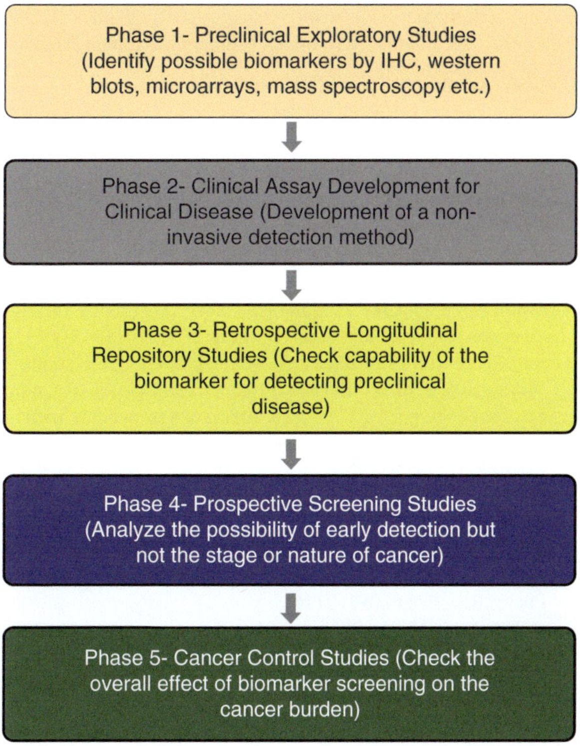

Fig. 1.3 Phases of development of novel biomarkers for early diagnosis of cancer (Pepe et al. 2001)

use in understanding the tumorigenesis process and in the identification of candidate tumor markers. However, tight thresholds, clear study design, and proper validation are essential to avoid any misinterpretation, bias, or false positives. Analytic validity, clinical validity, and clinical utility are three major factors to be considered during the development of novel tumor biomarkers (Henry and Hayes 2012; Pepe et al. 2001).

Conclusion

Most of the cancers are incurable as they are diagnosed in the advanced stage due to lack of good biomarkers. Several decades of research identified hundreds of tumor biomarkers; however their clinical usage is very limited due to their non-specificity and accuracy. As cancer is a multiple molecular disorder there is an urgent need for developing novel biomarkers for the diagnosis of cancers at an early stage and to identify different stages and grades of cancer to develop personalized treatment protocols. Unlike the conventional methods, the new era of cancer diagnosis shows promising results with specific biomarkers for the diagnosis and prognosis of cancer. However, more rigorous studies are required in this area to develop highly specific and accurate biomarkers for the better management of this disease.

Acknowledgments This work is supported by BT/P/ABK/01 Start-up Grant awarded to Dr. Ajaikumar B Kunnumakkara by the Ministry of Human Resource Development, Govt. of India.

Conflict of Interest: The authors express no conflict of interest.

References

Al-Hadiya BM, Bakheit AH, Abd-Elgalil AA (2014) Imatinib mesylate. Profiles Drug Subst Excip Relat Methodol 39:265–297

Alharbi RA, Pettengell R, Pandha HS, Morgan R (2013) The role of HOX genes in normal hematopoiesis and acute leukemia. Leukemia 27(5):1000–1008

Allegra CJ, Jessup JM, Somerfield MR, Hamilton SR, Hammond EH, Hayes DF, McAllister PK, Morton RF, Schilsky RL (2009) American Society of Clinical Oncology provisional clinical opinion: testing for KRAS gene mutations in patients with metastatic colorectal carcinoma to predict response to anti-epidermal growth factor receptor monoclonal antibody therapy. J Clin Oncol 27(12):2091–2096

Allred DC (2010) Issues and updates: evaluating estrogen receptor-alpha, progesterone receptor, and HER2 in breast cancer. Mod Pathol 23(Suppl 2):S52–S59

Anand P, Kunnumakkara AB, Sundaram C, Harikumar KB, Tharakan ST, Lai OS, Sung B, Aggarwal BB (2008) Cancer is a preventable disease that requires major lifestyle changes. Pharm Res 25(9):2097–2116

Bae YJ, Schaab M, Kratzsch J (2015) Calcitonin as biomarker for the medullary thyroid carcinoma. Recent Results Cancer Res 204:117–137

Bajetta E, Ferrari L, Martinetti A, Celio L, Procopio G, Artale S, Zilembo N, Di Bartolomeo M, Seregni E, Bombardieri E (1999) Chromogranin a, neuron specific enolase, carcinoembryonic antigen, and hydroxyindole acetic acid evaluation in patients with neuroendocrine tumors. Cancer 86(5):858–865

Ballesta AM, Molina R, Filella X, Jo J, Giménez N (1995) Carcinoembryonic antigen in staging and follow-up of patients with solid tumors. Tumour Biol 16(1):32–41

Bast RC Jr, Ravdin P, Hayes DF, Bates S, Fritsche H Jr, Jessup JM, Kemeny N, Locker GY, Mennel RG, Somerfield MR (2001) American Society of Clinical Oncology tumor markers Expert panel. 2000 update of recommendations for the use of tumor markers in breast and colorectal cancer: clinical practice guidelines of the American Society of Clinical Oncology. J Clin Oncol 19(6):1865–1878

Bernard DR, Delanghe JR, De Buyzere ML, Leroux-Roels GG (1996) Quantitative nephelometric assay for determining alpha-foetoprotein evaluated. Eur J Clin Chem Clin Biochem 34(10):847–852

Bethea M, Forman DT (1990) Beta 2-microglobulin: its significance and clinical usefulness. Ann Clin Lab Sci 20(3):163–168

Blackburn GF, Shah HP, Kenten JH, Leland J, Kamin RA, Link J, Peterman J, Powell MJ, Shah A, Talley DB et al (1991) Electrochemiluminescence detection for development of immunoassays and DNA probe assays for clinical diagnostics. Clin Chem 37(9):1534–1539

Blume-Jensen P, Hunter T (2001) Oncogenic kinase signalling. Nature 411(6835):355–365

Bottoni P, Scatena R (2015) The role of CA 125 as tumor marker: biochemical and clinical aspects. Adv Exp Med Biol 867:229–244

Bresler SC, Weiser DA, Huwe PJ, Park JH, Krytska K, Ryles H, Laudenslager M, Rappaport EF, Wood AC, McGrady PW, Hogarty MD, London WB, Radhakrishnan R, Lemmon MA, Mossé YP (2014) ALK mutations confer differential oncogenic activation and sensitivity to ALK inhibition therapy in neuroblastoma. Cancer Cell 26(5):682–694

Brutsaert EF, Gersten AJ, Tassler AB, Surks MI (2015) Medullary thyroid cancer with undetectable serum calcitonin. J Clin Endocrinol Metab 100(2):337–341

Burchardt M, Burchardt T, Shabsigh A, De La Taille A, Benson MC, Sawczuk I (2000) Current concepts in biomarker technology for bladder cancers. Clin Chem 46(5):595–605
Campeau PM, Foulkes WD, Tischkowitz MD (2008) Hereditary breast cancer: new genetic developments, new therapeutic avenues. Hum Genet 124(1):31–42
Cantile M, Franco R, Schiavo G, Procino A, Cindolo L, Botti G, Cillo C (2011) The HOX genes network in uro-genital cancers: mechanisms and potential therapeutic implications. Curr Med Chem 18(32):4872–4884
Chayvialle JA, Ganguli PC (1973) Radioimmunoassay of alpha-fetoprotein in human plasma. Lancet 1(7816):1355–1357
Clark G (2000) Prognostic and predictive factors. In: Harris JR (ed) Diseases of the breast, vol 2. Lippincott Williams & Wilkins, Philadelphia, pp 489–514
Compagni A, Christofori G (2000) Recent advances in research on multistage tumorigenesis. Br J Cancer 83(1):1–5
Costante G, Durante C, Francis Z, Schlumberger M, Filetti S (2009) Determination of calcitonin levels in C-cell disease: clinical interest and potential pitfalls. Nat Clin Pract Endocrinol Metab 5(1):35–44
Curry JL, Torres-Cabala CA, Tetzlaff MT, Bowman C, Prieto VG (2012) Molecular platforms utilized to detect BRAF V600E mutation in melanoma. Semin Cutan Med Surg 31(4):267–273
De Braekeleer E, Douet-Guilbert N, De Braekeleer MRARA (2014) Fusion genes in acute promyelocytic leukemia: a review. Expert Rev Hematol 7(3):347–357
de França Azevedo I, da Silva Júnior RM, de Vasconcelos AV (2014) Das Neves WB, de Barros Correia Melo FC, Melo RA. Frequency of p190 and p210 BCR-ABL rearrangements and survival in Brazilian adult patients with acute lymphoblastic leukemia. Rev Bras Hematol Hemoter 36(5):351–355
de Martel C, Franceschi S (2009) Infections and cancer: established associations and new hypotheses. Crit Rev Oncol Hematol 70(3):183–194
de Visser KE, Eichten A, Coussens LM (2006) Paradoxical roles of the immune system during cancer development. Nat Rev Cancer 6(1):24–37
Dean-Colomb W, Esteva FJ (2008) Her2-positive breast cancer: herceptin and beyond. Eur J Cancer 44(18):2806–2812
Debruyne EN, Delanghe JR (2008) Diagnosing and monitoring hepatocellular carcinoma with alpha-fetoprotein: new aspects and applications. Clin Chim Acta 395(1–2):19–26
Diamandis EP (2014) Present and future of cancer biomarkers. Clin Chem Lab Med 52(6):791–794
Duffy MJ (2001) Carcinoembryonic antigen as a marker for colorectal cancer: is it clinically useful? Clin Chem 47(4):624–630
Duffy MJ, van Dalen A, Haglund C, Hansson L, Klapdor R, Lamerz R, Nilsson O, Sturgeon C, Topolcan O (2003) Clinical utility of biochemical markers in colorectal cancer: European group on tumour markers (EGTM) guidelines. Eur J Cancer 39(6):718–727
Duffy MJ, McGowan PM, Harbeck N, Thomssen C, Schmitt M (2014) uPA and PAI-1 as biomarkers in breast cancer: validated for clinical use in level-of-evidence-1 studies. Breast Cancer Res 16(4):428
Early Breast Cancer Trialists' Collaborative Group (EBCTCG), Davies C, Godwin J, Gray R, Clarke M, Cutter D, Darby S, McGale P, Pan HC, Taylor C, Wang YC, Dowsett M, Ingle J, Peto R (2011) Relevance of breast cancer hormone receptors and other factors to the efficacy of adjuvant tamoxifen: patient-level meta-analysis of randomised trials. Lancet 378(9793):771–784
Easton DF, Ford D, Bishop DT (1995) Breast and ovarian cancer incidence in BRCA1-mutation carriers. Breast cancer linkage consortium. Am J Hum Genet 56(1):265–271
Eklöf V, Wikberg ML, Edin S, Dahlin AM, Jonsson BA, Öberg Å, Rutegård J, Palmqvist R (2013) The prognostic role of KRAS, BRAF, PIK3CA and PTEN in colorectal cancer. Br J Cancer 108(10):2153–2163
El-Bahrawy M (2010) Alpha-fetoprotein-producing non-germ cell tumours of the female genital tract. Eur J Cancer 46(8):1317–1322

Elledge RM, Fuqua SA (2000) Estrogen and progesterone receptors. In: Harris JR (ed) Diseases of the breast, vol 2. Lippincott Williams & Wilkins, Philadelphia, pp 471–488
Eychène A, Rocques N, Pouponnot C (2008) A new MAFia in cancer. Nat Rev Cancer 8(9):683–693
Ferla R, Calò V, Cascio S, Rinaldi G, Badalamenti G, Carreca I, Surmacz E, Colucci G, Bazan V, Russo A (2007) Founder mutations in BRCA1 and BRCA2 genes. Ann Oncol 18(Suppl 6):vi93–vi98
Filella X, Foj L (2016) Prostate cancer detection and prognosis: from prostate specific antigen (PSA) to Exosomal biomarkers. Int J Mol Sci 17(11):e1784
Fiore M, Mitchell J, Doan T, Nelson R, Winter G, Grandone C, Zeng K, Haraden R, Smith J, Harris K et al (1988) The Abbott IMx automated benchtop immunochemistry analyzer system. Clin Chem 34(9):1726–1732
Gadducci A, Guerrieri ME, Greco C (2013) Tissue biomarkers as prognostic variables of cervical cancer. Crit Rev Oncol Hematol 86(2):104–129
Gion M, Mione R, Leon AE, Dittadi R (1999) Comparison of the diagnostic accuracy of CA27.29 and CA15.3 in primary breast cancer. Clin Chem 45(5):630–637
Gion M, Mione R, Leon AE, Lüftner D, Molina R, Possinger K, Robertson JF (2001) CA27.29: a valuable marker for breast cancer management. A confirmatory multicentric study on 603 cases. Eur J Cancer 37(3):355–363
Granatowicz A, Piatek CI, Moschiano E, El-Hemaidi I, Armitage JD, Akhtari M (2015) An overview and update of chronic myeloid leukemia for primary care physicians. Korean J Fam Med 36(5):197–202
Grebe SKG (2009) Diagnosis and management of thyroid carcinoma: focus on serum thyroglobulin. Expert Rev Endocrinol Metab 4:25–43
Greene LA, Lee HY, Angelastro JM (2009) The transcription factor ATF5: role in neurodevelopment and neural tumors. J Neurochem 108(1):11–22
Grivennikov SI, Karin M (2011) Inflammatory cytokines in cancer: tumour necrosis factor and interleukin 6 take the stage. Ann Rheum Dis 70(Suppl 1):i104–i108
Grivennikov SI, Greten FR, Karin M (2010) Immunity, inflammation, and cancer. Cell 140(6):883–899
Grunnet M, Mau-Sørensen M (2014) Serum tumor markers in bile duct cancer--a review. Biomarkers 19(6):437–443
Grunnet M, Sorensen JB (2012) Carcinoembryonic antigen (CEA) as tumor marker in lung cancer. Lung Cancer 76(2):138–143
Gupta A, Roobol MJ, Savage CJ, Peltola M, Pettersson K, Scardino PT, Vickers AJ, Schröder FH, Lilja H (2010) A four-kallikrein panel for the prediction of repeat prostate biopsy: data from the European Randomized study of prostate cancer screening in Rotterdam, Netherlands. Br J Cancer 103(5):708–714
Guruvayoorappan C, Sakthivel KM, Padmavathi G, Bakliwal V, Monisha J, Kunnumakkara AB (2015) Cancer Preventive and therapeutic properties of fruits and vegetables: an overview. A Scientific Review, world scientific publishers, Anticancer Properties of Fruits and Vegetables
Hanahan D, Weinberg RA (2011) Hallmarks of cancer: the next generation. Cell 144(5):646–674
Harris L, Fritsche H, Mennel R, Norton L, Ravdin P, Taube S, Somerfield MR, Hayes DF, Bast RC Jr, American Society of Clinical Oncology (2007) American Society of Clinical Oncology 2007 update of recommendations for the use of tumor markers in breast cancer. J Clin Oncol 25(33):5287–5312
Henry NL, Hayes DF (2012) Cancer biomarkers. Mol Oncol 6(2):140–146
Hou MF, Chen YL, Tseng TF, Lin CM, Chen MS, Huang CJ, Huang YS, Hsieh JS, Huang TJ, Jong SB, Huang YF (1999) Evaluation of serum CA27.29, CA15-3 and CEA in patients with breast cancer. Kaohsiung J Med Sci 15(9):520–528
Hurt EM, Wiestner A, Rosenwald A, Shaffer AL, Campo E, Grogan T, Bergsagel PL, Kuehl WM, Staudt LM (2004) Overexpression of c-maf is a frequent oncogenic event in multiple myeloma that promotes proliferation and pathological interactions with bone marrow stroma. Cancer Cell 5(2):191–199

Isgrò MA, Bottoni P, Scatena R (2015) Neuron-specific Enolase as a biomarker: biochemical and clinical aspects. Adv Exp Med Biol 867:125–143

Janoueix-Lerosey I, Lequin D, Brugières L, Ribeiro A, de Pontual L, Combaret V, Raynal V, Puisieux A, Schleiermacher G, Pierron G, Valteau-Couanet D, Frebourg T, Michon J, Lyonnet S, Amiel J, Delattre O (2008) Somatic and germline activating mutations of the ALK kinase receptor in neuroblastoma. Nature 455(7215):967–970

Kakinoki Y, Hashiguchi J, Ishio T, Chiba K, Niino D, Ohshima K (2015) CD20-positive primary gastric T-cell lymphoma poorly responding to initial treatment with rituximab plus CHOP, and a literature review. Int J Hematol 102(6):702–708

Kamalov AA, Maksimov VA, Takhirzade TB, Gevorkian AR, Okhobotov DA, Avakian AI, Vasil'eva EG (2012) Detection of prostate cancer based on monitoring of prostate-specific antigen in outpatient clinic. Urologiia 6:58–60

Kannan MB, Solovieva V, Blank V (2012) The small MAF transcription factors MAFF, MAFG and MAFK: current knowledge and perspectives. Biochim Biophys Acta 1823(10):1841–1846

Karamouzis MV, Konstantinopoulos PA, Papavassiliou AG (2007) The role of STATs in lung carcinogenesis: an emerging target for novel therapeutics. J Mol Med (Berl) 85(5):427–436

Keen JC, Davidson NE (2003) The biology of breast carcinoma. Cancer 97(3 Suppl):825–833

Kelly ZL, Michael A, Butler-Manuel S, Pandha HS, Morgan RG (2011) HOX genes in ovarian cancer. J Ovarian Res 4:16

Kerr KM, Tsao MS, Nicholson AG, Yatabe Y, Wistuba II, Hirsch FR, IASLC Pathology Committee (2015) Programmed death-ligand 1. Immunohistochemistry in lung cancer: In what state is this art? J Thorac Oncol 10(7):985–989

Khasawneh J, Schulz MD, Walch A, Rozman J, Hrabe de Angelis M, Klingenspor M, Buck A, Schwaiger M, Saur D, Schmid RM, Klöppel G, Sipos B, Greten FR, Arkan MC (2009) Inflammation and mitochondrial fatty acid beta-oxidation link obesity to early tumor promotion. Proc Natl Acad Sci U S A 106(9):3354–3359

Kim S, Keku TO, Martin C, Galanko J, Woosley JT, Schroeder JC, Satia JA, Halabi S, Sandler RS (2008) Circulating levels of inflammatory cytokines and risk of colorectal adenomas. Cancer Res 68(1):323–328

Kontos CK, Scorilas A, Papavassiliou AG (2013) The role of transcription factors in laboratory medicine. Clin Chem Lab Med 51(8):1563–1571

Lasota J, Kowalik A, Wasag B, Wang ZF, Felisiak-Golabek A, Coates T, Kopczynski J, Gozdz S, Miettinen M (2014) Detection of the BRAF V600E mutation in colon carcinoma: critical evaluation of the imunohistochemical approach. Am J Surg Pathol 38(9):1235–1241

Lin K, Lipsitz R, Miller T, Janakiraman S, Preventive Services Task Force US (2008) Benefits and harms of prostate-specific antigen screening for prostate cancer: an evidence update for the U.S. Preventive services task force. Ann Intern Med 149(3):192–199

Liu G, Yang D, Sun Y, Shmulevich I, Xue F, Sood AK, Zhang W (2012) Differing clinical impact of BRCA1 and BRCA2 mutations in serous ovarian cancer. Pharmacogenomics 13(13):1523–1535

Locker GY, Hamilton S, Harris J, Jessup JM, Kemeny N, Macdonald JS, Somerfield MR, Hayes DF, Bast RC (2006) Jr; ASCO. ASCO 2006 update of recommendations for the use of tumor markers in gastrointestinal cancer. J Clin Oncol 24(33):5313–5327

Maletzki C, Emmrich J (2010) Inflammation and immunity in the tumor environment. Dig Dis 28(4–5):574–578

Mancal P, Srámek M, Malbohan I, Simek L (1988) The first clinical trial for determination of alpha 1 fetoprotein by means of Sevatest-ELISA AFP kit (micro I). J Hyg Epidemiol Microbiol Immunol 32(2):209–217

Masson-Lecomte A, Rava M, Real FX, Hartmann A, Allory Y, Malats N (2014) Inflammatory biomarkers and bladder cancer prognosis: a systematic review. Eur Urol 66(6):1078–1091

Mazoyer S (2005) Genomic rearrangements in the BRCA1 and BRCA2 genes. Hum Mutat 25(5):415–422

McGrath SE, Michael A, Morgan R, Pandha H (2013) EN2: a novel prostate cancer biomarker. Biomark Med 7(6):893–901
Miao P, Sheng S, Sun X, Liu J, Huang G (2013) Lactate dehydrogenase a in cancer: a promising target for diagnosis and therapy. IUBMB Life 65(11):904–910
Milde-Langosch K (2005) The Fos family of transcription factors and their role in tumourigenesis. Eur J Cancer 41(16):2449–2461
Molina R, Zanón G, Filella X, Moreno F, Jo J, Daniels M, Latre ML, Giménez N, Pahisa J, Velasco M et al (1995) Use of serial carcinoembryonic antigen and CA 15.3 assays in detecting relapses in breast cancer patients. Breast Cancer Res Treat 36(1):41–48
Molina R, Auge JM, Escudero JM, Marrades R, Viñolas N, Carcereny E, Ramirez J, Filella X (2008) Mucins CA 125, CA 19.9, CA 15.3 and TAG-72.3 as tumor markers in patients with lung cancer: comparison with CYFRA 21-1, CEA, SCC and NSE. Tumour Biol 29(6):371–380
Morris SW, Kirstein MN, Valentine MB, Dittmer KG, Shapiro DN, Saltman DL, Look AT (1994) Fusion of a kinase gene, ALK, to a nucleolar protein gene, NPM, in non-Hodgkin's lymphoma. Science 263(5151):1281–1284
Mossé YP (2016) Anaplastic lymphoma kinase as a cancer target in pediatric malignancies. Clin Cancer Res 22(3):546–552
National Cancer Institute BRCA1 and BRCA2: Cancer risk and genetic testing. https://www.cancer.gov/about-cancer/causes-prevention/genetics/brca-fact-sheet#r2
Nicolini A, Ferrari P, Rossi G (2015) Mucins and Cytokeratins as serum tumor markers in breast cancer. Adv Exp Med Biol 867:197–225
NIH-National Cancer Institute Tumor markers. https://www.cancer.gov/about-cancer/diagnosis-staging/diagnosis/tumor-markers-fact-sheet#q5
NIH-National Cancer Institute Cancer types. https://www.cancer.gov/types. Accessed 2 Mar 2017
O'Brien DP, Gough DB, Skehill R, Grimes H, Given HF (1994) Simple method for comparing reliability of two serum tumour markers in breast carcinoma. J Clin Pathol 47(2):134–137
Ordóñez NG (2012) Value of thyroid transcription factor-1 immunostaining in tumor diagnosis: a review and update. Appl Immunohistochem Mol Morphol 20(5):429–444
Ozanne BW, Spence HJ, McGarry LC, Hennigan RF (2007) Transcription factors control invasion: AP-1 the first among equals. Oncogene 26(1):1–10
Pal T, Permuth-Wey J, Betts JA, Krischer JP, Fiorica J, Arango H, LaPolla J, Hoffman M, Martino MA, Wakeley K, Wilbanks G, Nicosia S, Cantor A, Sutphen R (2005) BRCA1 and BRCA2 mutations account for a large proportion of ovarian carcinoma cases. Cancer 104(12):2807–2816
Park EJ, Lee JH, GY Y, He G, Ali SR, Holzer RG, Osterreicher CH, Takahashi H, Karin M (2010) Dietary and genetic obesity promote liver inflammation and tumorigenesis by enhancing IL-6 and TNF expression. Cell 140(2):197–208
Park JY, Dunbar KB, Vemulapalli R, Wang DH, Zhang PJ (2015) Human epidermal growth factor receptor 2 testing in gastric and gastroesophageal junction adenocarcinomas: role of the gastroenterologist. Gastrointest Endosc 81(4):977–982
Pepe MS, Etzioni R, Feng Z, Potter JD, Thompson ML, Thornquist M, Winget M, Yasui Y (2001) Phases of biomarker development for early detection of cancer. J Natl Cancer Inst 93(14):1054–1061
Peshkin BN, DeMarco TA, Brogan BM, Lerman C, Isaacs C (2001) BRCA1/2 testing: complex themes in result interpretation. J Clin Oncol 19(9):2555–2565
Petrelli F, Cabiddu M, Coinu A, Borgonovo K, Ghilardi M, Lonati V, Barni S (2015) Prognostic role of lactate dehydrogenase in solid tumors: a systematic review and meta-analysis of 76 studies. Acta Oncol 54(7):961–970
Piérard GE, Piérard-Franchimont C (2012) HOX Gene aberrant expression in skin melanoma: a review. J Skin Cancer 2012:707260
Pitot HC (1993) Multistage carcinogenesis--genetic and epigenetic mechanisms in relation to cancer prevention. Cancer Detect Prev 17(6):567–573
Porter AC, Vaillancourt RR (1998) Tyrosine kinase receptor-activated signal transduction pathways which lead to oncogenesis. Oncogene 17(11 Reviews):1343–1352

Pradhan MP, Prasad NK, Palakal MJ (2012) A systems biology approach to the global analysis of transcription factors in colorectal cancer. BMC Cancer 12:331
Rajkumar SV, Kumar S (2016) Multiple myeloma: diagnosis and treatment. Mayo Clin Proc 91(1):101–119
Rayet B, Gélinas C (1999) Aberrant rel/nfkb genes and activity in human cancer. Oncogene 18(49):6938–6947
Rhea JM, Molinaro RJ (2011) Cancer biomarkers: surviving the journey from bench to bedside. MLO Med Lab Obs 43(3):10–12
Richards MA (2009) The size of the prize for earlier diagnosis of cancer in England. Br J Cancer 101(Suppl 2):S125–S129
Robertson SC, Tynan J, Donoghue DJ (2000) RTK mutations and human syndromes: when good receptors turn bad. Trends Genet 16(8):368
Ruoslahti E, Seppälä M (1971) Studies of carcino-fetal proteins. 3. Development of a radioimmunoassay for -fetoprotein. Demonstration of -fetoprotein in serum of healthy human adults. Int J Cancer 8(3):374–383
Sankpal UT, Maliakal P, Bose D, Kayaleh O, Buchholz D, Basha R (2012) Expression of specificity protein transcription factors in pancreatic cancer and their association in prognosis and therapy. Curr Med Chem 19(22):3779–3786
Sawyers CL (1999) Chronic myeloid leukemia. N Engl J Med 340(17):1330–1340
Scatena R (2015) Cancer Biomarkers: a status quo. Adv Exp Med Biol 867:3–8
Schmetter BS, Habicht KK, Lamm DL, Morales A, Bander NH, Grossman HB, Hanna MG Jr, Silberman SR, Butman BT (1997) A multicenter trial evaluation of the fibrin/fibrinogen degradation products test for detection and monitoring of bladder cancer. J Urol 158(3 Pt 1):801–805
Shiota M, Nakamura S, Ichinohasama R, Abe M, Akagi T, Takeshita M, Mori N, Fujimoto J, Miyauchi J, Mikata A, Nanba K, Takami T, Yamabe H, Takano Y, Izumo T, Nagatani T, Mohri N, Nasu K, Satoh H, Katano H, Fujimoto J, Yamamoto T, Mori S (1995) Anaplastic large cell lymphomas expressing the novel chimeric protein p80NPM/ALK: a distinct clinicopathologic entity. Blood 86(5):1954–1960
Shukla VK, Gurubachan, Sharma D, Dixit VK, Usha (2006) Diagnostic value of serum CA242, CA 19-9, CA 15-3 and CA 125 in patients with carcinoma of the gallbladder. Trop Gastroenterol 27(4):160–165
Singh S, Law C (2012) Chromogranin a: a sensitive biomarker for the detection and post-treatment monitoring of gastroenteropancreatic neuroendocrine tumors. Expert Rev Gastroenterol Hepatol 6(3):313–334
Sisinni L, Landriscina M (2015) The role of human chorionic gonadotropin as tumor marker: biochemical and clinical aspects. Adv Exp Med Biol 867:159–176
Slattery ML, Lundgreen A, Kadlubar SA, Bondurant KL, Wolff RK (2013) JAK/STAT/SOCS-signaling pathway and colon and rectal cancer. Mol Carcinog 52(2):155–166
Sopik V, Phelan C, Cybulski C, Narod SA (2015) BRCA1 and BRCA2 mutations and the risk for colorectal cancer. Clin Genet 87(5):411–418
Stephens FO, Aigner KR (eds) (2009) Investigations that may be useful in detecting cancer. In: Basics of oncology. Springer, Berlin. doi: 10.1007/978-3-540-92925-3_1
Subramaniam A, Shanmugam MK, Perumal E, Li F, Nachiyappan A, Dai X, Swamy SN, Ahn KS, Kumar AP, Tan BK, Hui KM, Sethi G (2013) Potential role of signal transducer and activator of transcription (STAT)3 signaling pathway in inflammation, survival, proliferation and invasion of hepatocellular carcinoma. Biochim Biophys Acta 1835(1):46–60
Suzuki M (1988) Fundamental and clinical evaluation of an immunoradiometric assay procedure "Amerwell AFP" for estimation of serum alpha-fetoprotein. Radioisotopes 37(4):225–228
Swiderska M, Choromańska B, Dąbrowska E, Konarzewska-Duchnowska E, Choromańska K, Szczurko G, Myśliwiec P, Dadan J, Ladny JR, Zwierz K (2014) The diagnostics of colorectal cancer. Contemp Oncol (Pozn) 18(1):1–6
Szabo CI, King MC (1995) Inherited breast and ovarian cancer. Hum Mol Genet 4:1811–1817

Takahashi H, Ogata H, Nishigaki R, Broide DH, Karin M (2010) Tobacco smoke promotes lung tumorigenesis by triggering IKKbeta- and JNK1-dependent inflammation. Cancer Cell 17(1):89–97

Tan W, Li Q, Chen K, Su F, Song E, Gong C (2016) Estrogen receptor beta as a prognostic factor in breast cancer patients: a systematic review and meta-analysis. Oncotarget 7(9):10373–10385

Tetsumoto S, Kijima T, Morii E, Goya S, Minami T, Hirata H, Takahashi R, Kohmo S, Inoue K, Nagatomo I, Takeda Y, Kida H, Tachibana I, Kumanogoh A (2013) Echinoderm microtubule-associated protein-like 4 (EML4)-anaplastic lymphoma kinase (ALK) rearrangement in congenital pulmonary airway malformation. Clin Lung Cancer 14(4):457–460

Torre LA, Siegel RL, Ward EM, Jemal A (2016) Global cancer incidence and mortality rates and trends--an update. Cancer Epidemiol Biomark Prev 25(1):16–27

Tørring ML, Frydenberg M, Hansen RP, Olesen F, Vedsted P (2013) Evidence of increasing mortality with longer diagnostic intervals for five common cancers: a cohort study in primary care. Eur J Cancer 49(9):2187–2198

Tothill IE (2009) Biosensors for cancer markers diagnosis. Semin Cell Dev Biol 20(1):55–62

Trimboli P, Treglia G, Giovanella L (2015) Preoperative measurement of serum thyroglobulin to predict malignancy in thyroid nodules: a systematic review. Horm Metab Res 47(4):247–252

Tubiana M (1989) Carcinogenesis in humans: a multistage and multifactorial process. Bull Acad Natl Med 173(8):997–1002. discussion 1002–4, 1013–5

Vaiopoulos AG, Papachroni KK, Papavassiliou AG (2010) Colon carcinogenesis: learning from NF-kappaB and AP-1. Int J Biochem Cell Biol 42(7):1061–1065

Vakiani E (2015) HER2 testing in gastric and gastroesophageal adenocarcinomas. Adv Anat Pathol 22(3):194–201

Vickers AJ, Cronin AM, Aus G, Pihl CG, Becker C, Pettersson K, Scardino PT, Hugosson J, Lilja H (2008) A panel of kallikrein markers can reduce unnecessary biopsy for prostate cancer: data from the European Randomized study of prostate cancer screening in Göteborg, Sweden. BMC Med 6:19

Vickers A, Cronin A, Roobol M, Savage C, Peltola M, Pettersson K, Scardino PT, Schröder F, Lilja H (2010a) Reducing unnecessary biopsy during prostate cancer screening using a four-kallikrein panel: an independent replication. J Clin Oncol 28(15):2493–2498

Vickers AJ, Cronin AM, Roobol MJ, Savage CJ, Peltola M, Pettersson K, Scardino PT, Schröder FH, Lilja H (2010b) A four-kallikrein panel predicts prostate cancer in men with recent screening: data from the European Randomized study of screening for prostate cancer, Rotterdam. Clin Cancer Res 16(12):3232–3239

Vickers AJ, Gupta A, Savage CJ, Pettersson K, Dahlin A, Bjartell A, Manjer J, Scardino PT, Ulmert D, Lilja H (2011) A panel of kallikrein marker predicts prostate cancer in a large, population-based cohort followed for 15 years without screening. Cancer Epidemiol Biomark Prev 20(2):255–261

Voncken JW, Kaartinen V, Pattengale PK, Germeraad WT, Groffen J, Heisterkamp N (1995) BCR/ABL P210 and P190 cause distinct leukemia in transgenic mice. Blood 86(12):4603–4611

Wallace EM, Yeh TC, Laird ER, Blake JF, Lyssikatos J (2007) Inhibition of growth factor signaling by small-molecule inhibitors of ErbB, Raf, and MEK. In: Bradbury RH (ed) Cancer, vol 1. Springer, New York, pp 83–132

Waller CF (2014) Imatinib mesylate. Recent Results Cancer Res 201:1–25

Wang QH, Ji ZG, Chen ZG, Li HZ, Fan H, Fan XR, Shi BB, Fang Y (2015) Serum CA 19-9 as a good prognostic biomarker in patients with bladder cancer. Int J Surg 15:113–116

Whitley RJ, Ain KB (2004) Thyroglobulin: a specific serum marker for the management of thyroid carcinoma. Clin Lab Med 24(1):29–47

Wieskopf B, Demangeat C, Purohit A, Stenger R, Gries P, Kreisman H, Quoix E (1995) Cyfra 21-1 as a biologic marker of non-small cell lung cancer. Evaluation of sensitivity, specificity, and prognostic role. Chest 108(1):163–169

Wong VW, Yu J, Cheng AS, Wong GL, Chan HY, Chu ES, Ng EK, Chan FK, Sung JJ, Chan HL (2009) High serum interleukin-6 level predicts future hepatocellular carcinoma development in patients with chronic hepatitis B. Int J Cancer 124(12):2766–2770

Wong RJ, Ahmed A, Gish RG (2015) Elevated alpha-fetoprotein: differential diagnosis- hepatocellular carcinoma and other disorders. Clin Liver Dis 19(2):309–323

Yamaguchi U, Hasegawa T, Masuda T, Sekine S, Kawai A, Chuman H, Shimoda T (2004) Differential diagnosis of gastrointestinal stromal tumor and other spindle cell tumors in the gastrointestinal tract based on immunohistochemical analysis. Virchows Arch 445(2):142–150

Yasue M, Sakamoto J, Teramukai S, Morimoto T, Yasui K, Kuno N, Kurimoto K, Ohashi Y (1994) Prognostic values of preoperative and postoperative CEA and CA19.9 levels in pancreatic cancer. Pancreas 9(6):735–740

Ying M, Zhu XX, Zhao Y, Li DH, Chen LH (2015) KRAS mutation as a biomarker for survival in patients with non-small cell lung cancer, a meta-analysis of 12 Randomized trials. Asian Pac J Cancer Prev 16(10):4439–4445

Yip CH, Rhodes A (2014) Estrogen and progesterone receptors in breast cancer. Future Oncol 10(14):2293–2301

Zhang SY, Lin M, Zhang HB (2015) Diagnostic value of carcinoembryonic antigen and carcinoma antigen 19-9 for colorectal carcinoma. Int J Clin Exp Pathol 8(8):9404–9409

Zhang Z, Dou M, Yao X, Tang H, Li Z, Zhao X (2016) Potential biomarkers in diagnosis of human gastric cancer. Cancer Investig 34(3):115–122

Zhao X, Han RB, Zhao J, Wang J, Yang F, Zhong W, Zhang L, Li LY, Wang MZ (2013) Comparison of epidermal growth factor receptor mutation statuses in tissue and plasma in stage I-IV non-small cell lung cancer patients. Respiration 85(2):119–125

Zhao Z, Verma V, Zhang M (2015) Anaplastic lymphoma kinase: role in cancer and therapy perspective. Cancer Biol Ther 16(12):1691–1701

Transcription Factors as Detection and Diagnostic Biomarkers in Cancer

2

W.L. Goh, E. Assah, X.T. Zheng, D.P. Lane, F.J. Ghadessy, and Y.N. Tan

2.1 Introduction to Transcription Factors and Diseases

The survival of cellular life depends on the accurate and coordinated maintenance of biological processes at the single-cell level such as cell-cycle progression, differentiation, metabolism, development, and programmed cell death (Rudel and Sommer 2003; Hanahan and Weinberg 2011; DeBerardinis and Thompson 2012). Consequently, simultaneous regulation of complex intracellular programs is heavily reliant on the precision of gene expression at the transcriptional level. Eukaryotic gene expression begins typically with the assembly of transcription-related protein complexes and cofactors on DNA before genetic information is transcribed into messenger RNA molecules, through the recruitment of RNA polymerase and cofactors, allowing for downstream protein translation (Lee and Young 2000). Sequence-specific DNA-binding transcription factors (TFs) are an integral part of the transcriptional machinery that regulate gene expression rates through the recognition and binding to precise DNA motifs (enhancer regions or response elements) resulting in either transcriptional activation or repression (Robertson et al. 2006) through further interaction with co-regulators and histone modifiers (HATs, HDACs) (Schaefer et al. 2011). Whole-genome studies have predicted 2000–3000 TFs in the

W.L. Goh (✉) • D.P. Lane • F.J. Ghadessy
p53 Laboratory, Biomedical Sciences Institute, Agency of Science, Technology and Research, A*STAR, Singapore, Singapore
e-mail: lpgoh@p53lab.a-star.edu.sg

E. Assah • X.T. Zheng • Y.N. Tan
Institute of Materials Research and Engineering (IMRE), Agency of Science, Technology and Research, A*STAR, Singapore, Singapore

P. Chandra et al. (eds.), *Next Generation Point-of-care Biomedical Sensors Technologies for Cancer Diagnosis*, DOI 10.1007/978-981-10-4726-8_2

human genome (Babu et al. 2004; Kummerfeld and Teichmann 2006; Venter et al. 2001), and bioinformatics, transcriptome analysis estimates that TFs account for ~8–10% of human genes expressed (Messina et al. 2004; Kummerfeld and Teichmann 2006).

DNA-binding transcription factors are typically modular and generally contain a DNA-binding domain (DBD) which controls DNA binding and gene specificity, and a transactivation domain (TAD) to regulate transcription through interaction with protein factors of the transcriptional machinery. The basis of DNA selectivity lies within the DBD, which can be classified based on structure and function. Three classes are most prolifically expressed in humans: the C_2H_2 zinc finger, homeodomain, and helix-loop-helix families (Vaquerizas et al. 2009). TADs are generally disordered and less structured than DBDs within TF families, allowing for promiscuity in protein interaction and cofactor recruitment. Distinct categories of TADs are observed in different classes of TFs and are grouped based on the amino acid composition: acidic, isoleucine-rich, proline-rich, and glutamine-rich domains (Mermod et al. 1989; Okuda et al. 2016; Mognol et al. 2016; Hibino et al. 2016). In addition, the nine amino acid transactivation domain (9aaTAD) is a class common to eukaryotic transcription factors (Piskacek et al. 2007). Table 2.1 shows seven of the most cited TFs in the literature to date (Vaquerizas et al. 2009).

As expected, the deregulation of proper transcriptional activity has been associated with many human diseases. For example, mutations in the transcription factors HNF1beta, HNF1alpha, and HNF4alpha have been linked to maturity-onset diabetes of the young (MODY) by respectively affecting differentiation processes in the pancreas and decreasing glucose-dependent insulin secretion in beta-cells (Maestro et al. 2007). The autoimmune regulator AIRE, a transcription factor expressed in the thymus, is responsible for the identification and negative selection of self-reactive T-cells, and its inactivation causes type-I autoimmune polyendocrinopathy syndrome (APS-1) (Kyewski and Klein 2006). Aberrant gene expression from deregulated transcription factors can occur at the genetic level as a result of increased TF expression (increased copy number from gene duplication, epigenetic modifications, or chromosomal translocations), or at the protein level (post-translational modifications or a derailment in biochemical pathways like protein turnover rates). Examples include translocation of the AML1 (or RUNX1) transcription factor (resulting in oncogenic fusion proteins like AML1-ETO) commonly associated with several forms of leukemia (Licht 2001; Lukasik et al. 2002), and the HPV (human papillomavirus)-related viral protein E6 which, when present in cells during viral infections, facilitates the degradation of the tumor-suppressor p53 and promotes cervical carcinogenesis (Mantovani and Banks 2001). Additionally, mutations in cis-acting regulatory DNA elements as well as inactivating mutations within the reading frame of a gene can affect mRNA splicing, protein translation, or protein structure, and have all been described and linked to disease phenotypes (Lee and Young 2013).

Table 2.1 List of most cited transcription factors and their involvement in health and diseases

Transcription factor	Description	References
p53	Master tumor suppressor that regulates cellular programs that decide cell fate (e.g., growth arrest, senescence, apoptosis) in response to genetic aberrations. It is frequently mutated in many forms of human cancer.	Lane (1992), Vousden and Lane (2007), Vogelstein et al. (2000)
Estrogen receptor (ER)	Nuclear receptor family transcription factor activated by the steroid hormone estrogen. Responsible for the maintenance of reproductive, immune, cardiac, and skeletal systems. Overexpressed in many types of breast cancer, constituting a common prognostic and treatment target.	Ascenzi et al. (2006)
FOS	Forms the AP-1 complex with c-jun and regulates many cellular processes that govern differentiation, growth, and survival. Often implicated in cancer and also involved in cancer-related processes such as hypoxia, angiogenesis, and epithelial-mesenchymal transition.	Milde-Langosch (2005)
MYC	A potent oncogene frequently active during tumorigenesis through mutagenesis, chromosal translocation and deregulated protein biochemistry. Binds DNA through basic helix-loop-helix domain to regulate cell fate. Overexpressed in many cancers.	Meyer and Penn (2008)
JUN	Forms homo- or heterodimeric (c-fos) transcription factors. Often required in cellular transformation due to proliferative and anti-apoptotic signals. Associated with increased aggressiveness, tissue invasion, and metastatic cancer phenotypes.	Eferl and Wagner (2003)
Androgen receptor (AR)	Nuclear receptor transcription factor that responds to androgenic hormones. Often required for the development and maintenance of male sexual organs. AR activity is strongly associated with prostate cancer development and progression.	Heinlein and Chang (2004)
SP1	Zinc finger TF involved in many cellular processes such as cellular growth and differentiation, chromatin remodeling, immune system, and apoptosis.	Beishline and Azizkhan-Clifford (2015)
NF-κB	Modulates the inflammatory response of the innate immune system through dimerization with members within the NF-κB family in response to biochemical signals. Displays a double-edged role in cancer development depending on cellular context.	Gilmore (2006)

2.2 Transcription Factors in Cancer

Cellular transformation and the development of cancer have been acutely linked to numerous transcription factors responsible for distinct cellular processes. The tumor-promoting c-Myc transcription factor is the most frequently amplified oncoprotein in human cancers (Lin et al. 2012). c-Myc recognizes and binds DNA enhancer motifs (E-boxes) through heterodimerization with another TF, Max (myc-associated factor x) (Blackwood and Eisenman 1991), to elevate expression of genes involved in cell proliferation and metabolism, promoting cell growth (Ji et al. 2011). Furthermore, tumorigenic cellular programs have also been ascribed to simultaneous activation of transcriptional networks. Elevated transcriptional activity of TAL1 is observed in 40% of all T-ALL (T-cell acute lymphoblastic leukemia) cases and has been reported as a master regulator of transcriptional circuitries involving other TFs including RUNX, GATA3, HEB, and E2A (Sanda et al. 2012).

The rest of this chapter focuses on three transcription factors: the p53 tumor suppressor, estrogen receptor, and NF-κB; their roles in cancer; as well as past and current technologies designed at targeting them for diagnostic purposes.

2.2.1 The p53 Tumor Suppressor

The p53 tumor suppressor (also known as the guardian of the genome) is a master regulator that sits at a central node within a sophisticated network of cellular programs (Lane 1992). It functions primarily as a transcription factor that acts to safeguard the genomic integrity of an organism by inducing biochemical pathways that ultimately determine cell fate (including cell cycle arrest, apoptosis, senescence) p53 responds to upstream stress signals and prevents cellular transformation caused by genetic aberrations (Vogelstein et al. 2000). Stress signals that activate p53 take many forms but typically result in DNA mutations or chromosomal damage when left unchecked. Examples include DNA or chromosomal breakages, ionizing radiation, hypoxia, dNTP depletion, and glucose starvation (Bieging et al. 2014).

p53 shares significant homology with its family members, transcription factors p63 and p73, and they are each organized to carry several critical domains including an *N*-terminal transactivation domain (TAD), a proline-rich domain (PD), a well-ordered DNA-binding domain (DBD), an oligomerization domain (OD), and an unstructured carboxy-terminal domain (CTD) (Vousden and Lane 2007). In addition, p63 and p73 possess a sterile α-motif (SAM) domain that participates in protein-protein interactions (Thanos and Bowie 1999). The physiological functions of p53's unstructured CTD have been highly controversial and both early reports and in vitro experiments have suggested an auto-inhibitory role possibly through interacting with the DBD (Hupp et al. 1992; Goh et al. 2010). However, recent animal and biochemical studies reveal more evidence of the CTD's involvement in p53-DNA interaction, particularly in the selectivity and coordinated binding of p53 to DNA response elements and also in the precise induction of p53 response in cells (Laptenko et al. 2015, 2016). There are also postulations that the CTD can help mediate sequence-specific p53-DNA binding through weak interactions between the positively charged lysine-rich regions and the negatively charged phosphate DNA backbone (Friedler et al. 2005). This interaction may also facilitate the sensing of DNA damage and expedite DNA repair (Reed et al. 1995). The DNA-binding core of p53 is responsible for interacting with DNA in a sequence-specific manner through a highly ordered domain that is well conserved within the protein family (Belyi et al. 2010). The DBD core structure consists of a β-sandwich scaffold consisting of two antiparallel β-sheets projecting a loop-sheet-helix motif and two additional large loops that make DNA contacts (Cho et al. 1994). p53 functions as a TF by recognizing and binding cognate DNA elements known as p53 response elements (p53-RE) which contain two palindromic half-site decamers, each carrying the consensus sequence 5′-RRRC(A/T)(T/A)GYYY-3′ (where Y = pyrimidine and R = purine) separated by a spacer ranging from 0 to 13 base pairs (el-Deiry et al. 1992). A stable complex (dimer of dimers) is formed with each monomer contacting a 5-bp quarter site when p53 tetramerizes on DNA, resulting in a close to 100-fold increase in binding affinity over monomeric units alone (Balagurumoorthy et al. 1995). The DNA core motif C(A/T)(T/A)G within each decamer half-site, in particular, has been shown to have a profound influence on p53 DBD binding (Wang et al. 2009a). Wild-type p53 is known to regulate hundreds of gene targets, through transcriptional activation or repression, by interacting with DNA REs located across the entire genome. More than 200 RE sites have been established as empirically verifiable p53 response elements (Menendez et al. 2009; Riley et al. 2008; Zeron-Medina et al. 2013), with thousands more possible p53-binding sites identified through predictive algorithms and whole-genome studies (Tebaldi et al. 2015; Smeenk et al. 2008; Chang et al. 2014; Sammons et al. 2015). Furthermore, the low intrinsic thermodynamic stability of the p53 core (9-min half-life at body temperature), a likely result of evolutionary adaptations, has been linked to structural plasticity, allowing for interaction with diverse protein partners and DNA sequences (Joerger and Fersht 2010). Indeed, gene expression regulating p53-REs have shown considerable degeneracy in sequence and size, seen in noncanonical motifs like half- and three-quarter sites (Jordan et al. 2008; Tebaldi et al. 2015).

In the classical p53 response, cellular stress stimuli result in the activation of p53 modifiers like ATM (ataxia telangiectasia mutated), ATR (ataxia telangiectasia and Rad3-related protein), and CHK1/2 (checkpoint kinase 1/2) serine/threonine protein-kinase which phosphorylates p53 at key residues leading to the stabilization of intracellular protein levels, nuclear accumulation, and increased transcriptional activity on target genes (Cheng and Chen 2010). Acetylation of lysine residues found in the DBD and CTD, through the recruitment of histone or lysine acetyltransferases, can further contribute to this process (Dornan and Hupp 2001; Lambert et al. 1998). The precision of p53 gene target selection is regulated at many levels including p53 post-translational modifications (such as phosphorylation, acetylation, and ubiquitination of precise residues) (Meek and Anderson 2009), histone remodeling factors (HATs, HDACs), as well as p53-protein interaction (protein cofactors and p53 isoforms) (Khoury and Bourdon 2011; Beckerman and Prives 2010). The result is a downstream augmentation of canonical p53 cellular responses through the upregulation of classic gene targets including p21, GADD45, and 14–3-3σ which mediate growth arrest and DNA repair (Hermeking et al. 1997; Chin et al. 1997; el-Deiry et al. 1993), as well as Puma, Bax, and Noxa that induces apoptosis (Nakano and Vousden 2001; Miyashita and Reed 1995), all by virtue of high-affinity p53-REs (Weinberg et al. 2005).

Transcriptional upregulation is also achieved through the ablation of an autoregulated negative feedback mechanism mediated by Mdm2, a p53 target gene product. p53 activity is kept low under normal cellular conditions by Mdm2, a ubiquitin E3-ligase capable of inactivating p53 through TAD binding and sequestration, followed by cytosolic translocation and ubiquitin-dependent proteasome degradation through the modification of lysine residues with poly-ubiquitin chains (Kussie et al. 1996; Lohrum et al. 2001). Overexpression of Mdm2 leads to the attenuation of the p53 response and promotes cancer development. High levels of Mdm2 are found in several types of human malignancies and hence it constitutes a promising therapeutic and prognostic target in cancer (Rayburn et al. 2005; Andre et al. 2014).

The crucial role of p53 in cancer development is obvious when examining Li-Fraumeni patients who carry germline mutations in the p53 encoding *TP53* gene, resulting in cancer predisposition at a young age (particularly sarcomas and cancers of the breast, brain, and adrenal glands) (Malkin 2011). The derailment of p53 pathways leading to constitutive proliferative, or pro-survival, cellular signals can be seen in almost all human cancers, particularly in above 50% of cases where p53 exists in *TP53* mutations which results in mutation-inactivated forms with compromised transcriptional functions. In human breast cancer, *TP53* is one of the most frequently mutated genes (up to 80% in certain subtypes), and demonstrates a correlation between mutation type (e.g. insertational/deletion or missense mutations) and molecular subtype (Powell et al. 2014; Ciriello et al. 2015). In 75% of *TP53* mutations which results in fully translated proteins carrying a single-amino acid mutation, 95% reside in the DNA-binding core domain causing varying extents of structural perturbations, abating wild-type DNA binding (Bullock and Fersht 2001). Several particular mutations, known as the "hotspot" mutations, occur most frequently in human cancers (G245S, R273H, R248Q, R175H, R282W, and R249S). In addition to losing the ability to bind canonical p53 DNA REs and transcribing p53 target genes (loss of function), these mutants are also known to possess tumorigenic functions (oncogenic gain-of-function) and provide poor disease prognosis (Joerger and Fersht 2007; Powell et al. 2014). In particular, mutants R273H and R175H have been shown to associate with other DNA-binding transcription factors to increase tumor aggressiveness and metastasis through numerous mechanisms such as exerting a dominant-negative effect over tumor-suppressing TFs (p63, p73) or associating with oncogenic TFs (eg. ETS2) to induce pro-survival programs and promote chromosomal instability (Lu et al. 2013; Martinez et al. 2016; Solomon et al. 2012; Song et al. 2007). Indeed, the requirement for *TP53* mutation as an early initiation event in pathogenesis shows almost complete penetrance in high-grade serous ovarian carcinoma (Vang et al. 2016), where mutant p53 contributes to anoikis resistance (Cai et al. 2015) and tissue invasion (Iwanicki et al. 2016).

2.2.2 Estrogen Receptor (ER)

The intracellular estrogen receptors (ERα and ERβ) belong to the class of nuclear receptor (NR) superfamily of ligand-regulated transcription factors and respond to the sex steroid hormone estrogen. It functions primarily in the maintenance of the female reproductive system, but also in physiological processes including skeletal, neuroendocrine, cardiovascular, and immune systems (Swedenborg et al. 2009).

ER, like other members of the NR family, show highly conserved functional domains which comprise an N-terminus transactivation domain (AF-1), a DBD, a hinge domain containing the nuclear localization signal, and a C-terminus ligand-binding and transactivation domain (LBD and AF-2) (Le Romancer et al. 2011). ERα and ERβ have high homology within their DBD (~96%) and differ functionally by their N-termini transcriptional activity (AF-1 domain) which regulates hormone-independent transcription (Kuiper et al. 1996). Both forms of ER are expressed widely in many tissues and their relative expression in cells determines tissue-specific responsiveness to the presence of estrogen (Thomas and Gustafsson 2011). Additionally, numerous amino acid residues on ER are susceptible to post-translational modifications which influences transcriptional function, DNA selectivity, and interaction with ER-coregulators (Le Romancer et al. 2011).

Estrogen-dependent tumorigenesis has been linked to the development of many cancer types including breast, ovary, colon, and prostate (Shang 2007), and is largely ascribed to the transcriptional activity of ERα which is responsible for mediating pro-survival and proliferative signals (Liang and Shang 2013). In contrast ERβ has been reported to inhibit estrogen-dependent cell growth and also displays ERα antagonism when ectopically expressed in ERα-positive breast cancer cells, reducing cell proliferation (Strom et al. 2004). Estrogen exists predominantly as 17β-estradiol (E2) in cells, but can also take other forms like estrone (E1) and estriol (E3). Long-term exposure to estrogenic compounds, such as in hormone replacement therapy (HRT), constitutes a major risk factor in developing breast cancer (Narod 2011). Upon ligand binding, estrogen receptors are activated through homodimerization, further allowing binding to estrogen response elements (ERE) in the nucleus. Receptor dimerization has also been reported to result in interaction with other transcription factors (such as p53, NF-κB, RUNX1), allowing ER to regulate gene expression in the absence of estrogen REs (Jerry et al. 2010; Stender et al. 2010).

2.2.3 Nuclear Factor Kappa B (NF-κB)

The nuclear factor NF-κB, originally discovered as a transcriptional regulator of immunoglobulin kappa-light-chain-activated B cells, is now known to be widely expressed in many cell types and mediates the inflammatory response, as part of the innate immune system, as well as cellular growth and death (Wan and Lenardo 2009).

The family of NF-κB transcription factors contains five members (RelA, RelB, c-Rel, NF-κB1, and NF-κB2) and regulates transcription in a modular way by forming homo- and heterodimers with each other in virtually every possible permutation. All members are evolutionarily conserved and carry the approximately 300-residue Rel homology domain (RHD) responsible for dimerization, nuclear localization, and DNA interaction (Wan and Lenardo 2009). RelA, RelB, and c-Rel each contain one or more C-terminal transactivation domains (TA) (May and Ghosh 1997) but are usually inhibited in quiescent cells by interacting with members of the IκB family (inhibitors of NF-κB). In addition to obscuring the nuclear localization signals on NF-κB proteins and preventing nuclear translocation when bound (Jacobs and Harrison 1998), IκB family proteins (IκBα, IκBβ, IκBε, IκBγ, IκBζ, BCL-3, and precursors p105 and p100) (Wan and Lenardo 2009) all carry characteristic ankyrin repeats responsible for associating with RHD DNA-binding domains and render NF-κB proteins transcriptionally inactive (Verma et al. 1995). Release from IκB inhibition is required for NF-κB transcriptional activity and occurs in part through proteasome-dependent degradation of the inhibitors. The process is catalyzed through the poly-ubiquitination of lysine residues on IκB molecules by unique E3-ligases ($SCF^{\beta TrCP}$, beta-transducin repeat-containing protein) through the recognition of a specific double-phosphorylated substrate catalyzed by IKK complexes (IκB kinase) (Karin and Ben-Neriah 2000). The remaining two NF-κB family proteins, NF-κB1 (p105) and NF-κB2 (p100), lack a transactivation domain and are synthesized as precursors that remain inactive through the negative self-regulating ankyrin repeats they carry. These inhibitory domains are cleaved during the maturation process resulting in the active forms, p50 and p52 (Hayden and Ghosh 2012). Homodimers of p50 and p52 act as transcriptional repressors as they bind κB DNA elements but lack a transactivation domain. NF-κB TFs recognize and bind DNA motifs containing the κB consensus sequence 5′GGGRNYYYCC3′ (where R = purine, Y = pyrimidine, and N = any nucleotide) (Chen et al. 1998). Additionally, precise transcriptional response and DNA-binding selectivity are also achieved through post-translational modifications of NF-κB complexes and from different combinations of heterodimerization, resulting in cellular context-dependent activity. For instance, phosphorylation of serine 536 on RelA can result in IκBα dissociation, nuclear accumulation, and enhanced transcriptional activity (Sasaki et al. 2005), but also enhances RelB association, leading to decreased κB DNA sites binding in the nucleus (Jacque et al. 2005).

In the classical pathway of NF-κB activation, upstream receptor-mediated signals (for example Toll-like receptor (TLR) stimulation from bacterial cell wall lipopolysaccharides or tumor necrosis factor receptor (TNFR) activation) lead to activation of IKK complexes which phosphorylate and target IκB family members to the proteasome, hence liberating NF-κB for nuclear localization and activation of inflammatory responses. While acute inflammation can lead to activation of cytotoxic immunity against transformed cells, chronic inflammation has been associated with pro-tumorigenic outcomes. Stimulation of NF-κB pathways can result in pro-survival signals in cells as a response to withstand physiological causes of the inflammation (Hoesel and Schmid 2013). Furthermore, release of ROS (reactive oxygen species) by neutrophils can also cause DNA damage and propagate cancer-driving mutations (Liou and Storz 2010). RANK (receptor activator of NF-κB) belongs to the TNF receptor family and is etiologically linked to some forms of metastatic bone tumors and mammary carcinoma through the activation of NF-κB signaling (Hanada et al. 2011). Recent studies using mouse models have presented RANK receptor and its ligand RANKL as potential diagnostic and therapeutic targets of breast cancer in carriers of *BRCA1* mutations (Nolan et al. 2016). Furthermore, NF-κB signaling is also reported to drive cancer aggression by regulating EMT (epithelial mesenchymal transition) and can contribute to metastasis as well as angiogenesis by upregulating VEGF and increasing tumor vascularization (vascular endothelial growth factor) (Huber et al. 2004; Xie et al. 2010).

2.3 Detecting Transcription Factors in Cancer Diagnostics

As illustrated in the sections above, transcription factors in general (and in particular p53, ER, and NF-κB) are acutely linked to cellular transformation and cancer development by eliciting erroneous cellular programs or transcriptional functions through TF-DNA interaction. The ability to qualitatively and quantitatively assess DNA-binding functions of transcription factors using clinical samples will undoubtedly provide valuable information on disease prognosis or opportunities for early disease detection. For example, stabilization of ERα proteins associated with certain unique DNA-binding properties may result in the early detection of pro-survival cells (Fan et al. 2015). Oncogenic point mutations in NF-κB and RelA have also been detected in Hodgkin lymphomas, likely carrying altered but specific DNA-binding signatures (Hoesel and Schmid 2013).

Conventional and early methods for detecting TF-DNA binding are crude and often only semiquantitative. They are also labor intensive, have low throughput and sensitivity, and frequently require the use of radioactive materials to increase signal-to-noise detection. Examples include electrophoretic mobility shift assay (EMSA) which involves the electrophoretic separation of protein-DNA complexes using non-denaturing agarose or polyacrylamide gels followed by visualization. TF-DNA complexes can be further "super-shifted" when the overall molecular weight of the complex is increased by the addition of a TF-specific antibody. DNA fragments are often radioactively labeled to increase the detection limit and sensitivity of the assay (Fried 1989). In DNase footprinting assay, DNA fragments mixed with a protein of interest (a transcription factor) are later subjected to restriction endonuclease digestion. Binding of a protein to a specific region on the DNA provides protection from endonuclease activity resulting in different fragmentation patterns and allows the identification of DNA sequence involved in binding (Brenowitz et al. 1986). As before, ^{32}P-labeled DNA can be used (through PCR amplification using radioactively labeled primers) for increased signal detection. Enzyme-linked immunosorbent assay (ELISA) is a common serological diagnostic technique which involves the use of protein-specific antibodies. In the classic approach, antigens (or targets of interest) are first immobilized on a solid matrix (typically a polystyrene microtiter plate) through either adsorption or a "capture" antibody (sandwich ELISA). Next, an antigen-targeting "detection" antibody is added followed by an enzyme-linked (e.g., horseradish peroxidase, HRP) secondary antibody which produces a chromogenic or fluorogenic signal when mixed with the appropriate substrate solution, giving an indication of the amount of antigen present in the sample (Lequin 2005). In ELISA, the "capture" and "detection" moieties can be replaced by many protein-protein or protein-chemical interacting modules, including streptavidin, biotin, peptides, protein affinity tags, and nucleic acids, making this technique modular and flexible. Jagelska and colleagues reconfigured the classic ELISA format to measure p53-DNA binding by immobilizing biotin-conjugated p53 DNA response element onto streptavidin-treated plates.

p53-containing samples were then added and detected using a p53-specific antibody (Jagelska et al. 2002). Although ELISA is amenable to high-throughput applications and can be highly specific, the success is heavily reliant on the availability of good antibodies and faces caveats like moderate sensitivity and low signal-to-noise. In the following sections, we examine new technologies developed more recently to interrogate transcription factors functionally and their ability to bind DNA sequence—specifically through the unique integration of materials, reagents, and techniques.

2.4 Optical Biosensor for Detecting Transcription Factors

Optical biosensors are powerful tools for the functional study of transcription factors due to their high specificity, sensitivity, and cost-effectiveness as compared to conventional bioassays like EMSA and DNase footprinting (Garner and Revzin 1981; Galas and Schmitz 1978). Optical biosensors typically comprise optically labeled probes and optical transducers to facilitate the detection of protein functions. In the last decade, technological development of optical biosensors has experienced significant growth in studying sequence-specific TF-DNA interactions due to the increase demands for direct, real-time, and label-free sensing. Three different types of optical sensing techniques including colorimetric, fluorescence, and surface plasmon resonance (SPR) have been employed extensively for this purpose and are discussed here.

2.4.1 Fluorescence Assays

Fluorescence assays are one of the most widely applied optical techniques to study protein-DNA interactions. Noureddine and coworkers developed a fluorescent microsphere-based technique termed MAPD (microsphere assay for protein-DNA binding) that can measure p53-DNA binding in a multiplexed platform. Microspheres carrying individual fluorescent signatures are annealed with different p53 response elements of varying binding affinities (p21, PUMA, consensus sequence A and C, GADD45, and non-binding control DNA) and exposed to p53-activating drug (doxorubicin)-treated whole-cell lysates containing endogenous p53 in a multiplexed reaction. Using fluorescently tagged antibody to detect bead-bound p53 molecules, a profile of relative fluorescence intensity detailing p53 binding levels (from DNA binding) for each respective microsphere-RE is generated, showing the degree of sequence-specific DNA interactions (Noureddine et al. 2009). MAPD assay was highly sensitive and could accurately discern the binding affinities of wild-type and mutant p53 (R175H and S121F) towards different REs as well as sequences carrying single-nucleotide mutations (SNPs). Additionally, MAPD binding data also correlated to results from luciferase transactivation reporter assay in cells, demonstrating biological relevance (Noureddine et al. 2009). In another study that targets the p53 pathway, Goh and colleagues developed a biosensor using a conditionally fluorescing molecular rotor conjugate. Molecular rotors are a unique class of fluorescent chemicals that can undergo twisted intramolecular charge transfer (TICT), according to an optical property of conditional fluorescence when excited in a sterically restrictive molecular environment (due to a red-shifted fluorescence emission instead of non-radiative torsional relaxation (Grabowski et al. 2003)). Julolidine rotor, when conjugated to the 12.1 peptide sequence (p53 N-terminal analogue that binds Mdm2 protein), behaved as a switchable molecular sensor for the presence of Mdm2 proteins. Experimental and computational data shows that the binding-induced alpha helix of the peptide-rotor conjugate can be subtly altered through single-amino acid substitution to suit the modality of protein-protein interaction and fluorescence turn-on sensitivity (Goh et al. 2014). Through the use of a cell-penetrating fluorophore with the aggregation-induced emission (AIE) property, the application was further developed for detecting p53 transcriptional activity in live cells through microscopy imaging by visualizing the increase in Mdm2 production following p53 induction (Geng et al. 2015). In a separate microscopy technique, streptavidin-coated magnetic beads conjugated to different DNA-REs displayed preferential binding when exposed to different variants of the p53 transcription factor (binding-competent wild-type p53 or inactivated mutant p53) visualized through the use of a fluorescently labeled anti-p53 antibody. The authors further demonstrate a multiplexing function by attaching different fluorescent dyes with unique DNA sequences (Ong et al. 2012). In a similar concept from a seperate technolgy with higher throughput capabilities, biotinylated single-nucleotide polyphormic p53 proteins are microarrayed on neutravidin-dextran coated glass slides functionally assessed through binding to fluorescently, or radioactively labelled GADD45 DNA-RE (Boutell et al. 2004).

The basic concept of fluorescence resonance energy transfer (FRET) describes energy transfer between two chromophores where a donor chromophore in the ground state initially transfers energy to an acceptor chromophore through non-radioactive dipole-dipole coupling (Jares-Erijman and Jovin 2003). Here, the proximity between acceptor and donor chromophore plays an essential role in producing an effective energy transfer, typically in the range of 10–100 Å. Apart from distance, the spectral overlap integral (the effective overlap between acceptor chromophore's absorption/excitation spectrum and emission spectrum of the donor chromophore) is another key determinant of FRET efficacy. The efficiency of energy transfer (E) decreases very rapidly with increasing distance (r) between the donor and acceptor, according to the relationship $E \propto [1 + (r/R_0)^6]^{-1}$, where R_0 is the distance at which E is 50%.

Ambra and coworkers developed a FRET-based protein-DNA binding assay for the successful detection of an active form of NF-κB, p50 (Giannetti et al. 2006). FRET was harnessed to study TF-DNA binding interaction between p50 proteins and double-strand DNA (dsDNA) immobilized in a glass capillary. The complementary sequence of the single-stranded DNA (ssDNA) is labeled with a Cy5 dye, and the p50 protein with a black hole quencher (BHQ-3), constituting an effective FRET pair. A change in fluorescence intensity occurs when p50 interacts with the DNA duplex. Accordingly, the optimal emission wavelength of Cy5-labeled DNA (670 nm) overlaps effectively with the excitation wavelength of BHQ-3 quencher (636 nm) when p50 binds DNA, resulting in a 90% drop in fluorescence intensity relative to pure Cy5 alone. Despite the effectiveness of this assay, fluorescence

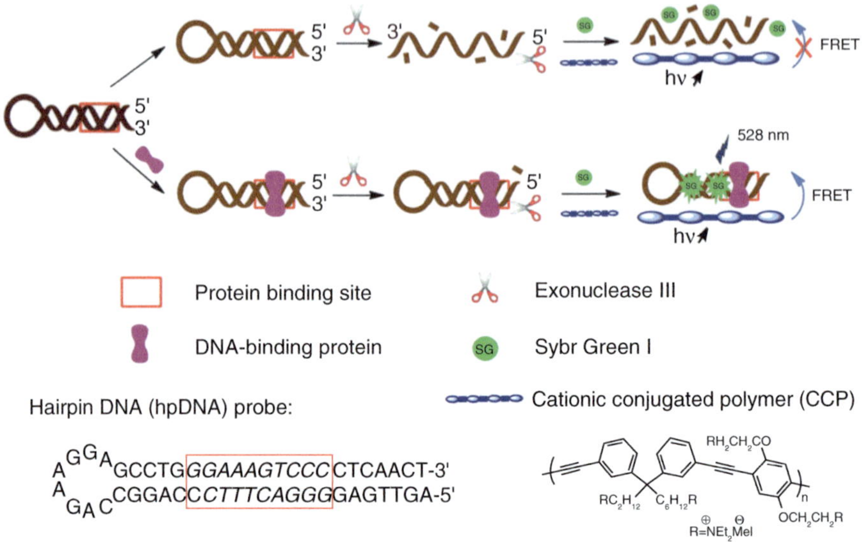

Fig. 2.1 Schematic illustration for the detection of DNA-binding protein (NF-κB) using hairpin DNA (hpDNA) and cationic conjugated polymer (CCP) (reproduced with permission from Liu et al. (2013), Copyright Elsevier, 2012)

labeling of proteins may limit its practical application, making label-free methods more attractive. Xingfen et al. developed a label-free FRET-based assay to study interactions between NF-κB and its target DNA (Liu et al. 2013). In this study, binding of the NF-κB protein to its DNA response elements shields it from digestion by exonuclease III (Fig. 2.1). The fluorescent cationic conjugated polymer (CCP) then interacts with the DNA duplex through strong electrostatic interactions with the DNA phosphate backbone, resulting in highly efficient FRET activity due to the presence of intercalated SYBR green by dsDNA that remains intact from p50 protection. In the absence of a sequence-specific binding protein, the enzyme digests the DNA duplex into single-stranded DNA fragments, preventing FRET activity. Furthermore, by using label-free hairpin DNA molecules containing two protein-binding site (PBS) as detection probes, an even lower detection limit of 1 pg/mL (with low error rates) has been achieved to detect NF-κB in HeLa cell nuclear extract. In another application, graphene oxide (GO) was used as the fluorochrome quencher. In this study, a FAM-labeled ssDNA carrying an NF-κB recognition site at the stem region of the hairpin conformation associates strongly onto the surface of the GO matrix (due to π-stacking interacting forces between the GO sheet and nucleotide bases) leading to fluorescence quenching. However, addition of NF-κB which binds to the κB consensus site on the hairpin leads to DNA desorption from the GO surface and FAM emission (Liu et al. 2012).

Certain fluorescence applications for sensing protein-DNA interactions can be labor intensive and unsuitable for complex biological samples. Molecular beacons are a class of facile, yet sensitive autonomous molecular sensors. In an early study, Heyduk and Heyduk developed a FRET-based molecular beacon for the detection of CAP proteins (a bacterial TF). The technique involves the use of a pair of DNA

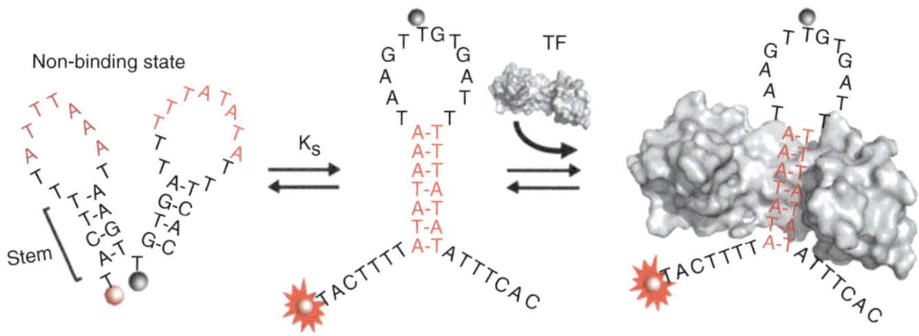

Fig. 2.2 Transcription factor (TF) beacons for the quantitative detection of DNA binding activity. DNA sequences containing the recognition site for a specific DNA-binding protein (here shown as red stem for TATA-binding protein (TBP)) are engineered into switches alternating between "non-binding" (*left*) and "binding-competent" (*right*) conformations. Binding of the protein to response element site present only in the "binding-competent" form shifts the switch's conformational equilibrium and stabilizes it in this form, which is associated with an increase in fluorescence due to the separation of dye and quencher (reproduced with permission from Vallée-Bélisle et al. (2011), Copyright American Chemical Society, 2011)

fragments each carrying a CAP-binding half-site conjugated to either an acceptor or a donor fluorochrome. In addition, each pair of half-sites carry a short complementary overhang which will only anneal when brought into proximity of each other during TF binding resulting in FRET and fluorescence activity. The background signal from spontaneous annealing of half-sites is kept low by making adjustments to probe concentration and complementary sequence (Heyduk and Heyduk 2002). Fang and colleagues developed a molecular beacon variant consisting of a hairpin-shaped single-stranded oligonucleotide labeled with a fluorophore/quencher pair at opposite ends. The oligonucleotide probe is designed to adopt a stem-and-loop structure in solution, bringing the fluorophore and quencher in close proximity which results in fluorescence quenching (Wang et al. 2009b; Fang et al. 2000). Additionally, the loop portion contains a sequence that is complementary to a target sequence which upon hybridization to target nucleic acids changes from a hairpin shape to the more rigid rodlike double helix. This conformational change forces the two arms of the hairpin to straighten, hence separating fluorophore and quencher and resulting in fluorescent activation (Vallée-Bélisle and Plaxco 2010; Tyagi 2009). The early utility of molecular beacons confined to ssDNA or ssDNA-binding proteins detection soon expanded to include more targets. Alexis and team developed a TF beacon strategy (Wang et al. 2009b; Stojanovic and Kolpashchikov 2004) based on the concept of structure-switching oligonucleotide probes. Accordingly, DNA probes conjugated to a fluorophore and quencher probe at two specific residues are designed with sequences that allow switching between two states of stem-loop structures in constant equilibrium. In the "non-binding" state, the formation of two smaller stem-loops results in the adjacent placement of dye and quencher leading to fluorescence quenching. In a second "binding-competent" conformation, the oligonucleotide probe takes a larger single stem-loop structure where the quencher is positioned distally from the fluorophore and also displays a TF-binding site at the stem region (Fig. 2.2). The addition of appropriate transcription factors

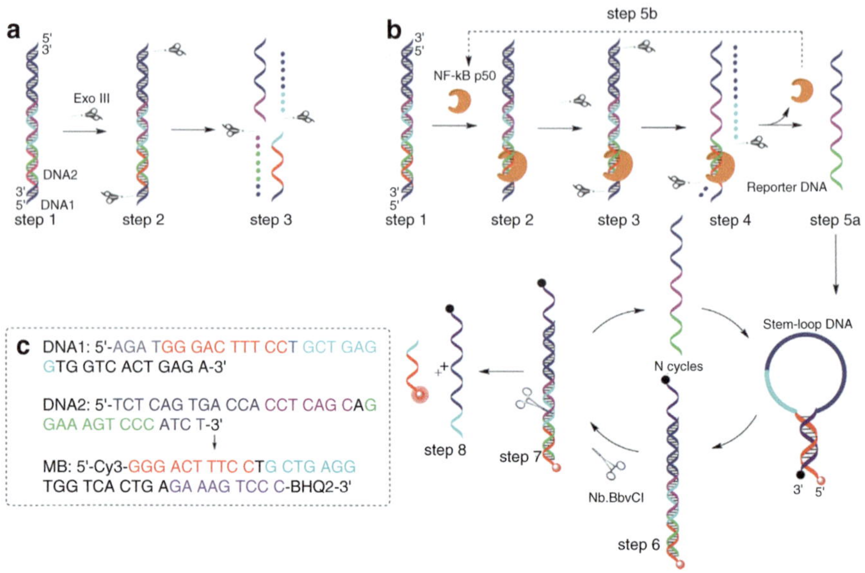

Fig. 2.3 Schematic representation of the TMDA fluorescence assay. Exonuclease III protection of DNA1/DNA2 duplex from NF-κB binding results in the release of reporter DNA fragments which hybridizes with stem-loop DNA (and fluorophore/quencher pair), constituting a Nb.BbvCI restriction site. Cleavage of this site releases free fluorophore into solution and returns reporter DNA fragment for hybridization process again (reproduced with permission from Zhang et al. (2016), Copyright Elsevier, 2016)

(TATA-binding protein, NF-κB, and Myc-Max heterodimers) stabilizes and shifts the equilibrium towards the 'binding-competent' form, leading to an increase in fluorescence activity. Additionally, the authors show that either conformation within the equilibrium can be stabilized by altering the probe's DNA sequence at the stem region (Vallee-Belisle et al. 2011).

In an even more sophisticated application of molecular beacon sensors, Zhang and coworkers describe a procedural method which generates a self-perpetuating signal amplification. The method begins with the protection of a specific DNA site from exonuclease III digestion through NF-κB (p50) binding. This leads to the liberation of a single-stranded 'reporter DNA' fragment which hybridizes with a stem-loop beacon probe (carrying a quencher/fluorophore pair) resulting in a dsDNA fragment containing a restriction endonuclease site not present before. Cleavage of this restriction site releases the fluorophore into the solution (hence increasing the fluorescence signal) and simultaneously releases the reporter DNA fragment, allowing it to target another stem-loop DNA probe, creating a self-perpetuating signal cycle (Fig. 2.3) (Zhang et al. 2016).

In another study, fluorescent readout from real-time qPCR cycling was used to sensitively and exponentially amplify detection signals from p53-DNA binding experiments. Double-stranded DNA probes each consisting of a different p53 response element (p21, PUMA, RGC, P2XM) placed adjacent to a qPCR quantifiable tag were used to detect sequence-specific DNA binding by immuno-capturing p53-DNA complexes before bound DNA are eluted and analyzed. Specific binding towards each RE was quantified by normalizing RE-binding signals against background signals from binding non-consensus DNA (qPCR tag alone), conveying absolute sequence-specific DNA binding values, and correlated well with published affinity constants. Furthermore, binding to different REs can be multiplexed in a single reaction by "barcoding" each RE with a unique qPCR tag that can be subsequently addressed with different primer sets (Goh et al. 2010). More recently, Sha et al. designed an elaborate sensor based on hairpin DNA cascade amplifier (HDCA). A dsDNA containing NF-kB p50 response element is first mixed with a specially designed ssDNA trigger in the presence of Ag+ to form a triplex. In the presence of p50, the triplex is destabilized leading to the release of the ssDNA trigger. The released trigger is then able to activate the HDCA, leading to the hybridization of specific hairpin probes, which in turn acts as an effective template for the formation of fluorescent CuNPs (Sha et al. 2016). This fluorescence-based biosensing strategy is ultrasensitive, achieving a detection limit of 0.096 pM with very high reproducibility.

2.4.2 Surface Plasmon Resonance

Surface plasmon resonance (SPR) relies on changes in refractive index at the surface/solution interface upon the binding of analyte for real-time measurements. SPR has been frequently applied for real-time monitoring of TF-DNA binding. To study the conformational effects of ligand binding on estrogen receptor alpha (ERα) and the induced selectivity towards different DNA elements (ERE or nonspecific DNA), SPR was combined with quartz crystal microbalance with dissipation monitoring (QCM-D) (Peh et al. 2007; Su et al. 2006). Here, it was observed that specific ERα-ERE complexes adopted a more compact conformation as compared to nonspecific complexes. QCM-D thus allowed for the study of conformational changes arising from ER-DNA interactions. Further evaluation of the binding capacity of ERα to ERE revealed that ligand binding affected viscoelasticity and structural conformations of protein-DNA complexes. SPR was used in this study to complement QCM as a tool for direct quantitative analysis of protein-DNA binding, as well as to elucidate ligand-dependent ERα binding capacity.

Apart from a direct detection of TF-DNA binding, additional surface modifications can allow for multiple detection modes and additional utility. As demonstrated by Wang et al., a sandwich assay format was adopted to achieve low detection limits and simultaneous measurement of total proteins using cancer cell lysates. In addition, wild-type p53 and mutant p53 were interrogated simultaneously by a dual-channel SPR technique. The surface of the SPR chip was co-immobilized with both the consensus dsDNA, to which wild-type p53 has high affinity, and monoclonal antibodies allowing the capture and quantitation of both wild-type and mutant p53 proteins (Wang et al. 2009c). This technique offers several advantages such as low

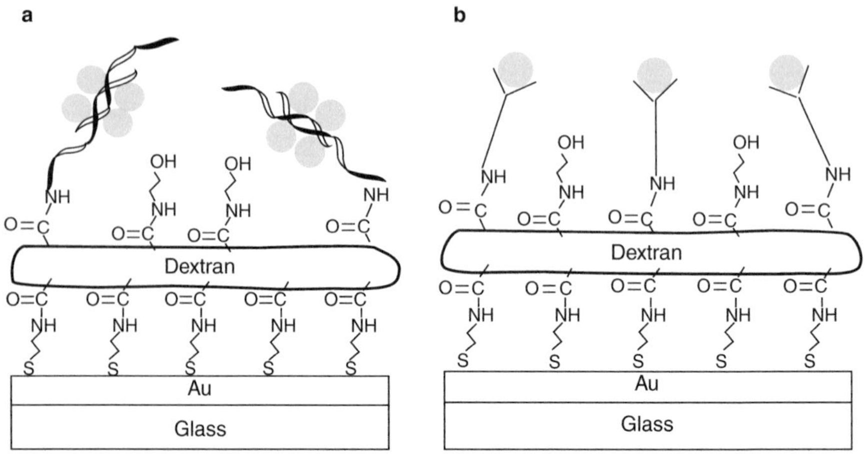

Fig. 2.4 Schematic diagram showing SPR-based detection of wild-type p53 through consensus DNA response elements (*left*), and total p53 proteins using monoclonal antibodies (*right*) in separate fluidic channels over a gold sensor chip functionalized with dextran (reproduced with permission from Wang et al. (2009c), Copyright American Chemical Society 2009)

detection limits for p53 proteins (10.6 pM for wild-type p53 and 1.06 pM for total p53 proteins), high specificity, and the feature of quantifying mutant p53 levels through signal differences between wild-type and total p53 proteins. Moreover, the dynamic range of the assay is impressive, allowing accurate measurement of p53 over a wide concentration range (Fig. 2.4).

2.4.3 Colorimetric Assay

Colorimetric assays are highly applicable as point-of-care diagnostics due to their instrument-free nature. Generally, noble metals such as gold or silver nanoparticles are suitable as colorimetric indicators due to their excellent extinction coefficients and strong distance-dependent optical properties (Wang et al. 2009c; Liu et al. 2009; Thaxton et al. 2006). Numerous colorimetric techniques have been developed for the sensitive and visually enabled analysis of metal ions, small molecules, proteins, as well as transcription factor-DNA binding. For example, colorimetric assays have been designed to sense estrogen receptor (ER) and specificity protein 1 (SP1) using metal nanoparticle probes (Tan et al. 2010a, b, 2011, 2013, 2014; Seow et al. 2015). One example is the measurement of ERα binding to its response elements (ERE), which for the purpose of this scheme, involved half-sites of the full response element conjugated on metal nanoparticles. Interaction of ER with nanoparticle-ERE probes leads to a decrease in aggregation (red spheres)

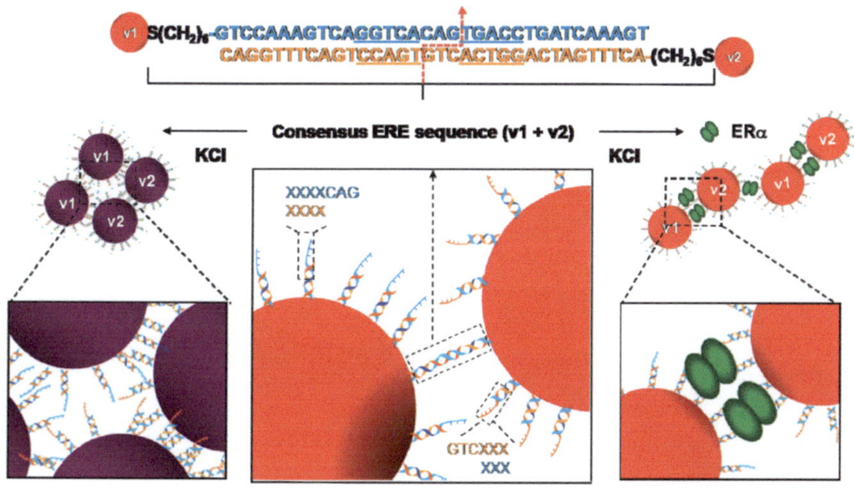

Fig. 2.5 Schematic diagram of AuNP colorimetric sensing of ER-DNA binding principle. Gold nanoparticles (AuNPs) are modified to carry either half of an ERE sequence (v1 and v2) with a 3-base complementary overhang and will aggregate spontaneously when mixed (*middle*). Addition of KCl salt reduces charge repulsion between DNA-AuNPs and promotes rapid particle aggregation and a consequent solution color change from red to purple (*left*). Addition of ERα results in DNA binding to full ERE sequence and exerts steric force to stabilize AuNPs resulting in solution color to remain red (*right*) (reproduced with permission from Tan et al. (2010b), Copyright American Chemical Society, 2010)

of DNA-metal nanoparticles from the introduction of steric protection forces between the nanoparticles in the presence of salt (Fig. 2.5).

Yan and group reported a user-friendly and sensitive colorimetric method to detect NF-kB p50 with an isothermal exponential amplification reaction (EXPAR) approach. Sequence-specific binding of p50 to a specially designed dsDNA results in the blocking of exonuclease III activity at a position which preserves and releases a ssDNA "DNA trigger" molecule to initiate the EXPAR cycle. DNA triggers anneal with an EXPAR ssDNA template allowing the synthesis of the antisense strand (in the presence of DNA polymerases) and the introduction of a nicking endonuclease site in between two copies of DNA triggers (Fig. 2.6). Endonuclease activity at this site leads to the release of a DNA trigger molecule which participates in another EXPAR cycle, creating an exponential increase in ssDNA trigger molecules which eventually serves as reporter oligonucleotides by aggregating AuNP probes through sequence complementarity and producing a color change (Fig. 2.6) (Zhang et al. 2012). However, in the absence of an appropriate DNA-binding protein, exonuclease III quickly degrades the DNA duplex preventing EXPAR amplification and resulting in no AuNP aggregation.

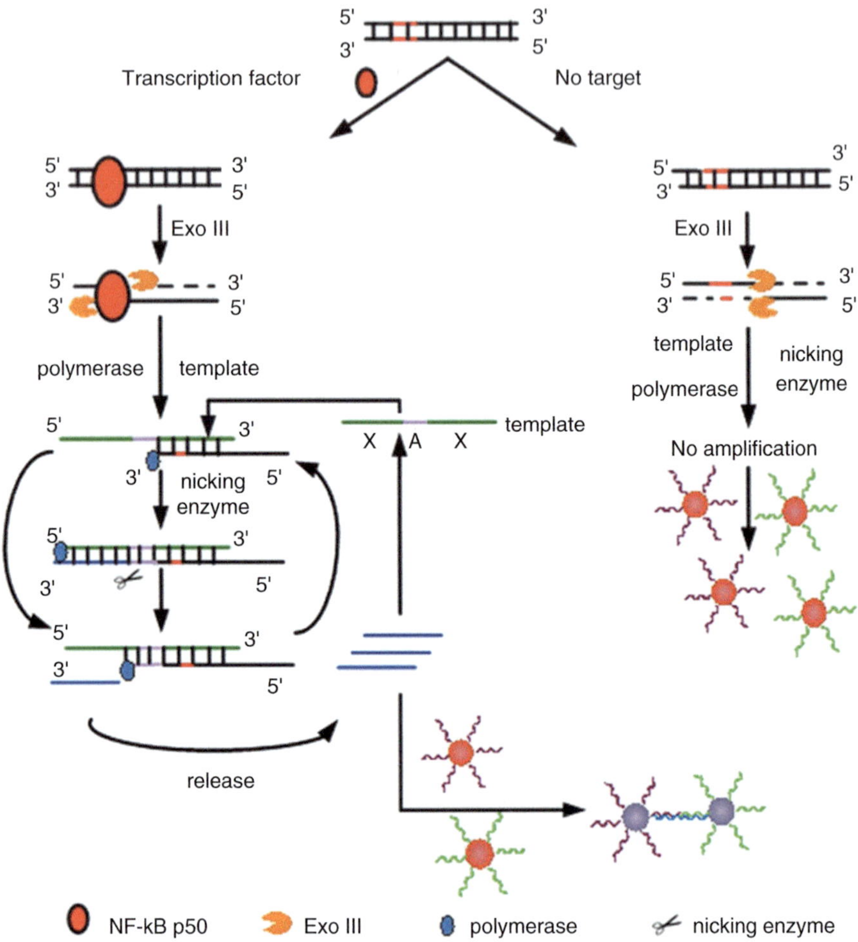

Fig. 2.6 Schematic diagram showing EXPAR-based colorimetric assay. Binding of p50 leads to the protection of dsDNA and subsequent release of trigger DNA molecule cycling and signal amplification in further EXPAR reactions. Trigger ssDNA also acts as reporter oligonucleotides that promote AuNP aggregation through DNA hybridization producing a colorimetric readout (reproduced with permission from Zhang et al. (2012), Copyright American Chemical Society, 2012)

2.5 Electrical Biosensors for Transcription Factor Detection

Electrical biosensors which include electrochemical sensors and electronic sensors are usually accurate, fast, and sensitive methods for molecular sensing. In addition, they provide operating simplicity, the option for miniaturization, cost-effectiveness, and have attracted much attention in the area of point-of-care diagnostics.

Electrochemical biosensors based on DNA-mediated charge transport offer an interesting approach to study transcription factor-DNA binding. Gorodetsky et al. demonstrated the use of DNA-modified microelectrodes to rapidly detect nanomolar concentrations of TATA-binding proteins (TBP), a ubiquitous transcription

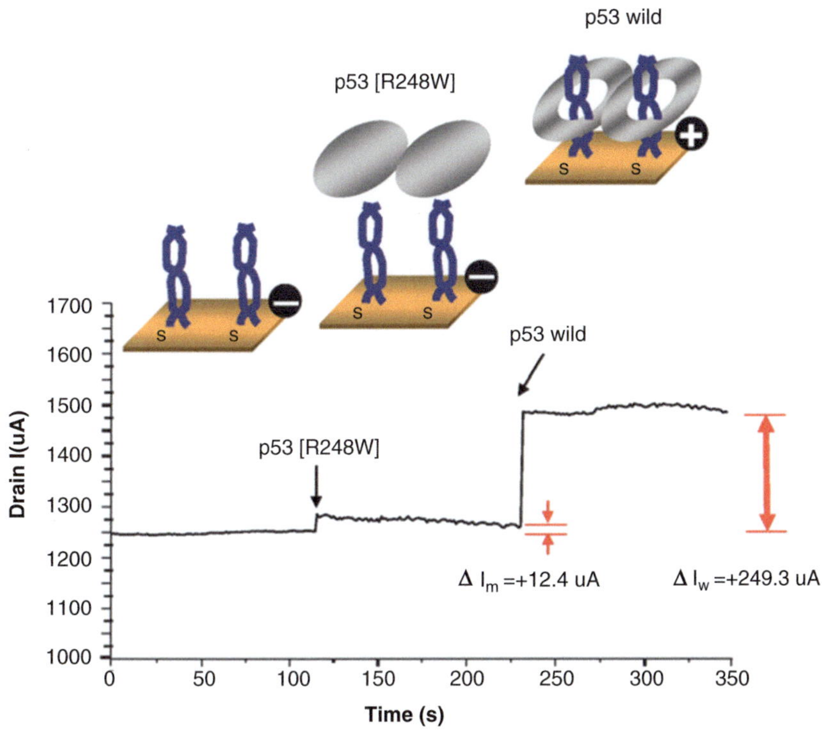

Fig. 2.7 Real-time monitoring of mutant p53 using MOSFET. Plot of drain current versus time in response to mutant p53 (R248W) and wild-type p53 on a single MOSFET device at an applied voltage of 2.0 V (reproduced with permission from Han et al. (2010), Copyright Elsevier, 2010)

factor (Gorodetsky et al. 2008). The double-stranded TBP-specific REs were immobilized on microelectrodes via Au-S chemistry and the distal ends of REs were modified with redox-active Nile Blue to give electrochemical signals. The binding of TBP bends the duplex RE and decreases the DNA-mediated reduction of Nile Blue, thus lowering the electrochemical signal. This electrochemical sensor can also be easily modified and applied to other TFs. It is also worth highlighting that the use of a microelectrode array can further allow the multiplexed detection of a panel of TFs on a single chip.

Besides electrochemical biosensors, metal oxide semiconductor field-effect transistor (MOSFET) is another popular electrical biosensor consisting of four terminals including the source, gate, drain, and substrate. Operation of the MOSFET depends on the electric field to control the size and shape of a channel from the source to the drain. Upon exposure to an analyte, a gate modulates the flow of electrons through the channel, thereby inducing changes in the drain current. Based on this principle, Han et al. pioneered the design of a field effect transistor (FET)-based biosensor to evaluate the DNA binding activity of wild-type and mutant p53 proteins. The MOSFET was immobilized with p53-specific GADD45 REs. As shown in Fig. 2.7, a significant increase in drain current was observed upon the addition of 100 nM wild-type p53, whereas addition of mutant p53 protein (R248W) gives no response (Han et al. 2010). This label-free method also allows real-time monitoring of p53-DNA binding. However, no detailed calibration has yet been performed to determine the limit of detection for this MOSFET; thus more studies are necessary to further apply this technology.

More recently, transcription factor biosensing at the level of single molecules has been successfully achieved by Squires et al. using a solid-state nanopore platform. Nanopores are label-free and ultrasensitive biosensors that are usually used to characterize biopolymers such as DNA, RNA, or proteins at the single-molecule level. An electrical field is applied to the nanopore to guide movement of the biopolymer into the nanopore, thus allowing the study of individual molecules. The ability to rapidly measure hundreds of samples and to resolve fine structural features alludes to the potential of nanopores in TF sensing. In this study, the DNA-binding domain of the early growth response protein 1 (EGR1), also known as zinc finger protein zif268, was used as the model TF (Squires et al. 2015). zif268-DNA binding was detected according to current blockage sublevels and duration of translocation through the nanopore. It was also demonstrated that different binding modes of zif268 will give rise to distinct current blockage patterns, demonstrating the feature of characterizing TF protein conformation. This unique nanopore technique provides a novel way to study transcription factor-DNA binding at the single-molecule level and will undeniably unveil new information about detailed molecular interaction.

2.6 Other Sensing Technology

In this section, we briefly visit TF sensing techniques using alternative detection methods as well as powerful protein sensing methods that could be repurposed for the functional sensing of transcription factors.

As an alternative to ELISA, Oberlander et al. developed a scintillation proximity assay (SPA) to measure total p53 protein in cell extracts. SPA beads are embedded scintillants, which give out light when they come close to radioactive compounds. The SPA beads are first immobilized with capture antibodies, and in the presence of p53 proteins, associate with biotinylated anti-53 antibodies. Addition of 35S-labeled streptavidin triggers the SPA, allowing photometric detection. This assay is sensitive enough to detect very low levels of p53 (50–300 pg) in small volumes of biological extracts, but requires the use of harmful radioactive labels (Oberlander et al. 2010). In a more recent EXPAR-based TF sensing technique, Ma and coworkers describe the detection of NF-κB p50 activity with remarkable sensitivity (10 fM) through the use of a dual-EXPAR scheme and G-quadruplex DNAzyme as reporter molecules. Binding of NF-κB p50 to a unique dsDNA provides protection from the nuclease activity of exonuclease I and III, sequentially added to disintegrate unbound DNA molecules. Intact DNA copies remaining serve as a template for RNA polymerase, producing RNA trigger molecules. RNA triggers then prime EXPAR ssDNA templates for DNA polymerization, producing a DNA duplex containing two trigger copies and an HRP-mimicking DNAzyme separated by nicking endonuclease sites. Cleavage of these sites releases more DNA trigger molecules (to initiate additional EXPAR cycles), a DNAzyme reporter molecule and the initial RNA trigger-EXPAR DNA template-bound fragment for elongation by DNA polymerase again; hence constituting a self-perpetuating signal amplification cycle. DNAzyme reporter molecules produced eventually catalyze a luminol-dependent chemiluminescence signal in the presence of hemin (Ma et al. 2014).

A powerful technique that is highly amenable for detecting TF-DNA interaction was developed by Langer et al. and involves a biochip with electrically actuated DNA levers able to sensitively detect hydrodynamic conformational changes. This biochip consists of four individually addressable flow channels on a glass substrate (Langer et al. 2013). Within each channel, six gold microelectrodes are immobilized with Cy3-labeled dsDNA. When positive potential is applied, the Au electrode attracts the negatively charged DNA molecules, leading to fluorescence quenching. When the potential is reversed, the DNA reverts back to an upright state, leading to fluorescence recovery. An epifluorescence setup is used to measure the change in fluorescence intensity during the DNA switching process. By applying designated capture sequences to the DNA fragment's distal ends, the DNA levers can specifically bind target proteins from solution. Protein binding slows down the DNA switching motion which is correlated to its hydrodynamic size. In addition, this method led to the development of an analytical model that predicts the hydrodynamic diameter of the bound protein from the kinetics of DNA-protein motion. This approach has also been successfully applied to detect post-translational modifications such as phosphorylation and glycosylation of proteins. Other notable advantages include microelectrode arrays for multiplexing and low sample consumption through the use of microfluidics.

2.7 Concluding Remarks

The last decade has witnessed a rapid advancement in the integration of biology, chemistry, and physics to yield novel, highly sensitive hybrid biosensors. This parallels the trend seen for latest-generation DNA-sequencing technologies that continue to push the limits of throughput and accuracy. The different detection methods reviewed here offer unique advantages each but present their own caveats. The popularity of fluorescence assays can be attributed to their high sensitivity, specificity, and multiplexing feasibility, but often require special labels. SPR techniques are label-free and provide real-time monitoring of binding kinetics but are less sensitive, and require extensive optimization and instrumentation. While colorimetric assays are more amenable as point-of-care detection tools due to their instrument-free and visually permissive detection modes, they often face the limitations of sample solution color. Electrical sensors can be highly sensitive down to the single-molecule level, but are prone to environmental interferences and may require expensive setups or complicated fabrication of sensor chips. Further integration, coupled with exciting advances in site-specific protein labeling (Proft 2010; Ravikumar et al. 2015), will advance the development of next-generation biosensors. These will find important use in multiplexed liquid biopsies to detect the ever-increasing number of clinically significant biomarkers.

References

Andre F, Bachelot T, Commo F, Campone M, Arnedos M, Dieras V, Lacroix-Triki M, Lacroix L, Cohen P, Gentien D, Adelaide J, Dalenc F, Goncalves A, Levy C, Ferrero JM, Bonneterre J, Lefeuvre C, Jimenez M, Filleron T, Bonnefoi H (2014) Comparative genomic hybridisation array and DNA sequencing to direct treatment of metastatic breast cancer: a multicentre, prospective trial (SAFIR01/UNICANCER). Lancet Oncol 15:267–274

Ascenzi P, Bocedi A, Marino M (2006) Structure-function relationship of estrogen receptor alpha and beta: impact on human health. Mol Asp Med 27:299–402

Babu MM, Luscombe NM, Aravind L, Gerstein M, Teichmann SA (2004) Structure and evolution of transcriptional regulatory networks. Curr Opin Struct Biol 14:283–291

Balagurumoorthy P, Sakamoto H, Lewis MS, Zambrano N, Clore GM, Gronenborn AM, Appella E, Harrington RE (1995) Four p53 DNA-binding domain peptides bind natural p53-response elements and bend the DNA. Proc Natl Acad Sci U S A 92:8591–8595

Beckerman R, Prives C (2010) Transcriptional regulation by p53. Cold Spring Harb Perspect Biol 2:a000935

Beishline K, Azizkhan-Clifford J (2015) Sp1 and the 'hallmarks of cancer'. FEBS J 282:224–258

Belyi VA, Ak P, Markert E, Wang H, Hu W, Puzio-Kuter A, Levine AJ (2010) The origins and evolution of the p53 family of genes. Cold Spring Harb Perspect Biol 2:a001198

Bieging KT, Mello SS, Attardi LD (2014) Unravelling mechanisms of p53-mediated tumour suppression. Nat Rev Cancer 14:359–370

Blackwood EM, Eisenman RN (1991) Max: a helix-loop-helix zipper protein that forms a sequence-specific DNA-binding complex with Myc. Science 251:1211–1217

Boutell et al (2004) Proteomics 4(7):1950–1958

Brenowitz M, Senear DF, Shea MA, Ackers GK (1986) Quantitative DNase footprint titration: a method for studying protein-DNA interactions. Methods Enzymol 130:132–181

Bullock AN, Fersht AR (2001) Rescuing the function of mutant p53. Nat Rev Cancer 1:68–76

Cai Q, Yan L, Xu Y (2015) Anoikis resistance is a critical feature of highly aggressive ovarian cancer cells. Oncogene 34:3315–3324

Chang GS, Chen XA, Park B, Rhee HS, Li P, Han KH, Mishra T, Chan-Salis KY, Li Y, Hardison RC, Wang Y, Pugh BF (2014) A comprehensive and high-resolution genome-wide response of p53 to stress. Cell Rep 8:514–527

Chen FE, Huang DB, Chen YQ, Ghosh G (1998) Crystal structure of p50/p65 heterodimer of transcription factor NF-kappaB bound to DNA. Nature 391:410–413

Cheng Q, Chen J (2010) Mechanism of p53 stabilization by ATM after DNA damage. Cell Cycle 9:472–478

Chin PL, Momand J, Pfeifer GP (1997) In vivo evidence for binding of p53 to consensus binding sites in the p21 and GADD45 genes in response to ionizing radiation. Oncogene 15:87–99

Cho Y, Gorina S, Jeffrey PD, Pavletich NP (1994) Crystal structure of a p53 tumor suppressor-DNA complex: understanding tumorigenic mutations. Science 265:346–355

Ciriello et al (2015) Cell 8;163(2):506–519. doi:10.1016/j.cell.2015.09.033

DeBerardinis RJ, Thompson CB (2012) Cellular metabolism and disease: what do metabolic outliers teach us? Cell 148:1132–1144

Dornan D, Hupp TR (2001) Inhibition of p53-dependent transcription by BOX-I phospho-peptide mimetics that bind to p300. EMBO Rep 2:139–144

Eferl R, Wagner EF (2003) AP-1: a double-edged sword in tumorigenesis. Nat Rev Cancer 3:859–868

el-Deiry WS, Kern SE, Pietenpol JA, Kinzler KW, Vogelstein B (1992) Definition of a consensus binding site for p53. Nat Genet 1:45–49

el-Deiry WS, Tokino T, Velculescu VE, Levy DB, Parsons R, Trent JM, Lin D, Mercer WE, Kinzler KW, Vogelstein B (1993) WAF1, a potential mediator of p53 tumor suppression. Cell 75:817–825

Fan D, Liu SY, van Hasselt CA, Vlantis AC, Ng EK, Zhang H, Dong Y, Ng SK, Chu R, Chan AB, Du J, Wei W, Liu X, Liu Z, Xing M, Chen GG (2015) Estrogen receptor alpha induces prosurvival autophagy in papillary thyroid cancer via stimulating reactive oxygen species and extracellular signal regulated kinases. J Clin Endocrinol Metab 100:E561–E571

Fang X, Li JJ, Tan W (2000) Using molecular beacons to probe molecular interactions between lactate dehydrogenase and single-stranded DNA. Anal Chem 72:3280–3285

Fried MG (1989) Measurement of protein-DNA interaction parameters by electrophoresis mobility shift assay. Electrophoresis 10:366–376

Friedler A, Veprintsev DB, Freund SM, von Glos KI, Fersht AR (2005) Modulation of binding of DNA to the C-terminal domain of p53 by acetylation. Structure 13:629–636

Galas DJ, Schmitz A (1978) DNAase footprinting a simple method for the detection of protein-DNA binding specificity. Nucleic Acids Res 5:3157–3170

Garner MM, Revzin A (1981) A gel electrophoresis method for quantifying the binding of proteins to specific DNA regions: application to components of the Escherichia coli lactose operon regulatory system. Nucleic Acids Res 9:3047–3060

Geng J, Goh WLP, Zhang C, Lane D, Liu B, Ghadessy FJ, Tan YN (2015) A highly sensitive fluoresce light-up probe for real-time detection of endogenous protein target and its antagonism in live cells. J Mater Chem B 3:5933–5937

Giannetti A, Citti L, Domenici C, Tedeschi L, Baldini F, Wabuyele MB, Vo-Dinh T (2006) FRET-based protein–DNA binding assay for detection of active NF-κB. Sensors Actuators B Chem 113:649–654

Gilmore TD (2006) Introduction to NF-kappaB: players, pathways, perspectives. Oncogene 25:6680–6684

Goh W, Lane D, Ghadessy F (2010) Development of a novel multiplex in vitro binding assay to profile p53-DNA interactions. Cell Cycle 9:3030–3038

Goh WL, Lee MY, Joseph TL, Quah ST, Brown CJ, Verma C, Brenner S, Ghadessy FJ, Teo YN (2014) Molecular rotors as conditionally fluorescent labels for rapid detection of biomolecular interactions. J Am Chem Soc 136:6159–6162

Gorodetsky AA, Ebrahim A, Barton JK (2008) Electrical detection of TATA binding protein at DNA-modified microelectrodes. J Am Chem Soc 130:2924–2925

Grabowski ZR, Rotkiewicz K, Rettig W (2003) Structural changes accompanying intramolecular electron transfer: focus on twisted intramolecular charge-transfer states and structures. Chem Rev 103:3899–4032

Han SH, Kim SK, Park K, Yi SY, Park H-J, Lyu H-K, Kim M, Chung BH (2010) Detection of mutant p53 using field-effect transistor biosensor. Anal Chim Acta 665:79–83

Hanada R, Hanada T, Sigl V, Schramek D, Penninger JM (2011) RANKL/RANK-beyond bones. J Mol Med (Berl) 89:647–656

Hanahan D, Weinberg RA (2011) Hallmarks of cancer: the next generation. Cell 144:646–674

Hayden MS, Ghosh S (2012) NF-kappaB, the first quarter-century: remarkable progress and outstanding questions. Genes Dev 26:203–234

Heinlein CA, Chang C (2004) Androgen receptor in prostate cancer. Endocr Rev 25:276–308

Hermeking H, Lengauer C, Polyak K, He TC, Zhang L, Thiagalingam S, Kinzler KW, Vogelstein B (1997) 14-3-3 sigma is a p53-regulated inhibitor of G2/M progression. Mol Cell 1:3–11

Heyduk T, Heyduk E (2002) Molecular beacons for detecting DNA binding proteins. Nat Biotechnol 20:171–176

Hibino E, Inoue R, Sugiyama M, Kuwahara J, Matsuzaki K, Hoshino M (2016) Interaction between intrinsically disordered regions in transcription factors Spl and TAF4. Protein Sci 25:2006–2017

Hoesel B, Schmid JA (2013) The complexity of NF-kappaB signaling in inflammation and cancer. Mol Cancer 12:86

Huber MA, Azoitei N, Baumann B, Grunert S, Sommer A, Pehamberger H, Kraut N, Beug H, Wirth T (2004) NF-kappaB is essential for epithelial-mesenchymal transition and metastasis in a model of breast cancer progression. J Clin Invest 114:569–581

Hupp TR, Meek DW, Midgley CA, Lane DP (1992) Regulation of the specific DNA binding function of p53. Cell 71:875–886
Iwanicki MP, Chen HY, Iavarone C, Zervantonakis IK, Muranen T, Novak M, Ince TA, Drapkin R, Brugge JS (2016) Mutant p53 regulates ovarian cancer transformed phenotypes through autocrine matrix deposition. JCI Insight 1:e86829
Jacobs MD, Harrison SC (1998) Structure of an IkappaBalpha/NF-kappaB complex. Cell 95:749–758
Jacque E, Tchenio T, Piton G, Romeo PH, Baud V (2005) RelA repression of RelB activity induces selective gene activation downstream of TNF receptors. Proc Natl Acad Sci U S A 102:14635–14640
Jagelska E, Brazda V, Pospisilova S, Vojtesek B, Palecek E (2002) New ELISA technique for analysis of p53 protein/DNA binding properties. J Immunol Methods 267:227–235
Jares-Erijman EA, Jovin TM (2003) FRET imaging. Nat Biotechnol 21:1387–1395
Jerry DJ, Dunphy KA, Hagen MJ (2010) Estrogens, regulation of p53 and breast cancer risk: a balancing act. Cell Mol Life Sci 67:1017–1023
Ji H, Wu G, Zhan X, Nolan A, Koh C, De Marzo A, Doan HM, Fan J, Cheadle C, Fallahi M, Cleveland JL, Dang CV, Zeller KI (2011) Cell-type independent MYC target genes reveal a primordial signature involved in biomass accumulation. PLoS One 6:e26057
Joerger AC, Fersht AR (2007) Structure-function-rescue: the diverse nature of common p53 cancer mutants. Oncogene 26:2226–2242
Joerger AC, Fersht AR (2010) The tumor suppressor p53: from structures to drug discovery. Cold Spring Harb Perspect Biol 2:a000919
Jordan JJ, Menendez D, Inga A, Noureddine M, Bell DA, Resnick MA (2008) Noncanonical DNA motifs as transactivation targets by wild type and mutant p53. PLoS Genet 4:e1000104
Karin M, Ben-Neriah Y (2000) Phosphorylation meets ubiquitination: the control of NF-[kappa]B activity. Annu Rev Immunol 18:621–663
Khoury MP, Bourdon JC (2011) p53 isoforms: an intracellular microprocessor? Genes Cancer 2:453–465
Kuiper GG, Enmark E, Pelto-Huikko M, Nilsson S, Gustafsson JA (1996) Cloning of a novel receptor expressed in rat prostate and ovary. Proc Natl Acad Sci U S A 93:5925–5930
Kummerfeld SK, Teichmann SA (2006) DBD: a transcription factor prediction database. Nucleic Acids Res 34:D74–D81
Kussie PH, Gorina S, Marechal V, Elenbaas B, Moreau J, Levine AJ, Pavletich NP (1996) Structure of the MDM2 oncoprotein bound to the p53 tumor suppressor transactivation domain. Science 274:948–953
Kyewski B, Klein L (2006) A central role for central tolerance. Annu Rev Immunol 24:571–606
Lambert PF, Kashanchi F, Radonovich MF, Shiekhattar R, Brady JN (1998) Phosphorylation of p53 serine 15 increases interaction with CBP. J Biol Chem 273:33048–33053
Lane DP (1992) Cancer. p53, guardian of the genome. Nature 358:15–16
Langer A, Hampel PA, Kaiser W, Knezevic J, Welte T, Villa V, Maruyama M, Svejda M, Jähner S, Fischer F, Strasser R, Rant U (2013) Protein analysis by time-resolved measurements with an electro-switchable DNA chip. Nat Commun 4:2099
Laptenko O, Shiff I, Freed-Pastor W, Zupnick A, Mattia M, Freulich E, Shamir I, Kadouri N, Kahan T, Manfredi J, Simon I, Prives C (2015) The p53 C terminus controls site-specific DNA binding and promotes structural changes within the central DNA binding domain. Mol Cell 57:1034–1046
Laptenko O, Tong DR, Manfredi J, Prives C (2016) The tail that wags the dog: how the disordered C-terminal domain controls the transcriptional activities of the p53 tumor-suppressor protein. Trends Biochem Sci 41:1022–1034
Le Romancer M, Poulard C, Cohen P, Sentis S, Renoir JM, Corbo L (2011) Cracking the estrogen receptor's posttranslational code in breast tumors. Endocr Rev 32:597–622
Lee TI, Young RA (2000) Transcription of eukaryotic protein-coding genes. Annu Rev Genet 34:77–137

Lee TI, Young RA (2013) Transcriptional regulation and its misregulation in disease. Cell 152:1237–1251
Lequin RM (2005) Enzyme immunoassay (EIA)/enzyme-linked immunosorbent assay (ELISA). Clin Chem 51:2415–2418
Liang J, Shang Y (2013) Estrogen and cancer. Annu Rev Physiol 75:225–240
Licht JD (2001) AML1 and the AML1-ETO fusion protein in the pathogenesis of t(8;21) AML. Oncogene 20:5660–5679
Lin CY, Loven J, Rahl PB, Paranal RM, Burge CB, Bradner JE, Lee TI, Young RA (2012) Transcriptional amplification in tumor cells with elevated c-Myc. Cell 151:56–67
Liou GY, Storz P (2010) Reactive oxygen species in cancer. Free Radic Res 44:479–496
Liu J, Cao Z, Lu Y (2009) Functional nucleic acid sensors. Chem Rev 109:1948–1998
Liu JJ, Song XR, Wang YW, Chen GN, Yang HH (2012) A graphene oxide (GO)-based molecular beacon for DNA-binding transcription factor detection. Nanoscale 4:3655–3659
Liu X, Ouyang L, Cai X, Huang Y, Feng X, Fan Q, Huang W (2013) An ultrasensitive label-free biosensor for assaying of sequence-specific DNA-binding protein based on amplifying fluorescent conjugated polymer. Biosens Bioelectron 41:218–224
Lohrum MA, Woods DB, Ludwig RL, Balint E, Vousden KH (2001) C-terminal ubiquitination of p53 contributes to nuclear export. Mol Cell Biol 21:8521–8532
Lu X, Liu DP, Xu Y (2013) The gain of function of p53 cancer mutant in promoting mammary tumorigenesis. Oncogene 32:2900–2906
Lukasik SM, Zhang L, Corpora T, Tomanicek S, Li Y, Kundu M, Hartman K, Liu PP, Laue TM, Biltonen RL, Speck NA, Bushweller JH (2002) Altered affinity of CBF beta-SMMHC for Runx1 explains its role in leukemogenesis. Nat Struct Biol 9(9):674
Ma F, Yang Y, Zhang CY (2014) Ultrasensitive detection of transcription factors using transcription-mediated isothermally exponential amplification-induced chemiluminescence. Anal Chem 86:6006–6011
Maestro MA, Cardalda C, Boj SF, Luco RF, Servitja JM, Ferrer J (2007) Distinct roles of HNF1beta, HNF1alpha, and HNF4alpha in regulating pancreas development, beta-cell function and growth. Endocr Dev 12:33–45
Malkin D (2011) Li-fraumeni syndrome. Genes Cancer 2:475–484
Mantovani F, Banks L (2001) The human papillomavirus E6 protein and its contribution to malignant progression. Oncogene 20:7874–7887
Martinez, LA. Mutant p53 and ETS2, a Tale of Reciprocity. Front Oncol 6, 35 (2016).
May MJ, Ghosh S (1997) Rel/NF-kappa B and I kappa B proteins: an overview. Semin Cancer Biol 8:63–73
Meek DW, Anderson CW (2009) Posttranslational modification of p53: cooperative integrators of function. Cold Spring Harb Perspect Biol 1:a000950
Menendez D, Inga A, Resnick MA (2009) The expanding universe of p53 targets. Nat Rev Cancer 9:724–737
Mermod N, O'Neill EA, Kelly TJ, Tjian R (1989) The proline-rich transcriptional activator of CTF/NF-I is distinct from the replication and DNA binding domain. Cell 58:741–753
Messina DN, Glasscock J, Gish W, Lovett M (2004) An ORFeome-based analysis of human transcription factor genes and the construction of a microarray to interrogate their expression. Genome Res 14:2041–2047
Meyer N, Penn LZ (2008) Reflecting on 25 years with MYC. Nat Rev Cancer 8:976–990
Milde-Langosch K (2005) The Fos family of transcription factors and their role in tumourigenesis. Eur J Cancer 41:2449–2461
Miyashita T, Reed JC (1995) Tumor suppressor p53 is a direct transcriptional activator of the human bax gene. Cell 80:293–299
Mognol GP, Carneiro FR, Robbs BK, Faget DV, Viola JP (2016) Cell cycle and apoptosis regulation by NFAT transcription factors: new roles for an old player. Cell Death Dis 7:e2199
Nakano K, Vousden KH (2001) PUMA, a novel proapoptotic gene, is induced by p53. Mol Cell 7:683–694
Narod (2011) Nat Rev Clin Oncol 8(11):669–676

Nolan E, Vaillant F, Branstetter D, Pal B, Giner G, Whitehead L, Lok SW, Mann GB, Rohrbach K, Huang LY, Soriano R, Smyth GK, Dougall WC, Visvader JE, Lindeman GJ (2016) RANK ligand as a potential target for breast cancer prevention in BRCA1-mutation carriers. Nat Med 22:933–939

Noureddine MA, Menendez D, Campbell MR, Bandele OJ, Horvath MM, Wang X, Pittman GS, Chorley BN, Resnick MA, Bell DA (2009) Probing the functional impact of sequence variation on p53-DNA interactions using a novel microsphere assay for protein-DNA binding with human cell extracts. PLoS Genet 5:e1000462

Oberlander S, Xie T, Chandrachud U, Gal S (2010) Scintillation proximity assay for total p53 protein as an alternative to ELISA. J Immunol Methods 360:173–177

Okuda M, Araki K, Ohtani K, Nishimura Y (2016) The interaction mode of the acidic region of the cell cycle transcription factor DP1 with TFIIH. J Mol Biol 428:4993–5006

Ong HJ, Siau JW, Zhang JB, Hong M, Flotow H, Ghadessy F (2012) Analysis of p53 binding to DNA by fluorescence imaging microscopy. Micron 43:996–1000

Peh WY, Reimhult E, Teh HF, Thomsen JS, Su X (2007) Understanding ligand binding effects on the conformation of estrogen receptor α-DNA complexes: a combinational quartz crystal microbalance with dissipation and surface plasmon resonance study. Biophys J 92:4415–4423

Piskacek S, Gregor M, Nemethova M, Grabner M, Kovarik P, Piskacek M (2007) Nine-amino-acid transactivation domain: establishment and prediction utilities. Genomics 89:756–768

Powell et al (2014) Cancer Discov 4(4):405–414

Proft T (2010) Sortase-mediated protein ligation: an emerging biotechnology tool for protein modification and immobilisation. Biotechnol Lett 32:1–10

Ravikumar Y, Nadarajan SP, Yoo TH, Lee CS, Yun H (2015) Unnatural amino acid mutagenesis-based enzyme engineering. Trends Biotechnol 33:462–470

Rayburn E, Zhang R, He J, Wang H (2005) MDM2 and human malignancies: expression, clinical pathology, prognostic markers, and implications for chemotherapy. Curr Cancer Drug Targets 5:27–41

Reed M, Woelker B, Wang P, Wang Y, Anderson ME, Tegtmeyer P (1995) The C-terminal domain of p53 recognizes DNA damaged by ionizing radiation. Proc Natl Acad Sci U S A 92:9455–9459

Riley T, Sontag E, Chen P, Levine A (2008) Transcriptional control of human p53-regulated genes. Nat Rev Mol Cell Biol 9:402–412

Robertson G, Bilenky M, Lin K, He A, Yuen W, Dagpinar M, Varhol R, Teague K, Griffith OL, Zhang X, Pan Y, Hassel M, Sleumer MC, Pan W, Pleasance ED, Chuang M, Hao H, Li YY, Robertson N, Fjell C, Li B, Montgomery SB, Astakhova T, Zhou J, Sander J, Siddiqui AS, Jones SJ (2006) cisRED: a database system for genome-scale computational discovery of regulatory elements. Nucleic Acids Res 34:D68–D73

Rudel D, Sommer RJ (2003) The evolution of developmental mechanisms. Dev Biol 264:15–37

Sammons MA, Zhu J, Drake AM, Berger SL (2015) TP53 engagement with the genome occurs in distinct local chromatin environments via pioneer factor activity. Genome Res 25:179–188

Sanda T, Lawton LN, Barrasa MI, Fan ZP, Kohlhammer H, Gutierrez A, Ma W, Tatarek J, Ahn Y, Kelliher MA, Jamieson CH, Staudt LM, Young RA, Look AT (2012) Core transcriptional regulatory circuit controlled by the TAL1 complex in human T cell acute lymphoblastic leukemia. Cancer Cell 22:209–221

Sasaki CY, Barberi TJ, Ghosh P, Longo DL (2005) Phosphorylation of RelA/p65 on serine 536 defines an I{kappa}B{alpha}-independent NF-{kappa}B pathway. J Biol Chem 280:34538–34547

Schaefer U, Schmeier S, Bajic VB (2011) TcoF-DB: dragon database for human transcription co-factors and transcription factor interacting proteins. Nucleic Acids Res 39:D106–D110

Seow N, Tan YN, Yung L-YL, Su X (2015) DNA-directed assembly of Nanogold dimers: a unique dynamic light scattering sensing probe for transcription factor detection. Sci Rep 5:18293

Sha L, Zhang X, Wang G (2016) A label-free and enzyme-free ultra-sensitive transcription factors biosensor using DNA-templated copper nanoparticles as fluorescent indicator and hairpin DNA cascade reaction as signal amplifier. Biosens Bioelectron 82:85–92

Shang Y (2007) Hormones and cancer. Cell Res 17:277–279

Smeenk L, van Heeringen SJ, Koeppel M, van Driel MA, Bartels SJ, Akkers RC, Denissov S, Stunnenberg HG, Lohrum M (2008) Characterization of genome-wide p53-binding sites upon stress response. Nucleic Acids Res 36:3639–3654

Solomon H, Buganim Y, Kogan-Sakin I, Pomeraniec L, Assia Y, Madar S, Goldstein I, Brosh R, Kalo E, Beatus T, Goldfinger N, Rotter V (2012) Various p53 mutant proteins differently regulate the Ras circuit to induce a cancer-related gene signature. J Cell Sci 125:3144–3152

Song H, Hollstein M, Xu Y (2007) p53 gain-of-function cancer mutants induce genetic instability by inactivating ATM. Nat Cell Biol 9:573–580

Squires A, Atas E, Meller A (2015) Nanopore sensing of individual transcription factors bound to DNA. Sci Rep 5:11643

Stender JD, Kim K, Charn TH, Komm B, Chang KC, Kraus WL, Benner C, Glass CK, Katzenellenbogen BS (2010) Genome-wide analysis of estrogen receptor alpha DNA binding and tethering mechanisms identifies Runx1 as a novel tethering factor in receptor-mediated transcriptional activation. Mol Cell Biol 30:3943–3955

Stojanovic MN, Kolpashchikov DM (2004) Modular aptameric sensors. J Am Chem Soc 126:9266–9270

Strom A, Hartman J, Foster JS, Kietz S, Wimalasena J, Gustafsson JA (2004) Estrogen receptor beta inhibits 17beta-estradiol-stimulated proliferation of the breast cancer cell line T47D. Proc Natl Acad Sci U S A 101:1566–1571

Su X, Lin C-Y, O'Shea SJ, Teh HF, Peh WY, Thomsen JS (2006) Combinational application of surface plasmon resonance spectroscopy and quartz crystal microbalance for studying nuclear hormone receptor-response element interactions. Anal Chem 78:5552–5558

Swedenborg E, Power KA, Cai W, Pongratz I, Ruegg J (2009) Regulation of estrogen receptor beta activity and implications in health and disease. Cell Mol Life Sci 66:3873–3894

Tan YN, Lai A, Su X (2014) Interrogating cooperative interactions of transcription factors with composite DNA elements using gold nanoparticles. Sci Adv Mater 6:1460–1466

Tan YN, Lee KH, Su X (2011) Study of single-stranded DNA binding protein–nucleic acids interactions using unmodified gold nanoparticles and its application for detection of single nucleotide polymorphisms. Anal Chem 83:4251–4257

Tan YN, Lee KH, Su X (2013) A study of DNA design dependency of segmented DNA-induced gold nanoparticle aggregation towards versatile bioassay development. RSC Adv 3:21604–21612

Tan YN, Su X, Liu ET, Thomsen JS (2010a) Gold-nanoparticle-based assay for instantaneous detection of nuclear hormone receptor– response elements interactions. Anal Chem 82:2759–2765

Tan YN, Su X, Zhu Y, Lee JY (2010b) Sensing of transcription factor through controlled-assembly of metal nanoparticles modified with segmented DNA elements. ACS Nano 4:5101–5110

Tebaldi T, Zaccara S, Alessandrini F, Bisio A, Ciribilli Y, Inga A (2015) Whole-genome cartography of p53 response elements ranked on transactivation potential. BMC Genomics 16:464

Thanos CD, Bowie JU (1999) p53 family members p63 and p73 are SAM domain-containing proteins. Protein Sci 8:1708–1710

Thaxton CS, Georganopoulou DG, Mirkin CA (2006) Gold nanoparticle probes for the detection of nucleic acid targets. Clin Chim Acta 363:120–126

Thomas C, Gustafsson JA (2011) The different roles of ER subtypes in cancer biology and therapy. Nat Rev Cancer 11:597–608

Tyagi S (2009) Imaging intracellular RNA distribution and dynamics in living cells. Nat Methods 6:331–338

Vallee-Belisle A, Bonham AJ, Reich NO, Ricci F, Plaxco KW (2011) Transcription factor beacons for the quantitative detection of DNA binding activity. J Am Chem Soc 133:13836–13839

Vallée-Bélisle A, Plaxco KW (2010) Structure-switching biosensors: inspired by nature. Curr Opin Struct Biol 20:518–526

Vang R, Levine DA, Soslow RA, Zaloudek C, Shih Ie M, Kurman RJ (2016) Molecular alterations of TP53 are a defining feature of ovarian high-grade serous carcinoma: a Rereview of cases lacking TP53 mutations in the cancer genome atlas ovarian study. Int J Gynecol Pathol 35:48–55

Vaquerizas JM, Kummerfeld SK, Teichmann SA, Luscombe NM (2009) A census of human transcription factors: function, expression and evolution. Nat Rev Genet 10:252–263

Venter JC, Adams MD, Myers EW, Li PW, Mural RJ, Sutton GG, Smith HO, Yandell M, Evans CA, Holt RA, Gocayne JD, Amanatides P, Ballew RM, Huson DH, Wortman JR, Zhang Q, Kodira CD, Zheng XH, Chen L, Skupski M, Subramanian G, Thomas PD, Zhang J, Gabor Miklos GL, Nelson C, Broder S, Clark AG, Nadeau J, McKusick VA, Zinder N, Levine AJ, Roberts RJ, Simon M, Slayman C, Hunkapiller M, Bolanos R, Delcher A, Dew I, Fasulo D, Flanigan M, Florea L, Halpern A, Hannenhalli S, Kravitz S, Levy S, Mobarry C, Reinert K, Remington K, Abu-Threideh J, Beasley E, Biddick K, Bonazzi V, Brandon R, Cargill M, Chandramouliswaran I, Charlab R, Chaturvedi K, Deng Z, Di Francesco V, Dunn P, Eilbeck K, Evangelista C, Gabrielian AE, Gan W, Ge W, Gong F, Gu Z, Guan P, Heiman TJ, Higgins ME, Ji RR, Ke Z, Ketchum KA, Lai Z, Lei Y, Li Z, Li J, Liang Y, Lin X, Lu F, Merkulov GV, Milshina N, Moore HM, Naik AK, Narayan VA, Neelam B, Nusskern D, Rusch DB, Salzberg S, Shao W, Shue B, Sun J, Wang Z, Wang A, Wang X, Wang J, Wei M, Wides R, Xiao C, Yan C et al (2001) The sequence of the human genome. Science 291:1304–1351

Verma IM, Stevenson JK, Schwarz EM, Van Antwerp D, Miyamoto S (1995) Rel/NF-kappa B/I kappa B family: intimate tales of association and dissociation. Genes Dev 9:2723–2735

Vogelstein B, Lane D, Levine AJ (2000) Surfing the p53 network. Nature 408:307–310

Vousden KH, Lane DP (2007) p53 in health and disease. Nat Rev Mol Cell Biol 8:275–283

Wan F, Lenardo MJ (2009) Specification of DNA binding activity of NF-kappaB proteins. Cold Spring Harb Perspect Biol 1:a000067

Wang B, Xiao Z, Ren EC (2009a) Redefining the p53 response element. Proc Natl Acad Sci U S A 106:14373–14378

Wang K, Tang Z, Yang CJ, Kim Y, Fang X, Li W, Wu Y, Medley CD, Cao Z, Li J (2009b) Molecular engineering of DNA: molecular beacons. Angew Chem Int Ed 48:856–870

Wang Y, Zhu X, Wu M, Xia N, Wang J, Zhou F (2009c) Simultaneous and label-free determination of wild-type and mutant p53 at a single surface plasmon resonance chip preimmobilized with consensus DNA and monoclonal antibody. Anal Chem 81:8441–8446

Weinberg RL, Veprintsev DB, Bycroft M, Fersht AR (2005) Comparative binding of p53 to its promoter and DNA recognition elements. J Mol Biol 348:589–596

Xie TX, Xia Z, Zhang N, Gong W, Huang S (2010) Constitutive NF-kappaB activity regulates the expression of VEGF and IL-8 and tumor angiogenesis of human glioblastoma. Oncol Rep 23:725–732

Zeron-Medina J, Wang X, Repapi E, Campbell MR, Su D, Castro-Giner F, Davies B, Peterse EF, Sacilotto N, Walker GJ, Terzian T, Tomlinson IP, Box NF, Meinshausen N, De Val S, Bell DA, Bond GL (2013) A polymorphic p53 response element in KIT ligand influences cancer risk and has undergone natural selection. Cell 155:410–422

Zhang K, Wang K, Zhu X, Xie M (2016) Sensitive detection of transcription factors in cell nuclear extracts by using a molecular beacons based amplification strategy. Biosens Bioelectron 77:264–269

Zhang Y, Hu J, Zhang C-y (2012) Sensitive detection of transcription factors by isothermal exponential amplification-based colorimetric assay. Anal Chem 84:9544–9549

Cancer Biomarker Immunosensing Monitoring Strategies via Graphene Surface-Engineered Materials

3

Shabi Abbas Zaidi

3.1 Introduction

Cancer is a group of associated diseases which involved abnormal cell growth possessing the potential to intrude or spread to other parts of the body. It is believed to be one of the most grave and fatal diseases responsible for large number of deaths worldwide. According to estimated data from the International Agency for Research on Cancer (IARC), 8.2 million cancer deaths were reported in addition to nearly 14.1 million new cancer cases worldwide in 2012. The situation is expected to be more aggravated by 2030 where roughly 1.5 times growth in new cancer cases and cancer deaths is estimated simply due to the growth and aging of the population (retrieved from Global Cancer Facts and Figures from American Cancer Society from http://www.cancer.org/research/cancerfactsstatistics/global, Cancer Facts and Figures 2016 from American Cancer Society, http://www.cancer.org/acs/groups/content/@research/documents/document/acspc-047079.pdf, Siegel et al. 2016). In one other report released by World Health Organization (WHO), the numbers of new cancer cases are expected to rise by about 70% over the next 20 years. It has also been reported that diagnostic and treatment costs for cancer patients cause about $263.8 billion annually in the USA only (retrieved from Cancer: Facts, Causes, Symptoms and Research from http://www.medicalnewstoday.com/info/cancer-oncology).

The cause of cancer diseases is dependent on various internal and external factors including inherited immune conditions, and genetic mutations and unhealthy diet, tobacco-chewing habit, and infections, respectively. These factors may act

S.A. Zaidi
Department of Chemistry, Kwangwoon University,
20 Kwangwoon-ro, Nowon-gu, Seoul 01897, South Korea
e-mail: shabizaidi79@gmail.com

P. Chandra et al. (eds.), *Next Generation Point-of-care Biomedical Sensors Technologies for Cancer Diagnosis*, DOI 10.1007/978-981-10-4726-8_3

together or in sequence to cause cancer. However, a careful consideration of our diet, and prevention of heavy consumption of alcohol and tobacco use, may curb the chances of cancer diseases completely. On the other hand, substantial proportion of certain cancers caused by infectious agents, such as human papillomavirus (HPV), hepatitis B virus (HBV), hepatitis C virus (HCV), human immunodeficiency virus (HIV), and *Helicobacter pylori* (*H. pylori*) could be evaded through vaccination, or by treating the infection. For instance, avoiding excessive sun exposure and indoor tanning devices may prove to be fruitful in skin cancer and which could save more than five million skin cancer patients that are diagnosed annually (retrieved from Cancer Facts and Figures 2016 from American Cancer Society, http://www.cancer.org/acs/groups/content/@research/documents/document/acspc-047079.pdf).

In many cancers, masses of tissues are formed called solid tumors unlike leukemia where generally there is no solid tumor formation. Generally, the tumors are categorized as benign and malignant tumors. The first type of tumors are stagnant and demonstrate limited growth with less harmful characteristics and symptoms, whereas the later ones are highly dangerous which manage to circulate throughout the body using the blood or lymphatic systems, damaging healthy tissue in a process called invasion, and feed on new blood vessels during the whole process which is termed as angiogenesis. In response to these conditions, certain benign (noncancerous) and cancerous cells produce special type of substances called as cancer markers (or tumor markers). In cancer cells, the production of tumor markers occurs at much higher levels in cancerous conditions which are secreted into the blood, urine, stool, tumor tissue, or other tissues or bodily fluids as compared to normal cells. Some tumor markers are associated with only one type of cancer, whereas others are associated with two or more cancer types. The tumor markers immediately stand out as superior prognostic or diagnostic tools for various types of cancer. There are some limitations to the use of tumor markers. Nevertheless, it is also sometimes possible that noncancerous conditions can cause the levels of certain tumor markers to increase. Therefore, it is usually recommended to combine the measurements of tumor markers with other tests, such as biopsies, to diagnose accurate condition of cancer (Bigbee and Herberman 2003; Sethi et al. 2013; Andriole et al. 2009).

There are nearly more than 100 types of cancers including carcinoma, sarcoma, lymphoma, leukemia, melanoma, multiple myeloma, brain and spinal cord tumors, neuroendocrine tumors, germ cell tumors, and carcinoid tumors. The types and names of cancer are usually originated for the organs or tissues where the cancers form. Nevertheless, the measurements of these tumor markers play critical roles in early determination of the type and the stage of the cancer in a patient which greatly improve the odds of successful treatment and survival (retrieved from What Is Cancer? From National Cancer Institute from https://www.cancer.gov/about-cancer/understanding/what-is-cancer). Most of the tumor markers are proteins and a complete list of these protein markers can be found elsewhere (Polanski and Anderson 2006). Besides protein tumor markers, some chemical substances have also been utilized to detect the cancer disease successfully. Cancer diagnostic science is currently undergoing an exemplary shift with the incorporation of molecular biomarkers as part of routine diagnosis panel via many detection techniques. Accordingly, the sensitive and reliable detection of cancer biomarkers provides an effective way for cancer screening and diagnosis, as well as evaluating the pathogenic processes, pharmacological responses to a therapeutic intervention, and prognosis of different cancers. Many different analysis techniques such as cytometric methods, polymerase chain reaction (PCR)-based methods, single-carbon nanotube field effect transistor, and fluorescence measurement have been developed (Jemal et al. 2005; Paterlini-Brechot and Benali-Furet 2007; Kang et al. 2007; Teker 2008).

However, one of the most sought criteria of any tumor marker detection technique is its high sensitivity, low cost, easy processing, and facile miniaturization of the diagnostic devices. Thus, immunosensors which use specific interaction between antigen and antibodies are emerging as a preferable approach in the clinical applications owing to their merits such as low cost and ease of miniaturization. Hence, until now, several immunoassay methods including fluorescence immunoassay (Matsuya et al. 2003; Cesaro-Tadic et al. 2004), enzyme-linked immunosorbent assay (ELISA) (Yates et al. 1999; Voller et al. 1978), chemiluminescence immunoassay (Fu et al. 2006), radioimmunoassay (Goldsmith 1975), mass spectrometric immunoassay (Hu et al. 2007; Aebersold and Mann 1997), electrophoretic immunoassay (Schmalzing and Nashabeh 1997), and polymerase chain reaction assay (Saito et al. 1999) have been carried out on the clinical serum sample measurements. Among several immunoassay techniques, electrochemical immunoassays have shown considerable potential owing to its intrinsic advantages including high sensitivity with antibody labeling, simple instrumentation, low cost, and easy portability. On the other hand, in some clinical applications, generally, a small tag is employed for antibody labeling, resulting in poor sensitivity. Thus, signal amplification strategies to improve the electrochemical response have been utilized. Furthermore, many approaches, especially the applications of various types of nanomaterials, are being utilized for greater sensitivity resulting from ultralow amount of target tumor markers (Zhong et al. 2010).

Since last two decades, nanomaterials (approximately less than 100 nm in size) have attracted tremendous interest among scientists owing to their extraordinary features such as magnetic, electrical, and optical properties; high specific surface area; and robust mechanical strength. There are various classes of nanomaterials which essentially depend on the number of dimensions lying within the nanometer range. Due to enormous influence of morphologies, and size of nanomaterials on their properties and applications, substantial efforts have been put forward for the controlled synthesis of nanostructured materials with novel morphologies (Chen and Chatterjee 2013; Zaidi and Shin 2015a, b, 2016a, b; Yusuf et al. 2015; Zaidi 2013; Ibrahim et al. 2012; Dar et al. 2012). Therefore, a wide variety of nanomaterials have also demonstrated their potential and appropriateness in the preparation of various nanocomposites (Umar et al. 2012; Zaidi and Shin 2014).

Among various nanostructured nanomaterials, graphene has attracted tremendous attention since its discovery, in fundamental and applied research because of its unique 2D honey comb structure with sp2 bonded carbon atoms, tunable surface chemistry, high surface area, and excellent electronic property (Pumera et al. 2010; Novoselov et al. 2004). The hybridization of graphene and other nanomaterials composites offers a fascinating platform for the utilization of electrochemical sensing applications with improved synergistic electro-catalytic properties (Liu et al. 2011). Hence, graphene and related nanocomposite have been employed in catalysts, sensors, energy storing, and many clinical studies such as cancer biomarker detection areas (Chen et al. 2012a).

In this review chapter, various graphene-based electrochemical sensors are discussed. We intend to summarize the selective work comprehensively which has been performed in this area. Hence, this review chapter is categorized into many sections based on various types of biomarkers and their detection methods exploited by using a wide variety of nanomaterials in combination of graphene. Each report is highlighted separately for the better understanding of synthesis strategy and its analytical performance.

3.2 Graphene-Engineered Surface and Their Applications for Protein-Based Biomarker Detection

3.2.1 Carcinoembryonic Antigen (CEA) Determination

Carcinoembryonic antigen (CEA), a glycoprotein, is one of the most extensively used clinical tumor markers. It is most often associated with colorectal cancer since its description in 1956; however, the CEA level in serum is usually being associated with different types of malignancies including ovarian carcinoma (Gould et al. 2000), breast cancer (Sahin et al. 1996; Kramer et al. 1998; Cameiro et al. 1998), lung cancer (Iwazawa et al. 2000; Hernandez et al. 2002), and cystadenocarcinoma (Kazuya et al. 1999). The determination of CEA amount is also an excellent indicator for tumor diagnoses of hollow organs such as gastrointestinal and respiratory. Thus, it is essential to develop a facile, sensitive, selective, and rapid method for CEA determination in serum in clinical research. Several works have appeared which employed many different sandwich-type immunosensor or simple immunosensor for the reliable detection of CEA as discussed below.

A label-free immunosensor for the susceptible detection of CEA using gold nanoparticle–thionine–reduced graphene oxide (AuNPs–TH–rGO) nanocomposite was synthesized (Kong et al. 2011). The nanocomposite was coated on glassy carbon electrode (GCE) and anti-CEA was immobilized over it. The sensor was characterized completely via scanning electron microscopy (SEM), ultraviolet–visible (UV–vis) spectrometry, electrochemical impedance spectroscopy (EIS), cyclic voltammetry (CV), and differential pulse voltammetry (DPV) studies. The formation of antigen–antibody complex displayed proportional decrease in the response current of CEA concentration in the range of 10–500 pg mL^{-1} and a detection limit (LOD) of 4 pg mL^{-1} was achieved. Su et al. (2011) studied an electrochemical immunosensor based on multiarmed starlike platinum nanowires (PtNWs) with biomolecular assembly as signal tags (CEA and horseradish peroxidase (HRP)) on an anti-CEA-functionalized graphene-sensing platform. The PtNWs provide better platform for conjugation of CEA and HRP. Two different supporting electrolytes, namely newborn cattle serum (NBCS) and acetate buffer solution (ABS, pH5.5), were used for competitive format assay. Both systems offered high LOD values of 5.0 pg mL^{-1} vs. 1.0 pg mL^{-1} in the linear concentration range of 0.01–60 ng mL^{-1} vs. 0.002–80 ng mL^{-1} toward CEA standards in the NBCS in comparison with ABS.

Tang et al. (2011) generated electrochemical signal for CEA when the catalytic recycling of the product for the antigen–antibody interaction by glucose oxidase

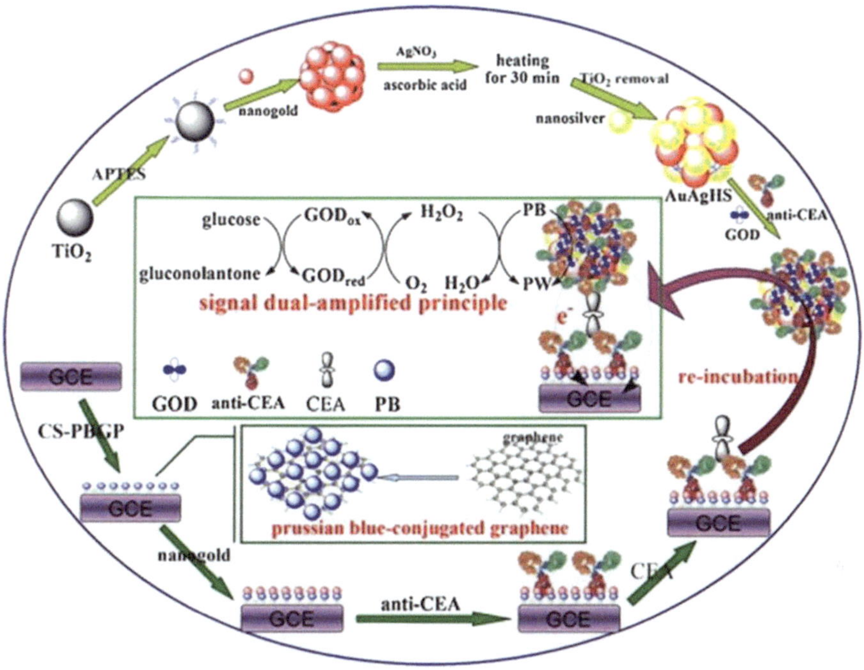

Fig. 3.1 (*Top*) Preparation process of anti-CEA-AuAgHS-GOD, (*bottom*) fabrication process of the electrochemical immunosensor, and (middle) the principle of signal dual-amplification (reproduced with permission from Elsevier publications from Tang et al. (2011))

(GOx)-conjugated gold–silver hollow microspheres (AuAgHSs) coupled with an artificial catalase and Prussian blue nanoparticles (PB) on a graphene-based immunosensing platform was employed in a dual-amplification approach as shown in Fig. 3.1. Firstly, catalytic oxidation of glucose is performed with the help of GOx which produced H_2O_2. Later on, the PB assisted in reducing the generated H_2O_2. This sensor worked in a wide linear dynamic range of 0.005–50 ng mL^{-1} and a LOD of 1.0 pg mL^{-1} for CEA was obtained. Chen et al. (2012b) demonstrated an assay for CEA based on a sandwich-type immunoassay protocol. In this work, horseradish peroxidase-labeled anti-CEA, as secondary antibodies (Ab_2), were immobilized on nanogold-patterned graphene oxide nanoscales (AuNPs-GO) in combination of biofunctionalized core-shell magnetic nanostructures. The magnetic nanocore had a shell composed of poly(o-phenylenediamine) (PPD) and metallic silver and exhibited enhanced adsorption properties for the attachment of anti-CEA antibody selective to CEA. The discussed immunosensor allowed the detection of CEA at a concentration as low as 1.0 pg mL^{-1}.

A similar dual-amplification approach was also employed by Zhou et al. (2012a) for the detection of CEA. Different types of monoclonal anti-CEA antibodies functionalized with nanoplatinum including multiarmed starlike platinum nanowires, hollow platinum nanospheres, and Pt nanostructures were used over gold nanocores (Pt@Au) as nanolables on the carbon nanospheres to prepare GOx-modified immunosensor. Using the functional Pt@Au nanolabels as molecular tags, the assay was implemented relative to glucose–hydroquinone system with a sandwich-type immunoassay. Among many nanolabels, Pt@Au nanostructures offered excellent analytical features with broad linear CEA concentration range from 0.001 to 120 ng mL^{-1} and the value of LOD was measured to be 0.5 pg mL^{-1}. Zhou et al. (2012b) designed a sandwich-type immunosensor for the detection of CEA using hollow platinum nanosphere (HPtNs)-labeled horseradish peroxidase-anti-CEA conjugates (HRP-anti-CEA) as molecular tags and anti-CEA-assembled carbon nanosphere-graphene hybrid nanosheets (CNS-GNS) as sensing probes. The direct electrolytic reduction and wet chemistry methods were employed to synthesize the probes, respectively. Among several labeling strategies, the improved dynamic concentration range between 0.001 ng mL^{-1} and 100 ng mL^{-1} with a LOD value of 1.0 pg mL^{-1} was obtained with HPtN labeling for CEA.

A simple immunoreaction for the determination of CEA was employed by Zhu and coworkers (Zhu et al. 2013). AuNP-decorated graphene composites (AuNPs–GN) were successfully synthesized followed by successful adsorption of HRP-labeled anti-CEA antibody (HRP-anti-CEA) and HRP over it. The synthesized materials were drop-coated on a GCE. The proposed sensor current change was proportional to the CEA concentration from 0.10 to 80 ng mL^{-1} with a LOD of 0.04 ng mL^{-1} under optimized experimental conditions. A highly sensitive sensor using a sandwich-type strategy was developed by Sun et al. (2013). In this work, primary antibody (Ab_1) was immobilized over a complex of graphene and 3-D AuNPs, whereas Ab_2 was functionalized on nanoporous silver (NPS) for detecting CEA. To prepare the immunosensor, HRP-Ab_2/TH/NPS signal label was used to capture CEA in the range between 0.001 and 10 ng mL^{-1}. A LOD value of 0.35 pg mL^{-1} and low limit of quantitation (LOQ) of 0.85 pg mL^{-1} were achieved by the proposed sensor.

Shi et al. (2014) utilized a label-free electrochemiluminescence (ECL) aptasensor using ECL property of cadmium sulfide-graphene (CdS–GR) nanocomposites with peroxydisulfate as the coreactant, for CEA. For easy immobilization of aptamer, fast electron transfer, and enhanced signal amplification, AuNPs were assembled on L-cysteine (L-Cys) during fabrication of sensor. The ECL intensity of proposed sensor was found proportional to the CEA concentration in the range of 0.01–10.0 ng mL^{-1} with a LOD of 3.8 pg mL^{-1}. The analytical application of sensor was shown for the recoveries of CEA in the real human serum samples in the range of 85.0–109.5% under RSD values of nearly 3.4%. Huang et al. (2015) used Ag/Au nanoparticle-modified graphene sheets (GS) for the immobilization of Ab_1 and as a tracer to label Ab_2 to fabricate electrochemical immunosensor for CEA monitoring. The 1,5-diaminonaphthalene (DN) molecules were employed for antibody immobilization onto GS. The sensor revealed a linear calibration range between 10 and 1.2×10^5 pg mL^{-1} with a LOD of 8 pg mL^{-1}.

Cryogel, a highly macroporous material, is fabricated under freezing and thawing process. The cryogenic conditions impart high loading capacity of immobilized agents such as enzyme owing to high surface area. Thus, Samanman et al. (2015) exploited the cryogenic conditions in order to prepare a composite of AuNPs, graphene, and chitosan (AuNPs-GP-CS) and coupled it to a silver deposition which acted as a redox mediator over Au electrode. Under optimized conditions, the decrease in the redox peak of silver was directly proportional to the CEA concentration in the range of 1.0×10^{-6}–1.0 ng mL^{-1} which offered a LOD value of

2.0×10^{-7} ng mL^{-1}. Unlike non-cryogel sensor, the proposed sensor exhibited 1.7 times higher sensitivity and 25 times lower detection limit which were attributed to high loading capacity of cryogels. Gao et al. (2015) studied the host–guest chemistry for nonenzymatic immunoassay for ultrasensitive electrochemical detection of CEA. For the preparation of sensor, β-CD-functionalized graphene nanosheet (β-CD-GS) was coated on electrode followed by immobilization of adamantine-modified Ab_1 (ADA-Ab_1) by supramolecular host–guest interaction. Finally, β-cyclodextrin-functionalized Cu@Ag (Cu@Ag-β-CD) core-shell nanoparticles were used as labels which could be captured by ADA-modified Ab_2 (ADA-Ab_2). This assembly resulted in high loading of Cu@Ag nanoparticles with enhanced electrical conductivity and high catalytic activity. This immunosensor exhibited worked quite well in broad CEA concentration range (0.0001–20 ng mL^{-1}) and offered a LOD of 20 fg mL^{-1}.

Pang et al. (2015) utilized a nanocomposite made up of GO/carboxylated multiwall carbon nanotubes/gold/cerium oxide nanoparticles (GO/MWCNTs-COOH/Au@CeO_2) as antibody carriers and sensing platforms over GCE, where the ECL property of CeO_2 NPs was exploited first time. The characterization of the sensor materials was carried out with various techniques as shown in Fig. 3.2. The proposed immunosensor displayed the wide linear concentration range of CEA (0.05–100 ng mL^{-1}) and the low LOD value of 0.02 ng mL^{-1}.

Feng et al. (2016) developed a sensitive electrochemical immunosensor for simultaneous detection of CEA and alpha-fetoprotein (AFP) using Cu_2O-GO-β-CD-anti-AFP and β-CD-graphene oxide-ferrocenecarboxylic acid (GO-β-CD-Fc-anti-CEA) as the distinguishable signal probes and GO-AuNPs as the sensor platform. The linear ranges from 0.001 ng mL^{-1} to 80 ng mL^{-1} for AFP and CEA with the LOD values of 0.0002 ng mL^{-1} for AFP and 0.0001 ng mL^{-1} for CEA were achieved. Wang et al. (2016) proposed a nonenzymatic sandwich-type immunosensor preparation approach by employing silver nanoclusters and GO nanocomposite (AgNCs/GO) as signal amplification tag for sensitive detection of CEA. This tag was functionalized with Ab_2 to capture Ab1 for the detection of CEA. The BET surface areas of the GO and the AgNCs-GO nanocomposites were estimated to be 107 and 327 m^2 g^{-1}, respectively, which revealed that inclusion of AgNCs offered high surface area for enhanced antibody immobilization and signal response. The designed immunosensor offered a LOD of 0.037 pg mL^{-1} with large linear range from 0.1 pg mL^{-1} to 100 ng mL^{-1} for CEA. Zhao et al. (2016) reported a novel ECL immunosensor for detection of CEA where poly (diallyldimethylammonium chloride) (PDDA), AuNPs, and anti-CEA were successively assembled on the surface of rGO-BaYF5:Yb, Er nanocomposites. The ECL assay revealed about high response of CEA in a linear range of 0.001–80 ng mL^{-1} with a LOD value of 0.87 pg mL^{-1}. Lee et al. (2017) also prepared a sandwich-type CEA electrochemical immunosensor by immobilizing anti-CEA between the Ab_1 and HRP-conjugated Ab_2 onto AgNPs-rGO-modified-SPEs. The proposed sensor exhibited a linear range of 0.05–0.50 mg mL^{-1} and a LOD of 0.035 mg mL^{-1} in comparison to non-sandwich counterpart, which provided a linear range of 0.05–0.40 mg mL^{-1}, with LOD value of 0.042 mg mL^{-1}.

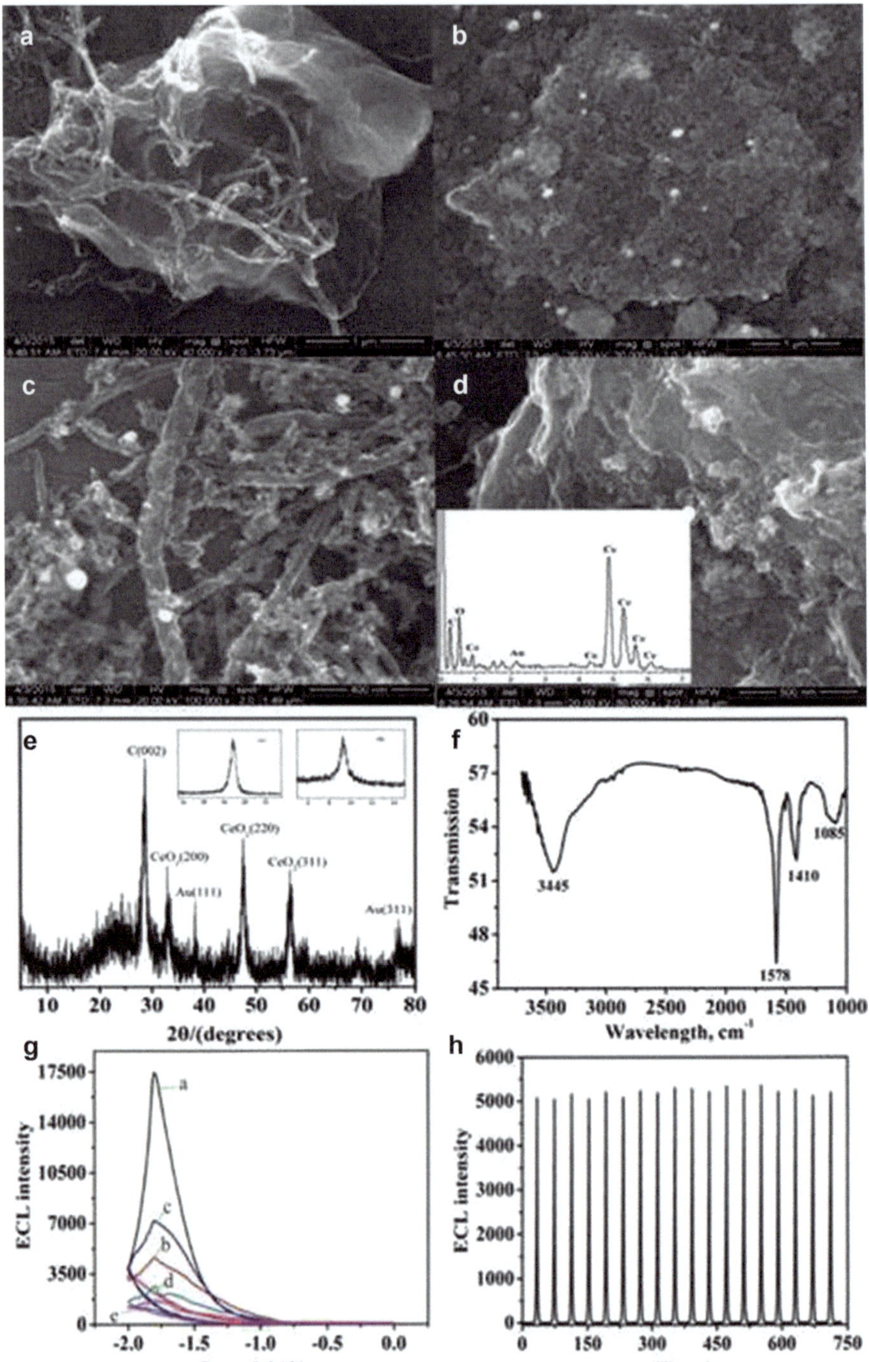
a
b
c
d
e
C(002)
CeO2(200)
Au(111)
CeO2(220)
CeO2(311)
Au(311)
10 20 30 40 50 60 70 80
2θ/(degrees)
f
Transmission
60 57 54 51 48 45
3445
1578
1410
1085
3500 3000 2500 2000 1500 1000
Wavelength, cm-1
g
ECL intensity
17500 14000 10500 7000 3500 0
a
b
c
d
e
-2.0 -1.5 -1.0 -0.5 0.0
Potential / V
h
ECL intensity
6000 5000 4000 3000 2000 1000 0
0 150 300 450 600 750
Time / s

3.2.2 Prostate-Specific Antigen (PSA) Determination

PSA is a protein produced by cells of the prostate gland. The elevated level of PSA in blood is associated with prostate cancer; however, a number of benign (not cancerous) conditions also cause a man's PSA level to rise. Thus, it is necessary to determine the accurate and selective method for PSA analysis.

Yang and Gong (2010) demonstrated an interesting strategy where the percolation threshold of the graphene film was utilized for the fabrication of facile and robust PSA detection immunosensor. The signals generated from resistance change before and after PSA binding onto anti-PSA antibody-modified graphene were recorded. It is known that the conductivity of the graphene film varies significantly with a small change of the graphene concentration; thus, the resistance change of the immunosensor before and after PSA binding onto anti-PSA antibody-modified graphene was recorded as the signal. The substantial resistance change in the signals was demonstrated owing to conductivity of graphene which varied with a small change of the graphene concentration, within the range of 0.1 ng to 100 ng mL^{-1} with a LOD of 0.08 ng mL^{-1}. Furthermore, the immunosensor was quite selective which yielded less than 10% of the resistance change in the presence of interfering species unlike the absence of interfering agents.

Li et al. (2011a) discussed the fabrication of ferrocene-functionalized iron oxide (Fe_3O_4) as label on GS for PSA detection. Firstly, the dopamine (DA) was anchored onto Fe_3O_4-functionalized GS sheets. Later, the ferrocene monocarboxylic acid (FC) and Ab_2 were conjugated through the exposed amine functionalities of DA to obtain labels. During analysis, the immunosensor showed large sensitivity and broad linear range (0.01–40 ng mL^{-1}), with low LOD (2 pg mL^{-1}) via redox properties of FC along with good reproducibility and stability. Additionally, this method was found suitable for other material immobilization such as fluorescence dyes of Fe_3O_4 in order to prepare various labels for immunosensing applications. Xu et al. (2011) proposed an ECL sandwich immunosensor for sensitive detection of PSA with a multiple signal amplification strategy from functionalized graphene and enzyme-antibody-conjugated gold nanorods as the sensor platform. The as-synthesized gold nanorods were serviced as carriers to load more Ab_2 and GOx. The CS-functionalized graphene-modified GCE was employed to increase the loading of Ab_1 and catalyzed the cathodic ECL reaction which was further amplified by the gold nanorods and the enzyme-catalyzed reaction. The gold nanorods were not only used as carriers of Ab_2 and GOx but also catalyzed the ECL reaction of luminol, which further amplified the ECL signal of

Fig. 3.2 SEM images of (**A**) GO/MWCNTs-COOH, (**B**) GO/Au@CeO_2, (**C**) MWCNTs-COOH/Au@CeO_2, and (**D**) GO/MWCNTs-COOH/Au@CeO_2 (the inset is EDS of GO/MWCNTs-COOH/Au@CeO_2); (**E**) XRD patterns of as-synthesized GO/MWCNTs-COOH/Au@CeO_2 (the insets are (*a*) MWCNTs-COOH and (*b*) GO); (**F**) FT-IR spectrometer analysis of GO/MWCNTs-COOH/Au@CeO_2; (**G**) the ECL intensity-potential curves of different materials: GO/MWCNTs-COOH/Au@CeO_2 (curve a), GO/MWCNTs-COOH/Au (curve b), GO/Au@CeO_2 (curve c), MWCNTs-COOH/Au@CeO_2 (curve d), MWCNTs-COOH/CeO_2 (curve e); (**H**) the ECL intensity–time curves of CeO_2 (reproduced with permission from ACS publications from Pang et al. (2015))

luminol in the presence of glucose and oxygen. The as-proposed low-potential ECL immunosensor exhibited high sensitivity and specificity on the detection of PSA in the range from 10 pg mL^{-1} to 8 ng mL^{-1} and the LOD was found to be 8 pg. mL^{-1}.

Yang et al. (2011) developed a sandwich-type electrochemical PSA immunosensor using primary anti-PSA antibody immobilization on quantum dot (QD)-functionalized GS. A broad linear concentration range from 0.005 to 10 ng. mL^{-1} with 3 pg. mL^{-1} LOD was obtained. In addition, the proposed sensor showed reproducibility, stability, selectivity, and satisfactory analysis of PSA in human serum samples. Yang et al. (2015) designed a sensitive sandwich ECL biosensor in order to determine the prostate PC-3 cancer cells by covalently immobilizing the anti-PSA antibody as capture probe and bis(2,2′-bipyridine)-4′-methyl-4-carboxybipyridine- ruthenium (*N*-succinimidyl ester-bis(hexafluorophosphate) (Ru1)-labeled wheat germ agglutinin (WGA) lectin as a signal probe on a GO-coated GCE. The WGA was utilized owing to its easy availability as compared to an antibody, and has an affinity for cells via the specific binding capacity to *N*-acetylglucosamine (GlcNAc) of *N*-glycans on the cell surface. The ECL response of this biosensor was found to be logarithmically proportional to the concentration of PC-3 cells over a range from 7.0×10^2 to 3.0×10^4 cells mL^{-1} and a LOD value of 2.6×10^2 cells mL^{-1} was calculated for PC-3 cells. Furthermore, the ECL biosensor also exhibited a LOD value of 0.1 ng mL^{-1} for the determination of PSA.

3.2.3 α-Fetoprotein (AFP) Determination

α-Fetoprotein (AFP), which is widely employed as a diagnostic biomarker for hepatocellular carcinoma, is a major plasma protein produced by the yolk sac and the liver (Tamura et al. 2009). The AFP expression is also often associated with hepatoma and teratoma.

Du et al. (2010) proposed a sandwich-type electrochemical immunosensor for the dual-amplified detection of cancer biomarker α-fetoprotein (AFP) by using a GS sensor platform and functionalized carbon nanospheres (CNSs) labeled with HRP-Ab_2. The incorporation of porous CNSs assisted in easy diffusion of guest molecules through interconnected micropores and a large amount of enzyme could be immobilized onto support platform. The application of graphene and CNS labeling were largely responsible for sevenfold increase in detection signals. The proposed biosensor exhibited good linearity in the concentration range from 0.05 to 6 ng mL^{-1} and offered a LOD of 0.02 ng mL^{-1} for AFP. Wang and Xue (2013) prepared a AFP-sensitive immunosensor where HRP-conjugated anti-AFP antibody was immobilized on graphene-modified GCE via diazonium chemistry. A linear relationship between current responses and concentrations of AFP was achieved between 0.1 ng mL^{-1} and 2 ng mL^{-1} and LOD was estimated to be 0.03 ng mL^{-1}.

3.2.4 Breast Cancer Biomarker Determination

In this section, determination methods for proteins related to breast cancer are discussed.

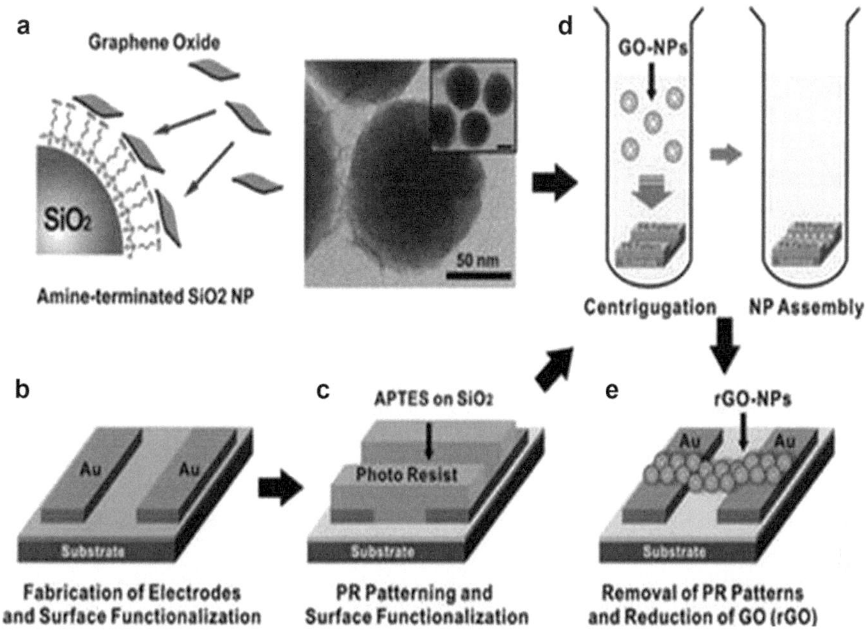

Fig. 3.3 Fabrication process of biomolecular sensor based on graphene-coated NPs. (**a**) Schematic diagram of GO assembly on amine-functionalized NPs and TEM image of NPs coated with GO. (**b**) Fabrication of a metal electrode on the oxide substrate and surface modification for the assembly of GO-NP. (**c**) Photoresist (PR) patterns on the metal electrodes. (**d**) GO-NP assembly in the centrifuge tube. (**e**) Removal of PR patterns and reduction of GO coated on the NP surface (reproduced with permission from Wiley publications from Myung et al. (2011))

In breast cancer patients, human epidermal growth factor receptor 2 (HER2) and epidermal growth factor receptor (EGFR) are two over-expressed biomarkers. For their accurate and selective detection, Myung and colleagues (Myung et al. 2011) prepared rGO-encapsulated NPs-based field-effect transistor (FET) biosensors using monoclonal antibodies (mAbs) against HER2 or EGFR, respectively. The fabrication of biosensor and real-time detection of cancer biomarkers have been shown in Figs. 3.3 and 3.4.

The sensitivity of the rGO-NP devices functionalized with HER2 mAbs was determined by measuring the changes in conductance as the solution concentration of HER2 was varied from 10 fM to 1 μM and LOD was measured to be 1 pM. The selectivity of proposed biosensor was investigated in the presence of bovine serum albumin (BSA). It was observed that conductance is not changed on addition of BSA and LOD of 100 pM was observed in the presence of BSA. For EGFR device, similar trends were obtained with HER2, with the LOD being 100 pM for EGFR and 10 nM in the presence of BSA.

Synthesis of graphene and subsequent doping with the heteroatoms (electron-donating or electron-receiving elements) to tailor electronic and surface properties have aroused great interest for material scientists in the last decade. Among many doping methods, *N*-doped graphene shows the possibility of opening the band gap

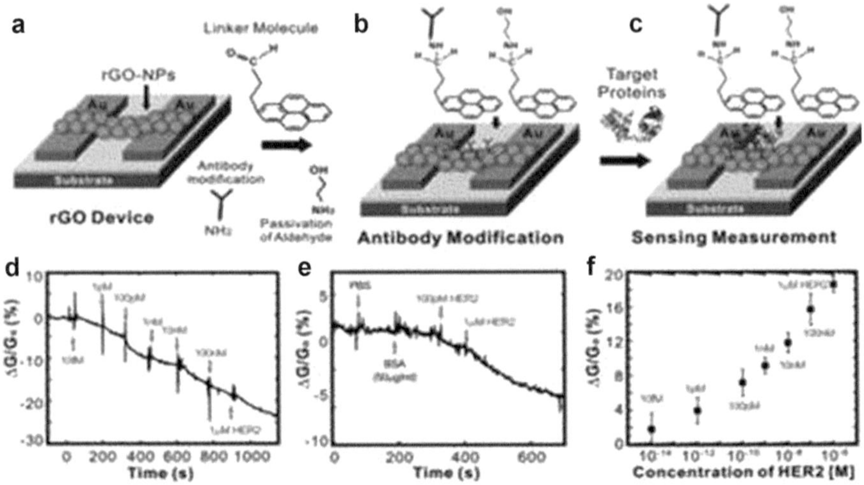

Fig. 3.4 Real-time detection of cancer marker, HER2. (**a**) The preparation of rGO-NP device. (**b**) Surface functionalization of rGO for immobilizing the antibody. (**c**) Measuring conductance of the devices when the target protein is introduced. (**d**) The sensitivity of the biosensor (relative conductance change, %) in response to the concentration of HER2 with V_{DS} (voltage drain to source) = 1 V and Vg (gate voltage) = 0 V. (**e**) The selectivity of the biosensor in response to PBS buffer, BSA with 50 μg mL^{-1} and HER2 (100 pM and 1 μM). (**f**) Sensor sensitivity (relative conductance change, %) as a function of the HER2 concentration with V_{DS} = 1 V and Vg = 0 V. All experiments were performed multiple times (sample number, n = 30) to collect statistical data (with error bars) and confirm the reproducibility and robustness of the biosensing system (reproduced with permission from Wiley publications from Myung et al. (2011))

and provides enhanced conducting influence by substitution of *N* atoms in the open vacancies and edges of GS that provide pathways for efficient electron transfer processes. Hence, the nitrogen-doped GS (*N*-GS) offered much higher electroconductibility than the GS. Considering the high conductivity and fast electron facilitation, *N*-GS have been increasingly utilized in electrochemical sensing (Wang et al. 2010; Shahzad et al. 2015).

The breast cancer susceptibility gene (BRCA1) is commonly expressed in various types of cancer such as pancreatic cancer, stomach cancer, and colon cancer. Thus, Ren et al. (2014) demonstrated dual-amplification sandwich immunosensor (DASI) design for ultrasensitive and precise detection of the breast cancer susceptibility gene (BRCA1) based on the combination of BRCA1 antigen-modified N-GS, hydroxypropyl chitosan, and Co_3O_4 mesoporous nanosheets. The immunosensor showed a wide linear response range of 0.001–35 ng mL^{-1} and a LOD value of 0.33 pg mL^{-1} was calculated. Abdul Rasheed and Sandhyarani (2014) also developed a graphene-based electrochemical DNA sensor for BRCA1 gene. To fabricate the immunosensor, capture probe (DNA-c) and reporter probe (DNA-r) DNAs were hybridized onto a target probe DNA (DNA-t) in a sandwich arrangement on a graphene-modified GCE. The sensor was found to be stable, reproducible, and sensitive and it could detect up to 1 fM BRCA1 gene (5.896 fg mL^{-1}).

Akter et al. (2016) designed a nanostructured immunosensor by immobilizing a monoclonal anti-CA 15–3 antibody on the GO-modified cysteamine (Cys) self-assembled monolayer (SAM) on an Au electrode (Au/Cys) through the amide bond formation between the carboxylic acid groups of GO/Py-COOH and amine groups of anti-CA 15–3. It was followed by Ab_2-conjugated MWCNT-supported ferritin labels (Ab_2-MWCNT-ferritin) which were prepared through the amide bond formation between amine groups of Ab_2 and ferritin and carboxylic acid groups of MWCNTs. The proposed nanoimmunosensor provided a LOD value of 0.01 U mL^{-1} in human serum samples in DPV analysis.

3.3 Chemical Marker Determination

Besides many valuable protein biomarkers, some simple chemical makers have also been detected for various cancer diseases. For instance, to prevent breast cancer and to maintain the genetic process, interaction of 17β-estradiol with tumor-suppressor genes BRCA1 and p53 is considered very crucial. Furthermore, the concentration of 17β-estradiol is also vital to reduce the risk of breast cancer. It has been reported that higher (more than pg) amount of 17β-estradiol in blood and urine samples of postmenopausal women affects the genetic cycle and increases the chance for breast cancer. Thus, Dharuman et al. (2013) constructed anti-estradiol antibody-modified electrical rGO-AuNP composite on the ITO surface for the label-free immunosensing of 17β-estradiol (E2) selectively in the presence of similar structured estrone (E1) and estriol (E3) successfully. The prepared composite was characterized completely using many techniques. Under tuned conditions, the LOD of anti-estradiol-antibody was calculated to be 0.1 fM and dynamic concentration range of 1×10^{-3}–0.1×10^{-12} M for 17β-estradiol was obtained without any signal amplifiers.

Nitric oxide (NO) is a naturally produced free radical in the human body. Besides its beneficial effects such as angiogenesis, apoptosis, cell cycle, invasion, and metastasis, it has been shown to exhibit tumoricidal effects also. Thus, it is necessary to create a specific probe for NO detection. Bai et al. (2017) used sonochemical approach for simultaneous exfoliation of graphite and the reduction of gold chloride to produce highly crystalline G-Au nanocomposite. The electrochemical detection of NO was investigated by linear sweep voltammetry analysis, utilizing the G-Au-modified GCE in a linear range of 10–5000 μM which exhibited a LOD of 0.04 μM without any interference.

Recently, Shahzad et al. (2017) proposed an interesting approach in which various sulfur-doped reduced graphene oxide (SrGO) products were synthesized via an eco-friendly biomass precursor "lenthionine" through a high-temperature doping process. The obtained products were employed for the sensitive determination of a cancer biomarker, 8-hydroxy-2′-deoxyguanosine (8-OHdG). The various sulfur-doped rGO nanomaterials were coated on GCE. The synthesis scheme of SrGO and CV analysis results obtained from various modified electrodes has been assembled in Figs. 3.5 and 3.6.

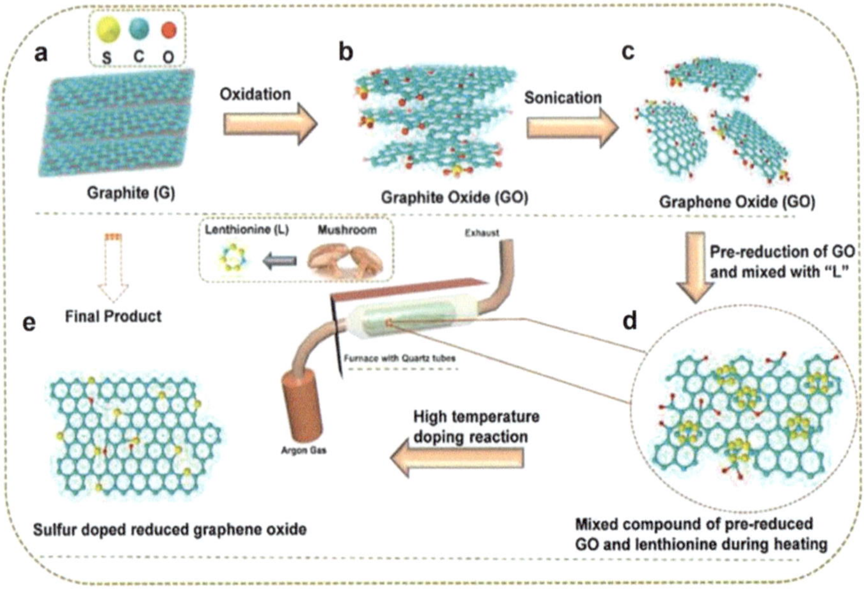

Fig. 3.5 Synthesis of *S*-doped reduced graphene oxide: (**a**) Graphite; (**b**) oxidation of graphite to get graphite oxide (GO); (**c**) sonication of GO to get exfoliated graphene oxide; (**d**) mixing of pre-reduced graphene oxide with lenthionine; (**e**) final product after high-temperature thermal treatment to obtain *S*-doped reduced graphene oxide (reproduced with permission from Elsevier publications from Shahzad et al. (2017))

It was found that high amount of sulfur-doped rGO (SrGO-HD) performed well as compared to mild amount of sulfur-doped rGO (SrGO-MD) which was attributed to the strong electron-donating ability of sulfur (*n*-type doping), strong catalytic activity of the doping sites in *S*-doped graphene, relatively high conductivity, high electrode surface area, and high adsorption capacity of 8-OHdG. The SrGO-HD-modified GCE sensor exhibited robust sensitivity (~1 nM), very wide detection window (20–0.002 μM), and reproducibility. Furthermore, the sensor was selective toward 8-OHdG in the presence of some common interfering species such as ascorbic acid (AA), uric acid (UA), guanine (G), and deoxyguanosine (dG) and excellent recoveries for the detection of 8-OHdG biomarker in various spiked 8-OHdG urine samples were also achieved under optimized experimental conditions.

Recently, it was reported that several types of tumor cells have generated more hydrogen peroxide (H_2O_2) than their normal counterparts due to their (1) higher ROS production or (2) lower ROS scavenging capacities originating from the abnormal growth of the tumor, and therefore, the level of H_2O_2 generated from living cancer cells can be utilized as a valuable biomarker of many types of cancer for early-stage recognition (Hu et al. 2005; Sun et al. 2016). For electrochemical detection of H_2O_2, Xi et al. (2016) presented a functionalized hollow-structured

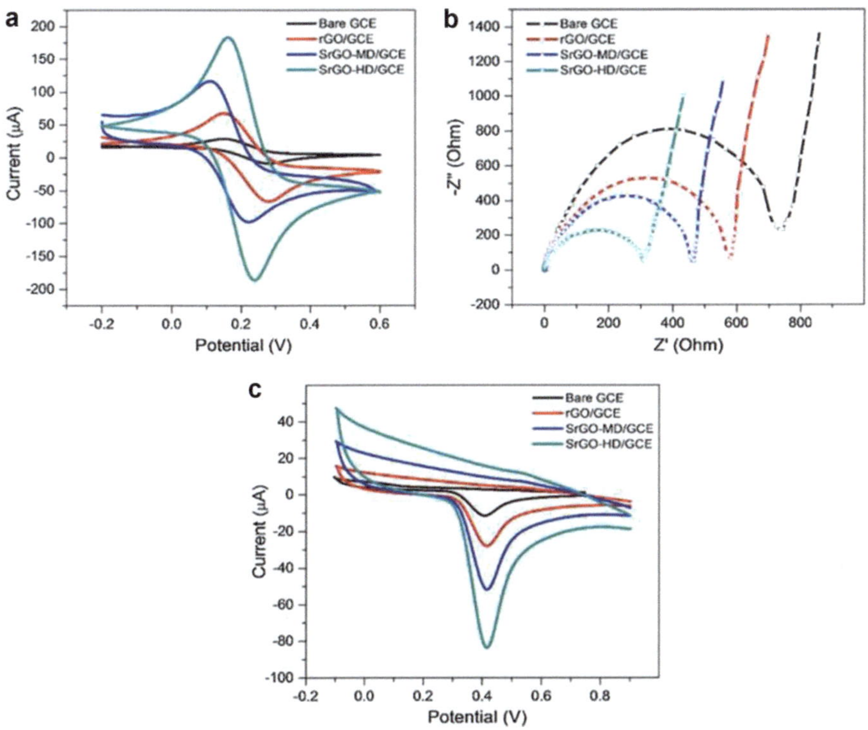

Fig. 3.6 (**a**) Typical cyclic voltammograms of bare GCE, rGO/GCE, SrGO-MD/GCE, and SrGO-HD/GCE in 1 mM $[Fe(CN)_6]^{3-/4-}$, (1:1) solution containing 0.1 M KCl at scan rate of 100 mV s^{-1}; (**b**) typical EIS analysis of various modified GCE. Experimental condition: 0.1 M PBS (pH-7.2) with 0.01 M $[Fe(CN)_6]^{3-/4-}$ system, frequency range: 100 mHz–100 kHz, potential: 0.2 V, AC voltage: 5 mV; (**c**) cyclic voltammograms of bare GCE, rGO/GCE, SrGO-MD/GCE, and SrGO-HD/GCE in the presence of 10 μM 8-OHdG in 0.1 M PBS (pH7.2) at a scan rate of 100 mV s^{-1} (reproduced with permission from Elsevier publications from Shahzad et al. (2017))

nanosphere (HNS) based on Pd NP-decorated double-shell-structured *N*-doped graphene quantum dots (NGQDs)@*N*-doped carbon (NC) HNSs, with ultrafine Pd NPs and "nanozyme" NGQDs as dual-signal-amplifying nanoprobes. The sensor showed an expanded linear response range up to 1.4 mM ($R^2 = 0.999$), with a sensitivity of 0.59 $mAcm^{-2}.mM^{-1}$ and a LOD of 20 nM.

For the detection of H_2O_2, Bai et al. (2016) studied a green and facile one-step synthesis approach for the preparation of rGO-Ag nanocomposite utilizing the extract of a medicinal mushroom, *Ganoderma lucidum*. It was shown that higher amount of polysaccharides in the extract is responsible for easy reduction of GO and $AgNO_3$ to produce rGO-Ag nanocomposite. The fabricated sensor exhibited two linear concentration ranges of 1–100 μM and 100–1100 μM with the LOD of 3 and 56 nM for H_2O_2, respectively, in amperometry analysis. On the other hand, LOD of 136 nM was obtained through LSV technique.

3.4 Miscellaneous Biomarker Determination

This section deals with some miscellaneous biomarker determination using graphene-based immunosensors.

A novel ECL immunosensor for cancer antigen 125 (CA 125) was fabricated by Zhang et al. (2013) using a 3D paper device where gold–silver nanocomposite-functionalized graphene (GN–Ag–Au) was employed as sensing substrate. For ECL signal amplification, CdTe quantum dot-coated carbon microspheres (QD@CMs) were used as bionanolabels. The ECL intensity of CA was found to be logarithmically proportional in the range of 0.008–50 U mL^{-1} with a LOD value of 2.5 mU mL^{-1}. Nucleolin, a multifunctional phosphoprotein, exists in the nucleolus, nucleus, and cytoplasm of the cell. It is usually correlated with the rate of functional activity of the nucleolus in exponentially growing cells. Its overexpression in cancer cell surface can distinguish cancer cells and normal ones in as low as 1000 cells. Therefore, Feng et al. (2011) realized a label-free electrochemical sensor for nucleolin using aptamer AS1411 and 3,4,9,10-perylene tetracarboxylic acid (PTCA)-functionalized graphene owing to high binding affinity and specificity of AS1411 to the overexpressed nucleolin on the cancer cell surface. The detection is ranged from 1×10^3 to 1×10^6 cells mL^{-1}, with a correlation coefficient (R^2) of 0.988, and the value of LOD was estimated to be 794 cells mL^{-1}.

The overexpression of cyclin A2 is considered another excellent prognostic biomarker in early-stage breast cancer, liver cancer, lung cancer, soft-tissue sarcoma, leukemia, and lymphoma. Feng et al. (2012) developed a label-free electrochemical impedance detection of cyclin A2 in cancer cells with outstanding sensitivity and selectivity by using porphyrin noncovalent-functionalized graphene-modified GCE, one specific hexapeptide P0 (RWIMYF) as detection probe, and Tween 20 for preventing nonspecific binding. The LOD of the proposed sensor was calculated to be 0.32 pM with a correlation coefficient (R^2) of 0.996. The fabricated sensor was capable not only to detect cyclin A2 in cancer cell extracts but also differentiate cancer cells from normal ones in many real sample analysis including complex starved cell extracts. Following similar strategy, Wu et al. (2014) proposed another label-free ECL immunosensor for the determination of cyclin A2 based on ECL graphene-upconversion nanoparticle hybrid (rGO-UCNPs). For the preparation of the immunosensor, rGO-UCNPs as the novel ECL emitter and a poly (ethylene glycol) (PEG)-modified specific hexapeptide P0 (PEG-RWIMYF) as the recognition probe were utilized. The proposed sensor worked well in the detection range from 100 fM to 10 nM for cyclin A2 with a LOD value of 10.5 fM (0.52 pg mL^{-1}).

Squamous cell carcinoma antigen (SCC-Ag) was first found in the uterine as cervical squamous cell carcinoma (SCC) by Kato and Torigoe (Kato and Torigoe 1977). The serum level of SCC-Ag is usually increased in parallel to the growth of the tumor size or the recurrence of the disease. Therefore, measurement of the serum level of SCC-Ag has been used clinically for the diagnosis and management of SCC in various organs. Therefore, Li et al. (2011b) adopted a magneto-controlled microfluidic device in which anti-SCC antibody (SCC-Ab)-functionalized magnetic mesoporous nanogold/TH/$NiCo_2O_4$ hybrid nanostructures as immunosensing probes (P1-Ab) and HRP-SCC-Ab conjugate-labeled nanogold/GS as signal tags (P2-Ab) were employed for sensitive electrochemical determination of SCC-Ag. The sandwich immunocomplex which was formed due to inclusion of analyte SCC-Ag between the probe and signal tags was attached to the microfluidic device with the aid of an external magnet. The assay was performed in newborn calf serum (NBCS) containing 2.5 mM H_2O_2 based on the labeled peroxidase on the P2-Ab toward the catalytic reduction of H_2O_2. Using proposed sensor, the current enhancement was shown to be proportional to the concentration of SCC-Ag from 2.5 pg mL^{-1} to 15 ng mL^{-1} with LOD of 1.0 pg mL^{-1}.

Gao et al. (2014) also fabricated a nonenzymatic sandwich-type electrochemical immunosensor for SCCA based on N-GS and carbon-supported Pd-Au binary NPs (Pd–Au/C) which were employed to increase the capacity of capturing Ab_1 and label Ab_2, respectively. The experimental results showed that the prepared Pd–Au/C exhibited high electrocatalytic activity toward H_2O_2 which was used to detect SCCA. The proposed sensor yielded a broad linear working range between 0.005 and 2 ng mL^{-1}, whereas a LOD of 1.7 pg mL^{-1} was obtained in addition to good reproducibility (4.2%) and stability (5.8%).

Recently, microRNAs (miRNAs) which are responsible for flow of biological information through DNA RNA-protein have drawn tremendous great attention due to their altered expression level in various cancers which can be either tumor suppressor whose downregulation brings about malignant cell phenotype or oncogenic (oncomirs) of which upregulation results in malignancies. Among many tumor suppressors and oncomirs, Kilic et al. (2015) developed a voltammetric and impedimetric detection of microRNA-21, mir-21, from cell lysates by using graphene-modified disposable pencil graphite electrodes (GME). The inosine-substituted antimicroRNA-21, antimir-21 probe, and InP (inosine-substituted Probe) were adsorbed onto the surface of GME followed by solid-phase hybridization of InP with mir-21. The DPV result showed a 2.77 times lower detection limit of 2.09 mg mL^{-1} (3.12 pM) as compared to unmodified GME. The proposed sensor was successfully utilized for analyzing mir-21 in the cell lysates of mir-21-positive breast cancer cell line (MCF-7) in contrast to mir-21-negative hepatoma cell line (HUH-7) without any purification steps.

Angiogenin, a 14.4-kDa polypeptide, belongs to the pancreatic ribonuclease family. It is capable to stimulate blood vessel formation strongly and its concentration in serum is elevated in patients affected by various types of cancers. Hence, a label-free and highly sensitive electrochemical aptasensor (ferrocene-5′-CGG ACG AAT GCT TTG ATG TTG TGC TGG ATC CAG CGT TCA TTC TCA-(CH2)6-(SH)-3′) for angiogenin detection based on a conformational change of aptamer and amplification by poly(diallyldimethylammonium chloride) (PDDA)-functionalized AuNP graphene modified on GCE was developed by Chen et al. (2015). The developed sensor was fully characterized and it offered outstanding response to angiogenin in a linear range from 0.1 pM to 5 nM with a LOD of 0.064 pM.

Nuclear matrix protein 22 (NMP22) is a typical biomarker for bladder cancer. The level of NMP22 in healthy person is less than 10 U mL^{-1} while it would increase among 80% of bladder cancer patients. For the sensitive and selective determination of NMP22, Ma and coworkers (Ma et al. 2015) fabricated a biosensor based on rGO–tetraethylene pentamine (rGO–TEPA) and trimetallic AuPdPt-NPs. In the biosensor preparation, rGO–TEPA worked as signal amplification and for antibody anchoring AuPdPt-NPs provided a suitable platform besides its excellent conductivity and large surface area. The proposed immunosensor offered a wide linear range from 0.040 to 20 U mL^{-1}, LOD of 0.01 U mL^{-1}, and short analysis time of 2 min.

3.5 Conclusion and Future Outlook

In conclusion, the potential applications of graphene and its derivatives for detection and quantification of cancer biomarkers have been discussed. Additionally, we also highlighted and described several works where graphene surface was tailored in various ways in order to achieve highly amplified sensing platforms for different types of biomarkers overexpressed by carcinoma cells. Based on the works discussed in abovementioned sections, it was observed that the biosensing of cancer biomarkers was dependent on the synergistic properties of the graphene-based sensing entities that are capable to act simultaneously for imaging as well as amplify the signal response generating from the biosensor properties. Additionally, the synthesis strategies and bio-quantification of cancer biomarkers/cells were described in details focusing the attention toward specific biomarkers responsible for numerous types of well-known cancer. However, it is worth mentioning that most of the fabrication strategies are quite complicated and their stability and reproducibility may be big challenges for researchers.

The science of graphene as a support material is growing rapidly beyond imagination; however, there are many challenges and hurdles that need to be overcome. The synthesis of GO and rGO is challenging, as there is always a need to look into more facile, robust, and efficient preparation methods for GO, graphene, and their composites. Furthermore, another issue may be related to the biocompatibility of graphene for its utilization in biosensing technologies, especially in vivo biosensor (Yusuf et al. 2015). Thus, as it was discussed, the ability of graphene-based materials to conjugate their outstanding chemical and physical features into remarkable sensing properties appears to be a solid milestone. Indeed, researchers around the world are continuing to explore the wide range of sensing strategies opened by graphene in the last few years due to its proven adaptability and functionality in different cancer microenvironments. For the near future, the ultimate goal is to develop graphene-based devices capable of simultaneously detect multiple cancer biomarkers (Cruz et al. 2016; Kierny et al. 2012).

Nevertheless, there are many hurdles which need to be overcome, yet we can expect that graphene-based cancer biomarker diagnostic system is going to be a powerful tool after tailoring the surface properties using nanostructured immunoassays for the individual based on their sequenced genome, their occupation, and their lifestyle which would definitely revolutionize the cancer-based biosensing field and unravel the complexity associated with early diagnosis of malignant tumors, and onset of inherited diseases, subsequently allowing for proper treatment. It can be envisioned that graphene-based immunosensors have bright future.

Acknowledgment This work was supported by Kwangwoon University in 2017.

References

Abdul Rasheed P, Sandhyarani N (2014) Graphene-DNA electrochemical sensor for the sensitive detection of BRCA1 gene. Sensors Actuators B Chem 204:777–782

Aebersold R, Mann M (1997) Mass spectrometry-based proteomics. Nature 422:198–207

Akter R, Jeong B, Choi JS, Rahman MA (2016) Ultrasensitive Nanoimmunosensor by coupling non-covalent functionalized graphene oxide platform and numerous ferritin labels on carbon nanotubes. Biosens Bioelectron 80:123–130

Andriole GL, Crawford ED, Grubb RL (2009) Mortality results from a randomized prostate-cancer screening trial. PLCO project team. N Engl J Med 360:1310–1319

Bai RG, Muthoosamy K, Shipton FN, Pandikumar A, Rameshkumar P, Huang NM, Manickam S (2016) Biogenic synthesis of reduced Graphene oxide-silver (RGO-Ag) Nanocomposite and its dual applications as antibacterial agent and cancer biomarker sensor. RSC Adv 6:36576–36587

Bai RG, Muthoosamy K, Zhou M, Ashokkumar M, Huang NM, Manickam S (2017) Sonochemical and sustainable synthesis of graphene-gold (G-au) nanocomposites for enzymeless and selective electrochemical detection of nitric oxide. Biosens Bioelectron 87:622–629

Bigbee W, Herberman RB (2003) Tumor markers and immunodiagnosis. In: R.C. Bast Jr., D.W. Kufe, R.E. Pollock, et al., editors. Cancer medicine, 6th edn. BC Decker Inc., Hamilton, Ontario, Canada

Cameiro C, Costa L, Melo M et al (1998) Serum tumor markers in metastatic breast cancer comparative study between CEA, CA 15-3 and MCA. Eur J Cancer A 34:s43

Cancer: Facts, causes, symptoms and research. http://www.medicalnewstoday.com/info/cancer-oncology; http://www.cancer.org/acs/groups/content/@research/documents/document/acspc-047079.pdf

Cancer: Facts and Figures 2016 from American Cancer Society. http://www.cancer.org/acs/groups/content/@research/documents/document/acspc-047079.pdf

Cesaro-Tadic S, Dernick G, Juncker D, Buurman G, Kropshofer H, Michel B, Fattinger C, Delamarche E (2004) High-sensitivity miniaturized immunoassays for tumor necrosis factor alpha using microfluidic systems. Lab Chip 4:563–569

Chen A, Chatterjee S (2013) Nanomaterials based electrochemical sensors for biomedical applications. Chem Soc Rev 42:5425–5438

Chen D, Feng H, Li J (2012a) Graphene oxide: preparation, functionalization, and electrochemical application. Chem Rev 112:6027–6053

Chen H, Tang D, Zhang B, Liu B, Cui Y, Chen G (2012b) Electrochemical immunosensor for carcinoembryonic antigen based on nanosilver-coated magnetic beads and gold-graphene nanolabels. Talanta 91:95–102

Chen Z, Zhang C, Li X, Ma H, Wan C, Li K, Lin Y (2015) Aptasensor for electrochemical sensing of angiogenin based on electrode modified by cationic polyelectrolyte-functionalized graphene/gold nanoparticles composites. Biosens Bioelectron 65:232–237

Cruz SM, Girão AF, Gonçalves G, Marques PA (2016) Graphene: the missing piece for cancer diagnosis? Sensors 16:137. (29). doi:10.3390/s16010137

Dar GN, Umar A, Zaidi SA, Baskoutas S, Hwang SW, Abaker M, Al-Hajry A (2012) Ultra-high sensitive ammonia chemical sensor based on ZnO nanopencils. Talanta 89:155–161

Dharuman V, Hahn JH, Jayakumar K, Teng W (2013) Electrochemically reduced graphene–gold nano particle composite on indium tin oxide for label free immuno sensing of estradiol. Electrochim Acta 114:590–597

Du D, Zou Z, Shin Y, Wang J, Wu H, Engelhard MH, Liu J, Aksay IA, Lin Y (2010) Sensitive Immunosensor for cancer biomarker based on dual signal amplification strategy of Graphene sheets and Multienzyme functionalized carbon Nanospheres. Anal Chem 82:2989–2995

Feng L, Chen Y, Ren J, Qu X (2011) A graphene functionalized electrochemical aptasensor for selective label-free detection of cancer cells. Biomaterials 32:2930–2937

Feng L, Wu L, Wang J, Ren J, Miyoshi D, Sugimoto N, Xiaogang Q (2012) Detection of a prognostic indicator in early-stage cancer using functionalized Graphene-based peptide sensors. Adv Mater 24:125–131

Feng T, Qiao X, Wang H, Sun Z, Qi Y, Hong C (2016) An electrochemical immunosensor for simultaneous point-of-care cancer markers based on the host–guest inclusion of β-cyclodextrin–graphene oxide. J Mater Chem B 4:990–996

Fu Z, Hao C, Fei X, Ju H (2006) Flow-injection Chemiluminescent immunoassay for alpha-fetoprotein based on Epoxysilane modified glass Microbeads. J Immunol Methods 312:61–67

Gao J, Du B, Zhang X, Guo A, Zhang Y, Wu D, Ma H, Wei Q (2014) Ultrasensitive enzyme-free immunoassay for squamous cell carcinoma antigen using carbon supported Pd–au as electrocatalytic labels. Anal Chim Acta 833:9–14

Gao J, Guo Z, Su F, Gao L, Pang X, Cao W, Du B, Wei Q (2015) Ultrasensitive electrochemical immunoassay for CEA through host–guest interaction of β-cyclodextrin functionalized graphene and cu@Ag core–shell nanoparticles with adamantine-modified antibody. Biosens Bioelectron 63:465–471

Goldsmith SJ (1975) Radioimmunoassay: review of basic principles. Semin Nucl Med 5:125–152

Global Cancer: Facts & Figures from American Cancer Society. http://www.cancer.org/research/cancerfactsstatistics/global

Gould DA, Moscoso GJ, Young MPA et al (2000) Human first trimester fetal ovaries express oncofetal antigens and steroid receptors. J Soc Gynecol Investig 7:131–138

Hernandez L, Espasa A, Fernandez C, Candela A, Martin C, Romero S (2002) CEA and CA 549 in serum and pleural fluid of patients with pleural effusion. Lung Cancer 36:83–89

Huang J, Tian J, Zhao Y, Zhao S (2015) Ag/au nanoparticles coated graphene electrochemical sensor for ultrasensitive analysis of carcinoembryonic antigen in clinical immunoassay. Sensors Actuators B Chem 206:570–576

Hu S, Zhang S, Hu Z, Xing Z, Zhang X (2007) Detection of multiple proteins on one spot by laser ablation inductively coupled plasma mass spectrometry and application to Immuno-microarray with element-tagged antibodies. Anal Chem 79:923–929

Hu Y, Rosen DG, Zhou Y, Feng L, Yang G, Liu J, Huang P (2005) Mitochondrial manganese-superoxide dismutase expression in ovarian cancer role in cell proliferation and response to oxidative. Stress J Biol Chem 280:39485–39492

Ibrahim AA, Dar GN, Zaidi SA, Umar A, Abaker M, Bouzid H, Baskoutas S (2012) Growth and properties of Ag-doped ZnO nanoflowers for highly sensitive phenyl hydrazine chemical sensor application. Talanta 93:257–263

Iwazawa T, Kanoh T, Matsui S, Monden T (2000) Diagnosis of lung cancer metastasis with CEA extracted from the dissected regional lymph nodes. Lung Cancer 29:254

Jemal A, Murry T, Ward E, Samuels A, Tiwari RC, Ghafoor A et al (2005) Cancer statistics. CA Cancer J Clin 55:10–30

Kang C, Chang C, Chang T, Liao L, Lou P, Xie W et al (2007) A handheld device for potential point-of-care screening of cancer. Analyst 132:745–749

Kazuya K, Yoshihiro K, Tsunekazu K et al (1999) Preoperative determination of several serum tumor markers in patients with primary epithelial ovarian carcinoma. Gynecol Obstet Investig 47:52–57

Kato H, Torigoe T (1977) Radioimmunoassay for tumor antigen of human cervical squamous cell carcinoma. Cancer 40:1621–1628

Kierny MR, Cunningham TD, Ka BK (2012) Detection of biomarkers using recombinant antibodies coupled to nanostructured platforms. Nano Reviews 3:17240. doi:10.3402/nano.v3i0.17240

Kilic T, Erdem A, Erac Y, Seydibeyoglu MO, Okur S, Ozsoz M (2015) Electrochemical detection of a cancer biomarker mir-21 in cell lysates using Graphene modified sensors. Electroanalysis 27:317–326

Kong FY, Xu MT, Xu JJ, Chen HY (2011) A novel lable-free electrochemical immunosensor for carcinoembryonic antigen based on gold nanoparticles–thionine–reduced graphene oxide nanocomposite film modified glassy carbon electrode. Talanta 85:2620–2625

Kramer S, Jager W, Lang N (1998) Early treatment of metastatic breast cancer patients after increase of CEA and CA 15-3 serum levels. Eur J Cancer A 34:s42

Lee SX, Lim HN, Ibrahim I, Jamil A, Pandikumar A, Huang NM (2017) Horseradish peroxidase-labeled silver/reduced graphene oxide thin film-modified screen-printed electrode for detection of carcinoembryonic antigen. Biosens Bioelectron 89:673–680

Li H, Wei Q, He J, Li T, Zhao Y, Cai Y, Du B, Qiane Z, Yang M (2011a) Electrochemical immunosensors for cancer biomarker with signal amplification based on ferrocene functionalized iron oxide nanoparticles. Biosens Bioelectron 26:3590–3595

Li Q, Tang D, Tang J, Su B, Chen G, Wei M (2011b) Magneto-controlled electrochemical immunosensor for direct detection of squamous cell carcinoma antigen by using serum as supporting electrolyte. Biosens Bioelectron 27:153–159

Liu XW, Mao JJ, Liu PD, Wei XW (2011) Fabrication of metal-graphene hybrid materials by electroless deposition. Carbon 2:477–483

Ma H, Zhang X, Li X, Li R, Du B, Wei Q (2015) Electrochemical immunosensor for detecting typical bladder cancer biomarker based on reduced graphene oxide–tetraethylenepentamine and trimetallic AuPdPt nanoparticles. Talanta 143:77–82

Matsuya T, Tashiro S, Hoshino N, Shibata N, Nagasaki Y, Kataoka K (2003) A Core−Shell-type fluorescent Nanosphere possessing reactive poly(ethylene glycol) tethered chains on the surface for Zeptomole detection of protein in time-resolved Fluorometric immunoassay. Anal Chem 75:6124–6132

Myung S, Solanki A, Kim C, Park J, Kim KS, Lee KB (2011) Graphene-encapsulated nanoparticle-based biosensor for the selective detection of cancer biomarkers. Adv Mater 23:2221–2225

Novoselov KS, Geim AK, Morozov SV, Jiang D, Zhang Y, Dubonos SV, Grigorieva IV, Firsov AA (2004) Electric field effect in atomically thin carbon films. Science 306:666–669

Pang X, Li J, Zhao Y, Wu D, Zhang Y, Du B, Ma H, Wei Q (2015) Label-free electrochemiluminescent immunosensor for detection of Carcinoembryonic antigen based on nanocomposites of GO/MWCNTs-COOH/au@CeO_2. ACS Appl Mater Interfaces 7:19260–19267

Paterlini-Brechot P, Benali-Furet NL (2007) Circulating tumor cells (CTC) detection: clinical impact and future directions. Cancer Lett 253:180–204

Polanski M, Anderson NL (2006) A list of candidate cancer biomarkers for targeted proteomics. Biomark Insights 1:1–48

Pumera M, Ambrosi A, Bonanni A, Khim Chng EL, Poh HL (2010) Graphene for electrochemical sensing and biosensing. Trends Anal Chem 29:954–965

Ren X, Yan T, Zhang S, Zhang X, Gao P, Wu D, Du B, Wei Q (2014) Ultrasensitive dual amplification sandwich immunosensor for breast cancer susceptibility gene based on sheet materials. Analyst 139:3061–3068

Sahin B, Paydak V, Paydas S (1996) 121 P-Hepatosteatosis and alterations of CA15.3 and CEA in patients with breast cancer receiving tamoxifen. Eur J Cancer 32:s24

Saito K, Kobayashi D, Sasaki M, Araake H, Kida T, Yagihashi A, Yajima T, Kameshima H, Watanabe N (1999) Detection of human serum tumor necrosis factor-alpha in healthy donors using a highly sensitive immuno-pcr assay. N Clin Chem 45:665–669

Samanman S, Numnuam A, Limbut W, Kanatharana P, Thavarungkul P (2015) Highly-sensitive label-free electrochemical carcinoembryonic antigen immunosensor based on a novel au nanoparticles–graphene–chitosan nanocomposite cryogel electrode. Anal Chim Acta 853:521–532

Schmalzing D, Nashabeh W (1997) Capillary electrophoresis based immunoassays: a critical review. Electrophoresis 18:2184–2193

Sethi S, Ali S, Philip AP, Sarkar FH (2013) Clinical advances in molecular biomarkers for cancer diagnosis and therapy. Int J Mol Sci 14:14771–14784

Shahzad F, Kumar P, Yu S, Lee S, Kim YH, Hong SM, Koo CM (2015) Sulfur-doped graphene laminates for EMI shielding applications. J Mater Chem C 3:9802–9810

Shahzad F, Zaidi SA, Koo CM (2017) Highly sensitive electrochemical sensor based on environmentally friendly biomass-derived sulfur-doped graphene for cancer biomarker detection. Sensors Actuators B Chem 241:716–724

Shi GF, Cao JT, Zhang JJ, Huang KJ, Liu YM, Chen YH, Ren SW (2014) Aptasensor based on tripetalous cadmium sulfidegraphene, electrochemiluminescence for the detection of carcinoembryonic antigen. Analyst 139:5827–5834

Siegel RL, Miller KD, Jemal A (2016) Cancer statistics, 2016. A Cancer J for Clinicians 66:7–30
Su B, Tang D, Tang J, Cui Y, Chen G (2011) Multiarmed star-like platinum nanowires with multienzyme assembly for direct electronic determination of carcinoembryoninc antigen in serum. Biosens Bioelectron 30:229–234
Sun G, Lu J, Ge S, Song X, Yu J, Yan M, Huang J (2013) Ultrasensitive electrochemical immunoassay for carcinoembryonic antigen based on three-dimensional macroporous gold nanoparticles/graphene composite platform and multienzyme functionalized nanoporous silver label. Anal Chim Acta 775:85–92
Sun Y, Zheng H, Wang C, Yang M, Zhou A, Duan H (2016) Ultrasonic-electrodeposition of Ptpd alloy nanoparticles on ionic liquid-functionalized graphene paper: towards a flexible and versatile nanohybrid electrode. Nanoscale 8:1523–1534
Tamura Y, Yamagiwa S, Aoki Y, Kurita S, Suda T, Ohkoshi S, Nomoto M, Aoyagi Y (2009) Serum alpha-fetoprotein levels during and after interferon therapy and the development of hepatocellular carcinoma in patients with chronic hepatitis C. Dig Dis Sci 54:2530–2537
Tang J, Tang D, Li Q, Su B, Qiu B, Chen G (2011) Sensitive electrochemical immunoassay of carcinoembryonic antigen with signal dual-amplification using glucose oxidase and an artificial catalase. Anal Chim Acta 697:16–22
Teker K (2008) Bioconjugated carbon nanotubes for targeting cancer biomarkers. Mater Sci Eng B 153:83–87
Umar A, Al-Hazmi F, Dar GN, Zaidi SA, Al-Ghamdi A (2012) Ultra-sensitive ethanol sensor based on rapidly synthesized Mg $(OH)_2$ hexagonal nanodisks. Sensors Actuators B Chem 166–167:97–102
Voller A, Bartlett A, Bidwell DE (1978) Enzyme immunoassays with special reference to ELISA techniques. J Clin Pathol 31:507–520
Wang J, Wang X, Wu S, Song J, Zhao Y, Ge Y, Meng C (2016) Fabrication of highly catalytic silver nanoclusters/graphene oxide nanocomposite as nanotag for sensitive electrochemical immunoassay. Anal Chim Acta 906:80–88
Wang R, Xue C (2013) Sensitive electrochemical immunosensor for alpha-fetoprotein based on covalently incorporating bio-recognition element onto graphene modified electrode via diazonium chemistry. Anal Methods 5:5195–5200
Wang Y, Shao Y, Matson DW, Li J, Lin Y (2010) Nitrogen-doped graphene and its application in electrochemical biosensing. ACS Nano 4:1790–1798
What Is Cancer? From National Cancer Institute. https://www.cancer.gov/about-cancer/understanding/what-is-cancer
Wu L, Wang J, Yin M, Ren J, Miyoshi D, Sugimoto N, Qu X (2014) Reduced graphene oxide upconversion nanoparticle hybrid for electrochemiluminescent sensing of a prognostic indicator in early-stage cancer. Small 10:330–336
Xi J, Xie C, Zhang Y, Wang L, Xiao J, Duan X, Ren J, Xiao F, Wang S (2016) Pd nanoparticles decorated N-doped Graphene quantum dots@NDoped carbon hollow Nanospheres with high electrochemical sensing performance in cancer detection. ACS Appl Mater Interfaces 8:22563–22573
Xu S, Liu Y, Wang T, Li J (2011) Positive potential operation of a Cathodic Electrogenerated Chemiluminescence Immunosensor based on Luminol and Graphene for cancer biomarker detection. Anal Chem 83:3817–3823
Yang H, Li Z, Shan M, Li C, Qi H, Gao Q, Wang J, Zhang C (2015) Electrogenerated chemiluminescence biosensing for the detection of prostate PC-3 cancer cells incorporating antibody as capture probe and ruthenium complex-labelled wheat germ agglutinin as signal probe. Anal Chim Acta 863:1–8
Yang M, Gong S (2010) Immunosensor for the detection of cancer biomarker based on percolated graphene thin film. Chem Commun 46:5796–5798
Yang M, Javadi A, Gong S (2011) Sensitive electrochemical immunosensor for the detection of cancer biomarker using quantum dot functionalized graphene sheets as labels. Sensors Actuators B Chem 155:357–360

Yates AM, Elvin SJ, Williamson DE (1999) The optimisation of a murine TNF-alpha ELISA and the application of the method to other murine cytokines. J Immunoass 20:31–44
Yusuf M, Elfghi FM, Zaidi SA, Abdullah EC, Khan MA (2015) Applications of graphene and its derivatives as an adsorbent for heavy metal and dye removal: a systematic and comprehensive overview. RSC Adv 5:50392–50420
Zhong Z, Wu W, Wang D, Wang D, Shan J, Qing Y, Zhang Z (2010) Nanogold-enwrapped graphene nanocomposites as trace labels for sensitivity enhancement of electrochemical immunosensors in clinical immunoassays: Carcinoembryonic antigen as a model. Biosens Bioelectron 25:2379–2383
Zaidi SA, Shin JH (2016a) Recent developments in nanostructure based electrochemical glucose sensors. Talanta 149:30–42
Zaidi SA, Shin JH (2016b) A review on the latest developments in nanostructure-based electrochemical sensors for glutathione. Anal Methods 8:1745–1754
Zaidi SA (2013) Graphene: a comprehensive review on its utilization in carbon paste electrodes for improved sensor performances. Int J Electrochem Sci 8:11337–11355
Zaidi SA, Shin JH (2014) Molecularly imprinted polymer electrochemical sensors based on synergistic effect of composites synthesized from graphene and other nanosystems. Int J Electrochem Sci 9:4598–4616
Zaidi SA, Shin JH (2015a) Graphene: functionalization with nanoparticles and applications, in book "chemical functionalization of carbon Nanomaterials: chemistry and applications", 844–867. Press, CRC
Zaidi SA, Shin JH (2015b) A novel and highly sensitive electrochemical monitoring platform for 4-nitrophenol on MnO_2nanoparticles modified graphene surface. RSC Adv 5:88996–89002
Zhao L, Li J, Liu Y, Wei Y, Zhang J, Zhang J, Xia Q, Zhang Q, Zhao W, Chen X (2016) A novel ECL sensor for determination of carcinoembryonic antigen using reduced graphene oxide-BaYF5:Yb, Er upconversion nanocomposites and gold nanoparticles. Sensors Actuators B Chem 232:484–491
Zhang Y, Li L, Yang H, Ding YN, Su M, Zhu J, Yan M, Yu J, Song X (2013) Gold–silver nanocomposite-functionalized graphene sensing platform for an electrochemiluminescent, immunoassay of a tumor marker. RSC Adv 3:14701–14709
Zhou J, Tang D, Houa L, Cui Y, Chen H, Chen G (2012a) Nanoplatinum-enclosed gold nanocores as catalytically promoted nanolabels for sensitive electrochemical immunoassay. Anal Chim Acta 751:52–58
Zhou J, Zhuang J, Miro M, Gao Z, Chen G, Tang D (2012b) Carbon nanospheres-promoted electrochemical immunoassay coupled with hollow platinum nanolabels for sensitivity enhancement. Biosens Bioelectron 35:394–400
Zhu L, Xu L, Jia N, Huang B, Tan L, Yang S, Yao S (2013) Electrochemical immunoassay for carcinoembryonic antigen using gold nanoparticle–graphene composite modified glassy carbon electrode. Talanta 116:809–815

Label-Free Biosensors for Early Diagnosis of Cancer Based on G-Quadruplex and Isothermal Amplification

4

Yahui Guo, Weirong Yao, and Renjun Pei

4.1 Introduction

Cancer is caused by the synthetic action of internal and external factors, including genetic mutations, environmental pollution, and unhealthy diet. According to the world cancer report published by the World Health Organization, cancer has become a leading cause of death worldwide with expected 21.7 million new cancer cases and 13 million cancer deaths in 2030.

Early diagnosis has a vital impact on improving disease-free survival and reducing the mortality of cancer patients. The commonly used tool for the diagnosis of cancer is imaging test, which includes CT scans, MRI, X-ray tests, nuclear medicine scans, ultrasound and endoscopy, etc. (Cao et al. 2017). However, imaging tests can only find big tumors or large groups of cancer cells. It takes millions of cancer cells to make a tumor big enough to show up on an imaging test. Besides, biopsy test is usually needed to avoid misjudgment and certify the statement of imaging test.

Compared with imaging tests and biopsy, biomedical sensors or biosensors with noninvasive sample collection play an increasingly important role in cancer diagnosis and prognosis. Biosensors bearing the capacity of detecting a few cancer cells or even a single have to be a robust technology for the early diagnosis of cancer. For instance, Sun et al. developed an excellent platform for the capture and identification of circulating tumor cells on an electrospray-fabricated chitosan nanoparticle surface (Sun et al. 2015b). This platform can be regarded as a biosensor with a soft

Y. Guo • W. Yao
State Key Laboratory of Food Science and Technology, School of Food Science and Technology, Jiangnan University, Wuxi 214122, China

R. Pei (✉)
Suzhou Institute of Nano-Tech and Nano-Bionics, Chinese Academy of Sciences, Suzhou 215123, China
e-mail: rjpei2011@sinano.ac.cn

P. Chandra et al. (eds.), *Next Generation Point-of-care Biomedical Sensors Technologies for Cancer Diagnosis*, DOI 10.1007/978-981-10-4726-8_4

hydrophilic interface by using DNA aptamer as the specific capture probe. As low as ten viable cancer cells (MCF-7) were successfully captured and detected in the sample solution containing 10^6 white blood cells. With biosensors, the identification and detection of the cancer-indicative biomarkers in body fluids is another way to get a view of a patient's cancer. Reported ultrasensitive biosensors for the detections of low-abundance biomarkers have demonstrated great potential for early diagnosis of cancers. Das et al. reported an electrochemical biosensor for the ultrasensitive detection of circulating tumor DNAs of mutated *KRAS* genes, which are associated with lung cancer, colorectal cancer, and ovarian cancer (Das et al. 2015). The biosensor was fabricated on a microchip with a 3D-nanostructured microelectrode, and employed peptide nucleic acid as the clamp probe for target DNAs. The biosensor was able to detect target mutant DNA directly in patient serum samples and produced the same accuracy as polymerase chain reaction (PCR). It responded rapidly (in 15 min) with high selectivity, and the use of a chip-based platform allowed a point-of-care test.

On the other hand, more than 70% of cancer death occurred in the economically developing countries, which contain 82% of the world's population. The availability and use of diagnostic tests is a major concern in these middle-income and low-income areas. Therefore, it is of great significance to develop economical biosensors when diagnosis can't be easily determined by imaging studies in the developing areas.

In this chapter, the content focused on the development of cost-effective and ultrasensitive biosensors for the early diagnosis of cancers by utilizing G-quadruplex-based output and isothermal amplification technology.

4.2 Label-Free Biosensor

Jonsson et al. firstly proposed the term "label-free" (Jonsson et al. 1991). In his study, surface plasmon resonance (SPR) was utilized for the study of real-time biospecific interaction, which was defining as "label-free" since SPR could monitor changes of the refractive index in the vicinity of reaction surface and was independent of the chemical nature of the sample being analyzed. So in SPR systems, no specific reporters were needed for the signal readout. Label-free biosensor (LFB) mentioned in this chapter, by definition, do not require the covalent labeling of reporter elements (fluorescent, colorimetric, or electrochemical) to facilitate measurements. There is a subtle difference between a *noun* and a *verb* in the same word "label." It refers to the *verb* in the term "label-free biosensor" throughout this review.

In a LFB system, the signal reporter or ligand dispersing in homogeneous solution can noncovalently bind to the probe-target conjugate and subsequently generate light-up or enhanced signal output. Compared with labeled systems such as enzyme-linked immunosorbent assay (ELISA), the main advantage of LFBs is that they avoid conjugation and purification of signal markers even hazardous radioactive materials (Guo et al. 2015c, 2016a). Such label-free systems are inexpensive, and

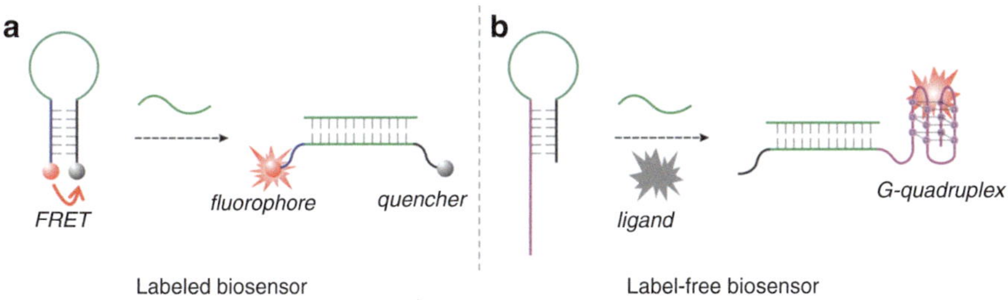

Fig. 4.1 (**a**) Molecular beacon labeled with fluorophore and quencher for the detection of nucleic acids. (**b**) Label-free molecular beacon based on G-quadruplex structure and binding-responsive reporting ligand

usually independent on high-price instruments. As illustrated in Fig. 4.1, there is a representative comparison between labeled and label-free nucleic acids sensor. Figure 4.1a depicts a typical molecular beacon probe with a hairpin structure and two terminal-modified functional moieties, a fluorophore and a quencher (Tyagi and Kramer 1996). The fluorophore is internally quenched due to the fluorescence resonance energy transfer; after binding to the nucleic acid of target sequence, the fluorophore and quencher is separated with restored fluorescence emission. While Fig. 4.1b depicts a label-free molecular beacon probe without modified moieties, which have been replaced with a partly complementarily stem structure (Guo et al. 2013). The hairpin structure will be opened when the probe is challenged with nucleic acid of target sequence, resulting in the liberation of guanine-rich sequence (pink color) and formation of G-quadruplex structure. A fluorescent ligand that can specifically bind to the G-quadruplex structure will generate light-up or enhanced intensity in fluorescence.

The two kinds of modes demonstrated equivalent performance, but the price of labeled molecular beacon probe is >10-fold higher than that of label-free probe. It also takes a longer turnaround time to synthesize the labeled probes. The LFBs displayed another advantage of separation-circumvented operation in some cases. ELISA is a conventional clinical technique for cancer biomarker assay, which involves at least one antibody with covalent-labeled reporter. As shown in Fig. 4.2, the basic steps of a typical ELISA include surface binding for separation and washes to remove unbound materials. Careful washes are required to get consistent results. ELISA can be quite complex when multiple layers of antibodies are employed with various intervening steps. In contrast, the LFBs based on aptamer and target-induced structural transformation performed a simple mix-and-detect procedure.

Aptamer is single-stranded DNA or RNA that specifically binds to the target, including small molecule, protein, and even whole cell (Liu et al. 2009). Due to its high affinity and specificity, aptamer has been considered as a great alternative of antibodies and has been employed in many analyses (Zhang et al. 2016). Our group has been devoting efforts to screening aptamers through the systematic evolution of ligands by exponential enrichment process, and developing label-free methodologies for the detections of small molecules, metal ions, nucleic acids, proteins, and

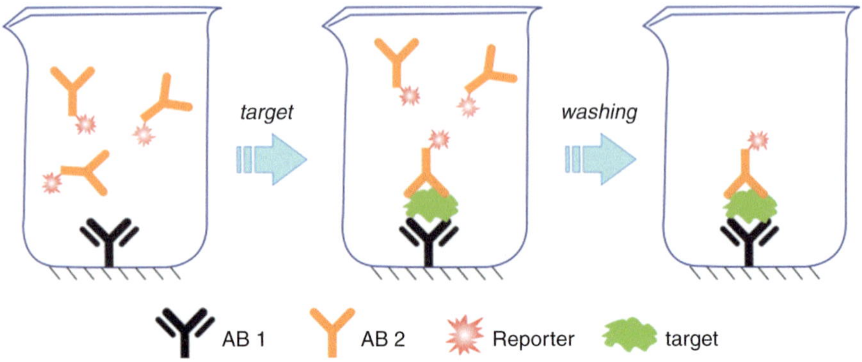

Fig. 4.2 Schematic procedure of ELISA for the detection of target

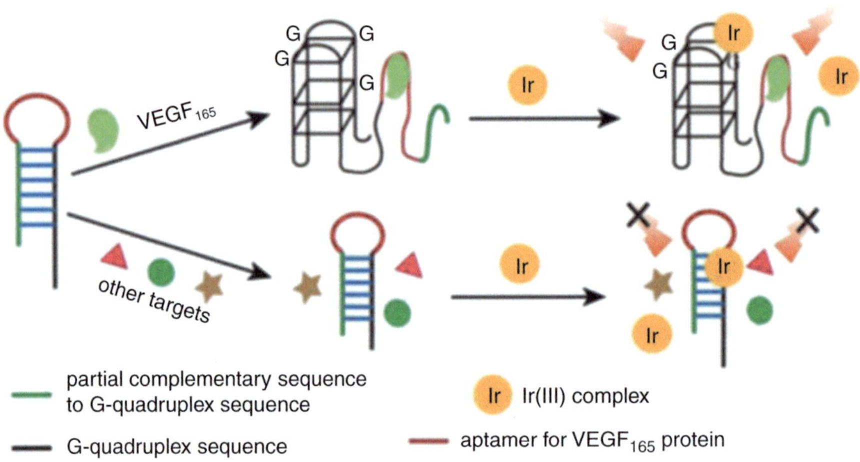

Fig. 4.3 The detection principle of the LFB for VEGF assay based on aptamer. Reprinted with permission from Lin et al. (2016a), © 2016 American Chemical Society

cancer cells using aptamers (Wang et al. 2016a, b; Yang et al. 2016). The works of Tan's group also greatly promoted the acknowledgement of aptamers and contributed to a broad scope of the biomedical applications of aptamers, which can be learned from their recent reviews (Lyu et al. 2016; Tan et al. 2013; Zheng et al. 2015). VEGF, known as vascular endothelial growth factor, is centrally involved in promoting tumor growth and metastasis. VEGF is overexpressed in neoplasms, and circulating VEGF levels have a significant correlation with the metastatic disease. Lin et al. designed an aptamer-based luminescent LFB for rapid and facile detection of VEGF in a microchip (Lin et al. 2016b). As shown in Fig. 4.3, in the absence of VEGF, the label-free probe was in an inactive state with a hairpin DNA structure that cannot interact with ligand reporter, a luminescent iridium (III) complex of selective binding affinity towards G-quadruplex; in the presence of VEGF, VEGF bound to the aptamer sequence and induced a structural transition to form a G-quadruplex structure, which could activate the ligand through noncovalent interaction with light-up luminescent signal.

Most of the reported LFBs depend on the principle of target-induced structural transition of nucleic acid probes because of the use of structural-specific signal ligands. In a label-free sensing system, signal ligand can noncovalently bind to a specific DNA structure, and enhanced fluorescence will be produced only after the formation of the specific DNA structure (Guo et al. (2016b). For example, the well-known SYBR Green I, an asymmetrical cyanine dye with negligible intrinsic fluorescence, but it exhibits thousand-fold enhancement of fluorescence when preferentially binding to double-stranded DNA (dsDNA) (Jin et al. 1994). SYBR Green I has been demonstrated a successful case for label-free DNA detections in quantitative PCR (Schneeberger et al. 1995) and loop-mediated isothermal amplification (Dhama et al. 2014) due to its excellent property.

Among the several nucleic structures, G-quadruplex is the most widely used signal transducer in label-free sensing applications and has stimulated significant improvement in LFBs in terms of vibrant binding activities, multiple signal output modes, and versatile roles in a variety of designing strategies (Guo et al. 2016c).

4.3 LFBs Based on G-Quadruplex

G-quadruplex (G4) possesses a kind of guanine-rich DNA structure that was originally found in telomeres as a biophysical oddity (Wright et al. 1997). As shown in Fig. 4.4a, G4 is formed by the stacking of two or more G-quartets, which are assembled with four guanines through Hoogsteen-type base pairing. Monovalent cations

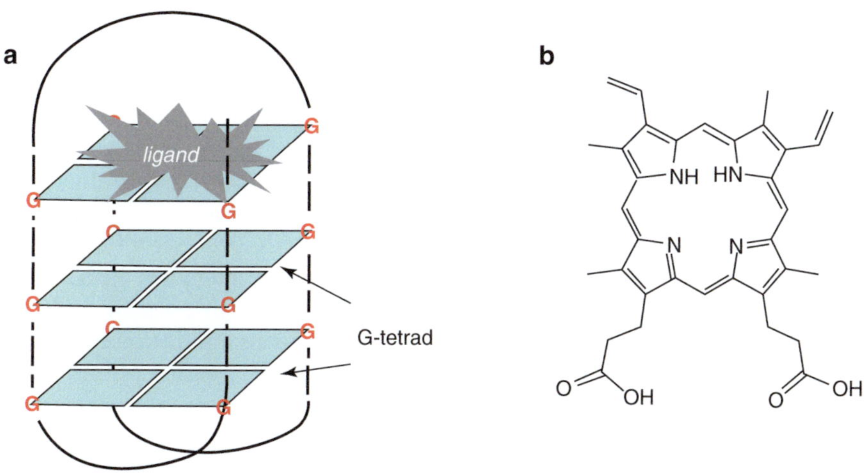

Fig. 4.4 (**a**) Depiction of G4 structure and its interaction with ligand through π-stacking binding mode. (**b**) Chemical structure of the G4-selective ligand PPIX

such as K^+/Na^+ can coordinate to the O_6 atoms of guanines and stabilize the negative potential in the central channel of G4. The converging arrangement of four guanines provides a large and exposed surface of G-tetrad, which accommodates many ligands in a well-known π-stacking mode (Patel et al. 2007). For example, in Fig. 4.4b, protoporphyrin IX (PPIX) is a kind of macrocyclic fluorophore with similar molecular dimension with G-tetrad, it can stack onto the terminal G-tetrad of G4 with 1:1 or 2:1 stoichiometry. Binding to G4 will result in a >16-fold increase in fluorescence (Li et al. 2010).

The vibrant binding activities of G4s contributed to multiple signal output modes and a variety of design strategies based on G4s, thereby promoting the applications of G4s in LFBs. There are mainly two kinds of G4-responsive ligands that are usually employed as signal reporters in LFBs.

The first kind of ligands is fluorescent compounds endowed with binding-responsive emission property towards G4. Since fluorescent dye was firstly employed to characterize the G4 structure (Sun et al. 1997), numerous fluorescent compounds with binding affinity to G4s have been found or synthesized. Fluorophores bind to G4s via noncovalent interactions including π-stacking, loop recognition, groove recognition, and intercalation (Largy et al. 2013). These dyes exhibited positive or negative responses in fluorescence intensity when binding to G4s, which have been employed as signal indicators in many G4-based LFBs. Representative groups of G4s-binding fluorescent ligands include cyanine dyes (thiazole orange and its derivatives, SYBR Green I, DODC, ETC, etc.) (Allain et al. 2006; Granotier et al. 2005; Monchaud et al. 2006; Paramasivan and Bolton 2008; Xu et al. 2010; Yang et al. 2009); benzimidazole derivatives (Maiti et al. 2003); EtBzEt derivatives (Jain and Bhattacharya 2011); ethidium derivatives (Guo et al. 1992); triphenylmethane dyes (Bhasikuttan et al. 2007; Kong et al. 2009); carbazoles (Chang et al. 2003; Dumat et al. 2011); porphyrins (Han et al. 1999; Li et al. 2010; Wei et al. 2008); aggregation-induced emission (AIE) luminogens (Hong et al. 2008; Luo et al. 2001); natural compounds (Arora et al. 2008); quercetin (Sun et al. 2006); isaindigotone derivatives (Tan et al. 2009); metal complexes (Ma et al. 2009; Shi et al. 2010); lanthanide complexes (Alzeer et al. 2009; Galezowska et al. 2007); heterocyclic derivatives (Mergny et al. 2001; Teulade-Fichou et al. 2003); anthracyclines (Manet et al. 2011a); and sabarubicin (Manet et al. 2011b). Wu et al. recently synthesized a new compound bis(methylpiperazinylstyryl)phenanthroline that showed strong binding affinity to G4s with up to 150-fold fluorescence enhancement (Wu et al. 2016). The good properties made it an excellent candidate serving as a label-free fluorescent indicator in G4-based LFBs. The abundant pool of G4-binding dyes has promoted the development of LFBs for the detection of small molecules, metal ions, nucleic acids, or proteins (Ma et al. 2013). Our group has devoted continuous efforts to the development of LFBs using G4-specific fluorescent ligands for the sensing of ions (Guo et al. 2015a, b), nucleic acids (Guo et al. 2013, 2014a), protein (Guo et al. 2014b), enzyme activity (Guo et al. 2015c; Zhou et al. 2015, 2016), etc.

Human platelet-derived growth factor BB (PDGF-BB) is considered as a potential biomarker for cancer diagnosis, since that it stimulated autocrine growth of different types of tumor cells and was often overexpressed in human malignant tumors.

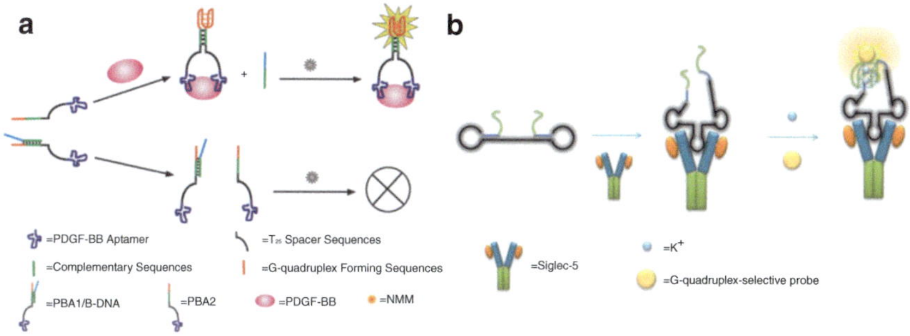

Fig. 4.5 (**a**) Illustration of the LFB for PDGF-BB assay based on target-induced split-G4 formation. Reprinted with permission from Wei et al. (2015), © 2015 Elsevier. (**b**) The G4-based LFB for *Siglec*-5 detection using aptamer as recognition unit and iridium(III) complex as signal ligand. Reprinted with permission from Lin et al. (2016a), © 2016 American Chemical Society

Wei et al. reported a fluorescence LFB for PDGF-BB detection (Wei et al. 2015). As shown in Fig. 4.5a, two DNA fragments integrating the sequences of split G4 and PDGF-BB aptamer were used as the recognition probe and the G4 selective ligand N-methyl mesoporphyrin IX (NMM) was employed as the label-free signal reporter. In the presence of target PDGF-BB, the aptamer parts of the two DNA fragments can simultaneously bind to PDGF-BB, resulting in the adjacency of the two split-G4 to form a complete G4 structure. Enhanced fluorescence emission signal was reached after the formation of complete G4 and activation of the G4-responsive dye NMM. As low as 3.2 nM PDGF-BB can be monitored by this LFB.

Sialic acid binding immunoglobulin-like lectin-5 (*Siglec*-5) was considered as a potential biomarker of granulocytic maturation and acute myeloid leukemia phenotype. Lin et al. developed a fluorescence LFB for sensitive detection of *Siglec*-5 based on target-induced structural transition and G4 formation (Lin et al. 2016a). A G4-binding luminescent iridium(III) complex was employed as the label-free signal ligand. As shown in Fig. 4.5b, the DNA probe of hairpin structure consisted of three parts of sequences: the *Siglec*-5-binding aptamer sequence (black), the G-rich sequences (green), and the blocking sequences (blue). *Siglec*-5 could bind to the aptamer sequence of the DNA probe, resulting in conformation change and formation of G4 structure with the assistance of monovalent cation K^+. The selective binding of the iridium complex to G4 brought out an enhancement in its luminescence. The LFB demonstrated a sensitive performance with a detection limit of 0.27 nM.

The second kind of ligands is a catalytic compound, the iron (III)-protoporphyrin IX complex (hemin). As shown in Fig. 4.6, hemin is an active cofactor of oxidative enzymes such as catalases and peroxidases. Hemin itself can also catalyze peroxidation reactions albeit at much lower levels compared to hemin-containing enzymes. However, it exhibits peroxidase activity larger by two orders of magnitude after noncovalently binding to G4s (Travascio et al. 1998). The G4/hemin complex can be regarded as a kind of catalytic DNAzyme, which has been employed as an amplifying reagent in many colorimetric and chemiluminescence methods as a label-free reporter.

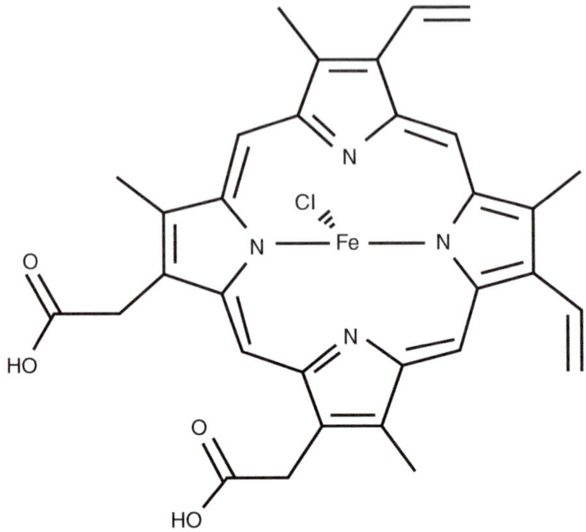

Fig. 4.6 The chemical structure of hemin

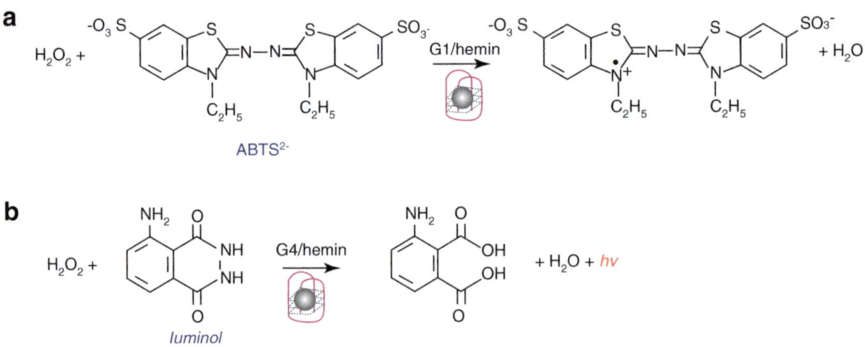

Fig. 4.7 Chemical reaction equations of (**a**) ABTS^{2-}; (**b**) luminol oxidization

As shown in Fig. 4.7a, G4/hemin can catalyze the oxidation of ABTS^{2-} by H_2O_2, and this is accompanied by a color change from colorless to green. The effect has been used in colorimetric assays for Hg^{2+} (Hao et al. 2014; Li et al. 2009), CEA (Zhou et al. 2014), lysozyme (Li et al. 2007), thrombin (Li et al. 2008), DNA (Jiang et al. 2014), enzyme activity (He et al. 2012), and others. The 3-3′-diaminobenzidine tetrahydrochloride and 3,3′,5,5′-tetramethylbenzidine (TMB) (Kolpashchikov 2008; Yin et al. 2009) can also be used as oxidizable substrates for colorimetric readout in G4-based LFBs. As shown in Fig. 4.7b, the G4/hemin DNAzyme can also catalyze the generation of chemiluminescence through oxidation of luminol by H_2O_2 (Xiao et al. 2004). Chemiluminescent sensors based on the use of G4/hemin have a large potential in analytical chemistry due to their ultrahigh sensitivity and label-free signal output mode (Freeman et al. 2011). On the other hand, the G4/hemin complex formed near the surface of electrodes may cause dielectric changes

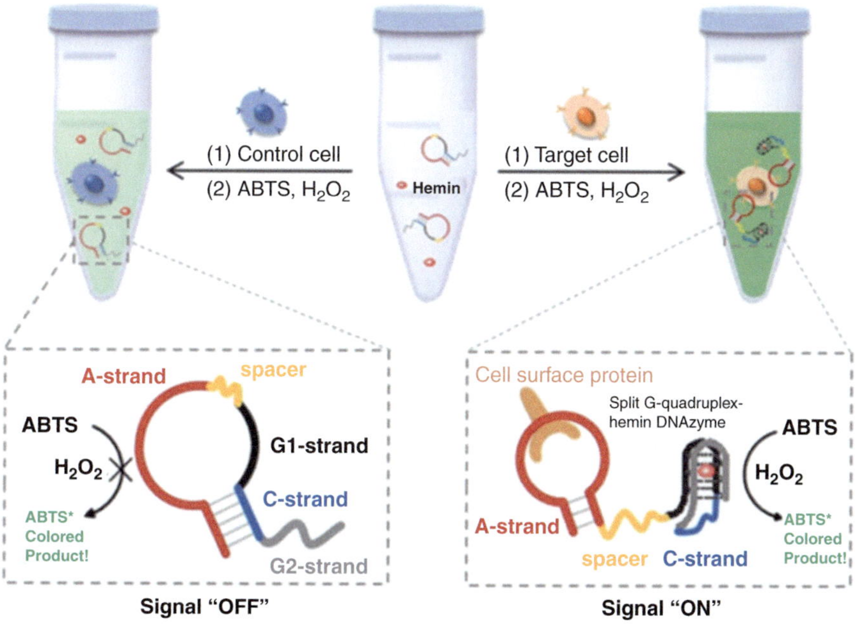

Fig. 4.8 A label-free and light-up probe for simple, rapid, and direct detection of cancer cells based on split-G4 and DNAzyme. Reprinted with permission from Shi et al. (2014), © 2014 Royal Society of Chemistry

due to surface plasmon resonance. It can act as a label-free electrocatalytic reporter in redox reactions for an electrochemical signal (Pelossof et al. 2012).

Shi et al. developed a LFB for the facile detection of human leukemic *CCRF-CEM* cells based on target-induced formation of G4 structure (Shi et al. 2014). Hemin and $ABTS^{2-}$ were used as the ligand and substrate, respectively, for the construction of DNAzyme system to give a label-free colorimetric readout. As shown in Fig. 4.8, light-up signal was realized through the assemblage of two G-rich fragment for a G4 architecture and subsequent activation of the DNAzyme system. An oligonucleotide of integrating cancer-binding aptamer and G-rich sequences was employed as the DNA probe. The DNA probe was tail-hairpin structured due to the self-complementary, resulting in the geographical segregation of the two G-rich fragments. While the presence of human leukemic *CCRF-CEM* cells induced the cell-aptamer conjugation and structural transformation of the DNA probe, resulting in the formation of G4 and construction of the DNAzyme system. A light-up colorimetric signal would be achieved as the consequence of the oxidation of $ABTS^{2-}$. The G4-based LFB achieved a simple, fast, washing-free, specific, and quantitative performance for colorimetric assay of human leukemic *CCRF-CEM* cells.

By employing aptamer and G4 as the recognition unit and output unit, respectively, LFBs have provided an economical and facile way for the detections of cancer biomarkers and even cancer cells. At the same time, urge requirement for early diagnosis of cancer demanded a more sensitive capacity of LFBs.

Isothermal amplification technique bearing the capacity of detecting ultralow abundance of biomarkers would make the early diagnosis possible, and had demonstrated tremendous potential for ultrasensitive detections in recent reported works (Zhao et al. 2015).

4.4 Ultrasensitive Detection Based on Isothermal Amplification

As a common and indispensable amplification technique in molecular biology, PCR is a thermal-cycling process with the assist of a heat-stable DNA polymerase (Saiki et al. 1988). PCR can exponentially amplify the target DNA but suffered dozen times of alternative heating and cooling steps. In a basic PCR protocol, defined series of temperature steps were set on a thermal cycler. It is time-consuming and instrument-dependent, which limits the applications in resource-limited regions for point-of-care assay.

It merits mention that the enzyme-assisted or catalytic isothermal amplification (ITA) possesses comparable sensitivity to PCR but does not require the lengthy thermo-cycling steps and special instruments. Moreover, ITA can be utilized not only for the biosensing of nucleic acids, but also for proteins, small molecules, ions, and even cells. As shown in Fig. 4.9, by searching "isothermal amplification" in the *PubMed* database, the number of published articles has increased steadily and an obvious increase was observed in the past decade.

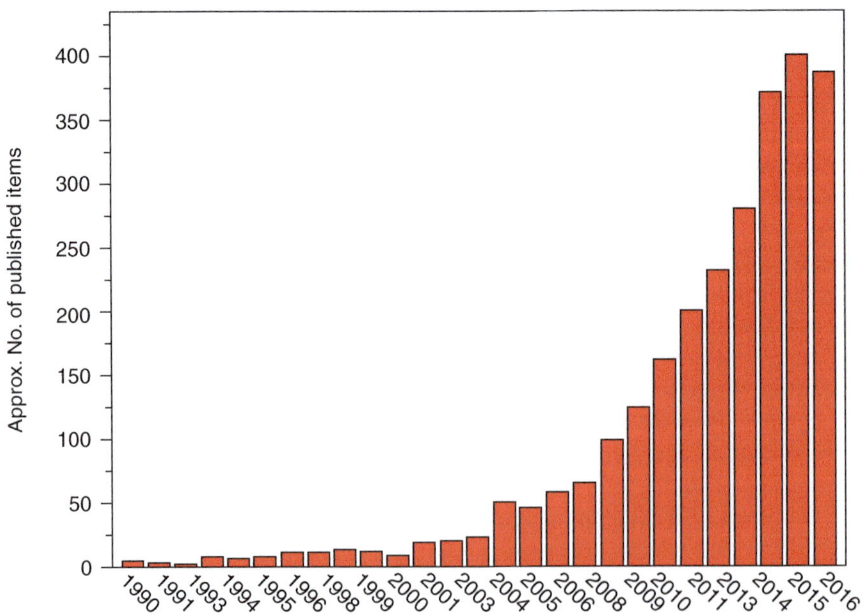

Fig. 4.9 The yearly number of published articles on the topic of ITA by searching "isothermal amplification" in the *PubMed* database

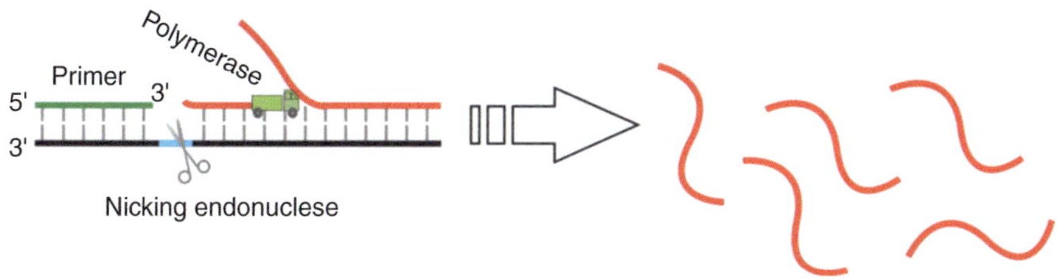

Fig. 4.10 Depiction of strand displacement amplification

This term "isothermal amplification" was first proposed by Gingeras et al. in 1990, when his group devised a self-sustained amplification system (3SR) under isothermal conditions by using three nucleases: reverse transcriptase, RNase H, and T7 RNA polymerase (Guatelli et al. 1990). When operated at 37 ° C consistently, 10^5-fold amplification would be achieved in 15 min, which should require more than 85 min in the PCR system. The pioneering work done by Gingeras et al. has inspired various ingenious strategies of ITA using the versatile polymerases and nucleases in molecular biotechnology. For example, restriction enzyme *Hinc*II bears the ability to nick the target single-stranded DNA (ssDNA) of recognition site at a dsDNA. The exo$^-$ Klenow polymerase is 5′–3′ exonuclease-deficient polymerase with strand displacement activity. It could extend the 3′-end of a partly complementary dsDNA and displace the downstream strand. By using the two specific enzymes, a strand displacement amplification (SDA) method was designed (Walker et al. 1992). As shown in Fig. 4.10, the target nucleic acid served as primer and bound to the complementary part of template DNA, forming a partly complementary strand with recognition sequence at the overhang. With the assistance of exo$^-$ Klenow polymerase, the primer was extended from 3′ end and a recognition site was formed. Then *Hinc*II nicked the primer strand, the exo$^-$ Klenow polymerase started work again from the nick and displaced the downstream ssDNA. Amplified amounts of ssDNAs can be produced after multiple polymerization, displacement, and nicking cycles.

In 2000, a more effective ITA, loop-mediated isothermal amplification, was developed by Notomi et al. using *Bst* DNA polymerase (Notomi et al. 2000; Tomita et al. 2008). The *Bst* DNA polymerase has a strong strand displacement activity and lacks 5′-3′ exonuclease activity. As schematically shown in Fig. 4.11, the elaborate insertion of the complementary sequence into the primer was the key point to form loop structure, which enabled anneal-free binding. And the *Bst* DNA polymerase with strong displacement activity enabled DNA elongation and release of new strands, which resulted in multiplied output of DNAs. Without tedious thermocycling process, it can accumulate more than 10^9 copies in less than an hour. Besides, due to the employment of six distinct probing sequences on the primers, the nucleic acids sensor based on loop-mediated isothermal amplification exhibited higher selectivity than PCR and other amplification methods.

As Zhao et al. summarized in his review (Zhao et al. 2015), ITA-based methods were usually designed based on the enzymatic activities of DNA replication or digestion. Theoretically, numerous isothermal amplifying methods could be developed by

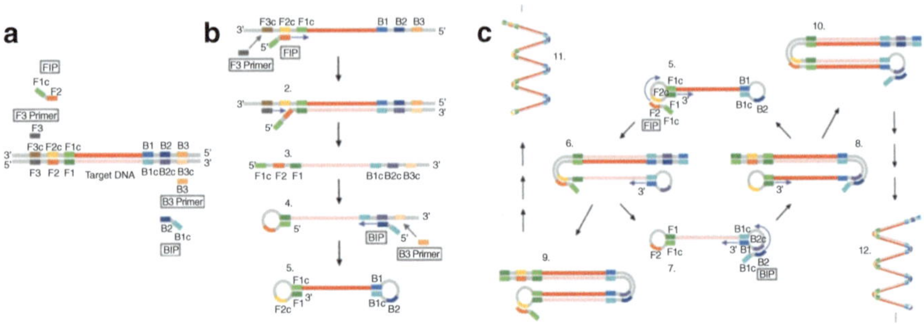

Fig. 4.11 Principle illustration of the loop-mediated isothermal amplification. (**a**) Design of the four primers with six probing regions towards target DNA. (**b**) Producing step of the dual-loop structure, which was resulted from DNA polymerase reaction and strand displacement. (**c**) Cycling amplification and elongation step for multiplied output. Reprinted with permission from Tomita et al. (2008), © 2008 Nature Publishing Group

combinational employment of polymerases and nucleases in the library of DNA enzymes. For examples, Guan et al. developed a ITA sensing strategy for the ultrasensitive detection of sequence-specific ssDNA using S1 nuclease (Guan et al. 2014). As low as 50 pM target ssDNA in 0.5 μg/mL DNA could be detected. Su et al. developed a low-background ITA-based method by using two enzymes, exonuclease I and DNA polymerase, for the detection of hepatitis RNA (Su et al. 2014). The linear range was 50 zM ~ 50 nM with a quantitative limit of 50 zM (10^{-21} mol/L), which was lower than general PCR method. Chen et al. developed an ultrasensitive electrochemiluminescence method based on rolling circle amplification (RCA) for the detection of microRNA (Chen et al. 2016). Three-way DNA junction and phi29 DNA polymerase were employed in his strategy and a detection limit of 22 aM (10^{-18} mol/L) was achieved. Min et al. developed a label-free ITA strategy for the detection of microRNA using exonuclease III (Exo III) and AIE fluorophore (Min et al. 2015). Samples in the urine of patient with bladder cancer were successfully detected and a detection limit of 1 pM was achieved in consistent temperature (37 °C).

ITA is more than a technique, it is a methodology.

The combination of G4 and ITA will definitely make for low-cost and ultrasensitive detections of biomarkers, and promote the early diagnosis of cancer in a more practical way in developing areas. In the following contents we try to introduce the latest works to illustrate the applications of G4&ITA-based methods for early cancer diagnosis.

4.5 Low-Cost and Ultrasensitive Detection of Biomarkers Based on G4 and ITA

4.5.1 Assay of Gene Mutation

Mutations in important genes can cause abnormal behaviors of cells, which always induces cancer (Bertram 2000). For examples, mutations in $BRCA_1$ and $BRCA_2$ genes indicate high risk of developing breast cancer; mutations in the *Rb* tumor

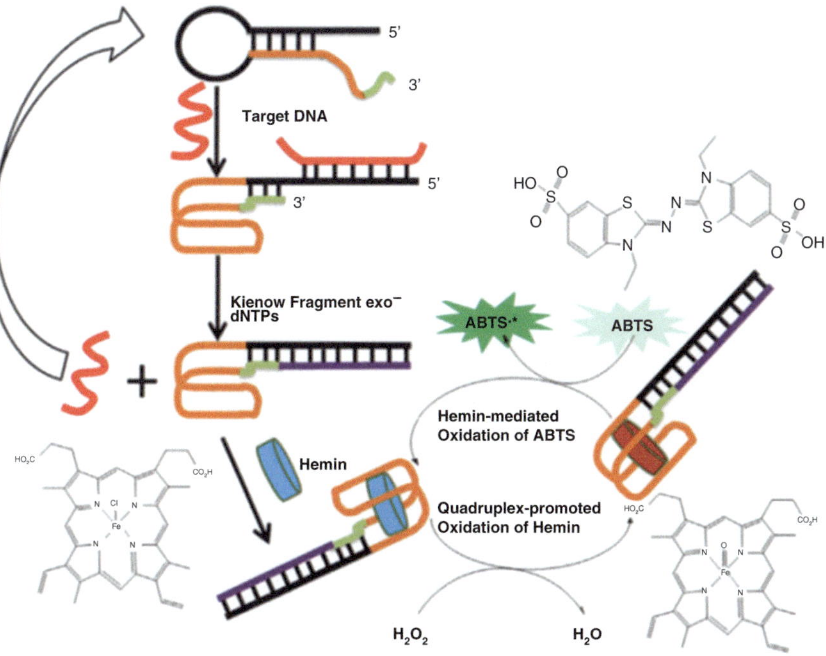

Fig. 4.12 Ultrasensitive detection of *TP53* DNA based on SDA and G4. Reprinted with permission from Li et al. (2013), © 2013 Elsevier

suppressor gene are related to eye cancer and several other types of cancers. Mutations in the *TP*53 tumor suppressor gene are the most frequent genetic alterations in human cancer and commonly compromise the gene's tumor suppressor activity (Vogiatzi et al. 2016). Li et al. developed an ITA-based method for the colorimetric detection of *TP*53 gene by employing SDA and G4 (Li et al. 2013). As shown in Fig. 4.12, the hybridization between target *TP*53 DNA and the stem-loop probe could induce the formation of G4 structure and initiation of polymerase-assisted strand displacement, thus liberating the target DNA for anther amplification circle. The specific binding of hemin to multiple numbers of G4 structure resulted in magnified colorimetric output by using $ABTS^{-}$ as catalytic substrate. A detection limit down to 25 fM was achieved by this ultrasensitive LFB.

Xu et al. developed a colorimetric method for ultrasensitive detection of *TP53* gene based on G4 and ITA (Xu et al. 2016). As shown in Fig. 4.13a, in step 1 and 2, target DNA hybridized with the stem-loop probe and opened its 3′-end for binding to a C-rich DNA template. Step 3 is polymerase and nickase-assisted SDA; in this step target DNA was also released for another amplification circle. The amplified G-rich DNAs in step 3 formed G4 structures in reaction buffer and catalyzed the generation of colorimetric signals by using hemin as binding ligand and $ABTS^{2-}$ as substrate. As low as 10 fM *TP53* DNA could be detected at consistent temperature (37 °C) in only 1 h. *K-ras* gene is a proto-oncogene, the mutation of which is usually found in the development of cancers, including leukemia, colorectal cancer, and lung cancer. The detection of *K-ras* mutation is of great importance for the

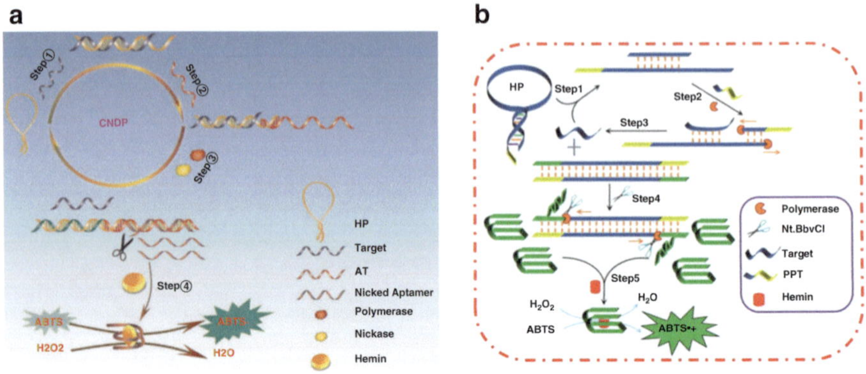

Fig. 4.13 Depiction of the operation principle based on G4 and SDA for the colorimetric detection of TP53 gene (**a**) and *K-ras* gene (**b**), respectively. Reprinted with permission from Xu et al. (2016), © 2016/2017 Elsevier

diagnosis and therapy of primary or metastatic cancer. Based on the similar strategy, Xu et al. developed an ultrasensitive LFB for the colorimetric detection of proto-oncogene *K-ras* based on G4 and SDA (Xu et al. 2017). As shown in Fig. 4.13b, the ssDNA of target gene can open the stem-loop probe, liberating the binding arm for the primer. Then in the presence of the nicking endonuclease and DNA polymerase with strand displacement capacity, large amount of G4 DNA were produced in the SDA process. By using hemin as ligand, G4/hemin DNAzyme was formed to catalyze the oxidation of $ABTS^{2-}$ for label-free and colorimetric signal output. The ultrasensitive method exhibited a linear range from 0.01 to 150 nM with a detection limit of 10 pM.

4.5.2 Assay of microRNA

MicroRNA (miRNA) is a class of noncoding small RNAs with 18–25 nucleotides, which functions as posttranscriptional regulator by hybridizing with the specific sequences (specifically the 3′-untranslated region) of the target mRNA to modulate gene expressions (Tavazoie et al. 2008). Aberrant expressions of miRNAs are implicated in tumorigenesis and proliferation. It has been universally considered that miRNA is an ideal biomarker for cancer diagnosis (Baker 2010).

For instance, the *lethal*-7 (*let-7*) family of miRNAs remains highly sequence-conserved and is one of the key miRNA regulators in the development of cancers (Bussing et al. 2008). The *let-7* family functions as a tumor suppressor and is down-regulated in human cancer cells including lung cancer (Takamizawa et al. 2004), hepatocellular carcinoma (Shimizu et al. 2010), and gastric cancer (Tang et al. 2016). The expression of *let-7* miRNAs can be regarded as indicators for cancer diagnosis. Zheng et al. designed a G4-based method for the label-free and ultrasensitive detection of *let-7d* miRNA by utilizing two ITA technologies, SDA and RCA (Zheng et al. 2016). As low as 1.5×10^{-13} M miRNA can be detected as a result of

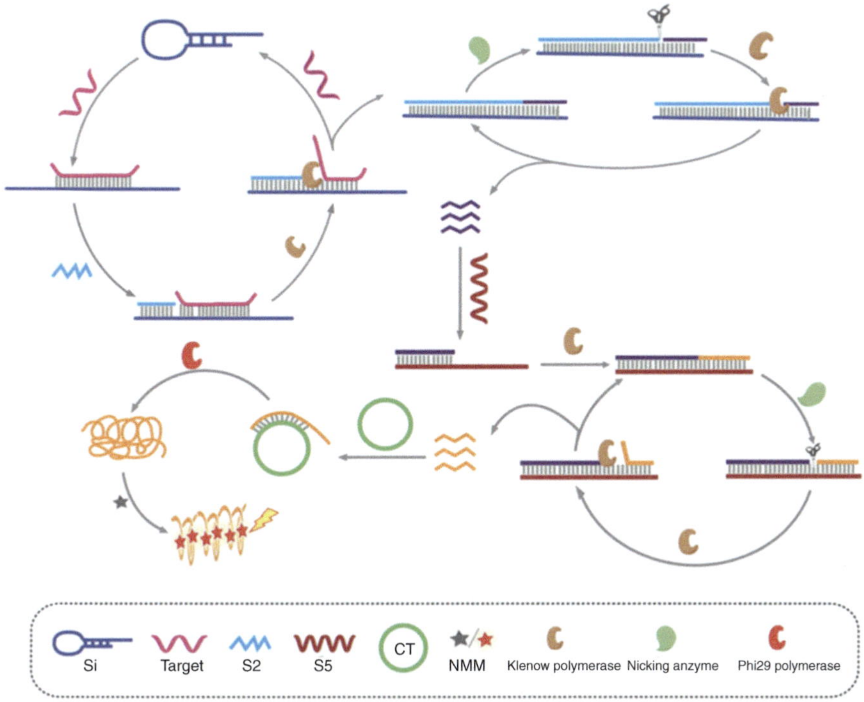

Fig. 4.14 Scheme of the LFB for the detection of let-7d miRNA based on G4 and SDA/RCA. Reprinted with permission from Zheng et al. (2016), © 2016 Nature Publishing Group

multiple amplifications. As shown in Fig. 4.14, after target *let-7d* miRNA opened the stem-loop structure through sequence-specific hybridization, three circles of SDA amplifications preceded for enlarged generation of the RCA primers (yellow color). Massive G-rich sequences could be produced by employing plenty primers and using a C-rich RCA template. G4-responsive NMM was employed as the fluorescent indicator for label-free signal output.

MiRNA-141 is affiliated with miRNA-200 family which is involved in the formation of cancer stem cells and regulation of epithelial-mesenchymal transition. Dysregulation of miRNA-141 is depending on the type of cancers: miRNA-141 is overexpressed in nasopharyngeal carcinoma, classic papillary thyroid carcinoma, bladder cancer, and colorectal cancers, while downregulated in gastric cancer, pancreatic cancer, osteosarcoma, hepatocellular, esophageal cancer, breast cancer, and renal cell carcinoma (Gao et al. 2016). The quantitative detection of the expression level of miRNA-141 can provide information for the diagnosis of one specific type of cancers (Zhao et al. 2013). Wang et al. developed a colorimetric method for label-free and ultrasensitive detection of miRNA-141 based on catalytic G4 DNAzyme and SDA technique (Wang et al. 2013). The detection principle was mainly based on a SDA strategy, but with an improved sensitivity since the elaborate design of the template. As shown in Fig. 4.15, there were two recognition sites for the nicking

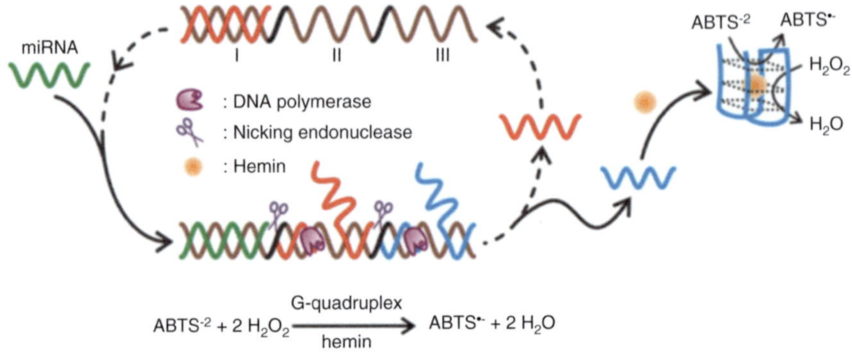

Fig. 4.15 Principle illustration of the label-free and ultrasensitive assay for miRNA-141 based on G4-based DNAzyme and SDA. Reprinted with permission from Wang et al. (2013), © 2013 Elsevier

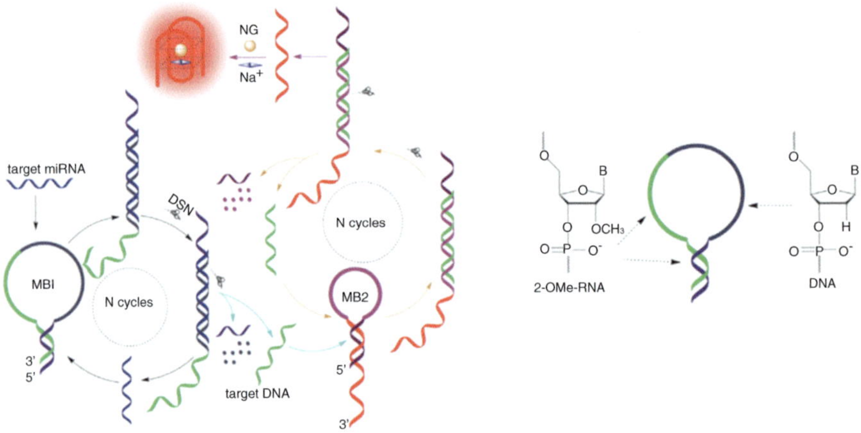

Fig. 4.16 Principle of the DSN-assisted ITA for miRNA-141 detection using G4/MG as label-free signal output. Reprinted with permission from Zhang et al. (2017), © 2017 Elsevier

endonuclease and an identical sequence of the complementary sequence for targeting miRNA-141. As a result, numerous short oligonucleotides of the same sequence of target miRNA-141 were synthesized and bound to the complementary sequence in the template, functioning as target in another amplification cycle. Finally, G-rich oligonucleotides were vastly produced in the C-rich part of the template during the SDA process, which can bind to hemin and form the catalytic G4/hemin DNAzyme to generate colorimetric signals by catalyzing the H_2O_2-mediated oxidation of $ABTS^{2-}$. The quantitative concentration by this label-free and ultrasensitive colorimetric method was in the range from 1 fM to 100 nM.

Zhang et al. developed an ultrasensitive assay for miRNA-141 based on G4 and duplex-specific nuclease (DSN)-assisted ITA (Zhang et al. 2017). As shown in Fig. 4.16, DSN displays an isothermal cleavage activity towards dsDNA and DNA in DNA-RNA hybrid duplexes. In this strategy, two stages of DSN-assisted recycling amplification were designed in this assay. Two types of molecular beacon probes were designed, MB1 and MB2. The MB1 contained RNA (green part) and DNA (black part) sequences, MB2 is a DNA probe. In the presence of miRNA-141, it opened MB1 and triggered the DSN-catalyzed digestion of the DNA part of MB1, resulting in release of miRNA-141 and the RNA part of MB1. Subsequently, miRNA-141 and RNA part initiated DSN-assisted amplification by opening MB1 and MB2, respectively, in the next amplification cycle. Finally, numerous G4 structures would form since G-rich DNAs were liberated from MB2. The G4-specific dye, malachite green (MG), was employed for label-free signal output. The method showed a wide dynamic linear range from 1 pM to 10 μM with a detection limit of 1.03 pM, and demonstrated a practical potential for cell lysate samples.

Wang et al. also reported a DSN-assisted amplification approach for ultrasensitive detection of miRNA-122 using catalytic G4/hemin for chemiluminescence signal output (Wang et al. 2016c). MiRNA-122 is a liver-specific miRNA, which accounts for 72% of total miRNAs in adult human liver and is frequently reduced in hepatocellular carcinoma (Coulouarn et al. 2009). MiRNA-122 was identified as a potential diagnostic biomarker in patients with hepatocellular carcinoma and hepatic disorders. As shown in Fig. 4.17, this approach comprises a G4-based label-free signal output mode, a polydopamine nanosphere (PDA)-based chemiluminescence resonance energy transfer (CRET) platform, and a DSN-assisted ITA. In the absence of target miRNA (pathway B), the chemiluminescence signal produced by the G4/hemin-luminol system would be quenched due to the CRET between PDA. In the CRET process, PDA acted as energy acceptor and the excited luminol by G4/hemin DNAzyme acted as energy donor. While in presence of target miRNA-122, miRNA-122 can drag the G4-contained DNA from the surface of PDA due to the selective hybridization between it and the complementary sequence of the probe. Then the hybridization was followed by a degradation process by DSN. The complementary DNA part in the

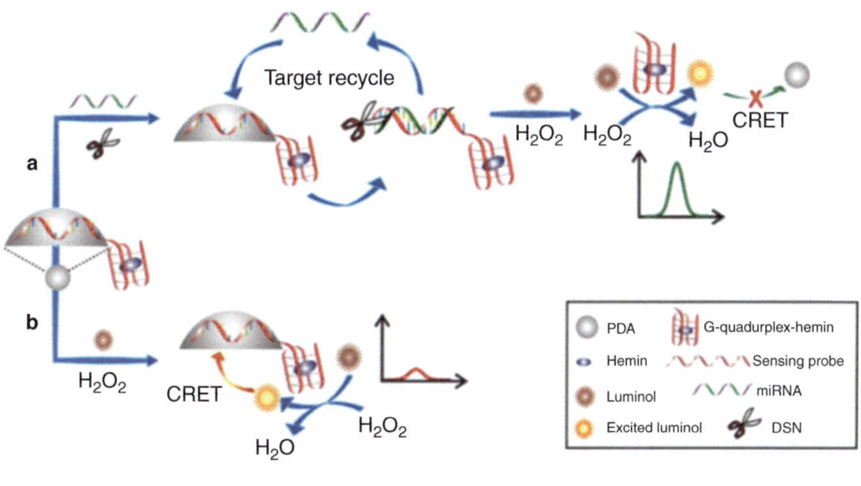

Fig. 4.17 Chemiluminescence approach for highly sensitive detection of miRNA-122 based on catalytic G4/hemin DNAzyme and DSN-based ITA. The sensing system in the presence (**a**) and absence (**b**) of target miRNA, respectively. Reprinted with permission from Wang et al. (2016c), © 2016 Elsevier

DNA–miRNA heteroduplex would then be degraded by DSN, releasing G4 DNAs and miRNA-122 for another amplification cycle. After binding to hemin, the G4/hemin DNAzyme catalyzed the H_20_2-mediated oxidation of luminol, generating enhanced chemiluminescence signals. The method allowed quantitative detection of target miRNA in the range of 80 pM–50 nM with a detection limit of 49.6 pM.

4.5.3 Assay of the Activity of Cancer-Related Enzyme

Telomerase is a ribonucleoprotein reverse transcriptase that catalyzes the addition of the telomeric repeats (TTAGGG)n onto the end of chromosomal DNA (Buys 2000). Telomerase activity is unregulated in greater than 85% of human tumors, such as breast, colon, lung, prostate, ovary, stomach, and skin, whereas it is found at relatively low levels in normal cells. The inhibition of telomerase activity has been proposed as a potential approach for the treatment of human cancers. This makes telomerase a promising biomarker in cancer diagnosis and therapy. Quach et al. developed a ITA method for rapid detection of telomerase activity based on elongation of G-rich DNA (Quach et al. 2013). As shown in Fig. 4.18, telomerase had polymerase activity and could synthesize TTAGGG repeat units from the 3′ end of the primer. The elongated G-rich ssDNA were able to form multiple G4 structures in the presence of K^+, resulting in fluorescence enhancement upon binding to G4-responsive dye. This method was simple, highly sensitive, and can produce results in only 5 min. Telomerase activity in 4 HeLa cells can be easily detected using this assay.

DNA methylation, a crucial epigenetic modification of the genome, plays a pivotal role in gene expression, cellular differentiation, and pathogenesis of various human diseases. This is closely associated with the activity of DNA methyltransferase. Alterations of the methyltransferase activity may lead to aberrant DNA methylation patterns. Abnormalities of DNA methylation normally occur before other signs of malignancy and can be used for early diagnosis. Recent researches have demonstrated that DNA methyltransferase can be a predictive biomarker in various types of cancer (Chen et al. 2017; Pronina et al. 2017). Xue et al. developed a Exo III-assisted ITA method for amplified detection of methyltransferase activity (Xue et al. 2015).

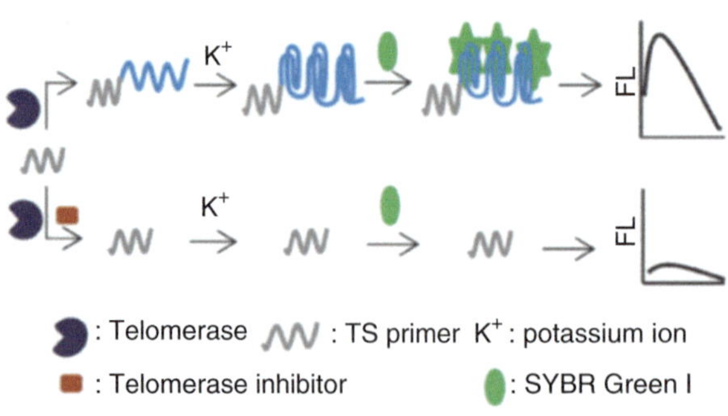

Fig. 4.18 Schematic illustration of the LFB for sensing of telomerase activity based on G4. Reprinted with permission from Quach et al. (2013), © 2013 Royal Society of Chemistry

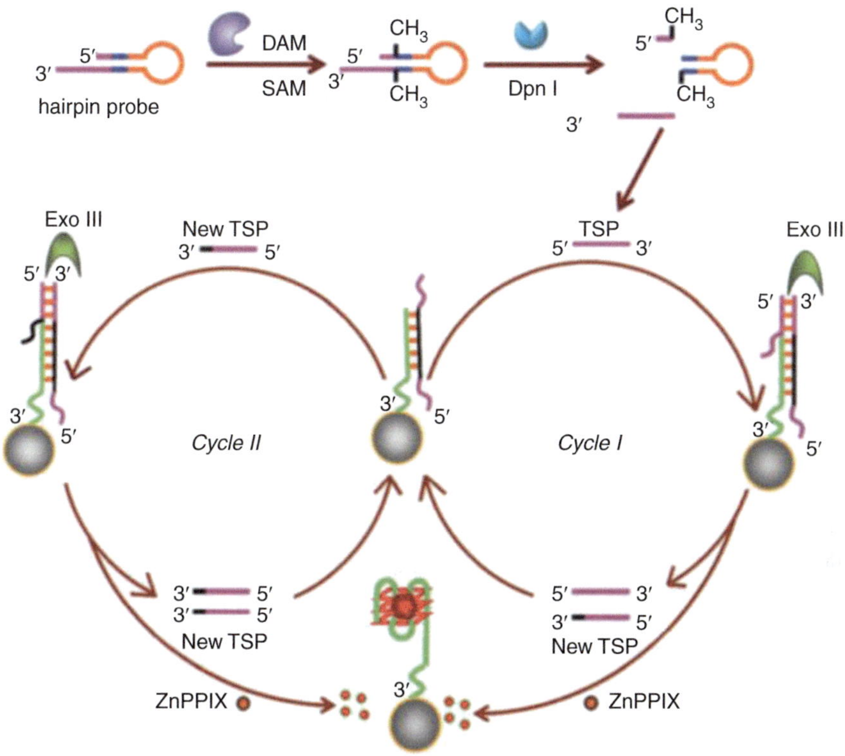

Fig. 4.19 Schematic depiction of the principle for the detection of methyltransferase activity based on G4 and ITA. Reprinted with permission from Xue et al. (2015), © 2015 Royal Society of Chemistry

As shown in Fig. 4.19, Exo III exhibits 3′ → 5′ exodeoxyribonuclease activity specific for dsDNA. It degrades dsDNA from blunt ends, 5′-overhangs or nicks, releasing 5′-mononucleotides from the 3′-ends of DNA strands. Once the stem-loop DNA was methylated by methyltransferase, the GATC sequence was then recognized and cleaved by the Dpn I that cleaved only when its recognition site was methylated. The released ssDNA acted as trigger to activate the Exo III-mediated ITA on the magnetic microbead, liberating numerous G4 DNAs. The G4-specific dye, ZnPPIX, was employed as the label-free signal reporter for fluorescence output. A very low detection limit down to 2.0×10^{-4} U/mL was obtained based on this method.

4.5.4 Assay of Protein Biomarker

Many different tumor protein biomarkers have been characterized and are in clinical use. Some are associated with only one type of cancer, whereas others are associated with two or more cancer types. Such as human epidermal growth factor receptor (EGFR) and transcription factor. EGFR is involved in tumor cell proliferation, adhesion, migration, apoptosis, and differentiation, and it is

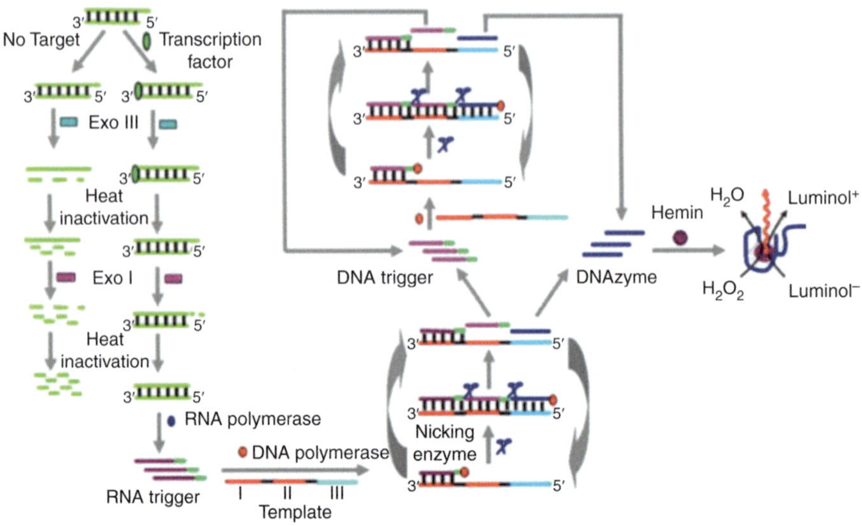

Fig. 4.20 Schematic depiction of the label-free chemiluminescence assay of transcription factor based on G4 and multiple enzyme-assisted ITA. Reprinted with permission from Ma et al. (2014), © 2014 American Chemical Society

overexpressed in gastric cancer, glioblastoma, anal cancers, and squamous-cell lung carcinoma (Gravalos and Jimeno 2008). Transcription factor modulates the gene expression by binding to the regulatory region of target DNA and regulating the transcription of target gene. Bours et al. observed high expression of the NF-κB inhibitor IκB in the ovarian carcinoma cell line *OVCAR*-3 (Bours et al. 1994). For ultrasensitive detection of the transcription factor (NF-κB), Ma et al. developed a label-free chemiluminescence method for the assay of NF-κB based on G4 and SDA (Ma et al. 2014). As shown in Fig. 4.20, NF-κB could bind to the recognition site of the dsDNA thus prohibiting the digestion by Exo III and exonuclease I. Then the dsDNA probe would be transcribed by T7 RNA polymerase to produce abundant RNA, which acted as the primer of SDA template. In the final stage, more DNA primers and G-rich DNAs can be produced in the SDA procedure. In the presence of hemin, the assembled G4/hemin DNAzyme catalyzed luminol for chemiluminescence signals. The detection limit of this method reached 6.03×10^{-15} M. And the linear range expanded from 10^{-14} to 10^{-9} M.

Platelet-derived growth factor (PDGF) is a mitogenic protein which is released from platelets, regulates cell growth and division in blood vessel formation. PDGF is frequently overexpressed in tumor cell lines, particularly glioma and sarcoma. Reported works have found that antagonism of PDGF resulted in inhibited growth of glioma cell lines (Westermark et al. 1995), suggesting that quantification of PDGF can be for the diagnosis of tumorigenesis and proliferation. Li et al. constructed a cascade amplification strategy for the sensitive detection of PDGF (Li et al. 2016). As shown in Fig. 4.21, the strategy contained three amplification stages: the target-activated SDA and the dual RCA-SDA process. A self-complementary DNA with the aptamer sequence was employed as the recognition probe. Upon binding to the target

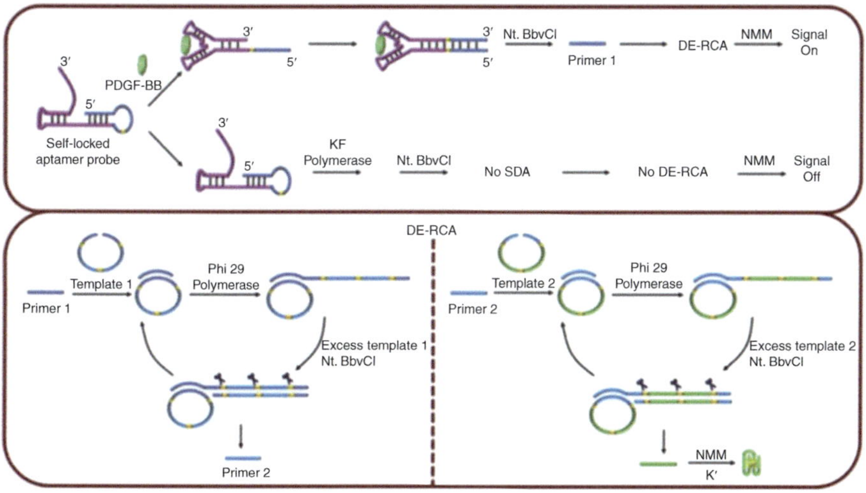

Fig. 4.21 Schematic illustration of the constructed three-stage amplification method for the detection of PDGF. Reprinted with permission from Li et al. (2016), © 2016 Elsevier

PDGF, the probe transformed to a three-way junction with 5′-overhang. Then the 3′-end of the three-way junction acted as the primer of SDA and released a large amount of primer 1. Next, the primer 1 activated the first RCA-SDA process. Since the template 1 contained the recognition site of Nt.BbvCI, as well as the complementary sequences of primer 1 and primer 2, a large amount of primer 1 and primer 2 can be yielded in the first RCA-SDA process. In the second RCA-SDA process, the primer 2 hybridizes with template 2 and triggered the amplification stage, generating numerous G4 DNAs. In the presence of G4-specific fluorescent dye NMM, enhanced fluorescence intensity would be detected. An ultralow detection limit of 3.8×10^{-16} mol/L was reached based on the cascade amplification strategy.

4.5.5 Assay of Cancer-Related Small Molecule

Evidences have shown that extracellular adenosine triphosphate (ATP) and adenosine receptors are overexpressed by tumor cells to enhance the tumor growth. And studies suggested that accumulation of extracellular ATP and adenosine can activate the receptors and induce pro-tumorigenesis, tumor immune escape, and cancer cell proliferation (Stagg and Smyth 2010). Therefore, there is positive correlation between the concentration of extracellular ATP/adenosine and status of cancer. By employing G4-specific dye ThT, Exo III and ATP aptamer, Wang et al. developed a fluorescence LFB for ultrasensitive detection of ATP (Wang et al. 2016e). As depicted in Fig. 4.22, the LFB was based on target-induced structure transition and Exo III-assisted ITA. The probe was a dual-strand complex with an overhanging part of ATP aptamer, which contributed steric hindrance and protected the DNA hairpin against the digestion by Exo III. After binding to the target ATP, the aptamer DNA will be released from the complex, and thus initiated the Exo III-assisted ITA

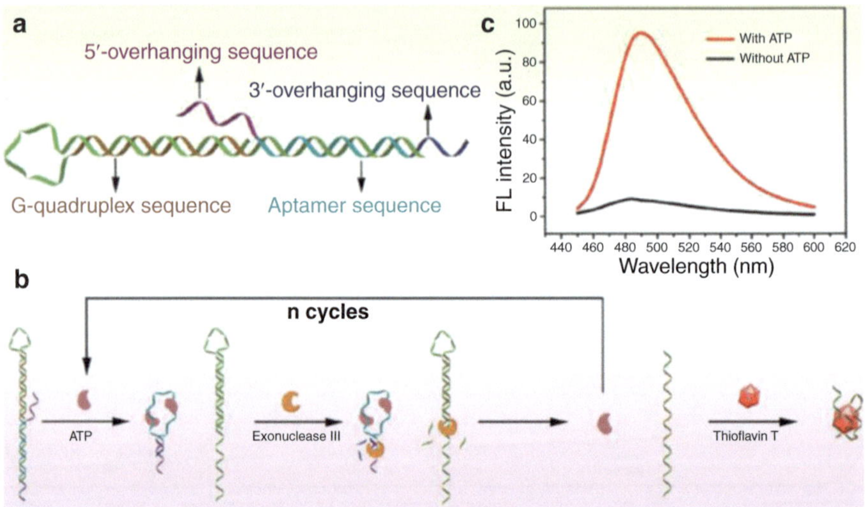

Fig. 4.22 Schematic illustration of the LFB for ultrasensitive detection of ATP based on G4 and Exo III-assisted ITA. (**a**) The sequence design of the G4-based probe. (**b**) The amplification cycles and detection procedure for the assay of ATP. (**c**) Fluorescence spectra of the LFB in the absence and presence of ATP, respectively. Reprinted with permission from Wang et al. (2016e), © 2016 Royal Society of Chemistry

process. The stem-loop probe will be hydrolyzed and the blocked G-rich sequences will be liberated from the hairpin DNA, resulting in formation of multiple G4 structures and enhanced fluorescence of ThT. The signal-to-noise ratio of the sensing system was calculated to be 22.4 with a detection limit of 280 pM.

Sun et al. constructed an amplified fluorescence LFB for the ultrasensitive analysis of adenosine based on Exo III-assisted ITA and hybridization chain reaction (HCR) (Sun et al. 2015a). The biosensor consisted three steps as shown in Fig. 4.23. In the first step, a streptavidin-magnetic mircobead functionalized with aptamer-contained dsDNA was employed to achieve a low background. After adding adenosine to the sensing system, adenosine bonded to the aptamer DNA, resulting in release of ssDNA. In the second step, the released ssDNA hybridized with the stem-loop DNA and formed dsDNA with blunt 3′-terminus. Exo III then catalyzed the digestion of the stem-loop DNA, releasing the trigger strand (pink). In the last step, the trigger strand initiated the assembling process of HCR, forming amplified amount of split G4 structures and generating enhanced intensity in the fluorescence of G4-specific dye NMM. This method exhibited a high sensitivity towards adenosine with a detection limit of 4.2×10^{-7} mol/L, and showed distinguish capacity for adenosine in urine samples from cancer patients.

4.6 Conclusion and Perspective

In the past decades, increasing efforts have been made to develop isothermal amplification technologies for the ultrasensitive and economic detections of biochemical samples. Furthermore, the employment of G4 in LFBs contributed to advantages of multiple signal outputs and a variety of design strategies. Prospectively, low cost,

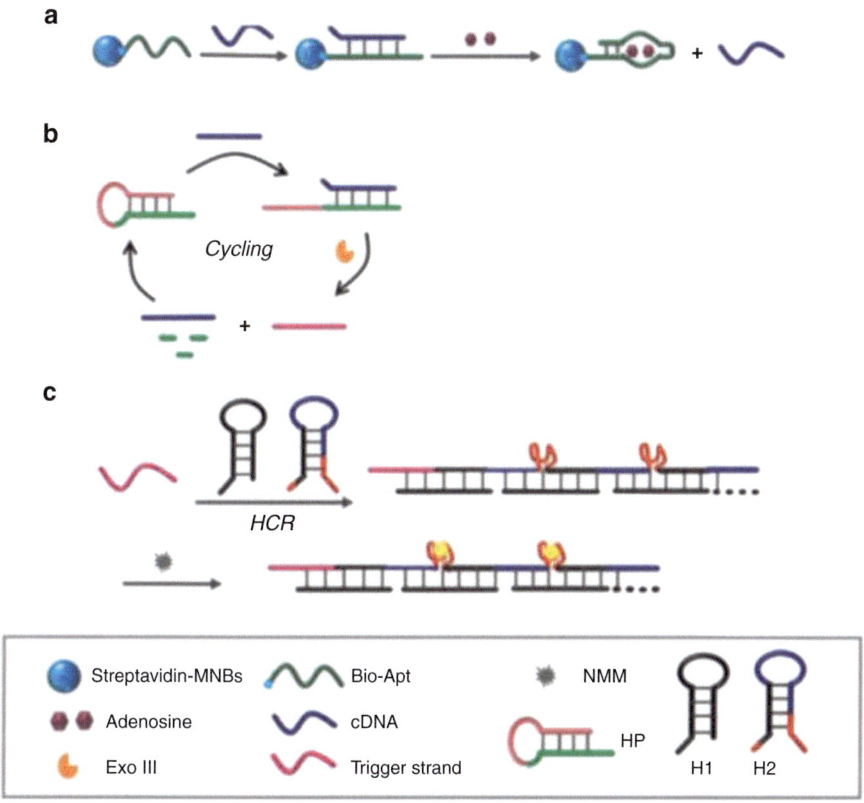

Fig. 4.23 Illustration of the detection principle of fluorescence biosensor for the ultrasensitive detection of adenosine based on aptamer, Exo III-assisted ITA and HCR. The recognizing (**a**), amplifying (**b**) and HCR assembling (**c**) steps of this ultrasensitive LFB. Reprinted with permission from Sun et al. (2015a) query, © 2015 Elsevier

high sensitivity, time-saving, as well as visual assay can be simultaneously realized based on G4 and ITA, which shows great potential for early diagnosis of cancer in developing areas.

However, there were also some inevitable drawbacks and limitations in the G4&ITA-based assays. It was generally known that many fluorescence dyes are susceptible to environment. The fluorescence emission property of some dyes can be affected by ionic strength, temperature, metal ions, proteins et al. in the environment. Especially in the label-free system, the dyes were homogeneously dispersed and separated with oligonucleotides probes. Meanwhile, the versatile G4 can also display a disruptive behavior due to its binding activities towards various kinds of ligands including metal ions, organic compounds, proteins, nanomaterials, and even cells. The respective binding of G4-responsive dyes and G4s to the interference compounds in the matrix might lower the sensitivity or cause false signals. Besides, the amplification strategies have become more and more sophisticated to get a higher sensitivity, which always suffered complicated steps and operations. So, the G4-based label-free system is more fragile than labeled system due to an increased uncertainty.

Except requirement for sample pretreatment, ratiometric detection may be an effective alternative to circumvent this vulnerability. As the two different signal indicators both have response to analyte, ratiometric methods have the advantages to eliminate the negative effect in complex environment and give more precise measurement. Increasing efforts are being devoted to develop ratiometric sensors for DNA (Guo et al. 2016a; Liu et al. 2016), proteins (Chang et al. 2016; Hu et al. 2016), and small compounds (Zhao et al. 2016; Zhu et al. 2016). For instance, our group developed a ratiometric LFB for fluorescence detection of DNA by using two kinds of binding-responsive dyes (Guo et al. 2016a). The ratio calculation of dual-signal intensities contributed to a better sensitivity and selectivity compared with single-signal intensity.

Though the cited works in this chapter showed excellent performances including low cost, ultrahigh sensitivity and selectivity, they all focused on single target. It is worth noting that, some biomarkers are associated with only one type of cancer, whereas others are associated with two or more cancer types. For example, miRNA-122 is a tumor suppressor specific in hepatocellular carcinoma, while miRNA-141 shows correlations with nasopharyngeal carcinoma, bladder cancer, colorectal cancers, gastric cancer, pancreatic cancer, osteosarcoma, breast cancer, and renal cell carcinoma. Squamous cell carcinoma antigen is mainly involved in the malignant behavior of squamous cell cancers, while carcinoembryonic antigen is a wide-spectrum tumor marker expressed in many types of cancer cells. Besides, there can be opposite correlation between the same biomarker with different types of cancer. For instance, miRNA-141 is overexpressed in colorectal cancers, but downregulated in gastric cancer. On the other hand, abnormal levels of some biomarkers are correlated to more than one kind of disease. For example, miRNA-29 exhibits dysregulation in lung cancer, chronic lymphocytic leukemia, Alzheimer disease, and cardiac ischemia. Abnormal AFP levels account to not only tumors, but also neural tube defects, omphalocele, and ataxia telangiectasia. Adenosine performs extremely important roles not only in tumor proliferation, but also in inflammation, brain ischemic damage, and Huntington's disease.

Therefore, a comprehensive and logical understanding of the combinational parameters is in urgent need for the clinical diagnosis and study of cancers. DNA logic gate and DNA computation provide an ideal solution to untangle the multiple and confused test parameters for a clear conclusion (Pei et al. 2010). In a logic gate-based assay strategy, a single conclusion can be obtained based on both negative and positive responses. For instance, Guo et al. proposed a three-target DNA logic circuit, which could be applied for intelligent assay of multiple DNA/RNA targets (Guo et al. 2014a). In a logic strategy, an "INHIBIT" system can be applied for the simultaneous assay of a positive biomarker and a negative biomarker. Undoubtedly, the employment of G4 and development of ITA were for the low-cost and time-efficient advantages for the early diagnosis of cancers. For an improved performance of more cost-saving and time-saving, we think intelligent assay based on DNA logic gates and DNA computation must be promising and useful.

4.7 Summary Table

Label-free and ultrasensitive biosensors for biomarker assay based on G4 and ITA

	Biomarker	Enzyme	G4 ligand/substrate and signal	Detection limit ($mol{\cdot}L^{-1}$)	Reference
DNA mutation	*TP53*	exo⁻ Klenow fragment	Hemin/ $ABTS^{2-}$ colorimetric	25×10^{-15}	Li et al. (2013)
	c-erbB-2	Exo III	Thioflavin T/— fluorescence	20×10^{-15}	Chen et al. (2014)
	T790M	DNA ligase phi29 DNA polymerase	Thioflavin T/— fluorescence	6.9×10^{-10}	Lee et al. (2016)
	TP53	exo⁻ Klenow fragment Nt.BbvCI	Hemin/ $ABTS^{2-}$ colorimetric	10×10^{-15}	Xu et al. (2016)
	K-ras	exo⁻ Klenow fragment Nt.BbvCI	Hemin/ $ABTS^{2-}$ colorimetric	10×10^{-12}	Xu et al. (2017)
miRNA	miRNA-141	Nt.Bst.NBI $Vent_R$ DNA polymerase	Hemin/$ABTS^{2-}$	1×10^{-15}	Wang et al. (2013)
	let-7a miRNA	T4 DNA ligase exo⁻ Klenow fragment Nt.Alwl	Hemin/ $ABTS^{2-}$ colorimetric	1.0×10^{-21}	Bi et al. (2013)
	miRNA-21	exo⁻ Klenow fragment Nb.BbvCI	Hemin/ $ABTS^{2-}$ colorimetric	1.7×10^{-12}	Yan et al. (2015)
	miRNA-21	$Vent_R$ DNA polymerase Nt.Bst.NBI 8–17 DNAzyme	Hemin/TMB chronoamperometry	0.5×10^{-18}	Cheng et al. (2016)
	miRNA-122	Duplex-specific nuclease	Hemin/luminol chemiluminescence	4.9×10^{-11}	Wang et al. (2016c)
	let-7b miRNA	exo⁻ Klenow fragment Nb.BbvCI T4 DNA ligase phi29 DNA polymerase	NMM/— fluorescence	3.2×10^{-12}	Wang et al. (2016d)
	let-7d miRNA	exo⁻ Klenow fragment Nb.BbvCI phi29 DNA polymerase	NMM/— fluorescence	1.5×10^{-13}	Zheng et al. (2016)
	miRNA-141	Duplex-specific nuclease	MG/— fluorescence	1.0×10^{-12}	Zhang et al. (2017)

(continued)

	Biomarker	Enzyme	G4 ligand/substrate and signal	Detection limit ($mol{\cdot}L^{-1}$)	Reference
Enzyme	Telomerase	Telomerase	SYBR Green I/— fluorescence	4 cells	Quach et al. (2013)
	Methyltransferase	Exo III Dpn I	ZnPPIX/— fluorescence	2.0×10^{-4} U/mL	Xue et al. (2015)
Protein	Transcription factor	Exo III exonuclease I T7 RNA polymerase $Vent_R$ DNA polymerase Nt.Bst.NBI	Hemin/luminol chemiluminescence	6.0×10^{-15}	Ma et al. (2014)
	PDGF	Nt.BbvCI exo^- Klenow fragment Phi29 polymerase	NMM/— fluorescence	3.8×10^{-16}	Li et al. (2016)
Small molecule	Adenosine	Exo III	NMM/— fluorescence	4.2×10^{-7}	Sun et al. (2015a)
	ATP	Exo III	ThT/— fluorescence	28×10^{-11}	Wang et al. (2016e)

References

Allain C, Monchaud D, Teulade-Fichou MP (2006) FRET templated by G-quadruplex DNA: a specific ternary interaction using an original pair of donor/acceptor partners. J Am Chem Soc 128:11890–11893

Alzeer J, Vummidi BR, Roth PJC, Luedtke NW (2009) Guanidinium-modified phthalocyanines as high-affinity G-quadruplex fluorescent probes and transcriptional regulators. Angew Chem Int Ed 48:9362–9365

Arora A, Balasubramanian C, Kumar N, Agrawal S, Ojha RP, Maiti S (2008) Binding of berberine to human telomeric quadruplex – spectroscopic, calorimetric and molecular modeling studies. FEBS J 275:3971–3983

Baker M (2010) RNA interference Micrornas as biomarkers. Nature 464:1227–1227

Bertram JS (2000) The molecular biology of cancer. Mol Asp Med 21:167–223

Bhasikuttan AC, Mohanty J, Pal H (2007) Interaction of malachite green with guanine-rich single-stranded DNA: preferential binding to a G-quadruplex. Angew Chem Int Ed 46:9305–9307

Bi S, Cui YY, Li L (2013) Dumbbell probe-mediated cascade isothermal amplification: a novel strategy for label-free detection of microRNAs and its application to real sample assay. Anal Chim Acta 760:69–74

Bours V, Dejardin E, Goujon-Letawe F, Merville MP, Castronovo V (1994) The NF-kappa B transcription factor and cancer: high expression of NF-kappa B- and I kappa B-related proteins in tumor cell lines. Biochem Pharmacol 47:145–149

Bussing I, Slack FJ, Grosshans H (2008) Let-7 microRNAs in development, stem cells and cancer. Trends Mol Med 14:400–409

Buys CH (2000) Telomeres, telomerase, and cancer. N Engl J Med 342:1282–1283

Cao Y, Liu M, Zhang K, Zu G, Kuang Y, Tong X, Xiong D, Pei R (2017) Poly(glycerol) used for constructing mixed polymeric micelles as T1 MRI contrast agent for tumor-targeted imaging. Biomacromolecules 18:150–158

Chang CC, Wu JY, Chien CW, Wu WS, Liu H, Kang CC, Yu LJ, Chang TC (2003) A fluorescent carbazole derivative: high sensitivity for quadruplex DNA. Anal Chem 75:6177–6183

Chang N, Lu Y, Mao J, Yang J, Li M, Zhang S, Liu Y (2016) Ratiometric fluorescence sensor arrays based on quantum dots for detection of proteins. Analyst 141:2046–2052

Chen J, Lin J, Zhang X, Cai S, Wu D, Li C, Yang S, Zhang J (2014) Label-free fluorescent biosensor based on the target recycling and Thioflavin T-induced quadruplex formation for short DNA species of c-erbB-2 detection. Anal Chim Acta 817:42–47

Chen A, Ma S, Zhuo Y, Chai Y, Yuan R (2016) In situ electrochemical generation of electrochemiluminescent silver naonoclusters on target-cycling synchronized rolling circle amplification platform for MicroRNA detection. Anal Chem 88:3203–3210

Chen HC, Huang HY, Chen YL, Lee KD, Chu YR, Lin PY, Hsu CC, Chu PY, Huang TH, Hsiao SH, Leu YW (2017) Methylation of the tumor suppressor genes HIC1 and RassF1A clusters independently from the methylation of Polycomb target genes in Colon cancer. Ann Surg Oncol 24:578–585

Cheng FF, Jiang N, Li X, Zhang L, Hu L, Chen X, Jiang LP, Abdel-Halim ES, Zhu JJ (2016) Target-triggered triple isothermal cascade amplification strategy for ultrasensitive microRNA-21 detection at sub-attomole level. Biosens Bioelectron 85:891–896

Coulouarn C, Factor VM, Andersen JB, Durkin ME, Thorgeirsson SS (2009) Loss of miR-122 expression in liver cancer correlates with suppression of the hepatic phenotype and gain of metastatic properties. Oncogene 28:3526–3536

Das J, Ivanov I, Montermini L, Rak J, Sargent EH, Kelley SO (2015) An electrochemical clamp assay for direct, rapid analysis of circulating nucleic acids in serum. Nat Chem 7:569–575

Dhama K, Karthik K, Chakraborty S, Tiwari R, Kapoor S, Kumar A, Thomas P (2014) Loop-mediated isothermal amplification of DNA (LAMP): a new diagnostic tool lights the world of diagnosis of animal and human pathogens: a review. Pak J Biol Sci 17:151–166

Dumat B, Bordeau G, Faurel-Paul E, Mahuteau-Betzer F, Saettel N, Bombled M, Metge G, Charra F, Fiorini-Debuisschert C, Teulade-Fichou MP (2011) N-phenyl-carbazole-based two-photon fluorescent probes: strong sequence dependence of the duplex vs quadruplex selectivity. Biochimie 93:1209–1218

Freeman R, Liu X, Willner I (2011) Chemiluminescent and chemiluminescence resonance energy transfer (CRET) detection of DNA, metal ions, and aptamer-substrate complexes using hemin/G-quadruplexes and CdSe/ZnS quantum dots. J Am Chem Soc 133:11597–11604

Galezowska E, Gluszynska A, Juskowiak B (2007) Luminescence study of G-quadruplex formation in the presence of Tb^{3+} ion. J Inorg Biochem 101:678–685

Gao Y, Feng B, Han S, Zhang K, Chen J, Li C, Wang R, Chen L (2016) The roles of MicroRNA-141 in human cancers: from diagnosis to treatment. Cell Physiol Biochem 38:427–448

Granotier C, Pennarun G, Riou L, Hoffschir F, Gauthier LR, De Cian A, Gomez D, Mandine E, Riou JF, Mergny JL, Mailliet P, Dutrillaux B, Boussin FD (2005) Preferential binding of a G-quadruplex ligand to human chromosome ends. Nucleic Acids Res 33:4182–4190

Gravalos C, Jimeno A (2008) HER2 in gastric cancer: a new prognostic factor and a novel therapeutic target. Ann Oncol 19:1523–1529

Guan Z, Liu J, Bai W, Lv Z, Jiang X, Yang S, Chen A, Lv G (2014) Label-free and sensitive fluorescent detection of sequence-specific single-strand DNA based on S1 nuclease cleavage effects. PLoS One 9:e108401

Guatelli JC, Whitfield KM, Kwoh DY, Barringer KJ, Richman DD, Gingeras TR (1990) Isothermal, in vitro amplification of nucleic acids by a multienzyme reaction modeled after retroviral replication. Proc Natl Acad Sci U S A 87:1874–1878

Guo Q, Lu M, Marky LA, Kallenbach NR (1992) Interaction of the dye ethidium bromide with DNA containing guanine repeats. Biochemistry 31:2451–2455

Guo Y, Xu P, Hu H, Zhou X, Hu J (2013) A label-free biosensor for DNA detection based on ligand-responsive G-quadruplex formation. Talanta 114:138–142

Guo Y, Cheng J, Wang J, Zhou X, Hu J, Pei R (2014a) Label-free logic modules and two-layer Cascade based on stem-loop probes containing a G-Quadruplex domain. Chem-Asian J 9:2397–2401

Guo Y, Zhou L, Xu L, Zhou X, Hu J, Pei R (2014b) Multiple types of logic gates based on a single G-quadruplex DNA strand. Sci Rep 4:7315

Guo Y, Sun Y, Shen X, Chen X, Yao W, Xie Y, Hu J, Pei R (2015a) Quantification of Zn(II) using a label-free sensor based on graphene oxide and G-quadruplex. Anal Methods 7:9615–9618

Guo Y, Sun Y, Shen X, Zhang K, Hu J, Pei R (2015b) Label-free detection of Zn2+ based on G-quadruplex. Anal Sci 31:1041–1045

Guo Y, Wang Q, Wang Z, Chen X, Xu L, Hu J, Pei R (2015c) Label-free detection of T4 DNA ligase and polynucleotide kinase activity based on toehold-mediated strand displacement and split G-quadruplex probes. Sensors Actuators B-Chem 214:50–55

Guo Y, Chen Q, Qi Y, Xie Y, Qian H, Yao W, Pei R (2016a) Label-free ratiometric DNA detection using two kinds of interaction-responsive emission dyes. Biosens Bioelectron 87:320–324

Guo Y, Xu L, Hong S, Sun Q, Yao W, Pei R (2016b) Label-free DNA-based biosensors using structure-selective light-up dyes. Analyst 141:6481–6489

Guo Y, Yao W, Xie Y, Zhou X, Hu J, Pei R (2016c) Logic gates based on G-quadruplexes: principles and sensor applications. Microchim Acta 183:21–34

Han FXG, Wheelhouse RT, Hurley LH (1999) Interactions of TMPyP4 and TMPyP2 with quadruplex DNA. Structural basis for the differential effects on telomerase inhibition. J Am Chem Soc 121:3561–3570

Hao Y, Guo Q, Wu H, Guo L, Zhong L, Wang J, Lin T, Fu F, Chen G (2014) Amplified colorimetric detection of mercuric ions through autonomous assembly of G-quadruplex DNAzyme nanowires. Biosens Bioelectron 52:261–264

He K, Li W, Nie Z, Huang Y, Liu Z, Nie L, Yao S (2012) Enzyme-regulated activation of DNAzyme: a novel strategy for a label-free colorimetric DNA ligase assay and ligase-based biosensing. Chemistry 18:3992–3999

Hong Y, Haussler M, Lam JW, Li Z, Sin KK, Dong Y, Tong H, Liu J, Qin A, Renneberg R, Tang BZ (2008) Label-free fluorescent probing of G-quadruplex formation and real-time monitoring of DNA folding by a quaternized tetraphenylethene salt with aggregation-induced emission characteristics. Chemistry 14:6428–6437

Hu Q, Zeng F, Wu S (2016) A ratiometric fluorescent probe for hyaluronidase detection via hyaluronan-induced formation of red-light emitting excimers. Biosens Bioelectron 79:776–783

Jain AK, Bhattacharya S (2011) Interaction of G-quadruplexes with nonintercalating duplex-DNA minor groove binding ligands. Bioconjug Chem 22:2355–2368

Jiang X, Zhang H, Wu J, Yang X, Shao J, Lu Y, Qiu B, Lin Z, Chen G (2014) G-quadruplex DNA biosensor for sensitive visible detection of genetically modified food. Talanta 128:445–449

Jin X, Yue S, Wells KS, Singer VL (1994) SYBR green-TM-1: a new fluorescent dye optimized for detection of picogram amounts of DNA in gels. Biophys J 66:A159

Jonsson U, Fagerstam L, Ivarsson B, Johnsson B, Karlsson R, Lundh K, Lofas S, Persson B, Roos H, Ronnberg I et al (1991) Real-time biospecific interaction analysis using surface plasmon resonance and a sensor chip technology. BioTechniques 11:620–627

Kolpashchikov DM (2008) Split DNA enzyme for visual single nucleotide polymorphism typing. J Am Chem Soc 130:2934–2935

Kong DM, Ma YE, Wu J, Shen HX (2009) Discrimination of G-quadruplexes from duplex and single-stranded DNAs with fluorescence and energy-transfer fluorescence spectra of crystal violet. Chemistry 15:901–909

Largy E, Granzhan A, Hamon F, Verga D, Teulade-Fichou MP (2013) Visualizing the quadruplex: from fluorescent ligands to light-up probes. Top Curr Chem 330:111–177

Lee IJ, Goo NI, Kim DE (2016) Label/quencher-free detection of single-nucleotide changes in DNA using isothermal amplification and G-quadruplexes. Analyst 141:6503–6506

Li D, Shlyahovsky B, Elbaz J, Willner I (2007) Amplified analysis of low-molecular-weight substrates or proteins by the self-assembly of DNAzyme-aptamer conjugates. J Am Chem Soc 129:5804–5805

Li T, Wang E, Dong S (2008) G-quadruplex-based DNAzyme for facile colorimetric detection of thrombin. Chem Commun 31:3654–3656

Li T, Dong S, Wang E (2009) Label-free colorimetric detection of aqueous mercury ion (Hg^{2+}) using Hg^{2+}-modulated G-quadruplex-based DNAzymes. Anal Chem 81:2144–2149

Li T, Wang E, Dong S (2010) Parallel G-quadruplex-specific fluorescent probe for monitoring DNA structural changes and label-free detection of potassium ion. Anal Chem 82:7576–7580
Li HB, Wu ZS, Qiu LP, Liu JW, Wang C, Shen GL, Yu RQ (2013) Ultrasensitive label-free amplified colorimetric detection of p53 based on G-quadruplex MBzymes. Biosens Bioelectron 50:180–185
Li W, Jiang W, Wang L (2016) Self-locked aptamer probe mediated cascade amplification strategy for highly sensitive and selective detection of protein and small molecule. Anal Chim Acta 940:1–7
Lin S, Lu L, Kang TS, Mergny JL, Leung CH, Ma DL (2016a) Interaction of an iridium(III) complex with G-Quadruplex DNA and its application in luminescent switch-on detection of Siglec-5. Anal Chem 88:10290–10295
Lin X, Leung KH, Lin L, Lin L, Lin S, Leung CH, Ma DL, Lin JM (2016b) Determination of cell metabolite VEGF(1)(6)(5) and dynamic analysis of protein-DNA interactions by combination of microfluidic technique and luminescent switch-on probe. Biosens Bioelectron 79:41–47
Liu J, Cao Z, Lu Y (2009) Functional nucleic acid sensors. Chem Rev 109:1948–1998
Liu Y, Ye M, Ge Q, Qu X, Guo Q, Hu X, Sun Q (2016) Ratiometric quantum dot-ligand system made by phase transfer for visual detection of double-stranded DNA and single-nucleotide polymorphism. Anal Chem 88:1768–1774
Luo J, Xie Z, Lam JW, Cheng L, Chen H, Qiu C, Kwok HS, Zhan X, Liu Y, Zhu D, Tang B (2001) Aggregation-induced emission of 1-methyl-1,2,3,4,5-pentaphenylsilole. Chem Commun 37:1740–1741
Lyu Y, Chen G, Shangguan D, Zhang L, Wan S, Wu Y, Zhang H, Duan L, Liu C, You M, Wang J, Tan W (2016) Generating cell targeting Aptamers for Nanotheranostics using cell-SELEX. Theranostics 6:1440–1452
Ma D, Che C, Yan S (2009) Platinum(II) complexes with dipyridophenazine ligands as human telomerase inhibitors and luminescent probes for G-quadruplex DNA. J Am Chem Soc 131:1835–1846
Ma DL, He HZ, Leung KH, Zhong HJ, Chan DS, Leung CH (2013) Label-free luminescent oligonucleotide-based probes. Chem Soc Rev 42:3427–3440
Ma F, Yang Y, Zhang CY (2014) Ultrasensitive detection of transcription factors using transcription-mediated isothermally exponential amplification-induced Chemiluminescence. Anal Chem 86:6006–6011
Maiti S, Chaudhury NK, Chowdhury S (2003) Hoechst 33258 binds to G-quadruplex in the promoter region of human c-myc. Biochem Biophys Res Commun 310:505–512
Manet I, Manoli F, Zambelli B, Andreano G, Masi A, Cellai L, Monti S (2011a) Affinity of the anthracycline antitumor drugs doxorubicin and Sabarubicin for human telomeric G-quadruplex structures. Phys Chem Chem Phys 13:540–551
Manet I, Manoli F, Zambelli B, Andreano G, Masi A, Cellai L, Ottani S, Marconi G, Monti S (2011b) Complexes of the antitumoral drugs doxorubicin and Sabarubicin with telomeric G-quadruplex in basket conformation: ground and excited state properties. Photochem Photobiol Sci 10:1326–1337
Mergny JL, Lacroix L, Teulade-Fichou MP, Hounsou C, Guittat L, Hoarau M, Arimondo PB, Vigneron JP, Lehn JM, Riou JF, Garestier T, Helene C (2001) Telomerase inhibitors based on quadruplex ligands selected by a fluorescence assay. Proc Natl Acad Sci U S A 98:3062–3067
Min X, Zhuang Y, Zhang Z, Jia Y, Hakeem A, Zheng F, Cheng Y, Tang BZ, Lou X, Xia F (2015) Lab in a tube: sensitive detection of MicroRNAs in urine samples from bladder cancer patients using a single-label DNA probe with AIEgens. ACS Appl Mater Interfaces 7:16813–16818
Monchaud D, Allain C, Teulade-Fichou MP (2006) Development of a fluorescent intercalator displacement assay (G4-FID) for establishing quadruplex-DNA affinity and selectivity of putative ligands. Bioorg Med Chem Lett 16:4842–4845
Notomi T, Okayama H, Masubuchi H, Yonekawa T, Watanabe K, Amino N, Hase T (2000) Loop-mediated isothermal amplification of DNA. Nucleic Acids Res 28:E63
Paramasivan S, Bolton PH (2008) Mix and measure fluorescence screening for selective quadruplex binders. Nucleic Acids Res 36:e106

Patel DJ, Phan AT, Kuryavyi V (2007) Human telomere, oncogenic promoter and 5′-UTR G-quadruplexes: diverse higher order DNA and RNA targets for cancer therapeutics Nucleic Acids Res 35:7429–7455

Pei RJ, Matamoros E, Liu MH, Stefanovic D, Stojanovic MN (2010) Training a molecular automaton to play a game. Nat Nanotechnol 5:773–777

Pelossof G, Tel-Vered R, Willner I (2012) Amplified surface Plasmon resonance and electrochemical detection of Pb^{2+} ions using the Pb^{2+}-dependent DNAzyme and Hemin/G-Quadruplex as a label. Anal Chem 84:3703–3709

Pronina IV, Loginov VI, Burdennyy AM, Fridman MV, Senchenko VN, Kazubskaya TP, Kushlinskii NE, Dmitriev AA, Braga EA (2017) DNA methylation contributes to deregulation of 12 cancer-associated microRNAs and breast cancer progression. Gene 604:1–8

Quach QH, Jung J, Kim H, Chung BH (2013) A simple, fast and highly sensitive assay for the detection of telomerase activity. Chem Commun 49:6596–6598

Saiki RK, Gelfand DH, Stoffel S, Scharf SJ, Higuchi R, Horn GT, Mullis KB, Erlich HA (1988) Primer-directed enzymatic amplification of DNA with a thermostable DNA polymerase. Science 239:487–491

Schneeberger C, Speiser P, Kury F, Zeillinger R (1995) Quantitative detection of reverse transcriptase-Pcr products by means of a novel and sensitive DNA stain. PCR Methods Appl 4:234–238

Shi S, Geng X, Zhao J, Yao T, Wang C, Yang D, Zheng L, Ji L (2010) Interaction of [Ru(bpy)(2)(dppz)](2+) with human telomeric DNA: preferential binding to G-quadruplexes over i-motif. Biochimie 92:370–377

Shi H, Li D, Xu F, He X, Wang K, Ye X, Tang J, He C (2014) A label-free activatable aptamer probe for colorimetric detection of cancer cells based on binding-triggered in situ catalysis of split DNAzyme. Analyst 139:4181–4184

Shimizu S, Takehara T, Hikita H, Kodama T, Miyagi T, Hosui A, Tatsumi T, Ishida H, Noda T, Nagano H, Doki Y, Mori M, Hayashi N (2010) The let-7 family of microRNAs inhibits Bcl-xL expression and potentiates sorafenib-induced apoptosis in human hepatocellular carcinoma. J Hepatol 52:698–704

Stagg J, Smyth MJ (2010) Extracellular adenosine triphosphate and adenosine in cancer. Oncogene 29:5346–5358

Su H, Meng X, Guo Q, Tan Y, Cai Q, Qin H, Meng X (2014) Label-free DNAsensor with PCR-like sensitivity based on background reduction and target-triggered polymerization amplification. Biosens Bioelectron 52:417–421

Sun D, Thompson B, Cathers BE, Salazar M, Kerwin SM, Trent JO, Jenkins TC, Neidle S, Hurley LH (1997) Inhibition of human telomerase by a G-quadruplex-interactive compound. J Med Chem 40:2113–2116

Sun H, Tang Y, Xiang J, Xu G, Zhang Y, Zhang H, Xu L (2006) Spectroscopic studies of the interaction between quercetin and G-quadruplex DNA. Bioorg Med Chem Lett 16:3586–3589

Sun J, Jiang W, Zhu J, Li W, Wang L (2015a) Label-free fluorescence dual-amplified detection of adenosine based on exonuclease III-assisted DNA cycling and hybridization chain reaction. Biosens Bioelectron 70:15–20

Sun N, Wang J, Ji LY, Hong SN, Dong JJ, Guo YH, Zhang KC, Pei RJ (2015b) A cellular compatible chitosan nanoparticle surface for isolation and in situ culture of rare number CTCs. Small 11:5444–5451

Takamizawa J, Konishi H, Yanagisawa K, Tomida S, Osada H, Endoh H, Harano T, Yatabe Y, Nagino M, Nimura Y, Mitsudomi T, Takahashi T (2004) Reduced expression of the let-7 microRNAs in human lung cancers in association with shortened postoperative survival. Cancer Res 64:3753–3756

Tan J, Ou T, Hou J, Lu Y, Huang S, Luo H, Wu J, Huang Z, Wong K, Gu L (2009) Isaindigotone derivatives: a new class of highly selective ligands for telomeric G-quadruplex DNA. J Med Chem 52:2825–2835

Tan W, Donovan MJ, Jiang J (2013) Aptamers from cell-based selection for bioanalytical applications. Chem Rev 113:2842–2862

Tang R, Yang C, Ma X, Wang Y, Luo D, Huang C, Xu Z, Liu P, Yang L (2016) MiR-let-7a inhibits cell proliferation, migration, and invasion by down-regulating PKM2 in gastric cancer. Oncotarget 7:5972–5984

Tavazoie SF, Alarcon C, Oskarsson T, Padua D, Wang QQ, Bos PD, Gerald WL, Massague J (2008) Endogenous human microRNAs that suppress breast cancer metastasis. Nature 451:147–U143

Teulade-Fichou MP, Carrasco C, Guittat L, Bailly C, Alberti P, Mergny JL, David A, Lehn JM, Wilson WD (2003) Selective recognition of G-quadruplex telomeric DNA by a bis(quinacridine) macrocycle. J Am Chem Soc 125:4732–4740

Tomita N, Mori Y, Kanda H, Notomi T (2008) Loop-mediated isothermal amplification (LAMP) of gene sequences and simple visual detection of products. Nat Protoc 3:877–882

Travascio P, Li Y, Sen D (1998) DNA-enhanced peroxidase activity of a DNA-aptamer-hemin complex. Chem Biol 5:505–517

Tyagi S, Kramer FR (1996) Molecular beacons: probes that fluoresce upon hybridization. Nat Biotechnol 14:303–308

Vogiatzi F, Brandt DT, Schneikert J, Fuchs J, Grikscheit K, Wanzel M, Pavlakis E, Charles JP, Timofeev O, Nist A, Mernberger M, Kantelhardt EJ, Siebolts U, Bartel F, Jacob R, Rath A, Moll R, Grosse R, Stiewe T (2016) Mutant p53 promotes tumor progression and metastasis by the endoplasmic reticulum UDPase ENTPD5. Proc Natl Acad Sci U S A 113:E8433–E8442

Walker GT, Little MC, Nadeau JG, Shank DD (1992) Isothermal in vitro amplification of DNA by a restriction enzyme/DNA polymerase system. Proc Natl Acad Sci U S A 89:392–396

Wang XP, Yin BC, Wang P, Ye BC (2013) Highly sensitive detection of microRNAs based on isothermal exponential amplification-assisted generation of catalytic G-quadruplex DNAzyme. Biosens Bioelectron 42:131–135

Wang H, Cheng H, Wang J, Xu L, Chen H, Pei R (2016a) Selection and characterization of DNA aptamers for the development of light-up biosensor to detect cd(II). Talanta 154:498–503

Wang K, Zhang R, Sun N, Li X, Wang J, Cao Y, Pei R (2016b) Near-infrared light-driven Photoelectrochemical Aptasensor based on the Upconversion nanoparticles and TiO2/CdTe Heterostructure for detection of cancer cells. ACS Appl Mater Interfaces 8:25834–25839

Wang Q, Yin BC, Ye BC (2016c) A novel polydopamine-based chemiluminescence resonance energy transfer method for microRNA detection coupling duplex-specific nuclease-aided target recycling strategy. Biosens Bioelectron 80:366–372

Wang R, Wang L, Zhao H, Jiang W (2016d) A split recognition mode combined with cascade signal amplification strategy for highly specific, sensitive detection of microRNA. Biosens Bioelectron 86:834–839

Wang Z, Zhao J, Dai Z (2016e) A label-free fluorescent adenosine triphosphate biosensor via overhanging aptamer-triggered enzyme protection and target recycling amplification. Analyst 141:4006–4009

Wei C, Han G, Jia G, Zhou J, Li C (2008) Study on the interaction of porphyrin with G-quadruplex DNAs. Biophys Chem 137:19–23

Wei Y, Zhou W, Liu J, Chai Y, Xiang Y, Yuan R (2015) Label-free and homogeneous aptamer proximity binding assay for fluorescent detection of protein biomarkers in human serum. Talanta 141:230–234

Westermark B, Heldin CH, Nister M (1995) Platelet-derived growth factor in human glioma. Glia 15:257–263

Wright WE, Tesmer VM, Huffman KE, Levene SD, Shay JW (1997) Normal human chromosomes have long G-rich telomeric overhangs at one end. Genes Dev 11:2801–2809

Wu S, Wang L, Zhang N, Liu Y, Zheng W, Chang A, Wang F, Li S, Shangguan D (2016) A Bis(methylpiperazinylstyryl)phenanthroline as a fluorescent ligand for G-Quadruplexes. Chemistry 22:6037–6047

Xiao Y, Pavlov V, Gill R, Bourenko T, Willner I (2004) Lighting up biochemiluminescence by the surface self-assembly of DNA-hemin complexes. Chembiochem 5:374–379

Xu H, Gao SL, Yang Q, Pan D, Wang LH, Fan CH (2010) Amplified fluorescent recognition of G-Quadruplex folding with a cationic conjugated polymer and DNA Intercalator. ACS Appl Mater Interfaces 2:3211–3216

Xu J, Qian J, Li H, Wu ZS, Shen W, Jia L (2016) Intelligent DNA machine for the ultrasensitive colorimetric detection of nucleic acids. Biosens Bioelectron 75:41–47

Xu H, Wu D, Li CQ, Lu Z, Liao XY, Huang J, Wu ZS (2017) Label-free colorimetric detection of cancer related gene based on two-step amplification of molecular machine. Biosens Bioelectron 90:314–320

Xue Q, Zhang Y, Xu S, Li H, Wang L, Li R, Zhang Y, Yue Q, Gu X, Zhang S, Liu J, Wang H (2015) Magnetic nanoparticles-cooperated fluorescence sensor for sensitive and accurate detection of DNA methyltransferase activity coupled with exonuclease III-assisted target recycling. Analyst 140:7637–7644

Yan Y, Shen B, Wang H, Sun X, Cheng W, Zhao H, Ju H, Ding S (2015) A novel and versatile nanomachine for ultrasensitive and specific detection of microRNAs based on molecular beacon initiated strand displacement amplification coupled with catalytic hairpin assembly with DNAzyme formation. Analyst 140:5469–5474

Yang Q, Xiang J, Yang S, Zhou Q, Li Q, Tang Y, Xu G (2009) Verification of specific G-quadruplex structure by using a novel cyanine dye supramolecular assembly: I. Recognizing mixed G-quadruplex in human telomeres. Chem Commun 9:1103–1105

Yang KA, Pei R, Stojanovic MN (2016) In vitro selection and amplification protocols for isolation of aptameric sensors for small molecules. Methods 106:58–65

Yin B, Ye B, Tan W, Wang H, Xie C (2009) An allosteric dual-DNAzyme unimolecular probe for colorimetric detection of copper(II). J Am Chem Soc 131:14624–14625

Zhang H, Zhou L, Zhu Z, Yang C (2016) Recent progress in Aptamer-based functional probes for Bioanalysis and biomedicine. Chemistry 22:9886–9900

Zhang K, Wang K, Zhu X, Xu F, Xie M (2017) Sensitive detection of microRNA in complex biological samples by using two stages DSN-assisted target recycling signal amplification method. Biosens Bioelectron 87:358–364

Zhao G, Wang B, Liu Y, Zhang JG, Deng SC, Qin Q, Tian K, Li X, Zhu S, Niu Y, Gong Q, Wang CY (2013) miRNA-141, downregulated in pancreatic cancer, inhibits cell proliferation and invasion by directly targeting MAP 4K4. Mol Cancer Ther 12:2569–2580

Zhao Y, Chen F, Li Q, Wang L, Fan C (2015) Isothermal amplification of nucleic acids. Chem Rev 115:12491–12545

Zhao W, Li Y, Yang S, Chen Y, Zheng J, Liu C, Qing Z, Li J, Yang R (2016) Target-activated modulation of dual-color and two-photon fluorescence of Graphene quantum dots for in vivo imaging of hydrogen peroxide. Anal Chem 88:4833–4840

Zheng J, Yang R, Shi M, Wu C, Fang X, Li Y, Li J, Tan W (2015) Rationally designed molecular beacons for bioanalytical and biomedical applications. Chem Soc Rev 44:3036–3055

Zheng X, Niu L, Wei D, Li X, Zhang S (2016) Label-free detection of microRNA based on coupling multiple isothermal amplification techniques. Sci Rep 6:35982

Zhou WJ, Su J, Chai YQ, Yuan R, Xiang Y (2014) Naked eye detection of trace cancer biomarkers based on biobarcode and enzyme-assisted DNA recycling hybrid amplifications. Biosens Bioelectron 53:494–498

Zhou L, Shen X, Sun N, Wang K, Zhang Y, Pei R (2015) Label-free fluorescence light-up detection of T4 polynucleotide kinase activity using the split-to-intact G-quadruplex strategy by ligation-triggered and toehold-mediated strand displacement release. Analyst 140:5450–5453

Zhou L, Cheng H, Wang JE, Pei RJ (2016) G-Quadruplex DNAzyme biosensor for quantitative detection of T4 polynucleotide kinase activity by using split-to-intact G-Quadruplex DNAzyme conversion. Chinese J Anal Chem 44:13–17

Zhu X, Xiao Y, Jiang X, Li J, Qin H, Huang H, Zhang Y, He X, Wang K (2016) A ratiometric nanosensor based on conjugated polyelectrolyte-stabilized AgNPs for ultrasensitive fluorescent and colorimetric sensing of melamine. Talanta 151:68–74

Point-of-Care and Implantable Biosensors in Cancer Research and Diagnosis

5

Christina G. Siontorou, Georgia-Paraskevi D. Nikoleli, Dimitrios P. Nikolelis, Stephanos Karapetis, Nikolaos Tzamtzis, and Spyridoula Bratakou

5.1 Introduction

Biosensors are compact analytical devices that mimic natural chemoreception schemes: biological components react with the analyte of concern to produce biochemical information, readily translated into an electric signal by a chemical transducer (Fig. 5.1). In context, the analytical characteristics of any device depend upon the intra-component properties and inter-component correlations: specificity is assigned by the biological system used, response times are determined by the transducer, miniaturization comes mostly inherent by the nanosize of the biological moieties, and intrinsic signal amplification capabilities are determined by the bioelement-transducer interface (Palchetti and Mascini 2010; Shruthi et al. 2014). Since 1960s, when Leland C. Clark, Jr. in 1960s used an oxygen probe as a glucose meter (Clark and Lyons 1962), the realization of the biosensor concept has been almost explicitly linked with the biomedical sector, where prospects, expectations, and deliverables could be readily translated into a worthwhile market-based rate of return in the portfolio of products.

C.G. Siontorou
Department of Industrial Management and Technology, Laboratory of Simulation of Industrial Processes, School of Maritime and Industry, University of Piraeus, Piraeus, Greece

G.-P. D. Nikoleli • S. Karapetis • N. Tzamtzis • S. Bratakou
Department of Chemical Sciences, Laboratory of Inorganic and Analytical Chemistry, School of Chemical Engineering, National Technical University of Athens, Athens, Greece

D.P. Nikolelis (✉)
Department of Chemistry, Laboratory of Environmental Chemistry, University of Athens, 15771 Athens, Greece
e-mail: nikolelis@chem.uoa.gr

P. Chandra et al. (eds.), *Next Generation Point-of-care Biomedical Sensors Technologies for Cancer Diagnosis*, DOI 10.1007/978-981-10-4726-8_5

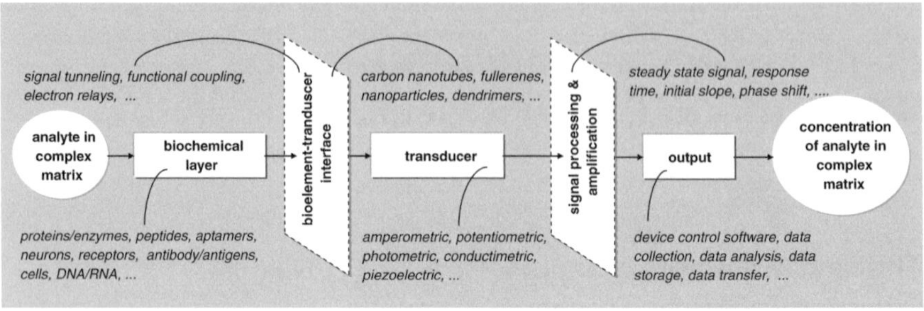

Fig. 5.1 Overview of biosensor architecture, general device assembly concepts, and basic mechanisms employed in the transduction of the biochemical information into a measurable signal

As diagnosis, monitoring, and therapy of diseases, especially cancer, shifts nowadays to a more molecular-based approach, biosensor technology may provide a suitable platform for real-time and personalized health monitoring. Fast responses, miniaturized sensor size, biocompatibility, rapid label-free detection, easy device tailoring, ultra-low detection limits, high reliability of measurements, and low development costs are appealing to patients, physicians, and the medical industry alike. The versatility of biosensor platforms offers a significant advantage in personalized and/or targeted monitoring: in concept, mush verified in practice, as well, any analyte can be correlated with a variety of suitable bioelements, which, in turn, can be paired to any transducer (and vice versa) and packaged according to any needs to yield a variety of devices with a larger variety of device characteristics. There exists the feasibility of engineering wearable or implantable point-of-care biosensors for monitoring clinical parameters, such as protein changes, biomarker concentrations, and drug targeting (Pantelopoulos and Bourbakis 2010; Jin et al. 2017). The development of unobtrusive, recurrent, and long-term nanomonitors can serve adequately early diagnosis of alarming health trends, while operating under strict medical specifications, several ergonomic constraints, and significant hardware resource limitations (Vasan et al. 2013).

The state-of-the-art in emerging concepts is presented herein, strategies and techniques in developing biosensor systems for cancer research and diagnosis. Critical issues, technology bottlenecks, and challenges are, also, discussed.

5.2 Construction of Biosensor Platforms

Most methods used for biosensor fabrication derive from the vast experience acquired in semiconductors and microelectromechanical systems. Briefly, bottom-up and top-down approaches are used (Prakash et al. 2017). The former involves the management of basic building blocks or materials. For example, self-assembly techniques use thermodynamic energy minimization processes to induce phase segregation and yield polymer structures (Ma et al. 2016; Prakash et al. 2009); more advanced tools such as optical tweezers (Song et al. 2010; Suei et al. 2015) or atomic force microscopy (Ozkan et al. 2016) enable greater accuracy for pick-and-place approaches. Top-down approaches rely on the machining of advanced materials through lithography and etching (Prakash et al. 2017). The processes used for the immobilization of the biological system on the transducer surface depend strongly on surface-species interactions; thus, the ability to control and manipulate surface properties (charge, stress, etc.) is a critical parameter in biosensor design.

The target analyte determines the biological system to be used. Apart from affinity, other criteria that may apply in bioelement selection include, inter alia (Siontorou and Batzias 2013): (a) kinetic parameters for the analyte-bioelement interaction; fast kinetics could provide fast response times in the event that the speed at which the biochemical information is transduced is equally fast (otherwise, the signal might be missed); (b) non-toxicity of interaction products, in order to avoid detector biofouling or patient intoxication; (c) reversible interaction in order to ensure the regeneration of the biochemical layer; (d) tight ligation of the bioelement onto the transducer surface to avoid leaching; (e) sufficient bioelement ruggedness to avoid denaturation. Evidently, matching the target analyte to a bioelement is not of critical concern; matching the bioelement to the conditions under which the biosensor will operate and to the aims and scopes of detection (i.e., the sensitivity and selectivity requisites) may prove problematic. These parameters should be taken into consideration when designing the diagnostic system, since any optimizations that will be applied at device testing might prove unsuccessful (Siontorou et al. 2010).

Nanotools now available can offer several alternatives for engineering biological moieties to suit any need, analytical or regulatory. Their coupling to a transducer may come in many forms, mostly as electrochemical, optical, or mass-based, depending on the type of biological response (Fig. 5.2). Frequently, hybrid transduction (e.g., electrochemical and optical) schemes may be used for signal optimization purposes. Many strategies have been proposed for enhancing the performance of the detectors, both material-based and instrumental. Some examples are given here below.

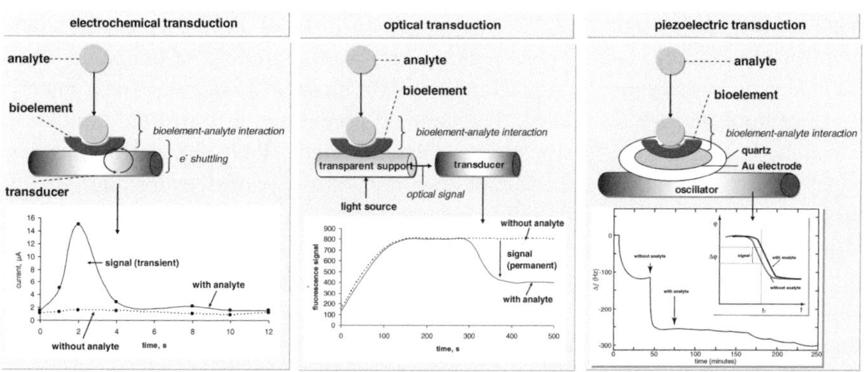

Fig. 5.2 Most commonly used transduction systems in biosensor for cancer diagnosis

Electrochemical sensors are compact devices of low resource settings; amenable to size reductions, they exhibit excellent linearity and repeatability and generally have a long life span, typically 1–3 years. Electrochemical sensors monitor changes in ion current, potential, conductance, or impedance (Bollella et al. 2017). Many devices have been suggested for targeting low molecular weight species, proteins, and cells. Recent advancements allow even for specific epitope targeting, such as the carbohydrate sites on cell surfaces; Cheng et al. (2009) developed a detection system for glycans on carcinoma cell surface. Signal enhancement may be achieved in a variety of ways. Strategies for accelerating the electron transfer with carbon nano-forms (nanotubes, nanofibers, nanosheets) have been proposed (Siontorou et al. 2016). Nanomaterials can be, also, used as tag molecules in hybrid transducers (Ding et al. 2008; Lai et al. 2011). Especially in electrochemiluminescent, these molecular tags behave as quantum dots and signals can be further amplified using ordered assembly or click chemistry, whereas the tags can be directly synthesized as dendrimers or polymers. If electrochemical transduction is preferred, gold nanoparticles can be added to produce conductive domains. Velev and Kaler (1999) have introduced this strategy when working on a conductivity immunoassay of proteins using antibody-functionalized latex spheres positioned between two interdigitated microelectrodes; the device could be miniaturized further to structure on-chip protein arrays with a picomolar detectablity. Silver-enhanced labelling may, also, prove quite useful. For example, Liu et al. (2010) used the silver enhancement technique in a conductimetric biochip, with a dual response: at a sub-threshold region, using electron hopping between silver islands and the electrolyte for conduction, and at an above-threshold region that employed direct flow of electrons. As the two regions use different conduction mechanisms and produce different slopes, the dynamic range of >40 dB produced gave a detection limit of 240 pg/mL. Single-walled carbon nanotubes can be easily fit into electrochemical systems to provide increased sensitivity to enzymatic reactions.

Optical transducers utilize light absorption, fluorescence, luminescence, total internal reflection, or surface plasmon resonance (SPR) for simple (Jeronimo et al. 2007) or multiplexed detection (Fan et al. 2008). Optical fibers and waveguide devices are used to improve sensitivity of the sensors by enhancing the interaction between the guided light and the sensor surface. Pu et al. (2010) proposed a new amplification strategy using hybrid nanomaterials (oligomeric silsesquioxane-based fluorescent nanoparticles) as signal amplifiers for biological imaging. These materials have fluorescent arms that can be chemically modified to adjust their emission wavelength, charge, and diameter according to needs; their signal amplification capabilities allow for the use of small quantities of indicator dyes for high-quality biological imaging. Similarly, semiconductor nanoparticles exhibit easily tunable absorbance and fluorescence. Jokerst et al. (2009) developed a microfluidic device for the multiplexed detection of cancer antigen 125 (CA125), carcinoembryonic antigen (CEA), and Her-2/Neu (C-erbB-2); the biosensor was based on fluorescence transduction of a quantum dot antibody probe immobilized on a microporous agarose bead array supported within a microfluidic system. The integration of semiconductor nanoparticles surpassed the response of standard molecular fluorophores by

30-fold. On the other hand, magnetic nanoparticles may offer certain advantages, especially in DNA-based platforms. Bi et al. (2009) used bio-barcode-functionalized magnetic nanoparticles as DNA hybridization platform to avoid cross-reactivity and lower detection limits; a femtomolar detection limit was achieved without any pre-concentration process.

Mass-based transducers detect the mass changes induced by the biochemical interaction. They consist of a piezoelectric crystal which oscillates at a particular frequency under an electric field. The mass of the crystal and the electrical frequency applied influence the frequency of oscillation of the crystal. Applications in cellular biology research involve mostly cell-surface interactions and morphological changes (Saitakis and Gizeli 2012). The process of transforming healthy cells to cancerous usually brings about changes in the morphology of cells and the arrangement of the cytoskeleton; these changes are expressed in dynamic cell adhesion processes and viscoelasticity modifications that can be monitoring in real time with a piezoelectric system (Zhou et al. 2011). The resistance vs. frequency changes provided a cell viscoelastic index that could be used to distinguish normal (HMEC) from malignant (MCF-7) mammary epithelial cells; during cell adhesion, malignant cells became softer, expressing a lower index compared to that of the healthy cells. The mechanical properties of cells were studied by applying centrifugal force during the interaction of cells on the surface of a quartz crystal microbalance, embedded in the rotor of centrifuge together with its driver (Webster et al. 2014). Apart from improving sensitivity, the viscoelastic properties of the cellular surfaces could be also measured. Su et al. (2013) developed a piezoelectric system for the direct detection of cancer biomarkers based on a lead titanate zirconate ceramic resonator as transducer. The dual sensing device had two resonators connected in parallel, one as the sensing unit and the other as the control unit; thus, they managed to minimize environment interference and compensate for temperature fluctuations. The device exhibited high sensitivity (0.25 ng/mL for prostate-specific antigen and α-fetoprotein) and fast analysis time (<30 min) of 1 μL samples. This ceramic resonator-based platform can be readily coupled to different chemical interfaces, for simple or multiplex detection.

Calorimetric biosensors are less common in cancer diagnostics, but nanotechnology-based modifications have broadened their range of applications. These systems measure enthalpy changes to monitor exothermic reactions, providing, indirectly, information about the concentration of the substrate (Bohunicky and Mousa 2011). Medley et al. (2008) developed a calorimetric biosensor based on aptamer-linked gold nanoparticles that could differentiate between acute leukemia cells and Burkitt's lymphoma cells. Their work demonstrated the feasibility of developing calorimetric platforms with aptamer-based recognition elements with the ability to discriminate between normal and cancer cells.

Microfluidic laboratory on-chip sensors may improve substantially patient care. Lab-on-chip technology integrates multiple steps of different analytical procedures, large variety of applications, sub-microliter consumption of reagents and samples, and portability (Gambari et al. 2003). Electrochemical detection based on paper-based microfluidic devices is also promising. Such devices could be developed as portable, easy-to-use, and low-cost point-of-care testing systems (Lu et al. 2012). Photolithography arranges microfluidic channels on cellulose fiber-based paper, while screen-printing fabricates electrodes on paper (Pires et al. 2014). The surface of the screen-printed electrodes can be functionalized with enzymes or DNA strands that serve as capture probes for the target analytes.

The use of luminescent nanocrystals (quantum dots) as molecular labels opened new horizons in cellular labeling and visualization (Tothill 2009). The nanocrystals can be attached to molecules for tracking intracellular components or used for antibody labeling. Their narrow emission peaks and spectroscopic properties support multiplexed analysis. Moreover, they exhibit high emission quantum yields that improve signal/noise ratios and increase the reliability of measurements.

Biosensor technology has indeed reached a level where state-of-the-art processes can offer amply a huge variety of engineering solutions for the manufacture of advanced micro- and nanosensors. Some examples are presented in the following sections. Still, physics present certain insurmountable constraints. The critical dimensions of micro- and nanofluidic-based systems are comparable to the scales of physical processes engaging small molecules. The minimization of detectable concentration levels and detection times are only limited by mass transport phenomena and reaction kinetics (Prakash et al. 2017; Siontorou and Batzias 2013). Reliability of detection is further reduced by nonspecific adsorption, matrix effects, Debye length, and streaming potential (Siontorou et al. 2010). Nanosensors exhibit ultralow detection limits because the screening of ions is reduced in packed spaces that are largely inaccessible by proteins, such as the corners between the nanowire and the substrate (Shoorideha and Chuia 2014). This corner effect exists in most biosensing structures, regardless of their scale; but at the nanoscale the effect becomes more important.

5.3 Biosensor Systems for Cancer

Using biosensors to monitor the levels of individual proteins secreted and/or expressed by cancerous cells may provide useful information to the health practitioner regarding cellular states. More than 160 types of biomarkers may be proven effective in diagnosing, staging, and treating early-stage cancer. For example, monitoring the levels of carcinoembryonic antigen (CEA) before and after treatment can be used to identify early recurrences or previously metastases (Kobayashi et al. 2012). Biosensor-based point-of-care monitoring could aid cancer management and facilitate earlier diagnosis. The systems developed are numerous, mostly on simple detection, although there are few platforms for multiplex analysis (Table 5.1). The detection limits achieved range between femto- and nano-scales, depending on the biosensor components used, such as carbon nanotubes, gold nanoparticles, quantum dots, and magnetic particles.

Antigen- and antibody-based biological systems are generally preferred due to the inherent specificity of antibody-antigen interactions. Kojima et al. (2003) developed an arrayed immunosensor with antibodies against α-fetoprotein immobilized

Table 5.1 Detection of tumor biomarkers with various biosensor platforms

Biomarker	Detection method	Biosensor principle	Detection limit	References
α-Fetoprotein (AFP)	Electrochemical	Arrayed immunosensor with antibodies immobilized in a plasma-polymerized film		Kojima et al. (2003)
		Prussian blue with screen-printed amperometric sensor	5 ng/mL	Guan et al. (2004)
α-Fetoprotein (AFP) and carcinoembryonic antigen (CEA)	Electrochemical	Dual immunosensor	1 ng/mL	Wilson (2005)
		Streptavidin-functionalized silver-nanoparticle-enriched carbon nanotube tag	0.093 pg/mL (AFP), 0.061 pg/mL (CEA)	Lai et al. (2011)
Breast cancer susceptibility gene (BRCA1)	Electrochemical	cDNA immobilized chitosan-co-polyaniline functionalized matrix	0.05 fM	Tiwari and Gong (2009)
		Mesoporous carbon nanospheres–toluidine blue nanocomposite	3.97 ng/mL	Fan et al. (2013)
Cancer antigen 125 (CA-125)	Electrochemical	Direct electrochemistry of horseradish peroxidase on titania sol-gel immunosensor	1.29 units/mL	Dai et al. (2003)
Cancer antigen 15–3 (CA15–3)	Optical	Gold nanorod -based plasmonic sensor	0.0249 units/mL	Chen et al. (2015)
Carcinoembryonic antigen (CEA)	Electrochemical	Direct electrochemistry of horseradish peroxidase on modified silica gel immunosensor	0.4 ng/mL	Tan et al. (2006)
		Thionine-doped magnetic gold nanospheres as labels and horseradish peroxidase as enhancer	0.01 ng/mL	Tang et al. (2008)
	Electrochemiluminescence	$Ru(bpy)_3^{2+}$-graphene-Nafion composite	0.002 pg/mL	Hao et al. (2012)
Ferritin	Piezoelectric	Gold chip immunosensor	0.1 ng/mL	Chou et al. (2002)

(continued)

Table 5.1 (continued)

Biomarker	Detection method	Biosensor principle	Detection limit	References
Human chorionic gonadotrophin (hCG)	Optical	Fluorescence immunosensor	25 units/mL	Nakamura et al. (2001)
Human epidermal growth factor receptor 2 (HER2)	Electrochemical	Label-free capacitive aptasensor coupled to non-Faradaic Impedance Spectroscopy	0.2 ng/mL	Qureshi et al. (2015)
Human prolactine biomarker (hPRL-3)	Electrochemical	Phage-modified light-addressable potentiometric sensor	0.04 nM	Jia et al. (2007)
Interleukin 6 (IL-6)	Electrochemical	Direct electrochemistry of horseradish peroxidase on carbon nanotubes gold-modified surfaces	0.5 pg/mL	Malhotra et al. (2010)
Mucin 1 (MUC1)	Electrochemical	Magnetic beads coupling screen-printed array	0.07 nM	Florea et al. (2015)
Prostate-specific antigen (PSA)	Piezoelectric	Microcantilever immunosensor	0.2 μg/mL	Wu et al. (2001)
	Optical	SPR with colloidal gold nanoparticles	0.15 ng/mL	Besselink et al. (2004)
		Gold layered dielectric-metal nanoparticles immunosensor	0.1 ng/mL	Hirsch et al. (2003)
		Micromechanical silicon nitride cantilevers	0.2 ng/mL	Wu et al. (2001)
	Electrochemical	Direct electrochemistry of horseradish peroxidase on carbon nanotubes gold-modified surfaces	0.5 pg/mL	Mani et al. (2009)
		Amine-terminated DNA aptamers were coupled to sulfo-betaine gold electrodes	1 ng/mL	Jolly et al. (2015)
	Electrochemi-luminescence	Carbon nanotubes-chitosan/gold nanoparticles	0.6 pg/mL	Zhang et al. (2012)

Table 5.1 (continued)

Biomarker	Detection method	Biosensor principle	Detection limit	References
Vascular endothelial growth factor (VEGF165)	Electrochemical	A label-free electrochemical aptasensor based on ordered mesoporous carbon-gold nanocomposite-modified screen-printed electrode	1 pg/mL	Tabrizi et al. (2015)

in a plasma-polymerized film. Dual systems for α-fetoprotein and carcinoembryonic antigen have been, also, proposed with either conventional platforms (Wilson 2005) or functionalized nanoparticles (Lai et al. 2011). Prostate-specific antigen (PSA) can be reliably detected with an anti-PSA antibody. The most successful platforms developed involve microcantilever-based transducers (Wu et al. 2001) and surface plasmon resonance (SPR)-based sensors (Hirsch et al. 2003), in which PSA antigen binding to antibody changes the vibrational frequency in an extend analogous to antigen concentration. Jia et al. (2007) developed a light-addressable potentiometric sensor using a phage recognition element for human prolactine biomarker (hPRL-3) and human breast cancer cell line MDA-MB-231; the results showed that the biosensor developed was more applicable to cancer cells detection. The major constraints of immunosensor platforms include the reduced thermal and physical ruggedness of the biological moieties and the difficulty in regenerating the antibody-based systems (Mittal et al. 2017); both limit considerably the reliability of the sensors, especially towards the limits of detection.

Aptamers and nucleic acids have been also proposed for cancer biosensing, offering almost endless different sequences that can express high affinities for their targets. A combinatorial chemistry-based technology that uses exponential enrichment for the systematic evolution of ligands can be used to generate specific nucleic acid probes from a library of RNA and DNA oligonucleotides. Despite the low success rates and time-consuming attributes of this technology (Mittal et al. 2017), many relevant biosensors have been developed, focused on the discovery of new cancer biomarkers for early diagnosis, such as the breast-specific protein NY-BR-1, and the cancer testis antigens CAGE-1 and NY-ESO-1 (Bohunicky and Mousa 2011). The latter are either detected by the cytotoxic T-lymphocytes of cancer patients or induce a serological immune response in the autologous host; these markers could be used for the development of anti-cancer vaccines (Balafoutas et al. 2013).

Aptasensors usually employ sandwich type methods, where the aptamers are attached to the transducer surface and analytes are attracted from liquid samples to yield high efficiencies. A second antibody with a measurable label is then bound to the attracted analytes; this label is readily detected by electrochemistry or other methods. For example, Mucin1 has been detected in real serum samples using a screen-printed array biosensor with magnetic beads and alkaline phosphatase labeling (Florea et al. 2015); the detection limit achieved was 0.07 nM within a linear range between 0 and 0.28 nM. Label-free schemes have been also reported. An aptasensor on carbon–gold nanocomposite-modified screen-printed electrode has been recently proposed for the detection of vascular endothelial growth factor VEGF165 in the serum of patients with lung cancer (Tabrizi et al. 2015). The sensor measures the changes in the interfacial charge transfer resistance of the electrode induced by the interaction of the immobilized anti-VEGF165 aptamer with the sample VEGF165 marker. In another study, a label-free capacitive aptasensor was developed for the human epidermal growth factor receptor 2 (HER2) protein using anti-HER2 DNA aptamers functionalized on interdigitated microelectrodes (Qureshi et al. 2015). The aptamer-protein complex induced concentration-dependent changes in the values of impedance/capacitance. Jolly et al. (2015) used impedimetric methods and an aptamer platform for detecting PSA in real blood samples. The authors compared two different methods in order to elucidate how the sensitivity and selectivity are impacted by surface chemistry. A thiolated DNA aptamer interacted with mercaptoethanol-modified gold electrodes; alternatively, amine-terminated DNA aptamers were coupled to anti-fouling sulfo-betaine gold electrodes. Although both fabrication processes were long and cumbersome, the detectability achieved was 1 ng/mL with sulfobetaine-probes and 10 μg/mL with mercaptoethanol-modified electrodes.

Light emission/absorption-based determination of biomarkers is a wide research field, mostly focused on nanoparticles, which involve photostable synthesis and provide noise-free fluorescence signals (Mittal et al. 2017). Chen et al. (2015) developed a combined detection assay for cancer antigen 15–3 (CA15–3) and copper level in serum using a gold nanorod-based plasmonic sensor. Manikandan et al. (2014) compared several surface-enhanced Raman spectroscopy substrates produced by in situ nucleation of gold nanohexagons on graphene nanosheets, gold nanoparticles, and gold-conjugated graphene nanomaterials; these nanomaterials enhanced Raman scattering to such a degree that human breast normal, cancer, and cancer stem cells could be discriminated. Cytotoxic studies indicated that graphene nanomaterials hardly enter cell; results on gold nanoparticles were inconclusive.

Some implantable electrochemical biosensors have been reported, designed to measure and transmit a specific response towards an analyte at the molecular level. A two or three electrode systems are commonly used, coupled to the appropriate enzyme. Apart from the use of nanomaterials to modify electrodes, some studies have been published on the development of devices with nanometric geometry (Goncalves et al. 2011), where one-dimensional structures serve as working electrodes for measuring femto- or pico-ampere activities. Various electrodes such as single-walled carbon nanotubes (Baughman et al. 2002) or boron-doped silicon nanowires (Goncalves et al. 2011) have been used in the construction of nanodevices. Hoeben et al. (2008) used a reduced scale of redox enzymes to electrochemically study a small amount of molecules. The measurements, made on lithographically fabricated 70 nm gold nanoelectrodes, showed successfully for the first time a distinct catalytic activity from less than 50 enzymes molecules. Cordeiro et al. (2015) developed an implantable biosensor for the continuous and simultaneous monitoring for glucose, lactate, and pyruvate. The sensor has been implanted in rats for evaluation; the brain levels of the carbohydrates could be monitored at the millimolar range. Zhang et al. (2016) developed a silicon-based 16-site implantable

25-mm long microelectrode array chip fabricated by standard lithography. The sensor was implanted in nonhuman primates for monitoring in real time the electrochemical activity of dopamine.

Flexible microelectrode arrays are expected to revolutionize point-of-care devices. Polyimide thin films have been proposed for implantable probe development. These films are deposited onto a carrier substrate; using anodic release, the carrier substrate discharges the polyimide structures in saline solution (Cheung and Renaud 2006). However, biocompatible interfaces between the implanted sensor and the surrounding tissue have not been demonstrated yet. The use of anti-inflammatory and biodegradable coating might interfere with analyte detection compromising the reliability of the measurements (Siontorou et al. 2010).

Optical platforms have been also proposed for in vivo sensing. Parameters such as fluorescence intensity and lifetime enhance sensitivity and offer long-term stability. Much work has been published on fluorescence resonant energy transfer-based biosensors for glucose, where the intensity of the signals is proportional to glucose levels (Khan et al. 2008). A transdermal system for continuous glucose monitoring has been reported by Ballerstadt et al. (2006). To further reduce invasiveness, transdermal glucose monitoring uses functionalized fluorescent microparticles injected in the patient (Shibata et al. 2010). Although promising, this approach is not suited for point-of-care continuous applications as a video camera and external light excitation are needed for image analysis. Valdastri et al. (2011) presented a miniaturized fluorescence biosensor suitable for long-term implantation. The device uses phototransistors as detectors and achieves fluorescence excitation and detection by driving a laser diode light source (Fig. 5.3). The signals are amplified and transmitted across the skin to a mobile device. Yet, the functionality of the sensor has been demonstrated only in vitro. Tong et al. (2016) studied the optical functionality, in vitro and in vivo, of a thermally hydrocarbonized porous silicon optical rugate filter, along with its stability and biocompatibility. The material proved to be cytotoxic, regardless of its surface chemistry, possibly due to the mitigation of reactive oxygen species levels during the pre-incubation of the film.

Magnetic resonance platforms have been also proposed. Harris et al. (2008) developed an in vivo sensor for measuring proteinase activity related to cancer. In

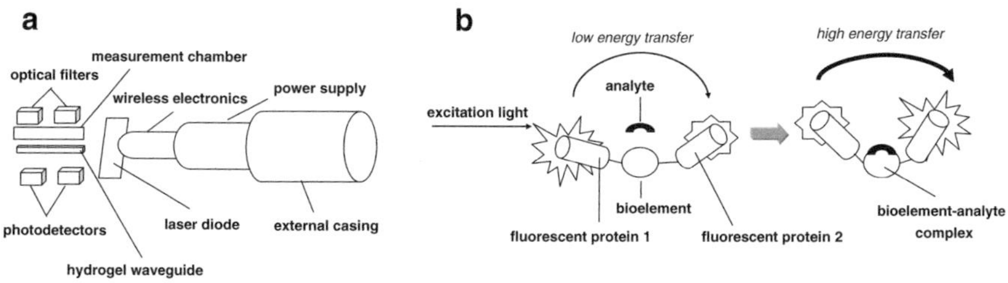

Fig. 5.3 (**a**) Architecture of the implantable fluorescent-based electrochemical biosensor. (**b**) The fluorescent resonant energy transfer concept: when two fluorescent proteins are covalently attached to the bioelement, a limited energy transfer from one protein to the other is recorded; the binding of the analyte to the bioelement induces conformation changes to the latter that result in bringing the two proteins closer and, thus, allowing for the transfer of a higher amount of energy (adopted from Valdastri et al. 2011)

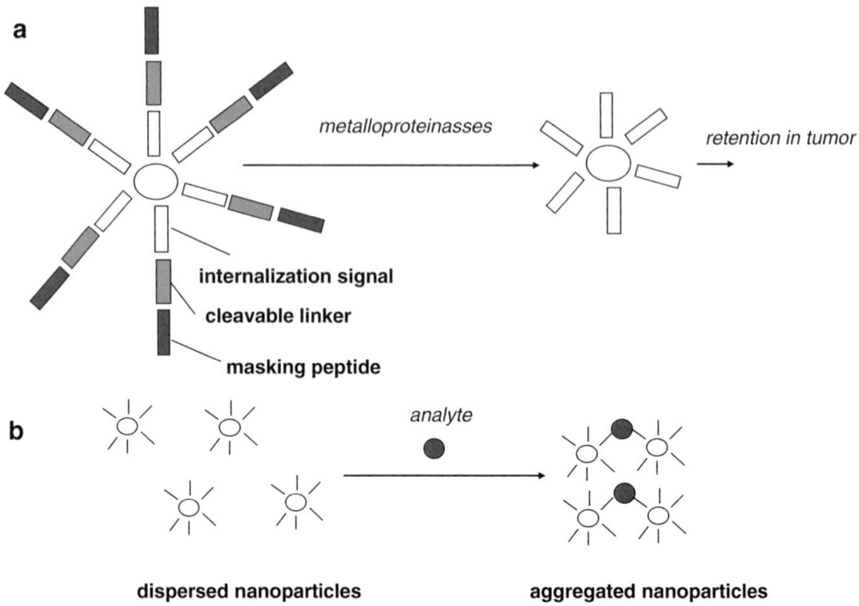

Fig. 5.4 Schematic representation of magnetic resonance platforms: (**a**) The magnetic nanoparticles are masked with protease-cleavable ligands to prevent internalization until the mask is removed by tumor-associated metalloproteins; the nanoparticles can then be efficiently internalized by the adjacent tumor cells. (**b**) Dispersed bifunctional particles exhibit a high relaxation time; when bound to the target analyte, they aggregate, quantitatively lowering the signal (adopted from Harris et al. 2008; Daniel et al. 2009)

this platform, protease-cleavable ligands coated a nanoparticle to mask a cell internalization signal embedded in the ligand (Fig. 5.4). When metalloproteinases are present in the matrix, the ligand is cleaved and the internalization signal is expressed. The presence of a tumor induces locally a high expression of proteases that drive the particles inside the tumor cells. Similar systems with other reporter proteases or with a combination of proteases and specific linkers could be developed for multiple cancer types. Daniel et al. (2009) developed an implantable biosensor that could sense the microenvironment. The sensor is built on a semi-permeable membrane containing nanoparticle magnetic relaxation switches. Ectopic tumors were produced in mice using a cell line that secreted a model cancer biomarker. After 1 day, tumor-bearing mice exhibited a transverse relaxation time that was 20 ± 10% lower than the healthy-control mice. The applicability of these devices in the verification of successful tumor resection may be realized quite soon.

5.4 Challenges in Implantable Sensor Development

In vivo monitoring has not been yet realized in a substantial extent. However, research in molecular therapy targets specific malfunctioning molecules and pathways in cancer. For example, kinase inhibitor imatinib proved promising for the management of chronic myeloid leukemia and gastrointestinal stromal tumors whose growth is related to the expression of specific kinase mutants (Sawyers 2004). The efficiency of the inhibitor needs to be evaluated at the level of protein interactions. Biosensors based on fluorescence resonance energy transfer (FRET) may provide the technology for monitoring kinase inhibition in live cells, even for in vivo applications. Numerous FRET-based biosensors have been recently published for the detection of oncogene-related kinase activities (Wang et al. 2005; Zhang and Allen 2007), and for other molecules that indicate cancer migration and invasion (Wang et al. 2008).

One major challenge in in vivo systems is powering. Inductive links for powering remotely devices has already reached the market. Size reductions in inductors for in vivo applications remain an open topic. The use of micro-fabricated inductors demonstrates the greatest potential (Olivo et al. 2014). Less power consuming and autonomous platforms have been reported. For example, nanoparticle magnetic relaxation switches have been developed for in vivo sensing (Daniel et al. 2009). The sensor is covered by a semi-permeable membrane that allows the selective diffusion of cancer biomarkers or drug molecules into the surface of the sensor.

Further, biocompatibility issues have not been adequately addressed. Their role in device engineering is inevitably dual: to prevent foreign body reaction and sensor fouling. Many polymeric materials, such as polyallylamines, horseradish peroxidase, or polyethylene glycol derivatives, have been suggested as coating materials but proven unsuccessful (Norton et al. 2007). Wang et al. (2013) have recently proposed the use of hydrogels from poly(lactic-co-glycolic) acid microsphere dispersed in poly(vinyl alcohol); preliminary in vivo testing results were very promising but more research is required with different biosensor systems in order to evaluate its efficiency.

Notwithstanding, a very interesting field has been recently introduced: nanobiolectronics. In brief, nanomaterials are integrated with biology and electronics in order to overcome existing challenges in biosensors. The downsizing of electronic transducers affords them a more nature-relevant and biocompatible character that is expected to bring sensitivity to near-nature levels (Zhang and Lieber 2016). Nanobioelectronic devices are used to study neural circuits at the cellular and subcellular level. Nanowire-nanotube heterostructures can penetrate cell membranes for minimally invasive recordings; when coupled with phospholipid functionalization, these nano-probes can facilitate spontaneous membrane penetration and a tight membrane seal of high resistance (Duan and Lieber 2015). Intracellular sensing becomes possible, opening new avenues in cancer diagnostics.

5.5 Conclusions and Future Prospects

Clinical biosensors have undoubtedly much to offer in cancer diagnosis. Recent progress in the development of multiplexed platforms is promising, while the sensitivity and selectivity of nanosensors might prove quite advantageous for novel approaches in early diagnosis and therapy monitoring. Lab-on-chip platforms show a steady potential towards rapid commercialization of point-of-care and implantable systems. Nanomaterials, particularly quantum dots, can facilitate the tracking of cancer cells or drug molecules. Integrating nanomaterials and biosensors might improve cancer imaging and drug delivery. Personalized health care systems might be a reality in the near future.

Biosensor technology presents the potential not only to serve the to-date cancer diagnostic strategy, but also to propose and support new, more efficient schemes. For example, cancer is usually expressed, at the molecular level, with a set of biomarkers; multiplexed platforms could be developed to provide reliable information for a wide dynamic range of many different biomarkers at ultra-low detectability. Further, the development of a diagnostic tool to inform on the borders of a tumor pre- or peri-operatively, could improve therapeutic success rates.

Notwithstanding, several issues need careful consideration when designing biosensor platforms. Despite progress in microfluidics, miniaturized transducers, and materials, the assembly of the biosensor components into a fully integrated device that could autonomously perform the analysis process has not been realized yet; possibly, the emerging nano-bioelectronics technology could support this goal. Also, as single-cell analysis is just started to post as requisite for early cancer diagnosis, nanoplatforms developed have not proved capabilities for detecting reliably just a limited number of biomolecules within a given cell. In addition, personalized medicine goes beyond disease diagnosis; more clinical information is required for a detailed molecular profiling, especially for the stage of tumorigenesis, the appropriate treatment regime, or in monitoring for disease recurrence. Thus, there exists the need for developing biosensors that could rapidly screen for DNA mutations and gene products.

Drug discovery and delivery may present another field where biosensing might prove beneficial and efficient. In vivo drug kinetics are affected by the properties of the active ingredient and how these properties are modified in vivo by transport, binding, or metabolism. This approach requires new strategies for reliably predicting drug delivery properties early in pharmaceutical development, so that the most efficient and suitable compounds move to clinical studies. This is especially true for the new therapeutic classes of gene-based drugs, although the proteomic information now available from gene expression data offers new prospects in both cancer management and biosensor development.

References

Balafoutas D, Hausen A, Mayer S, Hirschfeld M, Jaeger M, Denschlag D, Gitsch G, Jungbluth A, Stickeler E (2013) Cancer testis antigens and NY-BR-1 expression in primary breast cancer: prognostic and therapeutic implications. BMC Cancer 13:271

Ballerstadt R, Evans C, Gowda A, McNichols R (2006) In vivo performance evaluation of a transdermal near-infrared fluorescence resonance energy transfer affinity sensor for continuous glucose monitoring. Diabetes Technol Therapeut 8:296–311

Baughman RH, Zakhidov AA, De Heer WA (2002) Carbon nanotubes–the route toward applications. Science 297:787

Besselink GAJ, Kooyman RPH, van Os PJHJ, Engbers GHM, Schasfoort RBM (2004) Signal amplification on planar and gel-type sensor surfaces in surface plasmon resonance-based detection of prostate-specific antigen. Anal Biochem 333:165–173

Bi S, Zhou H, Zhang SS (2009) Bio-bar-code functionalized magnetic nanoparticle label for ultrasensitive flow injection chemiluminescence detection of DNA hybridization. Chem Commun 37:5567–5569

Bohunicky B, Mousa SA (2011) Biosensors: the new wave in cancer diagnosis. Nanotechnol Sci Appl 4:1–10

Bollella P, Fusco G, Tortolini C, Sanzò G, Favero G, Gorton L, Antiochia R (2017) Beyond graphene: electrochemical sensors and biosensors for biomarkers detection. Biosens Bioelectron 89:152–166

Chen S, Zhao Q, Zhang L, Wang L, Zeng Y, Huang H (2015) Combined detection of breast cancer biomarkers based on plasmonic sensor of gold nanorods. Sens Actuators B: Chem 221:1391–1397

Cheng W, Ding L, Ding S, Yin Y, Ju H (2009) A simple electrochemical cytosensor array for dynamic analysis of carcinoma cell surface glycans. Angew Chem Int Ed 48:6465–6468

Cheung KC, Renaud P (2006) BioMEMS for medicine: on-chip cell characterization and implantable microelectrodes. Solid State Electron 50:551–557

Chou SF, Hsu WL, Hwang JM, Chen CY (2002) Development of an immunosensor for human ferritin, a nonspecific tumor marker, based on a quartz crystal microbalance. Anal Chim Acta 453:181–189

Clark LC, Lyons C (1962) Electrode systems for continuous monitoring in cardiovascular surgery. Ann N Y Acad Sci 102:29–45

Cordeiro CA, de Vries MG, Ngabi W, Oomen PE, Cremers TIFH, Westerink BHC (2015) In vivo continuous and simultaneous monitoring of brain energy substrates with a multiplex amperometric enzyme-based biosensor device. Biosens Bioelectron 67:677–686

Dai Z, Yan F, Chen J, Ju H (2003) Reagentless amperometric immunosensors based on direct electrochemistry of horseradish peroxidase for rapid determination of carcinoma antigen-125. Anal Chem 75:5429–5434

Daniel KD, Kimb GY, Vassiliou CC, Galindo M, Guimaraes AR, Weisslederd R, Charest A, Langer R, Cima MJ (2009) Implantable diagnostic device for cancer monitoring. Biosens Bioelectron 24:3252–3257

Ding L, Cheng W, Wang X, Ding S, Ju H (2008) Carbohydrate monolayer strategy for electrochemical assay of cell surface carbohydrate. J Am Chem Soc 130:7224–7225

Duan X, Lieber CM (2015) Nanoscience and the nano-bioelectronics frontier. Nano Res 8:1–22

Fan X, White IM, Shopova SI, Zhu H, Suter JD, Sun Y (2008) Sensitive optical biosensors for unlabeled targets: a review. Anal Chim Acta 620:8–26

Fan H, Zhang Y, Wu D, Ma H, Li X, Li Y, Wang H, Li H, Du B, Wei Q (2013) Construction of label-free electrochemical immunosensor on mesoporous carbon nanospheres for breast cancer susceptibility gene. Anal Chim Acta 770:62–67

Florea A, Ravalli A, Cristea C, Sandulescu R, Marrazza G (2015) An optimized bioassay for Mucin1 detection in serum samples. Electroanalysis 2015(27):1594–1601

Gambari R, Borgatti M, Altomare L, Manaresi N, Medoro G, Romani A, Tartagni M, Guerrieri R (2003) Applications to cancer research of "lab-on-a-chip" devices based on dielectrophoresis. Technol Cancer Res Treat 2:31–40

Goncalves W, Lanfredi AJC, Crespilho FN (2011) Development of numerical methods for signal smoothing and noise modeling in single wire-based electrochemical biosensors. J Phys Chem C 115:16172–16179

Guan JG, Miao YQ, Chen JR (2004) Prussian blue modified amperometric FIA biosensor: one-step immunoassay for alpha-fetoprotein. Biosens Bioelectron 19:789–794

Hao T, Guo Z, Du S, Shi L (2012) Ultrasensitive detection of carcinoembryonic antigen based on electrochemiluminescence quenching of $Ru(bpy)_3^{2+}$ by quantum dots. Sens. Actuators B 171–172:803–809

Harris TJ, von Maltzahn G, Lord ME, Park JH, Agrawal A, Min DH, Sailor MJ, Bhatia SN (2008) Protease-triggered unveiling of bioactive nanoparticles. Small 4:1307–1312

Hirsch LR, Jackson JB, Lee A, Halas NJ, West JL (2003) A whole blood immunoassay using gold nanoshells. Anal Chem 75:2377–2381

Hoeben FJM, Meijer FS, Dekker C, Albracht SPJ, Heering HA, Lemay SJ (2008) Toward single-enzyme molecule electrochemistry:[NiFe]-hydrogenase protein film voltammetry at nanoelectrodes. ACS Nano 2:2497–2504

Jeronimo PC, Araujo AN, Conceicao BSMMM (2007) Optical sensors and biosensors based on sol-gel films. Talanta 72:13–27

Jia Y, Qin M, Zhang H, Niu W, Li X, Wang L, Li X, Bai Y, Cao Y, Feng X (2007) Label-free biosensor: a novel phage-modified light addressable potentiometric sensor system for cancer cell monitoring. Biosens Bioelectron 22:3261–3266

Jin H, Gui R, Yu J, Lv W, Wang Z (2017) Fabrication strategies, sensing modes and analytical applications of ratiometric electrochemical biosensors. Biosens Bioelectron 91:523–537

Jokerst JV, Raamanathan A, Christodoulides N, Floriano PN, Pollard AA, Simmons GW, Wong J, Gage C, Furmaga WB, Redding SW, McDevitt JT (2009) Nano-bio-chips for high performance multiplexed protein detection: determinations of cancer biomarkers in serum and saliva using quantum dot bioconjugate labels. Biosens Bioelectron 24:3622–3629

Jolly P, Formisano N, Tkac J, Kasak P, Frost CG, Estrela P (2015) Label-free impedimetric aptasensor with antifouling surface chemistry: a prostate specific antigen case study. Sensor Actuat B-Chem 209:306–312

Khan F, Gnudi L, Pickup JC (2008) Fluorescence-based sensing of glucose using engineered glucose/galactose-binding protein: a comparison of fluorescence resonance energy transfer and environmentally sensitive dye labelling strategies. Biochem Biophys Res Commun 365:102–106

Kobayashi E, Ueda Y, Matsuzaki S, Yokoyama T, Kimura T, Yoshino K, Fujita M, Kimura T, Enomoto T (2012) Biomarkers for screening, diagnosis, and monitoring of ovarian cancer. Cancer Epidemiol Biomark Prev 21:1902–1912

Kojima K, Hiratsuka A, Suzuki H, Yano K, Ikebukuro K, Karube I (2003) Electrochemical protein chip with arrayed immunosensors with antibodies immobilized in a plasma-polymerized film. Anal Chem 75:1116–1122

Lai GS, Wu J, Ju HX, Yan F (2011) Streptavidin-functionalized silver-nanoparticle-enriched carbon nanotube tag for ultrasensitive multiplexed detection of tumor markers. Adv Funct Mater 21:2938–2943

Liu Y, Zhang D, Evangelyn CA, Chakrabartty S (2010) Biomolecules detection using a silver-enhanced gold nanoparticle-based biochip. Nanoscale Res Lett 5:533–538

Lu J, Ge S, Ge L, Yan M, Yu J (2012) Electrochemical DNA sensor based on three-dimensional folding paper device for specific and sensitive point-of-care testing. Electrochim Acta 80:334–341

Ma W, Xu L, Wang L, Kuang H, Xu C (2016) Orientational nanoparticle assemblies and biosensors. Biosens Bioelectron 79:220–236

Malhotra R, Patel V, Vaque JP, Gutkind JS, Rusling JF (2010) Ultrasensitive electrochemical immunosensor for oral cancer biomarker IL-6 using carbon nanotube forest electrodes and multilabel amplification. Anal Chem 82:3118–3123

Mani V, Chikkaveeraiah BV, Patel V, Gutkind JS, Rusling JF (2009) Ultrasensitive immunosensor for cancer biomarker proteins using gold nanoparticle film electrodes and multienzyme-particle amplification. ACS Nano 3:585–594

Manikandan M, Abdelhamid HN, Talib A, Wu H-F (2014) Facile synthesis of gold nanohexagons on graphene templates in Raman spectroscopy for biosensing cancer and cancer stem cells. Biosens Bioelectron 55:180–186

Medley CD, Smith JE, Tang Z, Wu Y, Bamrungsap S, Tan W (2008) Gold nanoparticle-based colorimetric assay for the direct detection of cancerous cells. Anal Chem 80:1067–1072

Mittal S, Kaur H, Gautam N, Mantha AK (2017) Biosensors for breast cancer diagnosis: a review of bioreceptors. Biosens Bioelectron 88:217–231

Nakamura N, Lim TK, Jeong JM, Matsunaga T (2001) Flow immunoassay for detection of human chorionic gonadotrophin using a cation exchange resin packed capillary column. Anal Chim Acta 439:125–130

Norton LW, Koschwanez HE, Wisniswski NA, Klizman B, Reichert WM (2007) Vascular endothelial growth factor and dexamethasone release from nonfouling sensor coatings affect the foreign body response. J Biomed Mater Res B Appl Biomater 81A:858–869

Olivo J, Carrara S, Micheli G (2014) Micro-fabrication of high-thickness spiral inductors for the remote powering of implantable biosensors. Microelectr Eng 113:130–135

Ozkan AD, Topal AE, Dana A, Guler MO, Tekinay AB (2016) Atomic force microscopy for the investigation of molecular and cellular behaviour. Micron 89:60–76
Palchetti I, Mascini M (2010) Biosensor technology: a brief history. In: Malcovati P, Baschirotto A, d'Amico A, Natale C (eds) Sensors and microsystems. Lecture notes in electrical engineering, vol 54. Springer, Dordrecht
Pantelopoulos A, Bourbakis N (2010) A survey on wearable sensor-based systems for health monitoring and prognosis. IEEE Trans Syst Man Cybern Part C Appl Rev 40:1–12
Pires NMM, Dong T, Hanke U, Hoivik N (2014) Recent developments in optical detection technologies in lab-on-a-Chip devices for biosensing applications. Sensors 14:15458–15479
Prakash S, Karacor MB, Banerjee S (2009) Surface modification in microsystems and nanosystems. Surf Sci Rep 64:233–254
Prakash S, Pinti M, Bhushan B (2017) Theory, fabrication and applications of microfluidic and nanofluidic biosensors. Phil Trans R Soc A 370:2269–2303
Pu KY, Li K, Liu B (2010) Cationic oligofluorene-substituted polyhedral oligomeric silsesquioxane as light-harvesting unimolecular nanoparticle for fluorescence amplification in cellular imaging. Adv Mater 22:643–646
Qureshi A, Gurbuz Y, Niazi JH (2015) Label-free capacitance based aptasensor platform for the detection of HER2/ErbB2 cancer biomarker in serum. Sensor Actuat B-Chem 220:1145–1151
Saitakis M, Gizeli E (2012) Acoustic sensors as a biophysical tool for probing cell attachment and cell/surface interactions. Cell Mol Life Sci 69:357–371
Sawyers C (2004) Targeted cancer therapy. Nature 432:294–297
Shibata H, Heo YJ, Okitsu T, Matsunaga Y, Kawanishi T, Takeuchi S (2010) Injectable hydrogel microbeads for fluorescence-based in vivo continuous glucose monitoring. Proc Nat Acad Sci USA 107:17894–17898
Shoorideha K, Chuia CO (2014) On the origin of enhanced sensitivity in nanoscale FET-based biosensors. Proc Natl Acad Sci U S A 111:5111–5116
Shruthi GS, Amitha CV, Mathew BB (2014) Biosensors: a modern day achievement. J Instrument Technol 2:26–39
Siontorou CG, Batzias FA (2013) A methodological combined framework for roadmapping biosensor research: a fault tree analysis approach within a strategic technology evaluation frame. Crit Rev Biotechnol 34:31–55
Siontorou CG, Georgopoulos KN, Nikoleli G-P, Nikolelis DP (2016) Graphene tools in protein-based biosensing: optimizing artificial chemoreception in bilayer lipid membranes. Membranes (Basel) 6:43
Siontorou CG, Batzias FA, Tsakiri V (2010) A knowledge-based approach to online fault diagnosis of FET biosensors. IEEE Trans Instrum Measur 59:2345–2364
Song G, Chen M, Chen C, Wang C, Hu D, Ren J, Qu X (2010) Design of proton-fueled tweezers for controlled, multi-function DNA-based molecular device. Biochimie 92:121–127
Su L, Zou L, Fong CC, Wong WL, Wei F, Wong KY, Wu RS, Yang M (2013) Detection of cancer biomarkers by piezoelectric biosensor using PZT ceramic resonator as the transducer. Biosens Bioelectron 46:155–161
Suei S, Raudsepp A, Kent LM, Keen SAJ, Filichev VV, Williams MAK (2015) DNA visualization in single molecule studies carried out with optical tweezers: covalent versus non-covalent attachment of fluorophores. Biochem Biophys Res Commun 466:226–231
Tabrizi MA, Shamsipur M, Farzin L (2015) A high sensitive electrochemical aptasensor for the determination of VEGF(165) in serum of lung cancer patient. Biosens Bioelectron 74:764–769
Tan F, Yan F, Ju HX (2006) A designer ormosil gel for preparation of sensitive immunosensor for carcinoembryonic antigen based on simple direct electron transfer. Electrochem Commun 8:1835–1839
Tang DP, Yuan R, Chai YQ (2008) Ultrasensitive electrochemical immunosensor for clinical immunoassay using thionine-doped magnetic gold nanospheres as labels and horseradish peroxidase as enhancer. Anal Chem 80:1582–1588
Tiwari A, Gong S (2009) Electrochemical detection of a breast cancer susceptible gene using cDNA immobilized chitosan-co-polyaniline electrode. Talanta 77:1217–1222

Tong WY, Sweetman MJ, Marzouk ER, Fraser C, Kuchel T, Voelcker NH (2016) Towards a subcutaneous optical biosensor based on thermally hydrocarbonised porous silicon. Biomaterials 74:217–230
Tothill IE (2009) Biosensors for cancer markers diagnosis. Sem Cell Dev Biol 20:55–62
Valdastri P, Susilo E, Forster T, Strohhofer C, Menciassi A, Dario P (2011) Wireless implantable electronic platform for chronic fluorescent-based biosensors. IEEE Trans Biomed Eng 58:1846–1854
Vasan ASS, Mahadeo DM, Doraiswami R, Huang Y, Pech M (2013) Point-of-care biosensor system. Front Biosci S5:39–71
Velev OD, Kaler EW (1999) In situ assembly of colloidal particles into miniaturized biosensors. Langmuir 15:3693–3698
Wang Y, Botvinick EL, Zhao Y, Berns MW, Usami S, Tsien RY, Chien S (2005) Visualizing the mechanical activation of Src. Nature 434:1040–1045
Wang Y, Shyy JY, Chien S (2008) Fluorescence proteins, live-cell imaging, and mechanobiology: seeing is believing. Annu Rev Biomed Eng 10:1–38
Wang Y, Papadimitrakopoulos F, Burgess DJ (2013) Polymeric "smart" coatings to prevent foreign body response to implantable biosensors. J Control Release 169:341–347
Webster A, Vollmer F, Sato Y (2014) Probing biomechanical properties with a centrifugal force quartz crystal microbalance. Nat Commun 5:AR 5284
Wilson MS (2005) Electrochemical immunosensors for the simultaneous detection of two tumor markers. Anal Chem 77:1496–1502
Wu G, Datar RH, Hansen KM, Thundat T, Cote RJ, Majumdar A (2001) Bioassay of prostate-specific antigen (PSA) using microcantilevers. Nat Biotechnol 19:856–860
Zhang J, Allen MD (2007) FRET-based biosensors for protein kinases: illuminating the kinome. Mol BioSyst 3:759–765
Zhang A, Lieber CM (2016) Nano-bioelectronics. Chem Rev 116:215–257
Zhang M, Dai WJ, Yan M, Ge SG, Yu JH, Song XR, Xu W (2012) Ultrasensitive electrochemiluminescence immunosensor using PtAg@carbon nanocrystals composites as labels and carbon nanotubes-chitosan/gold nanoparticles as enhancer. Analyst 137:2112–2118
Zhang S, Song Y, Wang M, Zhang Z, Fan X, Song X, Zhuang P, Yue F, Chan P, Cai X (2016) A silicon based implantable microelectrode array for electrophysiological and dopamine recording from cortex to striatum in the non-human primate brain. Biosens Bioelectron 85:53–61
Zhou T, Marx KA, Dewilde AH, McIntosh D, Braunhut S (2011) Dynamic cell adhesion and viscoelastic signatures distinguish normal from malignant human mammary cells using quartz crystal microbalance. Anal Biochem 421:164–171

Electrochemical Redox Cycling Amplification Technology for Point-of-Care Cancer Diagnosis

6

Gorachand Dutta

6.1 Introduction

Biosensors are analytical devices with one or more biologic recognition elements that are specific for a certain target analyte of interest. Clinical diagnosis and disease prevention are the major applications of biosensor technology (Mascini and Tombelli 2008). It has several benefits over conventional diagnostic analysis including simplicity of use, specificity for the target analyte, and capability for continuous monitoring and multiplexing (Mascini and Tombelli 2008; Yu and Irudayaraj 2007). Point-of-care testing is known as medical diagnostic process which is conducted to the near patient and outside the clinical laboratory that don't involve the use of laboratory staff. The diagnosis device should be cheap and disposable to provide the benefits to the large part of the population in developing countries. Recently, with the printed electronics and roll-2-roll technologies tools have been developed that could potentially make diagnostics available to a much wider population (Gates et al. 2005). Electrochemical detection method is considered as one of the major sensing methods in microchip-based biosensors as it facilitates miniature sensing system and could be practically applied to a portable biosensing device for next-generation point-of-care biomedical testing (Gubala et al. 2012; Sharma et al. 2015; Rusling et al. 2010; Mahato et al. 2016). The tests need only simple instruction to use and detect multiple analytes or biomarkers in patient body fluids.

Enzyme labels are most commonly used in electrochemical biosensors for signal amplification (Hu et al. 2014; Nistor et al. 2002). Also nanomaterial-based highly sensitive biosensors have been newly developed (Chandra et al. 2015). The enzyme-based biosensors are most famous because of its high and reproducible signal

G. Dutta
Department of Mechanical Engineering, Michigan State University,
East Lansing, MI 48824, USA
e-mail: dutta.gorachand@gmail.com

P. Chandra et al. (eds.), *Next Generation Point-of-care Biomedical Sensors Technologies for Cancer Diagnosis*, DOI 10.1007/978-981-10-4726-8_6

amplification. But the direct electron transfer between enzyme label and electrode is a formidable challenge because of the large electron-hoping distance between the electrode and the redox center of the enzyme label and the signal amplification by enzymatic reaction not enough to obtain ultrasensitive biomolecule detection which is urgent for early and rapid diagnosis of diseases (Bourdillon et al. 1996 and Chen et al. 2007). To overcome this limitation, mediated electron transfer could be used where a redox mediator can help in fast transfer of electron from enzyme label to electrode, i.e., enzyme label-based electrochemical-enzymatic (EN) redox cycling. Chemical amplification can significantly enhance the signal level without significantly changing the background level (Park and Yang 2014). High signal-to-background ratios can be obtained using an enzymatic reaction or a nanocatalytic reaction plus a redox cycling reaction. Redox cycling is a process that can repetitively generate or consume signaling species in the presence of reversible redox species. The reversible two reactions, i.e., oxidation and reduction in a redox cycling process, can be observed either enzymatically, chemically, or electrochemically (Yang 2012). Electrochemical-electrochemical (EE) redox cycling, electrochemical-chemical (EC) redox cycling, or electrochemical-chemical-chemical (ECC) is effective in regenerating the signaling species and provides a very high and reproducible electrochemical signal (Yang 2012; Sen et al. 2012; Xia et al. 2014). In EE redox cycling, two close proximity electrodes are used for redox cycling of a signaling species, and in EC redox cycling, an extra chemical (i.e., reducing agent) is used to cycle the redox species which is generated on the electrode surface during electrochemical reaction. In ECC redox cycling two chemicals are used to cycle the electrochemically redox species and signal-to-noise ratio for ECC redox cycling is much better than EC redox cycling (Park and Yang 2014; Akanda et al. 2011).

A combination of redox cycling and electrochemical detection can play an important role in the development of ultrasensitive and reproducible biosensors for point-of-care testing. Point-of-care testing of biomarkers in clinical samples is of great importance for rapid and cost-effective diagnosis. However, up to now it is extremely challenging to develop an electrochemical POCT technique retaining both simplicity and very high sensitivity. However, many DNA or RNA sensors were reported with ultrahigh sensitivity but the detection procedures are quite complicated, which is unacceptable for next-generation point-of-care testing (Deng et al. 2017; Tiwari et al. 2016). Electrochemical biosensors are most suited to determine the biomarkers in "real-life" samples due to their low cost, high sensitivity, portable field-based size, and rapid diagnosis. But in most cases electrochemical signals are mass transfer limited of their signaling species to the electrode. Signaling species can be effectively regenerated and consumed by redox cycling and provides a reproducible and steady-state electrochemical signal and an ultrahigh detection scheme can be designed based on redox cycling technique. Accurate measurement of cancer biomarkers in their very early stage (benign) facilitates the categorization of tumor cells. Redox cycling-based electrochemical biosensors could be a promising tool for early-stage cancer diagnosis before the malignant stage and can save the patient life.

The present review pays attention to a novel electrochemical concept based on different redox cycling schemes to develop next-generation POC devices for early-stage detection of cancer biomarkers. Different redox cycling mechanisms are discussed that are used for over the past few years for cancer diagnosis. Also multiple detection of cancer biomarkers at single diagnostic platform is focused that may lead to a better interpretation (Table 6.1).

Table 6.1 Most reported electrochemical redox cycling techniques for highly sensitive cancer detection

Types of redox cycling	Schematic representation	Reason for signal amplification	References
Electrochemical-enzymatic (EN) redox cycling	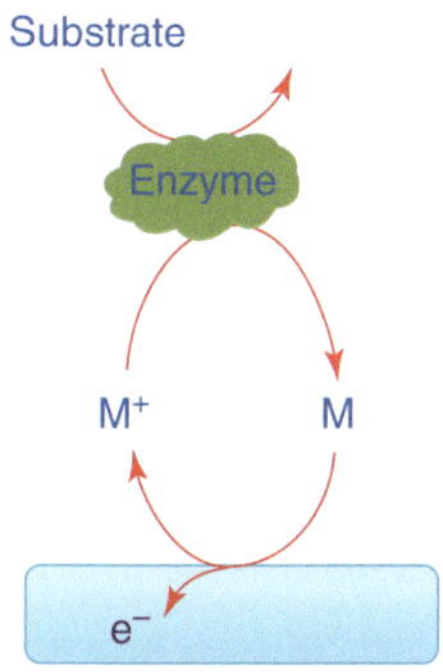	Redox mediator (M^+/M) helps in fast transfer of electron from enzyme label to electrode	Singh et al. (2013)
Electrochemical-electrochemical (EE) redox cycling	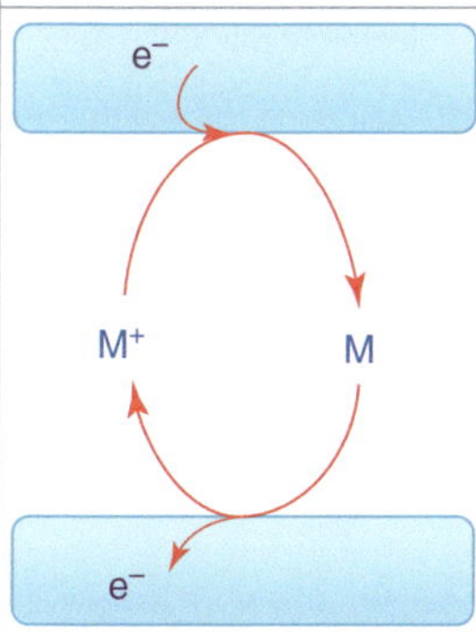	Repetitive generation and consumption of M and M^+ into the anode and cathode compartment, respectively, significantly enhanced the detection signal	Yasukawa et al. (2012), Han et al. (2014)
Electrochemical-chemical (EC) redox cycling	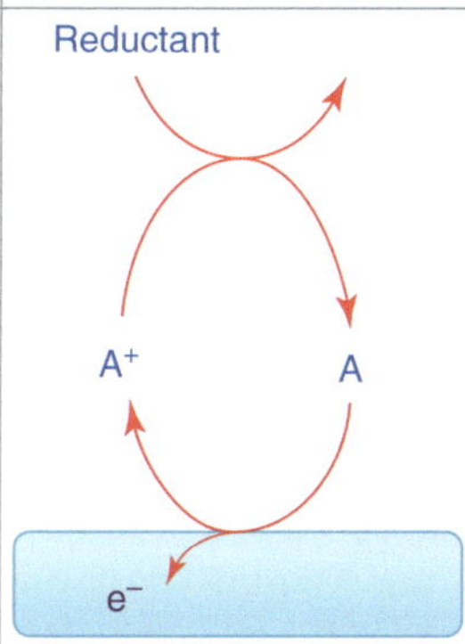	Reductant helps regeneration of signaling species (A) on the electrode surface and enhances the detection signal	Akanda and Ju (2016), Das et al. (2015), Cheng et al. (2014), Zhang et al. (2016a, b), Wang et al. (2014)

(continued)

Table 6.1 (continued)

Types of redox cycling	Schematic representation	Reason for signal amplification	References
Chemical-chemical (CC) redox cycling	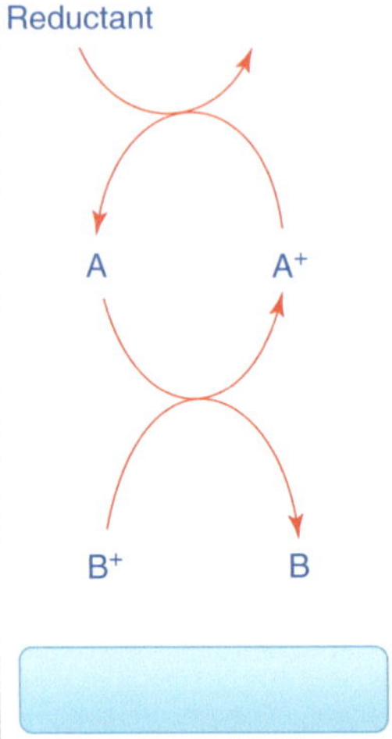	Two chemical reaction (one by reductant and another by substrate A) effectively regenerates the electroactive species B	Haque et al. (2015)
Electrochemical-chemical-chemical (ECC) redox cycling	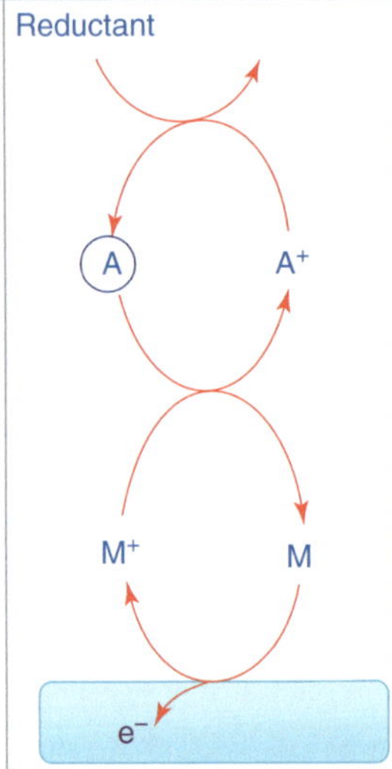	CC redox cycling regenerates the signaling species M which is electrochemically oxidized on the electrode surface	(i) Substrate (A) generated by nanomaterial, Das et al. (2006), Peng et al. (2016) (ii) Substrate (A) generated by enzyme, Akanda et al. (2012, 2014), Shuai et al. (2016)

6.2 Cancer Biomarkers

At the nineteenth century ectopic hormones and isoenzymes were known as cancer markers. In the last few decades due to the development of monoclonal antibody and recombinant DNA technology, there was a virtual explosion of discovery of cancer biomarkers. Some tumor markers are used as the indication of cancer in mass screening program. In China, alpha-fetoprotein (α-AFP) measurement is used for hepatocellular carcinoma screening. In Japan, urinary vanillylmandelic acid and homovanillic acid measurements have been used for neuroblastoma in children <1 year. Direct or indirect biomarker detection such as CA-125 for ovarian cancer, prostate-specific antigen (PSA) for prostate cancer, myoglobin creatinine kinase isoenzyme (CKMB), and cardiac troponins (cTnI and cTnT) for cardiac injury helps for effective screening, lower hospitalization rate, and cost saving.

6.3 Different Redox Cycling for Cancer Diagnosis

Over the past few years many types of redox cycling schemes have been used in biosensor for cancer diagnosis to improve the detection signal and to obtain an ultrahigh detection limit. Many enzyme-based sensors are introduced combined with redox cycling (Ertek et al. 2016; Park et al. 2014). Electrochemical-chemical (EC) redox cycling was used widely in biosensor for highly sensitive and selective protein detection. $NaBH_4$, hydrazine, or nicotinamide adenine dinucleotide (NADH) were used as reductant for EC redox cycling. But $NaBH_4$ and hydrazine can electroxidize readily on highly electrocatalytically active electrodes (i.e., gold electrode) and increases the background level. NADH electrooxidation on gold is slow and minimizes the background level (Kwon et al. 2008). A tyrosinase-responsive nonenzymatic redox cycling was developed for highly sensitive CEA detection (Fig. 6.1). A chitosan-coated immunosensor electrode was fabricated by covalently binding capture antibody. EC redox cycling scheme was incorporated for this study

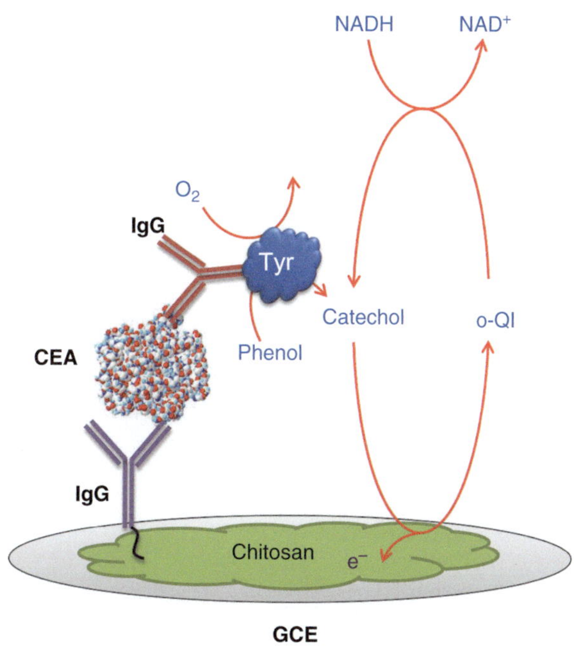

Fig. 6.1 Schematic representation of a tyrosinase-responsive electrochemical-chemical (EC) redox cycling for CEA detection (adapted from Akanda and Ju (2016))

using phenol as a substrate and NADH as a reducing agent. Tyrosinase converted low electroactive phenol to highly electroactive catechol and the catechol was electroxidized to its oxidized product ortho-quinoneimine (o-QI). o-QI can be chemically reduced by reducing agent NADH. The signal-to-background level was significantly high due to low electroactivity of phenol and high oxidation overpotential of NADH. The detection range of this EC redox cycling-based immunoassay was 1.0 pg/mL–0.1 μg/mL and the detection limit was 100 fg/mL. Because of very good reproducibility and high sensitivity, the proposed immunoassay method could be suitable in clinical diagnosis.

Enzyme-linked immunosorbent assay (ELISA) is a powerful tool for biomarker detection (Qureshia et al. 2012). However, the electron transfer rate between electrode and enzyme level is not so fast because of large electron hopping distance between electrode and redox center of enzyme. To overcome this limitation for highly sensitive protein detection, many electron-mediated enzymatic redox cycling schemes have been developed. Detection of cancer antigen 125 (CA-125) in human serum was studied by Singh et al. (2013) based on EN redox cycling using GOx as an enzyme label, $Ru(NH_3)_6^{3+}$ as a redox mediator, and glucose as a substrate (Fig. 6.2). $Ru(NH_3)_6^{3+}$ undergoes a fast outer-sphere electron transfer reaction at indium-tin oxide (ITO) electrodes and at the same time a fast electron transfer reaction with redox enzyme glucose oxidase (GOx) occurred. The EN redox cycling-based highly sensitive detection of cancer antigen 125 (CA-125) in human serum helped to detect 0.1 U/mL with 0-min incubation period and can be useful for near-patient cancer diagnosis.

A chemical-chemical (CC) and electrochemical-chemical-chemical (ECC) redox cycling for ultrasensitive detection of cardiac troponin-I (cTnI) in human serum was reported (Fig. 6.3). A highly OSR-philic (outer-sphere reactions) $Ru(NH_3)_6^{3+}$/$Ru(NH_3)_6^{2+}$ couple, highly ISR-philic (inner-sphere reactions) tris(2-carboxyethyl) phosphine (TCEP), and a OSR- and ISR-philic p-quinone imine/p-aminophenol (QI/AP) couple were used for the high, selective, and reproducible signal amplification (outer sphere to inner sphere). The formal potential of AP/QI is very close to

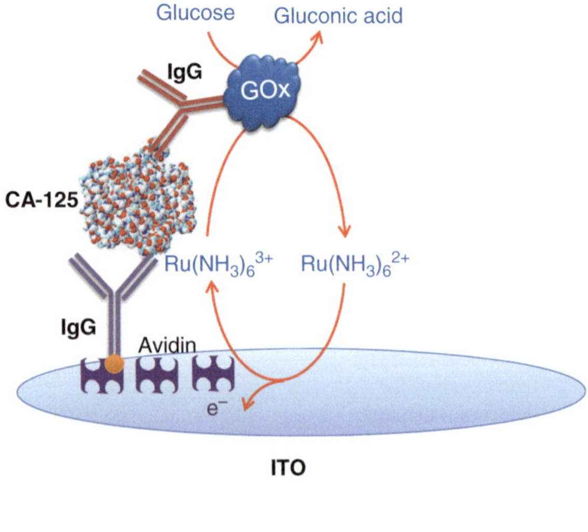

Fig. 6.2 Schematic view of a GOx label-based electrochemical enzymatic (EN) redox cycling for CA-125 detection (adapted from Singh et al. (2013))

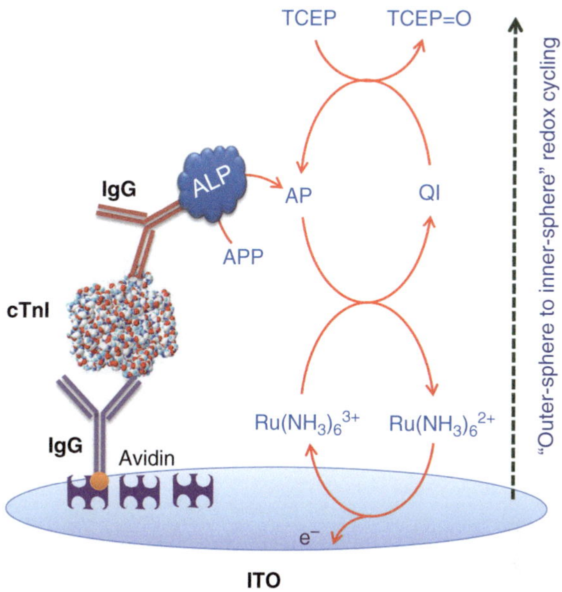

Fig. 6.3 Schematic illustration for cTnI detection based on enzymatic reaction and electrochemical-chemical-chemical (ECC) redox cycling (adapted from Akanda et al. (2012))

$Ru(NH_3)_6^{3+}$/$Ru(NH_3)_6^{2+}$couple; as a result $Ru(NH_3)_6^{3+}$ was reduced to $Ru(NH_3)_6^{2+}$ by AP and AP oxidized to QI. TCEP reduced QI to AP which can be reoxidized by $Ru(NH_3)_6^{3+}$ and regenerated $Ru(NH_3)_6^{2+}$. Many $Ru(NH_3)_6^{2+}$ were generated during 10-min incubation period and amplified the detection signal. Chronocoulometry and cyclic voltammetry were used for electrochemical signal measurements. The detection limit of cTnI was 10 fg/mL, which is quite lower than healthy persons and patients with myocardial infarction.

Development of a pure enzyme-free sensor with excellent stability is undoubtedly important for commercialization and next-generation point-of-care testing. Chemiluminescence immunoassay (Zhang et al. 2016a, b), radio immunoassay (Ledecky et al. 2013), and fluorescence immunoassay (Akter et al. 2016) without enzyme were developed. But many enzyme-free sensors suffer from sluggish electrode kinetics or poor signal stability and high detection limit. Also these sensors require complicated detection protocols or lack of ultrahigh detection limit in biological samples. Authors also reported sensitive protein detection using a transducer which was modified by a highly electroactive material to amplify the detection signals (Zhu et al. 2015). But in maximum, the modification procedures were complicated and background signal was very high due to highly electroactive materials.

To obtain very high signal amplification for early-stage prostate-specific antigen (PSA) detection, Das et al. (2006) first time established an enzyme-free nanocatalyst-based electrochemical redox cycling assay (Fig. 6.4). In this work, $NaBH_4$ was used for the catalytic reduction of p-nitrophenol (NP) to p-aminophenol (AP) and the reduction of electrochemically generated p-quinone imine (QI) to AP. The ferrocenium ion (Fc^+) was used as a redox mediator. The detection limit was 1 fg/mL which is comparable with bio-barcode assay.

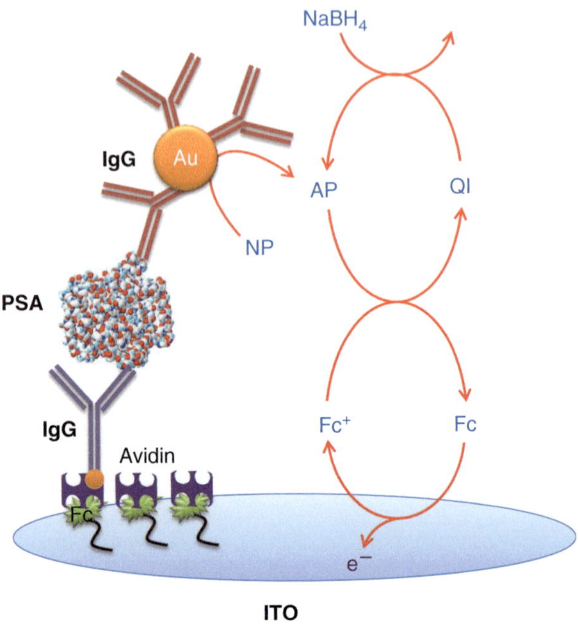

Fig. 6.4 Schematic view of a Au nanocatalyst label-based electrochemical-chemical-chemical (ECC) redox cycling assay for PSA detection (adapted from Das et al. (2006))

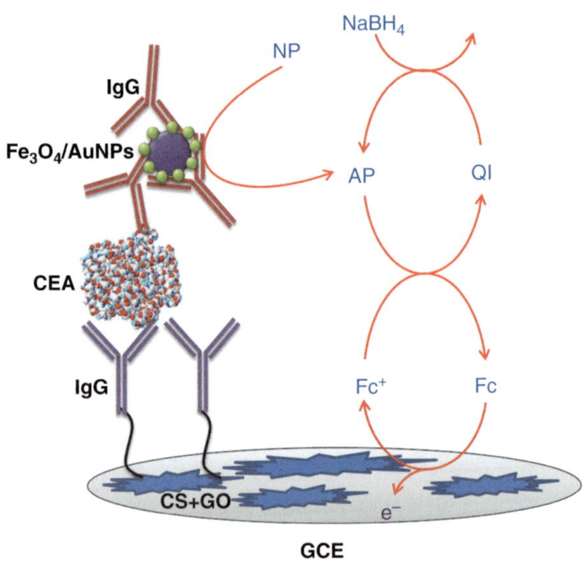

Fig. 6.5 Schematic representation of a Fe_3O_4/AuNP label-based electrochemical-chemical-chemical (ECC) redox cycling assay for CEA detection (adapted from Peng et al. (2016))

Recently, a similar study was made by Peng et al. (2016) to investigate a pure enzyme-free redox cycling immunoassay for the detection of carcinoembryonic antigen (CEA) based on Fe_3O_4/AuNPs label which revealed that p-nitrophenol (NP) can be catalytically reduced by reducing agent $NaBH_4$ to p-aminophenol (AP) and can take part in redox cycling with oxidant Fc^+ (Fig. 6.5). Based on this study, a

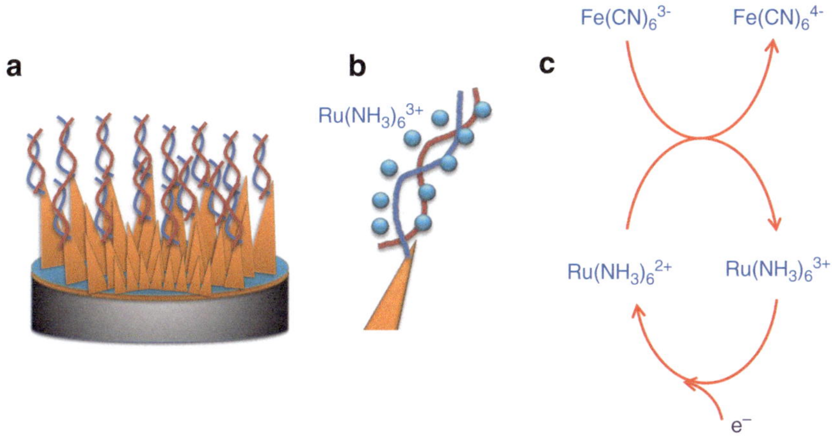

Fig. 6.6 Schematic view of (**a**) target mutant nucleic acid hybridization on PNA probe-modified nanostructured microelectrodes (NMEs), (**b**) negatively charged phosphate backbone of nucleic acids attracted electrostatically the positively charged $Ru(NH_3)_6^{3+}$, (**c**) electrochemical-chemical (EC) redox cycling for signal readout (adapted from Das et al. (2015))

sensitive and reproducible determination of CEA was done in the range of 0.001–30 ng/mL with a detection limit of 0.39 pg/mL.

Cell-free nucleic acids that originated from the mutated sequences shed by tumor cells which are present at very high levels in the blood of cancer patients could be detected and it is more challenging to differentiate the nucleic acids that originate from healthy cells and the mutated sequences shed by tumor cells. A EC redox cycling-based nanostructured microelectrode (NME) sensor was developed to detect the cell-free nucleic acids that originated from the mutated sequences shed by tumor cells (Fig. 6.6). The NMEs were functionalized with PNA probes which are specific to the mutant nucleic acid. An electrocatalytic reporter pair $Ru(NH_3)_6^{3+}$ and $Fe(CN)_6^{3-}$ was used for read out the signal. If target presents in the analyte solution, it would bind to the capture PNA probe and the negatively charged phosphate backbone of nucleic acids attracts electrostatically the positively charged $Ru(NH_3)_6^{3+}$. A negative bias potential was applied which reduced the $Ru(NH_3)_6^{3+}$ to $Ru(NH_3)_6^{2+}$ and at the same time the $Fe(CN)_6^{3-}$ which was present in solution chemically oxidized $Ru(NH_3)_6^{2+}$ back to $Ru(NH_3)_6^{3+}$ (EC redox cycling). The repetitive generation and consumption of $Ru(NH_3)_6^{3+}$ and $Ru(NH_3)_6^{2+}$ allowed a high electrocatalytic current. This new ultrahigh sensitive chip-based assay can detect selectively the presence of mutations within 15 min using a collection of samples taken from lung cancer and melanoma patients. The detection range was 1 fg/μL–100 pg/μL with 1 fg/μL detection limit.

MicroRNA (miRNA) detection plays an important role in various cancer disease diagnosis and in the research of miRNA function (Xia et al. 2015). The miRNA detection is important because of its complicated regulatory functions and plays a vital role in various life processes. Shuai et al. (2016) developed an ultrasensitive

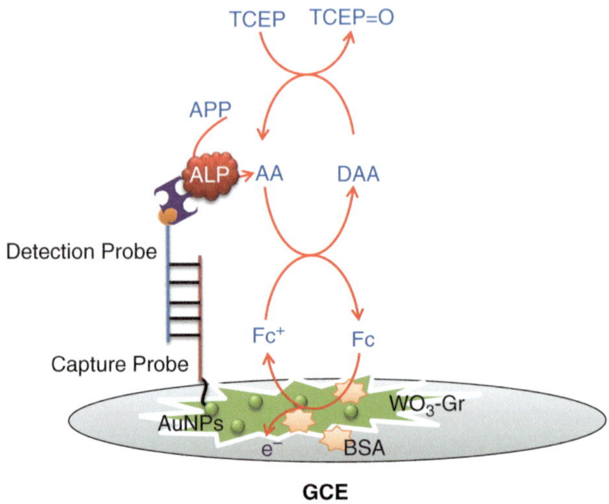

Fig. 6.7 Schematic view of an ultrasensitive electrochemical-chemical-chemical (ECC) redox cycling-based biosensor for miR-21 detection (adapted from Shuai et al. (2016))

electrochemical-chemical-chemical (ECC) redox cycling-based biosensor for miR-21 detection, a potential biomarker for breast cancer, based on tungsten oxide-graphene composites coupling with catalyzed hairpin assembly target recycling (Fig. 6.7). The enzyme substrate (APP) produced the ascorbic acid (AA) by enzymatic reaction with ALP and took part in the ECC redox cycling with ferrocene methanol (Fc) and tris (2-carboxyethyl) phosphine (TCEP). The regeneration of AA produced a significant electrochemical response. The detection range was 0 fM–100 pM and the detection limit was 0.05 fM. This proposed sensor was able to quantify the cancer miRNAs in clinical diagnostic and prognostic.

A novel electrochemical biosensor based on EC redox cycling with triple-signal amplification for the detection of human miRNAs from cancer cell-specific total RNA extracts was developed (Fig. 6.8). Redox cycling amplification, a bimetallic Pd–Pt-supported graphene-functionalized screen-printed gold electrode, and two stem-loop-structured DNAs as target capturers significantly enhance the detection signal. The enzyme substrate 1-naphthyl phosphate disodium salt (p-NPP) produced electroactive 1-naphthol (p-NP) by enzymatic reaction with ALP. The reducing agent TCEP reduced the electrochemically oxidized product (p-NP = O) and regenerated many p-NP. The linear range of detection was 10 fM–0.1 nM with a detection limit of 3.55 fM. The proposed EC redox cycling-based immunoassay was sufficiently selective and specific to discriminate the target cancer miRNA from homologous miRNAs and noncomplementary miRNAs which is suitable for the design of next-generation point-of-care electrochemical biosensors.

Over the last few years, the demands of the point-of-care chip-based immunosensors have steadily increased due to its ease of use and the fact that low volume of sample is required for diagnosis (Gubala et al. 2012). Recently, many electrochemical biosensors were developed based on electrochemical-electrochemical (EE) redox cycling, where two working electrodes are separated from each other between

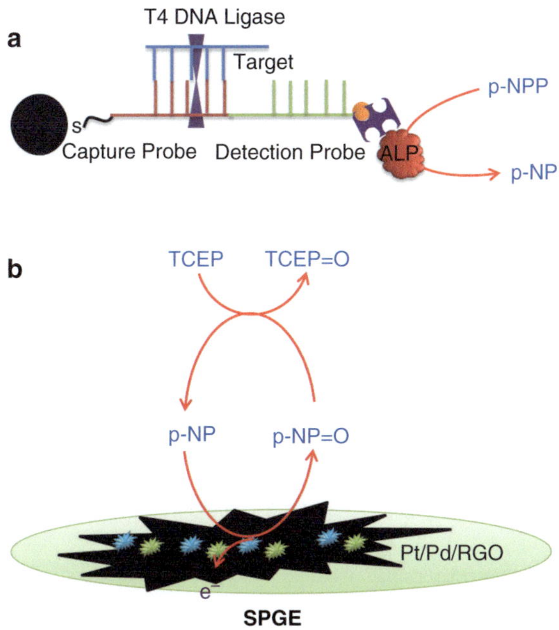

Fig. 6.8 Schematic illustration for (**a**) miRNA hybridization on Fe_3O_4@PDA–probe DNA1 and enzymatic conversion of 1-naphthyl phosphate disodium salt (p-NPP) into electroactive 1-naphthol (p-NP) by alkaline phosphatase (ALP) enzyme, (**b**) Pt/Pd/RGO-modified screen-printed gold electrode (SPGE) for electrochemical-chemical (EC) redox cycling for enhancement of the electrochemical signal (adapted from Cheng et al. (2014))

a nanogap and regenerate the electroactive species by redox reaction on anode and cathode electrodes and amplify the detection signal. Yasukawa et al. (2012) reported a highly sensitive immunosensor chip based on dual-amplification system combining an electrochemical-redox cycling and coulometric signal transduction using a galvanic cell (Fig. 6.9a–c). A sandwich immunocomplex was formed on a microparticle. In the events of antigen-antibody binding the β-galactosidase (β-gal)-conjugated secondary antibody bound to the immunosurface. β-Gal produced 4-aminophenol (AP) by enzymatic reaction with p-aminophenyl-β-D-galactopyranoside (APG) into the anode compartment consisting of a comb type of an interdigitated array (IDA) electrode. Because of the coupled reduction of silver ions the AP was oxidized to quinone imine (QI) resulting in the deposition of silver metal on the cathode GC electrode. Another comb of IDA reduced QI to AP and silver was deposited corresponding to the degree of AP oxidation by EE redox cycling which led to an enhancement of the stripping signal. The reproducible and highly sensitive immunoassay can detect 10 pg/mL CEA in human serum. Recently, Haque et al. (2015) demonstrated a similar enzymatic Ag-deposition scheme combined with redox cycling which is more simple than Yasukawa et al.'s study (Fig. 6.9d). A chemical-chemical (CC) redox cycling was used for the deposition of metallic silver by reduced β-nicotinamide adenine dinucleotide (NADH). The highly sensitive CC redox cycling-based sensor was effective for selective detection of a cardiac biomarker, creatine kinase-MB, in human serum.

Han et al. (2014) introduced an electrochemical redox cycling-based three-dimensional interdigitated array (3D IDA) electrode for the sensitive detection of

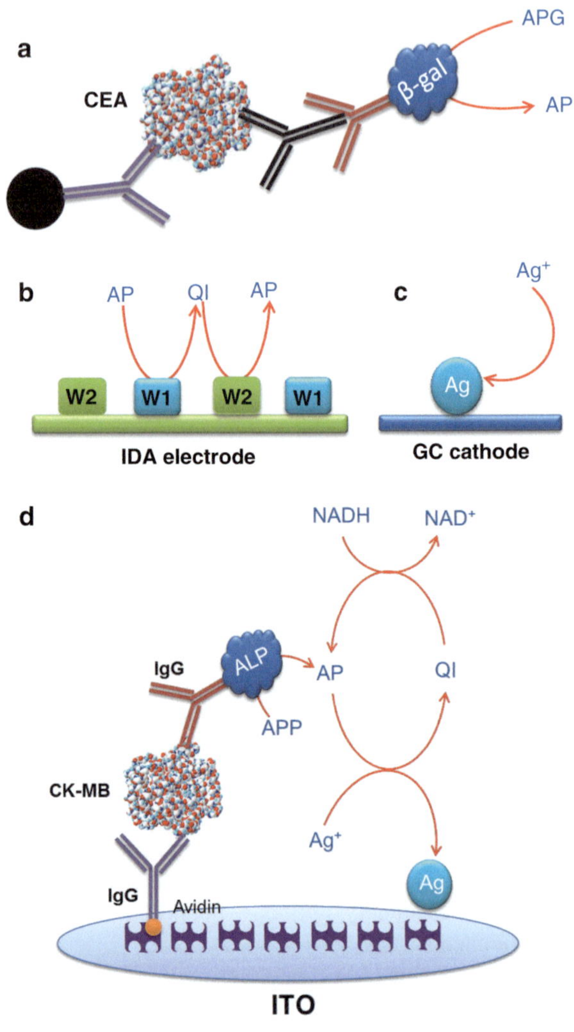

Fig. 6.9 Schematic illustration of (**a**) sandwich immunocomplex formation on a microparticle and enzymatic conversion of p-aminophenyl-β-D-galactopyranoside (APG) to p-aminophenol (AP), and (**b**) electrochemical-electrochemical (EE) redox cycling for regeneration of AP, (**c**) silver deposition on cathode glassy carbon electrode corresponding to the degree of AP oxidation by EE redox cycling which leads to an enhancement of the stripping signal (adapted from Yasukawa et al. 2012). (**d**) Schematic view of an electrochemical immunosensor using enzymatic silver deposition by chemical-chemical (CC) redox cycling and readout of the electrochemical signal by electrooxidation of the deposited silver (adapted from Haque et al. (2015))

cardiac troponin I in human serum (Fig. 6.10). The chip-based 3D IDA immunosensor had an indium tin oxide (ITO) working electrode modified with electroactive ferrocene (Fc). Alkaline phosphatase (ALP) and p-aminophenyl phosphate (APP) were used as an enzyme label and enzyme substrate, respectively. The enzyme product p-aminophenol (AP) electrochemically oxidized to p-quinone imine (QI) on the Fc-grafted generator electrodes and QI reduced back to AP on the surface of collector electrode. The repetitive generation of AP by EE redox cycling significantly enhanced the detection signal and 100 fg/mL cTnI could be detected in human serum. The parallel design of immunosensors could be suitable for multiple target detection on a single chip in "real-life" patient sample.

Cost-effective and sensitive diagnosis is the ultimate goal for the next-generation point-of-care testing of biomarkers (Vashist et al. 2015). Recently, many

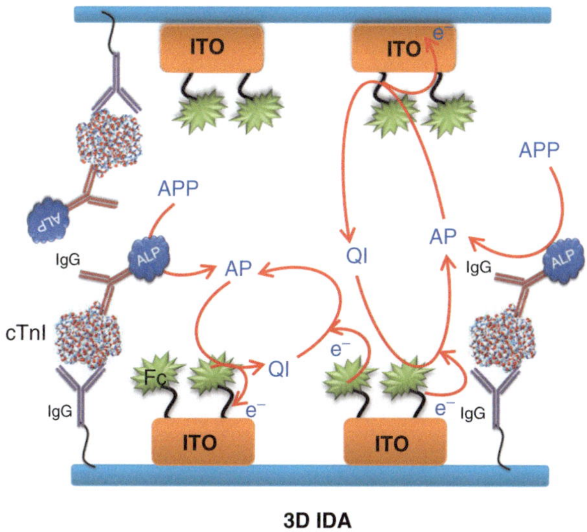

Fig. 6.10 Schematic representation of a chip-based three-dimensional interdigitated array (3D IDA) immunosensor for cTnI detection using enzymatic reaction and electrochemical-electrochemical (EE) redox cycling (adapted from Han et al. (2014))

lateral-flow strip-based biosensors are developed because of its simplicity of use and ease of miniaturizing for POCT devices (Toubanaki et al. 2016; Ang et al. 2016). But maximum sensors suffer from sensitive detection. A redox cycling-based lateral-flow immunostrip for one-step ultrasensitive detection with serum was developed for early-stage point-of-care cancer diagnosis (Fig. 6.11). In a noble ECC redox cycling technique β-galactosidase (Gal), 4-amino-1-naphthyl β-D-galactopyranoside (AN-GP), and 4-amino-1-naphthol (AN) were used as enzyme label, substrate, and electroactive product, respectively, with ultrahigh detection of cTnI (100 fg/mL after 11 min) and a minimum interference effect. This immunostrip was practically applied for real clinical samples with high accuracy.

An electrochemical-chemical redox cycling-based genosensor was developed using gold nanoparticle (Au NP)–DNA complex (Fig. 6.12). $Ru(NH_3)_6^{3+}$ was captured by Au NP–DNA complex via electrostatic force. The applied negative bias potential reduced the $Ru(NH_3)_6^{3+}$ to $Ru(NH_3)_6^{2+}$ and an externally added chemical $Fe(CN)_6^{3-}$ chemically oxidized $Ru(NH_3)_6^{2+}$ back to $Ru(NH_3)_6^{3+}$ (EC redox cycling) and increased significantly the cathodic signal. A 0.3 U/mL DNMT1 (human DNA (cytosine-5)–methyltransferase 1 from crude lysates of cancer cells) could be detected in human serum in the range of 1 U/mL–40 U/mL.

A highly sensitive and simple electrochemical aptasensor was introduced for the determination of mucin 1 (associated with colon, breast, ovarian, lung, and pancreatic cancers) based on p-aminophenol redox cycling (Fig. 6.13). A competitive assay was developed where biotinylated mucin 1 and streptavidin-alkaline phosphatase were captured by an anti-mucin 1 aptamer-modified electrode. Electrochemically active p-aminophenol was produced from the reaction between p-aminophenyl phosphate substrate and streptavidin-conjugated alkaline phosphatase. The resulting p-aminophenol was electrooxidized on electrode surface and reproduced by tris(2-carboxyethyl) phosphine. The repetitive generation of AP by EC redox cycling significantly enhanced the detection signal. The mucin 1 competed with the

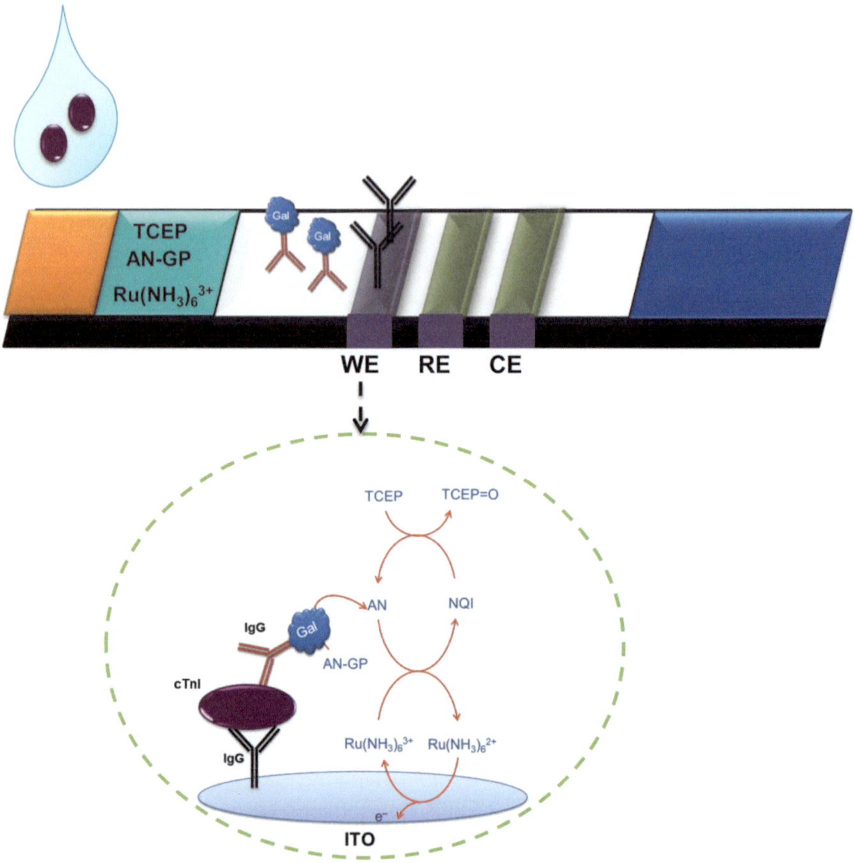

Fig. 6.11 Schematic view of an electrochemical lateral-flow immunostrip for cTnI detection using enzymatic reaction and electrochemical-chemical-chemical (ECC) redox cycling (adapted from Akanda et al. (2014))

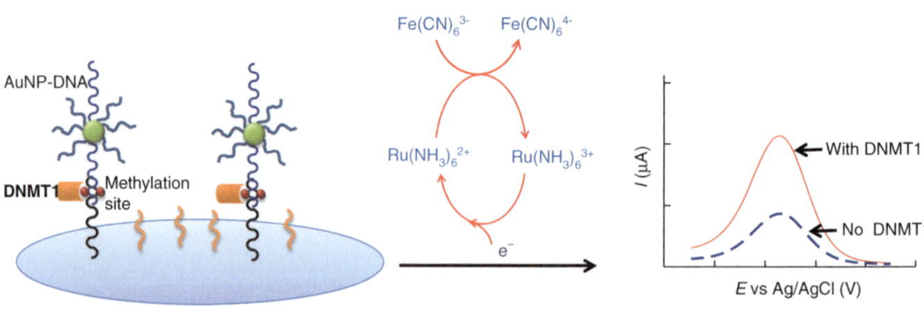

Fig. 6.12 Schematic representation for DNTMT1 detection based on recycling of Ru(III) redox using electrochemical-chemical (EC) redox cycling (adapted from Zhang et al. (2016a, b))

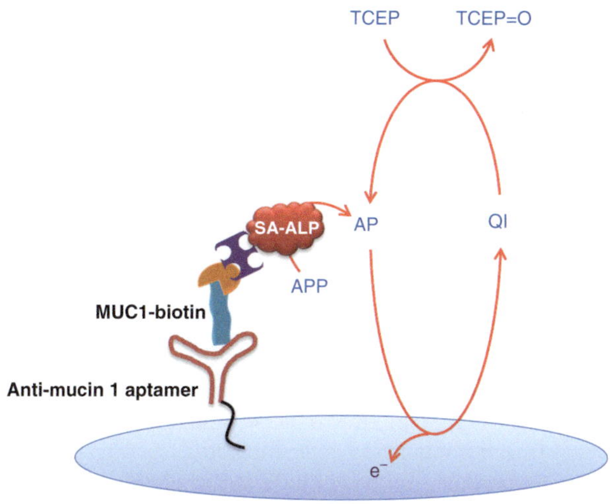

Fig. 6.13 Schematic view of a highly sensitive electrochemical aptasensor based on electrochemical-chemical (EC) redox cycling for mucin 1 detection (adapted from Wang et al. (2014))

conjugates which is already bound to the anchored aptamer and the detection signal was decreased with an increase in mucin 1 concentration. The detection range was 0.5–6 nM and the detection limit was 0.1 nM.

6.4 Multiplex Detection Using Redox Cycling

Recently, there has been more and more interest in electrochemical biosensor for detection of multianalytes with high sensitivity, low cost, and excellent selectivity for next-generation point-of-care testing (Spindel and Sapsford 2014; Chandra 2013). Redox cycling scheme could make it possible to perform highly sensitive detection of several biomarkers with printed electronics. With the rise of printed electronics and roll-2-roll technologies tools have been developed that could bring the redox cycling-based sensor technology to a much wider population. Multianalyte detection immunosensors have been developed to facilitate the detection of several biomarkers from one sample. Different capture probes could be immobilized on a single chip to allow the multiple detection within same sample for fast and sensitive point-of-care testing.

Zheng et al. (2013) developed a simultaneous multiplex detection and classification of both acute myeloid leukemia and acute lymphocytic leukemia cells based on nanobiotechnology electrochemical approach using redox cycling reaction for signal enhancement. Dual-aptamer-functionalized graphene–Au multilayered nanostructures on a glassy carbon electrode surface were used for high detection sensitivity and selectivity. The hybrid electrochemical nanoprobes were conjugated with distinguishable cell-targeting aptamers and redox tag, and signal-amplifying enzyme. This simple and unique method has several advantages including operational simplicity, low cost, high sensitivity, excellent selectivity, and ease of miniaturization.

6.5 Wash-Free Detection of Cancer Biomarkers Based on Redox Cycling Immunoassay

Immunoassays using affinity binding between antigen and antibody have been widely used in bioassays because of their high sensitivity and selectivity. In recent years, point-of-care testing (POCT) of biomarkers in clinical samples has been of great importance for rapid and cost-effective diagnosis. Many methods have been developed to this end; many of them have limitations in terms of simplicity, rapidness, cost-effectiveness, and ultrasensitivity (Milligan and Ghindilis 2002; Rishpon and Ivnitski 1997; McNeil et al. 1995; Ramón-Azcón et al. 2010). Washing processes are essential in most heterogeneous labeled assays; if a wash-free scheme is combined with the assays, this could significantly simplify the detection procedure and reduce the assay time. Recently, Dutta et al. developed a redox cycling-based wash-free immunoassay that allowed fast, sensitive, and single-step detection of biomarkers in serum with low interference using electrochemical-enzymatic (EN) redox cycling (Dutta et al. 2014, 2015). After a sample solution was mixed with a solution for enzymatic reaction, no operation was required prior to the electrochemical measurement. The discrimination between a bound and an unbound label was obtained with single-enzyme-based proximity-dependent electron mediation: a bound labeled probe allowed faster electron mediation between an enzyme label and an electrode than an unbound one (Fig. 6.14).

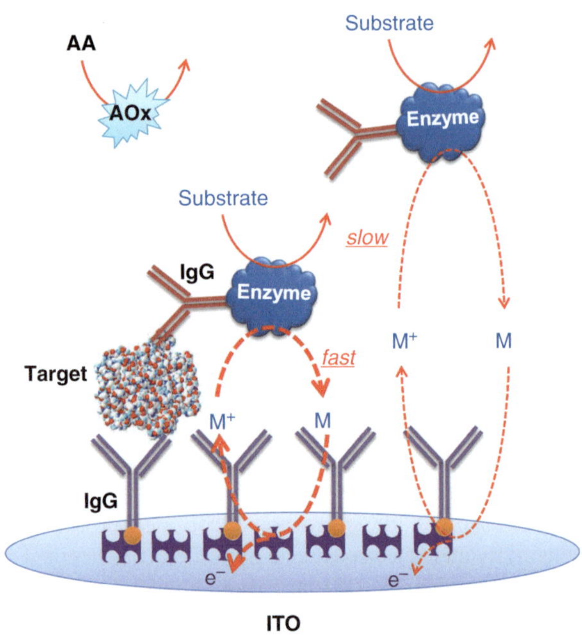

Fig. 6.14 Schematic illustration of a washing-free electrochemical immunosensor using proximity-dependent electron mediation and electrochemical-enzymatic (EN) redox cycling. The annotations M^{+}/M and AOx represent electron mediator and ascorbate oxidase, respectively (adapted from Dutta et al. (2014, 2015))

After immobilization of biotinylated IgG on the electrode surface a sample solution was prepared with enzyme-conjugated IgG and enzyme-substrate, which was mixed with serum or blood spiked with different concentrations of target antigen. The mixed solution was then injected into the electrochemical cell. The sensing electrode was already modified with capture antibody. The main interfering species ascorbic acid effect was eliminated by ascorbate oxidase (AOx) during the 10-min solution incubation. Concentration-dependent signal was obtained after a 10-min incubation period. The signal data was increased with increase in the target concentration, although the concentration of injected enzyme-conjugated IgG was the same. The surface concentration of bound enzyme-conjugated IgG was increased with increase in target concentration. Therefore, the increased signal indicated that a surface-bound enzyme allowed faster electron mediation than an unbound enzyme. The main advantage of the immunoassay was that the single solution-based wash-free detection can significantly simplify the assay procedure and reduce the assay time (Table 6.2).

Conclusions

Biosensors offered fast, simple, and cost-effective detection capabilities for cancer diagnosis. Redox cycling-based biosensor significantly improved the signal amplification and opened a new avenue for early-stage cancer detection. The main challenges of point-of-care testing require implementing complex analytical methods into low-cost technologies. The main future prospects of redox cycling technology are to develop a simple and cost-effective electrochemical sensor for real sample analysis before the malignant stage and that can save the patient life. Also, the principle of redox cycling-based electrochemical sensors becomes appealing if we can develop an assay for low-cost disposable systems that can be possibly integrated with a mobile electronic device. This is particularly true for countries with less developed health care infrastructure. A low-cost disposable chip and redox cycling amplification approach could make it possible to perform highly sensitive detection of cancer biomarkers with printed electronics. The redox cycling technology helps to detect several interesting targets at the same time on a printed chip by amplifying the signal 10^2 to 10^5 times. The electrochemical-chemical (EC) redox cycling or electrochemical-chemical-chemical (ECC) redox cycling method combined with a washing-free technique could generate a stronger impact on point-of-care cancer diagnosis.

Acknowledgments Dr. Gorachand Dutta gratefully acknowledges Prof. Dr. Peter B. Lillehoj for his support in Michigan State University. The author would also like to thank Prof. Dr. Pranjal Chandra, Indian Institute of Technology, Guwahati, for the valuable discussion in this chapter.

Table 6.2 Specification of redox cycling-based electrochemical biosensor for cancer detection

Detection method	Biomarker	Detection matrix	Detection range	Limit of detection (LOD)	Reference
Electrochemical-chemical (EC) redox cycling based on tyrosinase-converted phenol	CEA	Serum	1.0–100,000 pg/mL	0.1 pg/mL	Akanda and Ju (2016)
Electrochemical enzymatic (EN) redox cycling using GOx as an enzyme label	CA-125	Serum	0.1 U/mL–100 U/mL	0.1 U/mL	Singh et al. (2013)
Electrochemical – chemical – chemical (ECC) redox cycling based on ALP-converted p-aminophenol	cTnI	Serum	10 fg/mL–10 ng/mL	10 fg/mL	Akanda et al. (2012)
Electrochemical-chemical-chemical (ECC) redox cycling based on Au nanocatalyst	PSA	PBS	1 pg/mL–30 ng/mL	1 fg/mL	Das et al. (2006)
Electrochemical-chemical-chemical (ECC) redox cycling based on Fe_3O_4/AuNPs	CEA	Tris	1 pg/mL–30 ng/mL	0.39 pg/mL	Peng et al. (2016)
Electrochemical-chemical (EC) redox cycling based on nanostructured microelectrode	cfRNA	Serum	1 fg/µl–100 pg/µl	1 fg/µl	Das et al. (2015)
Electrochemical-chemical-chemical (ECC) redox cycling based on tungsten oxide-graphene composites	miR-21	Tris	0 fM–100 pM	0.05 fM	Shuai et al. (2016)
Electrochemical-chemical (EC) redox cycling based on bimetallic Pd–Pt-supported graphene	miRNA let-7b	Tris	10 fM–0.1 nM	3.55 fM	Cheng et al. (2014)
Electrochemical-electrochemical (EE) redox cycling and coulometric signal transduction using a galvanic cell	CEA	Serum	10 pg/mL–100 ng/mL	10 pg/mL	Yasukawa et al. (2012)
Chemical-chemical (CC) redox cycling based on metallic silver deposition	CK-MB	Serum	100 pg/mL–1 µg/mL	100 pg/mL	Haque et al. (2015)
Electrochemical-electrochemical (EE) redox cycling based on ALP-converted p-aminophenol	cTnI	Serum	100 fg/mL–100 ng/mL	100 fg/mL	Han et al. (2014)
Electrochemical-chemical-chemical (ECC) redox cycling using a lateral-flow immunostrip	cTnI	Serum	100 fg/mL–100 ng/mL	100 fg/mL	Akanda et al. (2014)

Electrochemical-chemical (EC) redox cycling based on gold nanoparticle (Au NP)–DNA complex	DNMT1	Serum	1 U/mL–40 U/mL	0.3 U/mL	Zhang et al. (2016a, b)
Electrochemical aptasensor based on electrochemical-chemical (EC) redox cycling	MUC1	PBS	0.5 nM–6 nM	0.1 nM	Wang et al. (2014)
Washing-free immunosensor based on electrochemical-enzymatic (EN) redox cycling	cTnI	Serum	10 pg/mL–100 ng/mL	10 pg/mL	Dutta et al. (2015)
	PSA	Serum	10 pg/mL–100 ng/mL	10 pg/mL	Dutta et al. (2014)

References

Akanda MR, Aziz MA, Jo K, Tamilavan V, Hyun MH, Kim S, Yang H (2011) Optimization of phosphatase- and redox cycling-based immunosensors and its application to ultrasensitive detection of troponin I. Anal Chem 83:3926–3933

Akanda MR, Choe YL, Yang H (2012) "Outer-sphere to inner-sphere" redox cycling for ultrasensitive immunosensors. Anal Chem 84:1049–1055

Akanda MR, Joung HA, Tamilavan V, Park S, Kim S, Hyun SMH, Kim MG, Yang H (2014) An interference-free and rapid electrochemical lateral-flow immunoassay for one-step ultrasensitive detection with serum. Analyst 139:1420–1425

Akanda MR, Ju H (2016) A tyrosinase-responsive nonenzymatic redox cycling for amplified electrochemical immunosensing of protein. Anal Chem 88:9856–9861

Akter S, Vehniäinen M, Spoof L, Nybom S, Meriluoto J, Lamminmäki U (2016) Broad-spectrum noncompetitive immunocomplex immunoassay for cyanobacterial peptide hepatotoxins (microcystins and nodularins). Anal Chem 88:10080–10087

Ang SH, Rambeli M, Thevarajah TM, Alias YB, Khor SM (2016) Quantitative, single-step dual measurement of hemoglobin A1c and total hemoglobin in human whole blood using a gold sandwich immunochromatographic assay for personalized medicine. Biosens Bioelectron 78:187–193

Bourdillon C, Demaille C, Moiroux J, Savéant JM (1996) From homogeneous electroenzymatic kinetics to antigen-antibody construction and characterization of spatially ordered catalytic enzyme assemblies on electrodes. Acc Chem Res 29:529–535

Chandra P (2013) Miniaturized multiplex electrochemical biosensor in clinical bioanalysis. Bioanal Biomed 5:e122

Chandra P, Noh HB, Pallela R, Shim YB (2015) Ultrasensitive detection of drug resistant cancer cells in biological matrixes using an amperometric nanobiosensor. Biosens Bioelectron 70:418–425

Chen D, Wang G, Li J (2007) Interfacial bioelectrochemistry: fabrication, properties and applications of functional nanostructured biointerfaces. J Phys Chem C 111:2351–2367

Cheng FF, Zhang JJ, He TT, Shi JJ, Abdel-Halimc ES, Zhu JJ (2014) Bimetallic Pd–Pt supported graphene promoted enzymatic redox cycling for ultrasensitive electrochemical quantification of microRNA from cell lysates. Analyst 139:3860–3865

Das J, Aziz MA, Yang H (2006) A nanocatalyst-based assay for proteins: DNA-free ultrasensitive electrochemical detection using catalytic reduction of p-nitrophenol by gold-nanoparticle labels. J Am Chem Soc 128:16022–16023

Das J, Ivanov I, Montermini L, Rak J, Sargent EH, Kelley SO (2015) An electrochemical clamp assay for direct, rapid analysis of circulating nucleic acids in serum. Nature Chem 7:569–575

Deng K, Li C, Huang H, Li X (2017) Rolling circle amplification based on signal-enhanced electrochemical DNA sensor for ultrasensitive transcription factor detection. Sensors Actuators B Chem 238:1302–1308

Dutta G, Kim S, Park S, Yang H (2014) Washing-free heterogeneous immunosensor using proximity-dependent electron mediation between an enzyme label and an electrode. Anal Chem 86:4589–4595

Dutta G, Park S, Singh A, Seo J, Kim S, Yang H (2015) Low-interference washing-free electrochemical immunosensor using glycerol-3-phosphate dehydrogenase as an enzyme label. Anal Chem 87:3574–3578

Ertek B, Akgul C, Dilgin Y (2016) Photoelectrochemical glucose biosensor based on a dehydrogenase enzyme and NAD^+/NADH redox couple using a quantum dot modified pencil graphite electrode. RSC Adv 6:20058–20066

Gates BD, Xu Q, Stewart M, Ryan D, Willson CG, Whitesides GM (2005) New approaches to nanofabrication: molding, printing, and other techniques. Chem Rev 105:1171–1196

Gubala V, Harris LF, Ricco AJ, Tan MX, Williams DE (2012) Point of care diagnostics: status and future. Anal Chem 84:487–515

Han D, Kim YR, Kang CM, Chung TD (2014) Electrochemical signal amplification for Immunosensor based on 3D Interdigitated Array electrodes. Anal Chem 86:5991–5998
Haque AMJ, Kim J, Dutta G, Kim S, Yang H (2015) Redox cycling-amplified enzymatic Ag deposition and its application in the highly sensitive detection of creatine kinase-MB. Chem Commun 51:14493–14496
Hu C, Yang DP, Zhu F, Jiang F, Shen S, Zhang J (2014) Enzyme-labeled Pt@BSA nanocomposite as a facile electrochemical biosensing interface for sensitive glucose determination. ACS Appl Mater Interfaces 6:4170–4178
Kwon SJ, Yang H, Jo K, Kwak J (2008) An electrochemical immunosensor using p-aminophenol redox cycling by NADH on a self-assembled monolayer and ferrocene-modified au electrodes. Analyst 133:1599–1604
Ledecky V, Valencakova-Agyagosova A, Lepej J, Frischova Z, Hornak S, Nagy V (2013) Determination of carcinoembryonic antigen and cancer antigen values with the radioimmunoassay method in healthy females dogs. Vet Med Czech 58:277–283
Mahato K, Prasad A, Maurya PK, Chandra P (2016) Nanobiosensors: next generation point-of-care biomedical devices for personalized diagnosis. J Anal Bioanal Tech 7:e125
Mascini M, Tombelli S (2008) Biosensors for biomarkers in medical diagnostics. Biomarkers 13:637–657
McNeil CJ, Athey D, Ho WO (1995) Direct electron transfer bioelectronics interfaces: application to clinical analysis. Biosens Bioelectron 10:75–83
Milligan C, Ghindilis A (2002) Laccase based sandwich scheme immunosensor employing mediatorless electrocatalysis. Electroanalysis 14:415–419
Nistor C, Rose A, Wollenberger U, Pfeiffer D, Emnéus J (2002) A glucose dehydrogenase biosensor as an additional signal amplification step in an enzyme-flow immunoassay. Analyst 127:1076–1081
Park S, Singh A, Kim S, Yang H (2014) Electroreduction-based electrochemical-enzymatic redox cycling for the detection of cancer antigen 15-3 using graphene oxide modified indium–tin oxide electrodes. Anal Chem 86:1560–1566
Park S, Yang H (2014) Sensitive and selective trypsin detection using redox cycling in the presence of L-ascorbic acid. Analyst 139:4051–4055
Peng D, Liang RP, Huang H, Qiu JD (2016) Electrochemical immunosensor for carcinoembryonic antigen based on signal amplification strategy of graphene and Fe_3O_4/au NPs. J Electroanal Chem 761:112–117
Qureshia A, Gurbuzb Y, Niazi JH (2012) Biosensors for cardiac biomarkers detection: a review. Sensors Actuators B Chem 171–172:62–76
Ramón-Azcón J, Yasukawa T, Lee HJ, Matsue T, Sánchez-Baeza F, Marco MP, Mizutani F (2010) Competitive multi-immunosensing of pesticides based on the particle. Biosens Bioelectron 25:1928–1933
Rishpon J, Ivnitski D (1997) An amperometric enzyme-channeling immunosensor. Biosens Bioelectron 12:195–204
Rusling JF, Kumar CV, Gutkinde JS, Patele V (2010) Measurement of biomarker proteins for point-of-care early detection and monitoring of cancer. Analyst 135:2496–2511
Sen M, Ino K, Shiku H, Matsue T (2012) Accumulation and detection of secreted proteins from single cells for reporter gene assays using a local redox cycling-based electrochemical (LRC-EC) chip device. Lab Chip 12:4328–4335
Sharma S, Zapatero-Rodríguez J, Estrela P, O'Kennedy R (2015) Point-of-care diagnostics in low resource settings: present status and future role of microfluidics. Biosensors 5:577–601
Shuai HL, Huang KJ, Xing LL, Chen YX (2016) Ultrasensitive electrochemical sensing platform for microRNA based on tungsten oxide-graphene composites coupling with catalyzed hairpin assembly target recycling and enzyme signal amplification. Biosens Bioelectron 86:337–345
Singh A, Park S, Yang H (2013) Glucose-oxidase label-based redox cycling for an incubation period-free electrochemical immunosensor. Anal Chem 85:4863–4868
Spindel S, Sapsford KE (2014) Evaluation of optical detection platforms for multiplexed detection of proteins and the need for point-of-care biosensors for clinical use. Sensors 14:22313–22341

Tiwari JN, Vij V, Kemp KC, Kim KS (2016) Engineered carbon-nanomaterial-based electrochemical sensors for biomolecules. ACS Nano 10:46–80
Toubanaki DK, Margaroni M, Karagouni E (2016) Dual enhancement with a nanoparticle-based lateral flow biosensor for the determination of DNA. Anal Lett 49:1040–1055
Vashist SK, Luppa PB, Yeo LY, Ozcan A, Luong JHT (2015) Emerging technologies for next-generation point-of-care testing. Trends Biotechnol 33:692–705
Wang Z, Xia N, Shi J, Li S, Zhao Y, Wang H, Liu L (2014) Electrochemical Aptasensor for determination of Mucin 1 by p-aminophenol redox cycling. Anal Lett 47:2431–2442
Xia N, Liu L, Wu R, Liu H, Li SJ, Hao Y (2014) Ascorbic acid-triggered electrochemical–chemical–chemical redox cycling for design of enzyme-amplified electrochemical biosensors on self-assembled monolayer-covered gold electrodes. J Electroanal Chem 731:78–83
Xia N, Zhang YJ, Wei X, Huang YP, Liu L (2015) An electrochemical microRNAs biosensor with the signal amplification of alkaline phosphatase and electrochemical chemical–chemical redox cycling. Anal Chim Acta 878:95–101
Yang H (2012) Enzyme-based ultrasensitive electrochemical biosensors. Curr Opin Chem Biol 16:422–428
Yasukawa T, Yoshimoto Y, Goto T, Mizutani F (2012) Highly-sensitive electrochemical immunosensing method based on dual amplification systems. Biosens Bioelectron 37:19–23
Yu C, Irudayaraj J (2007) Multiplex biosensor using gold nanorods. Anal Chem 79:572–579
Zhang H, Dong H, Yang G, Chen H, Cai C (2016a) Sensitive electrochemical detection of human methyltransferase based on a dual signal amplification strategy coupling gold nanoparticle–DNA complexes with Ru(III) redox recycling. Anal Chem 88:11108–11114
Zhang Q, Wang CR, Yu JP, Ma Q, Xu WW (2016b) The establishment of an HE4-CLIA method and the combined analysis of HE4 and CA125 in ovarian cancer. J Clin Lab Anal 30:709–718
Zheng T, Tan T, Zhang Q, Fu JJ, Wu JJ, Zhang K, Zhu JJ, Wang H (2013) Multiplex acute leukemia cytosensing using multifunctional hybrid electrochemical nanoprobes at a hierarchically nanoarchitectured electrode interface. Nanoscale 5:10360–10368
Zhu C, Yang G, Li H, Du D, Lin Y (2015) Electrochemical sensors and biosensors based on nanomaterials and nanostructures. Anal Chem 87:230–249

Hyperbolic Metamaterial-Based Ultrasensitive Plasmonic Biosensors for Early-Stage Cancer Detection

7

G. Strangi, K.V. Sreekanth, and M. Elkabbash

7.1 Introduction

7.1.1 Biosensors: Background, Significance, and Challenges

Optical biosensors are analytical devices that detect bioanalytes immobilized on the device surface via monitoring a given characteristic of light, e.g., intensity, wavelength, and phase. Commonly, optical biosensors detect changes in the refractive index of their environment; that is, they are refractometers (Homola et al. 1999). In order to properly attribute changes in refractive index to bioanalytes, the sensing domain should be restricted to the immobilized bioanalytes only. To do so, refractometers based on guided waves have been used as biosensors, where the superstrate (upper surface) of the waveguide is functionalized to immobilize a certain bioanalyte. Guided waves are waves that are forced to propagate in a given plane of propagation. These waves have an evanescently decaying field in the superstrate and thus allow for selective detection to immobilized biomolecules.

Coupling to a waveguided mode usually require total internal reflection (TIR) inside an optical waveguide due to momentum (or phase) mismatch between the propagation constants inside and outside the guide. The momentum mismatch is

G. Strangi (✉)
Department of Physics, Case Western Reserve University,
10600 Euclid Avenue, Cleveland, OH 44106, USA

Department of Physics and CNR-NANOTEC UOS di Cosenza, University of Calabria,
Rende 87036, Italy
e-mail: giuseppe.strangi@case.edu

K.V. Sreekanth • M. Elkabbash
Department of Physics, Case Western Reserve University,
10600 Euclid Avenue, Cleveland, OH 44106, USA

P. Chandra et al. (eds.), *Next Generation Point-of-care Biomedical Sensors Technologies for Cancer Diagnosis*, DOI 10.1007/978-981-10-4726-8_7

directly related to the refractive index of both the waveguide and the superstrate. The field evanescently decays due to the momentum mismatch. The evanescent field is the sensing element, i.e., transducer. Particularly, changes in the refractive index of the superstrate can be sensed by the evanescent field which in turn changes in the propagation condition of the field inside the guide. This is because the propagation conditions inside the waveguide are dependent on the refractive index of the waveguide and the superstrate. By using the evanescent field for detection, it is possible to constrain the sensing domain to the immobilized analytes (Gauglitz and Proll 2008).

An important class of guided waves are surface plasmon polaritons (SPPs). In this case, the wave is guided at the interface between a metal and a dielectric under strict conditions. In general, a wave can be confined (and guided) at an interface between two media if the field is evanescently decaying in both media. For an SPP, the field is evanescently decaying inside the metal because of its high imaginary refractive index component which rejects the penetration of an electromagnetic wave. In order for the field to be evanescently decaying inside the dielectric, an SPP can only exist if the wave momentum is mismatched with that of the dielectric such that the field is totally internally reflected. Such SPPs are also refractometers that are extremely sensitive (Zeng et al. 2014; Borisoy and Wolfbeis 2008) to changes in the refractive index of the superstrate. Changes in the refractive index of the superstrate result in a change in the surface wave propagation constant, i.e., momentum component at the interface, as its interface confinement depends on the momentum mismatch with the dielectric superstrate.

In order to excite SPPs, a momentum matching scheme is employed due to its momentum mismatch with its superstrate. This can be achieved using a prism illuminated at its critical angle of TIR. In such case, the evanescent field from the total internally reflected beam is capable of exciting the surface wave due to its higher momentum which caused its TIR in the first place. Another method is to fabricate a diffraction grating on the metal supporting SPPs. The diffracted light of nonzero order has higher momentum than the incident light and is capable of exciting an SPP. In both cases, once an SPP is created a drop-in reflection (absorption mode) exists within a relatively narrow spectral range for a given angle of incidence. This mode is due to the excitation of an SPP where light is absorbed due to the losses inside the metal. Although SPP momentum is mismatched with the incident medium light momentum, the momentum coupler which couples the momentum to the SPP can out-couple it to the free space. For that reason, SPP excited via a momentum coupler is a leaky wave. To achieve complete absorption of light which provides a strong mode, light must be critically coupled to the SPP. Critical coupling takes place when the absorption rate (due to SPP losses in the metal) is equal to the leakage rate effectively canceling any leakage to free space. However, strong coupling of light to an SPP mode using a grating is not possible because for a given wavelength and angle of incidence only one diffraction order can couple to the surface wave and thus critical coupling is difficult to achieve.

The previous discussion sheds the light on the current problems facing traditional SPP biosensors.

First of all, using momentum coupling via Kretschmann configuration requires bulky optical elements, a prism, which makes it not suitable for point-of-care (POC) applications. While grating-coupled SPP sensors are not bulky, their sensitivity is an order of magnitude lower than prism-coupled SPPs. Additionally, while SPPs are only sensitive to their immediate superstrate as they employ their evanescent field for detection, an SPP sensor is a label-free sensor which cannot distinguish between the target molecules and parasitic molecules resulting in possible false-positive signals. Finally, while SPP sensors have shown their ability to detect highly diluted analytes, the sensitivity of these sensors is not high enough to detect dilute analytes of low molecular weights (<500 Da). If the sensitivity of SP sensors is increased, POC clinical evaluation, early cancer screening, and real-time diagnosis of diseases will be possible.

A further improvement in sensitivity was demonstrated using phase-sensitive interferometry techniques (Wu et al. 2004). Recent progress in microfabrication and nanofabrication has encouraged the development of novel label-free plasmonic biosensors, particularly metamaterials, which can overcome the limitations of conventional plasmonic sensors (Brolo 2012; Kravets et al. 2013).

7.1.2 Hyperbolic Metamaterials: Background and Opportunities

To address these concerns, a new sensing platform is needed. In order to circumvent the limits of traditional materials, research in metamaterials has intensified in the past decade. Metamaterials are a class of engineered materials that do not exist in nature and exhibit exotic and unusual electromagnetic properties. One of the most promising applications of metamaterials is in biosensing (Anker et al. 2008). Recently, metamaterial-based plasmonic biosensor that exhibits hyperbolic dispersion, commonly known as hyperbolic metamaterials (HMMs), has shown extreme sensitivity for low concentrations of low-molecular-weight bio-analytes (Kabashin et al. 2009; Sreekanth et al. 2016a, b). The dispersion relation of homogenous, isotropic materials is $k_x^2 + k_y^2 + k_z^2 = \frac{\omega^2}{c^2}$ which implies an elliptical dispersion. For a uniaxial anisotropic material, the dielectric response is given by the tensor $\ddot{\varepsilon} = \left[\varepsilon_{xx}, \varepsilon_{yy}, \varepsilon_{zz}\right]$, yielding a hyperbolic dispersion relation $\frac{k_x^2 + k_y^2}{\varepsilon_{zz}} + \frac{k_z^2}{\varepsilon_{xx}} = \frac{\omega^2}{c^2}$. HMMs are a class of artificial materials that enjoy hyperbolic dispersion (Fig.7.1a) because the out-of-plane dielectric component $\varepsilon_{zz}=\varepsilon_{\perp}$ has an opposite sign to the in-plane dielectric components $\varepsilon_{xx}=\varepsilon_{yy}=\varepsilon_{\parallel}$ (Poddubny et al. 2013; Hoffman et al. 2007; Sreekanth et al. 2013a, b). If ε_{zz} is negative, the material is type I HMM (Fig. 7.1(i) and (ii)), and if it is positive, the material is type II HMM (Fig. 7.1 (iii) and (iv)). To realize hyperbolic dispersion in different wavelength ranges of interest from the UV, visible, near-IR to mid-IR, appropriate choice of metal and metallic filling fraction should be used. For different HMM structures, the optical phase diagram in which effective medium response with the wavelength and the fill fraction of metal are shown in Fig. 7.1b (Cortes et al. 2013). It is evident that all structures show both type I and type II hyperbolic dispersion behavior in a broadband wavelength region. Importantly, Harish et al. experimentally realized the

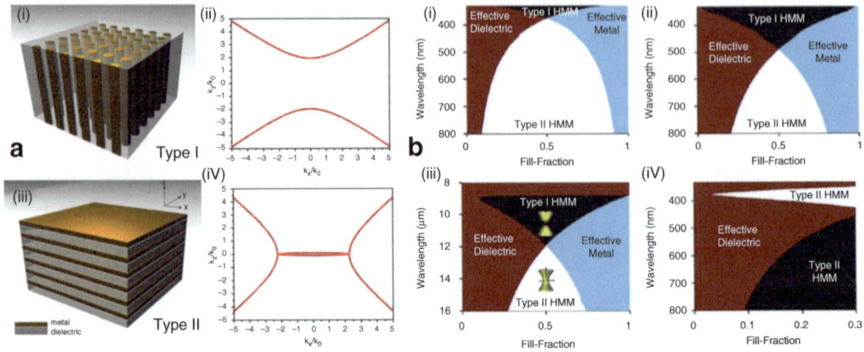

Fig. 7.1 (**a**) Schematic diagram of HMMs and their isofrequency diagrams: (i) 2D metal nanorod HMM (type I), (ii) the isofrequency contour of type I HMM ($\varepsilon_{xx}=\varepsilon_{yy}>0, \varepsilon_{zz}<0$), (iii) metal/dielectric planar HMM (type II), and (iv) the isofrequency contour of type II HMM ($\varepsilon_{xx}=\varepsilon_{yy}<0, \varepsilon_{zz}>0$). (**b**) Optical phase diagrams for (i) Ag/Al_2O_3 multilayer system, (ii) Ag/TiO_2 multilayer system, (iii) AlInAs/InGaAs multilayer system in the mid-IR region, and (iv) silver nanowires in an alumina matrix. (Fig. 7.1b is reproduced with permission from Cortes et al. (2013))

topological transition behavior (elliptical to hyperbolic dispersion) in HMM, by studying the short-living excitonic states of the chromophores placed in close proximity to the HMM (Krishnamoorthy et al. 2012). An important consequence of hyperbolic dispersion is that these materials can support waves propagating inside them with infinitely large momentum in the effective medium limit; on the other hand such waves are evanescent and decay away exponentially in superstrate (Cortes et al. 2013). Since these modes enjoy really high momentum, they can only be excited using a momentum coupler just like the case of SPPs. Coincidently, the first HMM used in biosensing was type I HMM consisting of an assembly of Au nanorods electrochemically grown into a substrate-supported, thin-film porous aluminum oxide template (Kabashin et al. 2009). By employing prism coupling scheme, they could excite a guided mode propagating that exhibited a refractive index sensitivity of 32,000 nm/RIU, exceeding the sensitivity of traditional SPP sensors by two orders of magnitudes. The origin of such high sensitivity is due to the modification of the coupling condition as the effective dielectric constant of the nanorods modifies by the presence of bioanalytes. In addition, the strength of the field associated with that mode is at maximum within the metamaterial itself. Given the nanoporous nature of the nanorods, bioanalytes are able to intercalate inside the metamaterial itself allowing for augmented sensitivity.

A more recent work used type II HMM that consists of multi-stack of metal/dielectric bilayers (Sreekanth et al. 2016a, b). Each bilayer supports a propagating SP wave coupled evanescently to its adjacent bilayer which in turn supports a propagating SP, allowing propagation inside the material as well as on the interface of each bilayer (Avrutsky et al. 2007). This high-momentum guided mode is called bulk plasmon polaritons (BPPs) indicating that it is propagating inside the bulk of the material while maintaining the properties of a propagating surface wave that polarize its dielectric superstrate. Such BPP enjoys extremely high momentum and is shown experimentally to be very sensitive to any change in the dielectric constant within the range of its evanescently decaying field in the superstrate. The excitation of these high-k modes was achieved by grating coupling method (Sreekanth et al. 2013a, b, 2014). The extremely high photonic local density of states (LDOS) of BPPs overcomes the leakage problem of coupled light. While a grating allows for coupling and out-coupling light to and from the HMM, light will preferentially propagate inside the HMM in the form of BPPs due to its high LDOS.

Furthermore, exciting BPPs by a grating coupler can excite multiple modes which satisfy the momentum matching condition. This is because a HMM can support multiple BPP modes. Particularly, a HMM with n bilayers supports $n-1$ BPP modes. Furthermore, a grating can efficiently diffract light over several diffraction orders. These modes have different sensitivity for bioanalytes with different molecular weights and thus provide an extra step towards transducer-based specificity.

7.2 Grating-Coupled HMM

7.2.1 Dispersion Analysis of HMM

In this section, we investigate the dispersion properties of surface transvers magnetic (TM) modes of HMM. As illustrated in the inset of Fig. 7.2a, the studied HMM geometry consisted of five bilayers of Au/SiO_2 stack. The calculation assumes that the light is incident from an isotropic medium, with a permittivity, $\varepsilon_1 = 1$, and a permeability, $\mu_1 = 1$ (region I), to an indefinite medium, HMM (region II). The principal dielectric tensor components of HMM (ε_α, ε_β and ε_γ) are in the ($\alpha\beta\gamma$) plane. The coordinate system of the HMM is considered in such a way that in-plane components are $\varepsilon_\alpha = \varepsilon_\beta = \varepsilon_{//}$ and out-of-plane component is $\varepsilon_\gamma = \varepsilon_\perp$. The TM mode surface wave dispersion relation is analyzed by considering a rectangular coordinate system (xyz). The magnetic field components of two regions are solved at the boundary, $z = 0$, and we then directly obtain the dispersion relation for surface TM mode, which is given by (Yan et al. 2007)

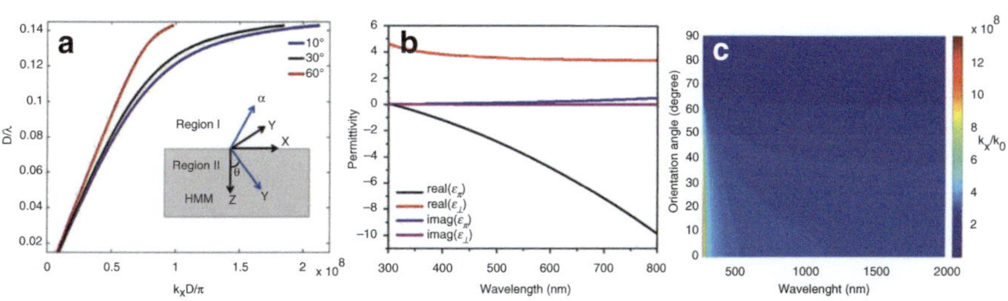

Fig. 7.2 (**a**) Controlling the dispersion diagram of HMM by changing the orientation of the boundary surface of HMM. Both axes are normalized with D, the period of the stack. A schematic of studied configuration is shown in the inset of (**a**). (**b**) Real and imaginary uniaxial dielectric permittivities of HMM obtained using EMT and (**c**) plot of the effective index of HMM with orientation angles and wavelengths

$$k_x = \omega\sqrt{\frac{\varepsilon_1\left(\varepsilon_1\varepsilon_{zz}\mu_y - \varepsilon_\perp\varepsilon_{//}\mu_1\right)}{\varepsilon_1^2 - \varepsilon_\perp\varepsilon_{//}}} \quad (7.1)$$

Since the HMM is assumed as nonmagnetic with $\mu_y = 1$ and unit cell dimension of the HMM belongs to effective medium approximations, we use effective medium theory (EMT) to calculate the principal dielectric tensor components of HMM (Cortes et al. 2013),

$$\varepsilon_{//} = \frac{t_m\varepsilon_m + t_d\varepsilon_d}{t_m + t_d} \quad (7.2)$$

$$\varepsilon_\perp = \frac{\varepsilon_m\varepsilon_d\left(t_m + t_d\right)}{t_m\varepsilon_d + t_d\varepsilon_m} \quad (7.3)$$

In Eqs. (7.2) and (7.3), (t_d, ε_d) and (t_m, ε_m) represent the thickness and dielectric permittivity of SiO_2 and gold, respectively. Drude free electron model $\varepsilon_m = 1 - \left(\frac{\omega_p^2}{\omega\left(\omega + i/\tau\right)}\right)$ was used to obtain the optical constants of gold, with ω_p, ω, and τ being the plasma frequency of gold, excitation frequency, and relaxation time, respectively. In the calculation, the dielectric permittivity of SiO_2 was set to be 2.1 (Palik 1985) and used thicknesses of thin films were 15 nm (gold) and 28 nm (SiO_2). The term ε_{zz}, in Eq. (7.1), is given by $\varepsilon_{zz} = \varepsilon_\perp\cos^2\theta + \varepsilon_{//}\sin^2\theta$, with θ being the orientations of the boundary surfaces of HMM (in Fig. 7.2a, the angle made by the z- and y-axes), and θ can be varied from $-\pi/2$ to $\pi/2$. There is a special condition for the existence of surface TM mode in the designed configuration, which is given as (Yan et al. 2007)

$$\left(\varepsilon_\perp/\varepsilon_1\right)\left(\mu_y/\mu_1\right) > 1 \text{ and } |\theta| < \theta_c \quad (7.4)$$

$$\text{where } \theta_c = \cos^{-1}\left(\left(\frac{\left(1 - \left(\varepsilon_{//}/\varepsilon_1\right)\left(\mu_y/\mu_1\right)\right)}{\left(\left(\mu_y/\mu_1\right)\left(\left(\varepsilon_\perp - \varepsilon_{//}\right)/\varepsilon_1\right)\right)}\right)^{1/2}\right) \quad (7.5)$$

The uniaxial dielectric permittivity components ($\varepsilon_{//}$ and $\varepsilon_\perp$) of HMM (five bilayers of gold and SiO_2) acquired using EMT are displayed in Fig. 7.2b. According to Fig. 7.2b, five bilayers of gold and SiO_2 show hyperbolic dispersion at $\lambda \geq 308$ nm. This is because the real parts of $\varepsilon_{//}$ and $\varepsilon_\perp$ provide negative and positive values at $\lambda \geq 308$ nm, respectively. In addition, the imaginary values of both components are positive. This response predicts the existence of surface TM mode in the designed HMM, which is the first mandatory condition as given in Eq. (7.4). We then plot (using Eq. (7.1)) the dispersion diagram of inverse wavelength versus wavevector projected into the interface plane of the HMM. As shown in Fig. 7.2c, the dispersion diagram is plotted for different orientations of the boundary surfaces of HMM. One can see that the orientation angle has a significant effect on the dispersion of

HMM. In particular, the wavevectors are independent of orientation angle at longer wavelengths, whereas wavevectors show extensive change with orientation angles at shorter wavelengths. Also, note that smaller angles provide higher wavevectors. In Fig. 7.2c, we plot the 2D color map of the effective index (n_{eff}=real(k_x)/k_0) of HMM, as a function of orientation angle and wavelength. It further confirms that higher effective index values are possible for shorter wavelengths and at smaller orientation angles. It should be noted that the effective index values are almost zero after 60°. This is because no surface TM modes exist in the HMM above 60°, which is the second condition, mentioned in Eq. (7.4). At 300 nm, the calculated critical angle (θ_c), an angle at which no more surface TM modes exist in HMM, is 65°. Our analysis shows that the dispersion properties of the HMM can be controlled by changing the orientation of the boundary surface of HMM.

7.2.2 Excitation of BPP Modes of HMM

The bulk plasmon modes of HMM were excited using a grating coupling technique. An essential condition for the excitation of surface mode is that the wavevectors of the grating diffraction orders must be greater than those of the incident light. In our case, multiple modes are possible for a given grating period that satisfy the momentum matching condition to bulk plasmon modes. Once the required condition is satisfied, energy is transferred to bulk plasmon modes. According to grating coupling condition, $k_{spp}^2 n_0^2 k_0^2 \sin^2\theta \pm 2n_0 m k_g k_0 \sin\theta \cos\phi + \left(mk_g\right)^2$, where θ, ϕ, n_0, $k_0 = 2\pi/\lambda$, and m are the incident grazing angle, azimuthal angle, refractive index of the incident medium, vacuum wavevector, and grating diffraction orders, respectively. $k_g = 2\pi/\Lambda$ is the grating wavevector with grating period, Λ. We considered p-polarized beam and $\phi = 0$ in our all experiments. Alternatively, this mechanism can be explained that it is possible to diffract light and produce a wide range of wavevectors entering into the HMM, by introducing a subwavelength periodic diffraction grating on top of the HMM. The coupling of generated wavevectors with the surface modes can happen due to the existence of impedance mismatch at the various openings (Yan et al. 2007).

By following this excitation mechanism, we designed and fabricated a grating-coupled HMM (GC-HMM), to excite the surface as well as bulk plasmon modes of HMM. GC-HMM is a combined structure of subwavelength metallic diffraction grating and HMM. As shown in Fig. 7.3a, the fabricated multilayer is an Au-TiO_2 HMM with individual layer thickness of 32 nm and 16 nm for Au and TiO_2, respectively. HMM was made by sequential deposition of the thin films of TiO_2 and Au on a glass substrate by means of RF sputtering of TiO_2 target and thermal evaporation of Au pellets, respectively. Electron beam lithography was used to pattern subwavelength 1D diffraction lines on top of the PMMA spin-coated HMM. We then directly deposited a 20 nm thick silver layer on the patterned PMMA grating. A SEM image of as-prepared pattern is shown in Fig. 7.3b. The measured average period and slit width of the grating pattern are 500 nm and 160 nm, respectively.

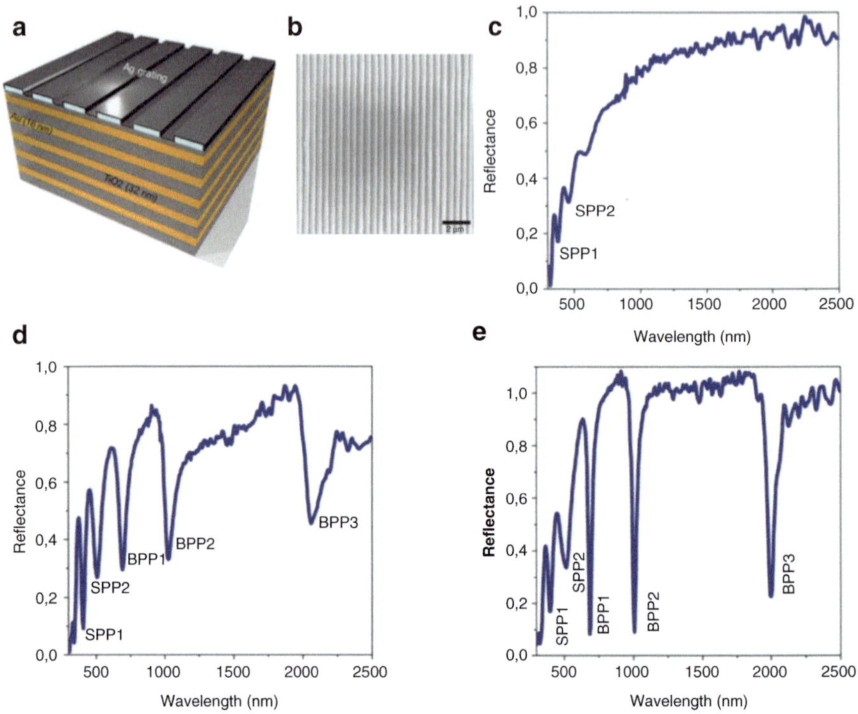

Fig. 7.3 Fabrication and characterization of GC-HMM. (**a**) A schematic of fabricated silver-diffraction grating-coupled Au-TiO2 HMM, (**b**) SEM image of sub-wavelength 1D diffraction grating, (**c**) reflectance spectrum of reference sample, (**d**) reflectance spectrum of control sample, and (**e**) reflectance spectrum of HMM

A very thin (10 nm) layer of TiO_2 was deposited between HMM and grating, in order to avoid the direct contact of grating with HMM.

By using a variable angle spectroscopic ellipsometer, the reflectance spectra of the samples as a function of excitation wavelength were acquired. In our measurements, the incident-grazing angle was fixed at 50° and the polarization of the beam was set to be TM polarized. A detailed experimental analysis is performed by comparing different samples. Result of a reference sample (Ag grating on a TiO_2/glass substrate) is shown in Fig. 7.3c. In this case, two low-quality factor resonances (reflectance minima) are observed at visible wavelengths (from 350 to 450 nm), which confirms the existence of surface plasmon polaritons in the sample. We then plot the reflectance spectrum of a control sample (Ag grating on a single bilayer of Au/TiO_2) in Fig. 7.3d. In comparison with reference sample, the control sample shows extra three resonances at longer wavelengths such as 700, 1000, and 2000 nm, in addition to SPP resonances. These three resonances confirm the existence of gap plasmon modes supported by the control sample. Our main result is shown in Fig. 7.3e, which is the reflectance spectrum of an Au-TiO_2 GC-HMM. In contrast to

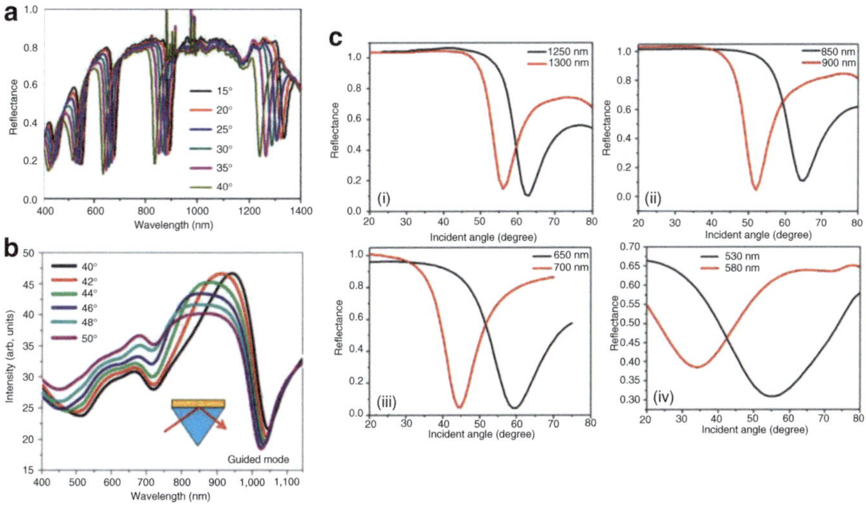

Fig. 7.4 Characterization of HMM-based biosensors. (**a**) Excitation of the BPP modes of Au-Al_2O_3 HMM (type II) via grating coupling, (**b**) Excitation of the BPP modes of Au nanorod HMM (type I) via prism coupling, and (**c**) angular reflectance spectrum of BPP mode band: (a) mode band 1 (1250–1300 nm), (b) mode band 2 (850–900 nm), (c) mode band 3 (650–700 nm), and (d) mode band 4 (530–580 nm) (Fig. 7.4b is reproduced with permission from Kabashin et al. (2009))

Fig. 7.3d, the resonance at wavelengths 700, 1000, and 2000 nm is very narrow and deeper. The high-quality factor mode at these wavelengths represents the highly confined bulk plasmon polariton (BPP) modes of GC-HMM. The resonance wavelength of these modes is red shifted when the number of bilayer is increased from 1 to 6, which is due to the coupling of individual gap plasmon modes. In order to prove that these modes are guided modes, we measured the reflectance of GC-HMM for different angles of incidence (Fig. 7.4a). We noticed that resonance wavelength of all modes was blue shifted as the incident angle was increased. This is because the modal index (effective index) of the modes varies with incident angle. Specifically, the excitation of the modes occurred at higher resonance angle, as the excitation wavelength in all BPP mode wavelength bands decreases. That means, resonance wavelength blue shifted with increase in angle of incidence. However, with decrease in excitation wavelength, the blue shift is decreased. Therefore, for higher incident angles, the reflectance dip became deeper and narrow. Also note that a smaller resonance angle variation is required to excite the BPP modes of longer wavelength band modes whereas a higher resonance angle variation is necessary for shorter wavelength band modes (Fig. 7.4c). This is the reason why blue shift decreased when the BPP mode wavelength is deceased from longer wavelength band mode to shorter wavelength band mode. In short, the excitation of the modes in type II HMM via a grating coupler results in multiple modes that satisfy the momentum matching condition.

7.3 Development of Plasmonic Biosensors

7.3.1 Design and Fabrication

Since the BPP modes show unique spectral and angular features in each BPP mode band, the spectacular properties of GC-HMM could be utilized to develop a potential plasmonic biosensor platform, working for a broad wavelength range from visible to near infrared. Therefore, we developed a miniaturized biosensor platform, which is a microfluidic channel-integrated GC-HMM. The refractive index sensing mechanism is based on the coupling condition between the grating surface modes and BPP modes. As the refractive index of the surrounding medium changes, one can observe a shift in resonance wavelength and resonance angle.

As illustrated in Fig.7.3a, the sensor device is a combination of a metallic diffraction grating and a HMM. For the realization of a biosensor, we fabricated a gold-Al_2O_3 HMM, which consists of 16 alternating thin films of gold and aluminum dioxide (Al_2O_3) with thickness 16 nm and 30 nm, respectively. HMM shows type II hyperbolic dispersion at $\lambda \geq 520$ nm. For exciting the BPP modes of HMM, a 2D sub-wavelength metallic hole array grating was designed and fabricated (Sreekanth et al. 2016a, b). To improve the coupling between HMM and grating, a very thin (10 nm) Al_2O_3 spacer layer was deposited on the HMM before making the grating. The patterned sub-wavelength hole array in PMMA has an average period of 500 nm and hole diameter of around 160 nm. Then, we sputtered a 20 nm thick gold on the sample to make gold diffraction grating so that a thiol-based surface chemistry can be used for biosensing.

7.3.2 Spectroscopic Characterizations

We acquired the reflectance spectra as a function of excitation wavelength as well as incident angle to determine the bulk plasmon modes in the fabricated Au-Al_2O_3 HMM. As discussed above, bulk plasmon modes have large modal indices because they are the entire family of gap plasmon modes of a multilayer. Therefore, strong mode confinement and shorter propagation length are possible, so that these modes can provide high quality (Q) factor. In Fig. 7.4a, we plot the reflectance spectra as a function of excitation wavelength for different angles of incidence. The deepest narrow resonance observed at 1120, 755, and 580 nm wavelength represents the highly confined BPP modes of the HMM and the calculated Q-factors for longer to shorter wavelength modes are 29.5, 26, and 23, respectively. It shows that Q-factor increases with the increase in excitation wavelength. This system would be useful for the design and development of multi-analyte biosensors since the mode at longest wavelength provides the maximum Q-factor and the mode at shortest wavelength provides the minimum Q-factor. However, type 1 HMM based on Au nanorods exciting BPPs by a prism coupler results in a single mode at near-infrared frequencies (Fig. 7.4b) (Kabashin et al. 2009). We further acquired the reflectance spectra of the HMM as a function of incident angle. As shown in Fig. 7.4c, with an increase of excitation wavelength in each BPP band, a decrease in coupling angle variation with a maximum for the shorter wavelength band and a minimum for the longer wavelength band is observed.

7.4 Calibration of Sensor Device

7.4.1 Interrogation Schemes

Since surface plasmon resonance sensors operate mainly on prism and grating coupling techniques, two interrogation schemes are commonly employed, which are spectral and angular scans. Therefore, we use both spectral and angular scan interrogation schemes to study the detection limit and sensitivity of the GC-HMM sensors.

7.4.2 Spectroscopic and Angular Scan

By injecting aqueous solutions of glycerol with different weight ratios into the sensor microchannel (sample volume $14 \times 2 \times 0.05\ mm^3$), first we determined the detection limit of the sensor in both spectral and angular scan interrogation schemes. We monitored the wavelength and angular shift as the sensing parameters to record the corresponding extremely small bulk refractive index changes. In Fig. 7.5a, c, we plot the

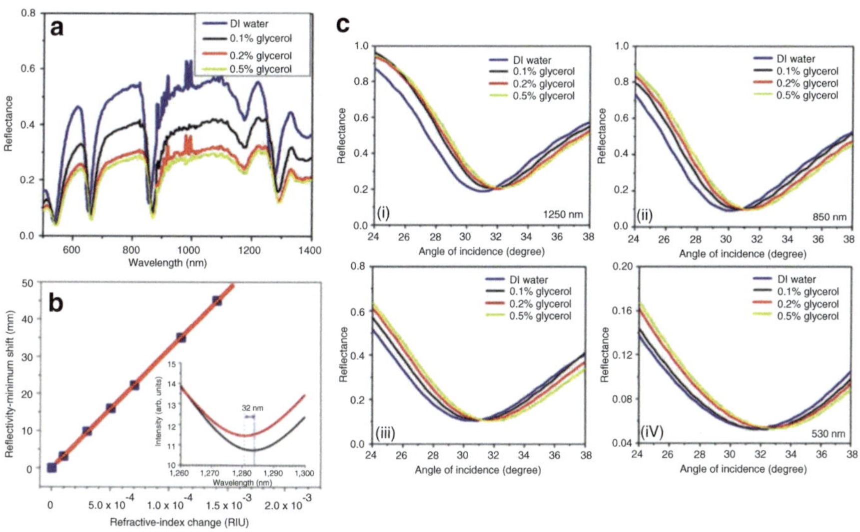

Fig. 7.5 (**a**), (**b**) Calibration test for spectral interrogation scheme. (**a**) Sensor (type II HMM) response by injecting different weight percentage of glycerol (0.1 to 0.5% w/v) in the channel. (**b**) Response of type I HMM-based sensor with changes of the refractive index of the environment using different glycerol–water solutions and inset shows the reflectivity spectrum modifications with the changes of the refractive index by 10^{-4} RIU. (**c**) Sensor (type II HMM) response in angular interrogation scheme by injecting different weight percentage of glycerol (0.1–0.5% w/v): (i) at 1250 nm, (ii) at 850 nm, (iii) at 650 nm, and (iv) at 530 nm (Fig. 7.5b is reproduced with permission from Kabashin et al. (2009))

reflectance spectra of the sensor with different weights of glycerol in distilled water (0.1–0.5% w/v), for spectral and angular scan interrogation schemes, respectively. As it is clear from Fig. 7.5a, resonance wavelength is red shifted and the quality factor of the resonance declined with increase in the weight ratio of glycerol concentration. Also, in Fig. 7.5c, a positive angular shift with increase in glycerol weight ratio was obtained. These results indicate the ability of the sensor to detect extremely small refractive index changes. One can also see that both spectral and angular shift varied between the different modes. This is because the transverse decay of the field in the superstrate strongly varies from one mode to another (Sreekanth et al. 2016a, b). The sensor has its highest performance at the lowest concentration because the shift varies nonlinearly with glycerol concentrations. In order to do a detailed sensitivity analysis, the spectral and angular sensitivity of the device at each BPP mode band was calculated, which is useful for the determination of the spectral and angular detection limit of the sensor device. For this purpose, 0.5% glycerol concentration is considered as a reference with bulk refractive index difference between DI water and 0.5% (w/v) glycerol in DI water is around 0.0006 (Weast 1987). A maximum spectral and angular sensitivities are recorded at 1300 nm, which are around 30,000 nm/RIU and 2500°/RIU, respectively. However, the minimum values are obtained for the mode at 530, which are 13,333 nm/RIU and 2333°/RIU for spectral and angular sensitivity, respectively. The figure of merit (FOM) of the sensor can be calculated using the expression, $(\Delta\lambda/\Delta n)\ (1/\Delta\omega)$, where $\Delta\lambda$ is the wavelength shift, Δn is the refractive index change, and $\Delta\omega$ is the full width of the resonant dip at half-maximum. This is an important biosensing parameter that determines the sensitivity with which very small wavelength changes can be measured, by considering the sharpness of the resonance. The FOM of the sensor for the mode from visible to NIR wavelengths are 206, 357, 535, and 590, respectively. In addition, the reported refractive index sensitivity and FOM of type 1 HMM-based sensor are 32,000 nm/RIU and 330, respectively (Fig. 7.5b) (Kabashin et al. 2009). Interestingly, there is a flexibility in the selection of a mode for the identification of specific biomolecules because the sensor shows different sensitivities and FOM for each mode. On the other hand, there is an option that higher sensitivity mode can be used for the detection of smaller molecular weight biomolecules and lower sensitivity mode can be used for the detection of heavier molecular weight biomolecules.

7.5 Biomolecule Sensing

In this section, we demonstrate the proof-of-concept biosensing experiments using both spectral and angular interrogation schemes. Here, we only consider the mode at 1300 nm because it shows maximum sensitivity.

7.5.1 Smaller Molecule Detection

Small molecules such as D-biotin (molecular weight = 244 Da) are selected as a bioanalyte to demonstrate the capabilities of the sensor device for the detection of lower molecular weight biomolecules (<500 Da). Importantly, it is a model system for small-molecule compounds such as cancer-specific proteins, hormones, and therapeutics. Another motivation is that circulating tumor cells overexpress small proteins (<500 Da) very early that are extremely difficult to detect with currently available sensing technologies. To study the binding kinetics of biotin, the well-known streptavidin-biotin affinity model was used (Sreekanth et al. 2016a, b). The refractive index change caused by the capture of biotin at the sensor surface was monitored as the resonant wavelength and angular shift in the reflectance spectra. We used a single-injection procedure (in batch mode) in which different concentrations (100 pM–10 μM) of biotin in phosphate-buffered saline (PBS) were injected into the sensor microchannel. Experiment is performed in such a way that the reflectance spectra of the sensor were recorded after a reaction time of 40 min for each concentration and PBS was then introduced into the channel to remove the unbound and weakly attached biotin molecules. In Fig. 7.6a, b, we show the responses of the sensor in spectral and angular interrogation schemes, by injecting different concentrations of biotin. Due to the increase in refractive index by the capture of biomolecules, a red wavelength shift in wavelength scan and a positive angular shift in angular scan were obtained. A nonlinear variation of the response with biotin concentration was observed in both interrogation schemes (Fig. 7.6c, d). In Fig. 7.6e, we provide the binding kinetics of 10 pM biotin in PBS with time progress. A discrete step in resonance wavelength over time is evident from the plot, which confirms the binding of biotin molecules on the sensor surface with time progress. Note that the variability in the step size is due to statistical

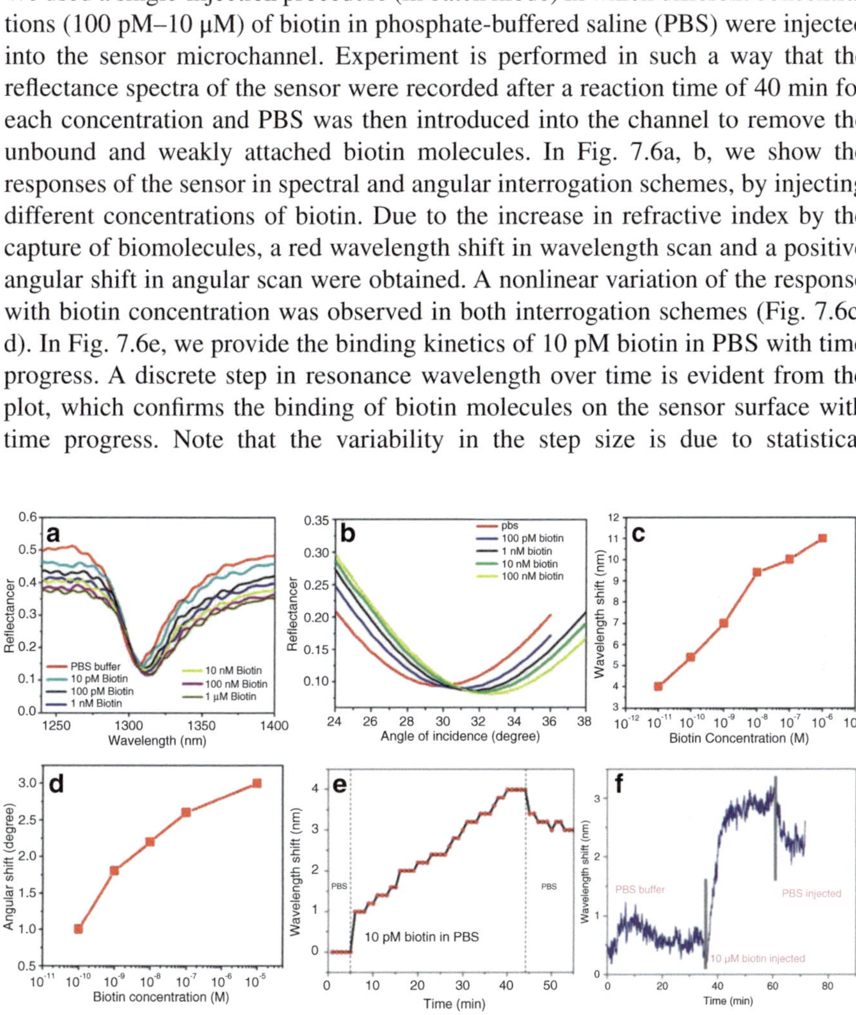

Fig. 7.6 Response of the HMM-based biosensors for small molecules binding in spectral and angular interrogation schemes: (**a**) Reflectance spectra as a function wavelength for different concentrations of biotin in PBS with angle of incidence of 30° (**b**) Reflectance spectra as a function of incident angle at 1250 nm for different concentrations of biotin in PBS. The variation of (**c**) wavelength shift and (**d**) angular shift with different concentrations of biotin. (**e**) The variation of wavelength shift in the presence of 10 pM biotin in PBS over time for type II HMM. (**f**) The variation of wavelength shift in the presence of 10 μM biotin in PBS over time for type I HMM (Fig. 7.6f is reproduced with permission from Kabashin et al. (2009))

fluctuations where larger or smaller numbers of binding events occurred. It is also reported that type I HMM-based sensor can detect biotin at concentrations as low as 10 μM using a standard analytical chemistry protocol (Fig. 7.6f). It shows that sensitivity can be increased further by six orders of magnitude by using type II HMM.

7.5.2 Heavier Molecule Detection

To demonstrate close to single-molecule sensitivity of the sensor device, we then performed experiments to detect high-molecular-weight macromolecules such as cowpea mosaic virus (CPMV, molecular weight = 5.6×10^6 Da), in the absence of surface functionalization. CPMV is a plant virus, which was prepared by following the method described in Wen et al. (2015). The importance of the detection of CPMV is that it is a safe model system to mimic infectious disease. We used varying concentrations (1 fM–1 nM) of CPMV prepared in PBS. The response of the device is monitored by recording the reflectance spectrum after a reaction time of 20 min and PBS was injected into the channel to remove unbound and weakly attached CPMV. As shown in Fig. 7.7a, b, a large wavelength and angular shift even for 1 fM CPMV were recorded, and the sensor saturated very quickly due to the very high molecular weight of CPMV. By recording the reflectance spectrum over time, the binding of 100 fM CPMV on the sensor surface is investigated (Fig. 7.7c). An increase in angular shift with time was obtained, which is due to statistical fluctuations where larger or smaller number of binding events occur. In this study, the angular resolution of the instrument was set to be 0.5° and the spectrum was recorded every 60 s. For simplicity, only limited spectra are shown in Fig. 7.7c. It is evident that the angular shift almost saturated at 15–20 min. No considerable shift was noticed after 20 min. A maximum angular shift of 3° was obtained for 100 fM CPMV after 20 min. Our future work direction is to detect viruses in complex media (blood or other body fluids and tissue samples), which is important for the commercial implementation of the device.

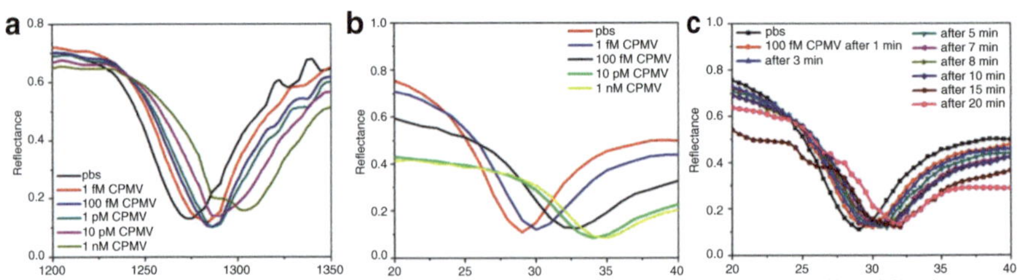

Fig. 7.7 Response of higher molecular weight macromolecules to the binding of CPMV: (**a**) Response of the sensor in wavelength interrogation scheme for different concentrations of CPMV in PBS. (**b**) Response of the sensor in angular interrogation scheme for different concentrations of CPMV in PBS. (**c**) Proof-of-concept heavier molecule binding experiment using CPMV

7.5.3 Quantification of Absorbed Molecule on the Sensor Surface

The quantification of the absorbed biomolecules on the sensor surface is very important to study the binding kinetics of molecules. Therefore, we have analyzed the sensitivity of the wavelength and angular shift to the number of adsorbed molecules on the sensor surface. For different concentrations c of biotin in PBS, we considered the saturation values of the wavelength ($\Delta\lambda$) and angular ($\Delta\theta$) shift, which occurs approximately after 45 min. In the sensing region, the number of bound molecules $N(c)$ depends on the shift of the resonance wavelength and angle. Note that we cannot directly measure the precise value of $N(c)$, but we can reliably estimate an upper bound $N_{max}(c)$ based on the sensor parameters, such that the actual $N(c) \leq N_{max}(c)$ at any concentration. The estimated upper bound is $N_{max}(c) = 8.4\ c \times 10^{15}\ M^{-1}$. In Sreekanth et al. (2016a, b), we showed that there was a nonlinear relationship between $N_{max}(c)$ and $\Delta\lambda(c)$, and $N_{max}(c)$ and $\Delta\theta(c)$, the shift in resonance wavelength and angle at concentration c. By using a phenomenological double-exponential fitting function, this behavior could be accurately represented, and we find an analogous functional relationship between $N_{max}(c)$ and $\Delta\theta(c)$, and $N_{max}(c)$ and $\Delta\lambda(c)$, which is as follows:

$$N_{max} = A_1\left(e^{\Delta\lambda/\beta_1} - 1\right) + A_2\left(e^{\Delta\lambda/\beta_2} - 1\right) \tag{7.6}$$

$$N_{max} = A_1\left(e^{\Delta\theta/\beta_1} - 1\right) + A_2\left(e^{\Delta\theta/\beta_2} - 1\right) \tag{7.7}$$

where A_1, A_2, β_1, and β_2 are fitting parameters. In the case of very few adsorbed molecules, this function could show limiting behavior at small $\Delta\lambda$ and $\Delta\theta$, which should be linear. In Fig. 7.8, we plot the experimental N_{max} versus $\Delta\lambda$ and $\Delta\theta$ results for biotin and solid lines represent the best fits of Eqs. (7.6) and (7.7). It should be noted that at the smallest probed concentration (1 fM) of CPMV, the estimated value for N_{max} is approximately eight adsorbed particles, which indicates that our sensor is close to the single-particle detection regime.

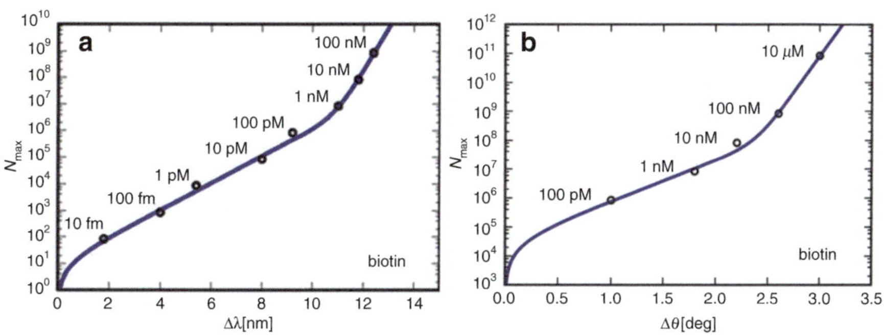

Fig. 7.8 Experimentally estimated maximum number of biotin molecules for the mode located at 1280 nm. (**a**) N_{max} versus wavelength shift $\Delta\lambda(c)$ and (**b**) N_{max} versus angular shift $\Delta\theta(c)$

7.6 Summary and Outlook

In summary, we discussed the basic principles of optical biosensing via SPPs and its shortcoming. Hyperbolic metamaterials which are artificial materials that enjoy hyperbolic dispersion of light support a different type of SPPs called BPPs which are surface waves that can also propagate through the bulk of the material and enjoy extremely high momentum and local density of photonic states. Such HMMs have reported to exhibit extreme sensitivity for angular and spectral sensitivities reaching 32,000 nm/RIU and 2500°/RIU, respectively, and figure of merit of 590, allowing for detection of low-molecular-weight biomolecules at low concentration.

Type I HMM biosensor based on Au nanorod arrays is a promising system as it can be grown on approximately large areas (~ 2 cm^2) and allow it to enjoy high fabrication tolerance due to the sub-wavelength nature of its components. In addition, there is a lot of room for improving the metamaterial parameters such as the height and surface area of the nanorods in order to increase the number of attached bioanalytes. In addition, transducer-based enhanced selectivity is possible by modifying the nanorod period such that only bioanalytes with certain sizes can permeate through the material and experience the enhanced local field.

On the other hand, due to the thin metal film nature of type II HMMs based on metal-dielectric bilayers it is possible to excite its bulk modes using the grating coupling configuration which overcomes the bulkiness problem of biosensors relying on prisms stated in the introduction. Furthermore, the existence of multiple BPP modes and multiple diffraction orders allows for multimodal excitation of resonant modes at different angles of incidence and wavelength ranges.

Such multimodal differential response offers a promising opportunity to design an assay for the selective detection of higher and lower molecular weight biomolecules in highly diluted solutions, i.e., transducer-based selectivity. Both miniaturization and transducer-based selectivity are two important features for new-generation lab-on-a-chip sensors.

Furthermore, embedding of the HMM in biocompatible matrices (for example, silk) could also allow for transcutaneous implantation. The transducer-based selectivity is important in sensing applications because it brings specificity properties to ligand-free sensors and adds specificity features to ligand-functionalized sensors. Although the reported sensitivity of HMMs is extremely high, it is still capable of operating within a wide range of molecule concentrations for low-molecular-weight analytes: from highly diluted concentrations (10 fM) to the more concentrated multi-analyte samples (1 μM). A further improvement in sensitivity of lower molecular weight biomolecules is possible by using differential phase shift (Zeng et al. 2015) and Goos-Hanchen shift (Yin and Hesselink 2006) interrogation schemes. These will be the future work directions.

An effective point-of-care biosensor should also be affordable. One main drawback is the reliance on nanofabrication techniques to create the grating. In future, efforts will be dedicated to lithographically free nanophotonic platforms. Not only would the fabrication costs be dramatically reduced, but also the sensing area could be significantly enlarged, allowing for a large-area biosensor and multichannel sensing platform.

Acknowledgments We acknowledge the support of the Ohio Third Frontier Project "Research Cluster on Surfaces in Advanced Materials (RC-SAM) at Case Western Reserve University." We also acknowledge the support of the MORE Center at Case Western Reserve University and the GU malignancies program of the Case Comprehensive Cancer Center.

References

Anker JN, Hall WP, Lyandres O, Shah NC, Zhao J, Van Duyne RP (2008) Biosensing with plasmonic nanosensors. Nat Mater 7:442–453

Avrutsky I, Salakhutdinov I, Elser J, Podolski V (2007) Highly confined optical modes in nanoscale metal-dielectric multilayers. Phys Rev B 75:241402

Borisoy SM, Wolfbeis OS (2008) Optical biosensors. Chem Rev 108:423

Brolo AG (2012) Plasmonics for future biosensors. Nat Photon 6:709–713

Cortes CL, Newman W, Molesky S, Jacob Z (2013) Quantum nanophotonics using hyperbolic metamaterials. J Opt 14:063001

Gauglitz G, Proll G (2008) Strategies for label-free optical detection. Adv Biochem Eng Biotechnol 109:395–432

Hoffman AJ, Alekseyev L, Howard SS, Franz KJ, Wasserman D, Podolskiy VA, Narimanov EE, Sivco DL, Gmachi C (2007) Negative refraction in semiconductor metamaterials. Nat Mater 6:946–950

Homola J, Yee SS, Gauglitz G (1999) Surface plasmon resonance sensors: review. Sensors Actuators B 54:3

Kabashin AV, Pastkovsky S, Hendren W, Wurtz GA, Atkinson R, Pollard R, Podolskiy VA, Zayats AV (2009) Plasmonic nanorod metamaterials for biosensing. Nat Mater 8:867–871

Kravets VG, Schedin F, Jalil R, Britnell L, Gorbachev RV, Ansell D, Thackray B, Novoselov KS, Geim AK, Kabashin AV, Grigorenko AN (2013) Singular phase nano-optics in plasmonic metamaterials for label-free single-molecule detection. Nat Mater 12:304–309

Krishnamoorthy HNS, Jacob Z, Narimanov E, Kretzschmar I, Menon VM (2012) Topological transitions in metamaterials. Science 336:205–209

Palik ED (1985) Handbook of optical constants of solids. Academic, Orlando

Poddubny A, Iorsh I, Belov P, Kivshar Y (2013) Hyperbolic metamaterials. Nat. Photon. 7:948–957

Sreekanth KV, De Luca A, Strangi G (2013a) Negative refraction in graphene-based hyperbolic metamaterials. Appl Phys Lett 103:023107

Sreekanth KV, De Luca A, Strangi G (2013b) Experimental demonstration of surface and bulk plasmon polaritons in hypergratings. Sci Rep 3:3291

Sreekanth KV, De Luca A, Strangi G (2014) Excitation of volume plasmon polaritons in metal-dielectric metamaterials using 1D and 2D diffraction gratings. J Opt 16:105103

Sreekanth KV, Alapan Y, Elkabbash M, Ilker E, Hinczewski M, Gurkan UA, De Luca A, Strangi G (2016a) Extreme sensitivity biosensor platform based on hyperbolic metamaterials. Nat Mater 15:621–627

Sreekanth KV, Alapan Y, Elkabbash M, Wen AM, Ilker E, Hinczewski M, Gurkan UA, Steinmetz NF, Strangi G (2016b) Enhancing the angular sensitivity of plasmonic sensors using hyperbolic metamaterials. Adv Opt Mater 4:1659

Weast RC (1987) CRC handbook of chemistry and physics. CRC Press, Boca Raton

Wen AM, Infusino M, De Luca A, Keernan DL, Czapar AE, Strangi G, Steinmetz NF (2015) Interface of physics and biology: engineering virus-based nanoparticles for biophotonics. Bioconjug Chem 26:51–62

Wu SY, Ho HP, Law WC, Lin C, Kong SK (2004) Highly sensitive differential phase sensitive surface plasmon resonance biosensor based on the Mach-Zehnder configuration. Opt Lett 29:2378–2380

Yan W, Shen L, Ran L, Kong JA (2007) Surface modes at the interfaces between isotropic media and indefinite media. J Opt Soc Am A 24:530–535

Yin X, Hesselink L (2006) Goos-Hanchen shift surface plasmon resonance sensor. Appl Phys Lett 89:261108

Zeng S, Baillargeat D, Hod HP, Yong KT (2014) Nanomaterials enhanced surface plasmon resonance for biological and chemical sensing applications. Chem Soc Rev 43:3426–3452

Zeng S, Sreekanth KV, Shang J, Yu T, Chen CK, Yin F, Baillargeat D, Coquet P, Ho HP, Kabashin AV, Yong KT (2015) Graphene-gold metasurface architectures for ultrasensitive plasmonic biosensing. Adv Mater 27:6163

SERS-Based Biosensors as Potential Next-Generation Point-of-Care Cancer Diagnostic Platforms

8

Shounak Roy and Amit Jaiswal

Abbreviations

4MBA	4-Mercaptobenzoic acid
A1AT	Alpha-1-antitrypsin
Ab	Antibody
AFP	Alpha-fetoprotein
AgNP	Silver nanoparticle
AuNP	Gold nanoparticle
AuNR	Gold nanorod
AuNR@Ag	Silver-coated Gold nanorod
AuNS	Gold nanosphere
CEA	Carcinoembryonic antigen
CTC	Circulating tumour cells
Cy5	Cyanine 5
DNA	Deoxyribonucleic acid
DTNB	5,5-Dithiobis(2-nitrobenzoic acid)
EGF	Epidermal growth factor
ELISA	Enzyme-linked immunosorbent assay
EpCAM	Epithelial cell adhesion molecule
ErbB2	Erythroblastic leukaemia viral oncogene homolog 2
ERL	Extrinsic Raman label
HCPCF	Hollow core photonic crystal fibre
HER2	Human epidermal growth factor receptor 2
HGNs	Hollow gold nanosphere
IGF-1	Insulin-like growth factor 1

S. Roy • A. Jaiswal (✉)
BioX Center and School of Basic Sciences, Indian Institute of Technology Mandi, Kamand, Mandi 175005, Himachal Pradesh, India
e-mail: j.amit@iitmandi.ac.in

P. Chandra et al. (eds.), *Next Generation Point-of-care Biomedical Sensors Technologies for Cancer Diagnosis*, DOI 10.1007/978-981-10-4726-8_8

iMS	Inverse molecular sentinel
LDA	Linear discriminate analysis
LDR	Ligase detection reaction
LOD	Limit of detection
LSPR	Localized surface plasmon resonance
mAb	Monoclonal antibody
MB	Magnetic beads
MCH	Mercaptohexane
MGITC	Malachite green isothiocyanate
miRNA	microRNA
MNP	Magnetic nanoparticle
MS	Molecular sentinel
MSC	Molecular sentinel-on-chip
MSTP	Methyl sulphanyl thiophenol
MUC4	Mucin protein 4
NiFe@Au	Gold shell-coated nickel-iron alloy magnetic nanoparticle
pAb	Polyclonal antibody
PATP	p-Aminothiophenol
PCA	Principal component analysis
PCI	Plasmonic coupling interference
PEG	Polyethylene glycol
PNTP	p-Nitrothiophenol
PSA	Prostate-specific antigen
RNA	Ribonucleic acid
RR	Raman reporter
SEHGNS	Silica-encapsulated hollow gold nanosphere
SERRS	Surface-enhanced resonance Raman scattering
SERS	Surface-enhanced Raman scattering/spectroscopy
Si	Silica
SNP	Single-nucleotide polymorphism
TSNW	Triangular shaped nanowire
VEGF	Vascular endothelial growth factor
WHO	World Health Organization

8.1 Introduction

Cancer, which is hailed as the "*Emperor of all maladies*," is the leading cause of millions of deaths worldwide every year. Approximately, 1.7 million new cancer cases are diagnosed each year in the United States (American Cancer Society. Cancer Facts & Figures 2015). According to the WHO, the world witnessed around 8.2 million deaths in 2012 due to cancer and related illness (Cancer, Fact sheet No: 297; WHO). For the past 5–6 decades, governments, pharma companies and private institutions throughout the world have invested billions and trillions of dollars for carrying out extensive research to understand the basics of cancer and to develop new technologies for diagnosing and treating cancer, with the sole goal of achieving a universal "cure" for cancer. Remarkable progress in basic cancer research over the years has strengthened our understanding of what cancer is and presently the research is focussed on developing diagnostic tools to detect cancer at its nascent stage. Despite the success that has been achieved over the past few years in developing new-generation anticancer therapeutics and treatment strategies, the mortality rate due to cancer and cancer-related diseases has not decreased significantly. This led scientists worldwide to switch the focus of research more towards developing novel tools and techniques for screening and detection of cancer at an early stage when the disease has not metastasized. It is very important to understand that the earlier the detection of a disease, the higher is the chance of survival and recovery. The outcome of a specific cancer treatment strategy greatly depends on the stage of the cancer at the time of diagnosis. In most cases, majority of the patients coming to the clinic turn up with late or terminal stages of cancer, where even the most advanced of the treatment strategies fail to save the patient. According to the Cancer Stat Fact Sheet of National Institute of Health, USA, patients diagnosed with stage 3 or stage 4 breast cancers, cervical cancers, ovarian cancers, prostate cancers and colon cancers have a 5-year survival rate of 28%, 16%, 28%, 28% and 13%, respectively, while for those diagnosed at stage 1 or stage 2 of the same five cancers, the 5-year survival rates are 99%, 91%, 92%, 100% and 90%, respectively (Cancer Stat Fact Sheets, NIH). This clearly shows the importance for developing novel screening technologies for specific, sensitive, cost-effective and accurate early detection of cancer in a short period. The tremendous potential of the various novel cancer treatment modalities that have been mentioned before can only be realized to its full extent when the cancers are detected early while it is still localized. This will not only decrease the mortality rate, but will also reduce morbidity and costs (Etzioni et al. 2003).

Thus, scientists have come up with new technologies for early diagnosis of cancer. Detection of biomarkers in body fluids such as blood, plasma, serum, urine and saliva, and also in cells and tissues, has emerged as a promising early cancer detection technology (Satish et al. 2016). According to the National Cancer Institute, a biomarker is "a biological molecule found in blood, other body fluids, or tissues that is a sign of a normal or abnormal process, or of a condition or disease" (NCI Dictionary of Cancer Terms, National Cancer Institute). The biochemical nature of a biomarker is diverse ranging from proteins, nucleic acids, peptides and antibodies up to collection of proteomic and metabolomic signatures that play very important roles for detection, monitoring and subsequent treatment of a variety of diseases including cancer (Henry and Hayes 2012; Mayeux 2004). This whole new field of early cancer diagnosis based on detection of circulating biomarkers in body fluids is referred to as "liquid biopsies" (Cree 2015). With progress in the field of nanotechnology and sensor technology, novel nanobiosensors have been developed for more specific and selective detection of cancer biomarkers (Altintas and Tothill 2015; Brian and Shaker 2011). Several technologies have been developed for designing biosensors based on fluorescence (Vuori et al. 1991; Xu et al. 1992), chemiluminescence (Brown et al. 1985), electrochemical assay (Hayes et al. 1994; Pallela et al. 2016), gel-based techniques, protein microarrays (Cho et al. 2006) and enzymatic methods (Butler 2000), which can be used as diagnostic tools for the detection of disease biomarkers (Brian and Shaker 2011). However, most of these biosensor platforms have certain limitations. These techniques suffer from low reproducibility, large sample volume requirement, poor specificity, low sensitivity, elaborate and tedious procedures, time consumption and expensiveness and above all most of these techniques fail to get translated into point-of-care diagnostic platforms for rapid and early detection of diseases. Colorimetric and absorbance-based assays such as ELISA, which is the most favoured and readily used immunoassay, suffer from low detection limits, limited multiplexing capabilities and issues related to variations in results with changes in pH, buffer and ionic strength (Butler 2000). Fluorescence-based detection platforms provide high sensitivity as compared to other methods, but the broad spectral overlaps and frequent quenching of the fluorescence significantly affect the performance and output of these assays (Strianese et al. 2012). To overcome these limitations scientists are working on the development of reliable, cheap, easy-to-perform diagnostic platform that has the potential to simultaneously detect multiple target analytes (multiplexing) with high sensitivity and specificity in a short period of time so that it can be used as a potential point-of-care diagnostic tool. Surface-enhanced Raman scattering or SERS has emerged as a powerful alternative spectroscopy-based detection technique having the advantages of high sensitivity, excellent multiplexing capabilities, ease of performance and reliability (Kyle et al. 2011). In the past 2–3 decades, SERS has been extensively studied and researched and SERS-based techniques have been developed for detection and monitoring of a range of target analytes starting from environmental pollutants ranging up to biomarkers for diagnostic purposes (Stephen and George 2009). Detection of any disease at an early stage increases the chance of recovery and also provides the physician an opportunity to properly treat and monitor the disease. Further, simultaneous detection of multiple biomarkers is considered to be crucial for an accurate and specific diagnosis of a complex disease such as cancer. The high specificity and excellent multiplexing capabilities of SERS have proved to be highly beneficial for early detection of cancer and thus a large number of scientific reports and articles have come up in the past few years focussing on the development of novel diagnostic platforms based on SERS (Marc et al. 2013). This chapter discusses about several SERS-based diagnostic methodologies developed in the last few years for detection of cancer.

8.2 Raman Spectroscopy and SERS

The wide application of Raman spectroscopy in the field of biosensing and diagnostics (Mahadevan and Richards 1996) that have been observed in the last few decades is due to the unique property of this optical spectroscopic technique to rapidly detect and identify different chemical compounds and materials based on the interaction of electromagnetic radiations with the material. When light (monochromatic radiation) is incident on a sample, it interacts with the sample in a manner such that the incident photons can either get reflected or absorbed or scattered. In case of scattering, the incident photons can get scattered elastically or inelastically. Majority of the photons get scattered elastically in which there is no change in energy or frequency between the incident photon and scattered photon (Rayleigh scattering) (Amer 2010). However, there is a small fraction of photons (1×10^{-7}) which gets scattered at frequencies or energies different from that of the incident photons (Amer 2010). Such type of scattering is referred to as inelastic scattering and the phenomenon is known as Raman effect, named after its discoverer Sir C. V. Raman. The Raman-scattered photons can have either lower energy (Stokes Raman scattering) or higher energy (anti-Stokes Raman scattering) as compared to incident photons depending on the nature of the vibrational state of the molecule. This difference in the energy between the incident photons and the inelastically scattered photons corresponds to the vibrational energy levels of the scatterer and is characteristic of the nature or type of bonds present in the scattering molecule, thereby unique for each molecule. As a result, every molecule that is capable of inelastic scattering has its own unique vibrational spectrum referred to as the Raman fingerprint, which gives information regarding the identity and molecular structure of a sample (Gremlich and Yan 2001). In comparison to fluorescence spectrum, a Raman spectrum consists of peaks that are discrete, highly resolved and 10^1–10^2 times narrower in width as compared to fluorescent emission peaks, which makes Raman spectroscopy an excellent and powerful detection tool for high-level multiplexing operations. In spite of all these properties and advantages over other conventional detection platforms, Raman spectroscopy found very few practical applications specifically in the field of biological sciences. This is because Raman scattering is a very weak scattering process with only one in 10^6–10^{10} incident photons getting scattered inelastically (Willard et al. 1974). The Raman scattering cross sections of molecules are usually 14 orders of magnitude lower than that of fluorescent molecules, thereby resulting in very weak scattering intensity (Willard et al. 1974). However, this limitation of Raman spectroscopy was overcome by the discovery of a new optical phenomenon known as surface-enhanced Raman scattering or spectroscopy (SERS), when in 1974 scientists for the first time observed the intensification of Raman signals of pyridine adsorbed on roughened silver electrodes (Fleischmann et al. 1974). Further research into this field along with development of new disciplines like nanotechnology and plasmonic science revealed that the weak intensity of Raman signals could be significantly enhanced to a detectable range by placing the molecular species of interest on or near the surface of nanoroughened metals or plasmonic nanostructures, especially made up of gold or silver (Nie 1997; Ko et al. 2008; Jaiswal et al. 2014). This process of enhancing the intensity of Raman spectra of a molecule by using a metal surface is known as SERS (Doering et al. 2007; Haynes et al. 2005; Sharma et al. 2012). This enhancement in the intensity of Raman spectrum is the outcome of the enhanced electric field caused due to the localized surface plasmon resonance (LSPR) of metal nanostructures. On exposure to electromagnetic radiation, the conduction band electrons at the surface of the metal nanostructure start to oscillate due to absorption of energy. This oscillation is referred to as localized surface plasmon. When the frequency of localized surface plasmons matches the frequency of the incident electromagnetic radiation, it results in resonance which is known as LSPR. This LSPR increases the absorbance as well as scattering properties, and also enhances the electromagnetic field at the surface of the nanostructure (Jeanmaire 1977). The target molecules or Raman labels that are adsorbed on the surface of the plasmonic nanostructure or are placed in close proximity to the surface experience this enhanced electromagnetic field, thereby resulting in increase in the intensity of the Raman scattering (Willets and Van Duyne 2007; Jensen et al. 2008). In addition to enhanced electromagnetic field, charge transfer mechanisms (Albrecht and Creighton 1977) and surface-enhanced resonance Raman scattering (SERRS) (Schatz et al. 2006) also contribute towards signal intensification by SERS. SERS have the potential to cause signal enhancements by 10^8–10^{14} orders of magnitude which is extremely helpful in the detection of target analytes at nano- or picomolar range.

8.3 SERS-Based Biosensor Platforms

Basically, there are two modes or platforms for SERS-based biosensors: an extrinsic/indirect SERS platform and an intrinsic/direct SERS platform (Driscoll et al. 2013). Extrinsic or indirect SERS-based biosensing refers to the detection of one or multiple target analytes via excitation of strong Raman scattering reporter molecules (Raman tags) that are coupled to the surface of target analytes (Driscoll et al. 2013). Detection of the strong Raman signals of the Raman tags indirectly indicates the presence of the target analyte to which it is bound. On the other hand, intrinsic or direct SERS-based biosensing involves detection of target analytes directly by measuring the excited Raman signals of the target analytes without the use of a Raman label. Plasmonic nanostructures made up of gold or silver are used as the label-free SERS-active substrates which significantly enhance the Raman signal of the analytes adsorbed on its surface, thereby facilitating direct detection of the analytes (Driscoll et al. 2013).

Immunoassay is the most preferred and widely used assay technique for development of biosensor platforms. An immunoassay consists of antibodies as capture or detection agents that specifically bind to its cognate antigen (a target protein) and help in its detection. Majority of the SERS-based biosensors are based on immunoassay platforms that can be performed either as a solution-based colloidal immunoassay or as a solid substrate/chip-based immunoassay (Porter et al. 2008). In a solution-based colloidal immunoassay, the target analyte or protein of interest is detected using gold nanoparticles functionalized with a specific antibody and a Raman label. This complex is then bound by a metallic nanoparticle conjugated to another antibody specific to the same target protein, thereby forming sandwich immunocomplex aggregates, which can then be probed with a laser directly in solution or can be pelleted down first and then excited with a laser for measurement of the Raman spectra of the reporter molecule. On the other hand, solid substrate/chip-based sandwich immunoassay platforms (Fig. 8.1) consist of a solid substrate or chip fabricated with a plasmonic material such as gold or silver. This plasmonic substrate is then immobilized with capture antibodies that specifically bind to the target antigen. Upon antigen binding to the immobilized capture antibodies, these complexes are probed using SERS-active reporter plasmonic nanoparticles labelled with a Raman tag and conjugated with an

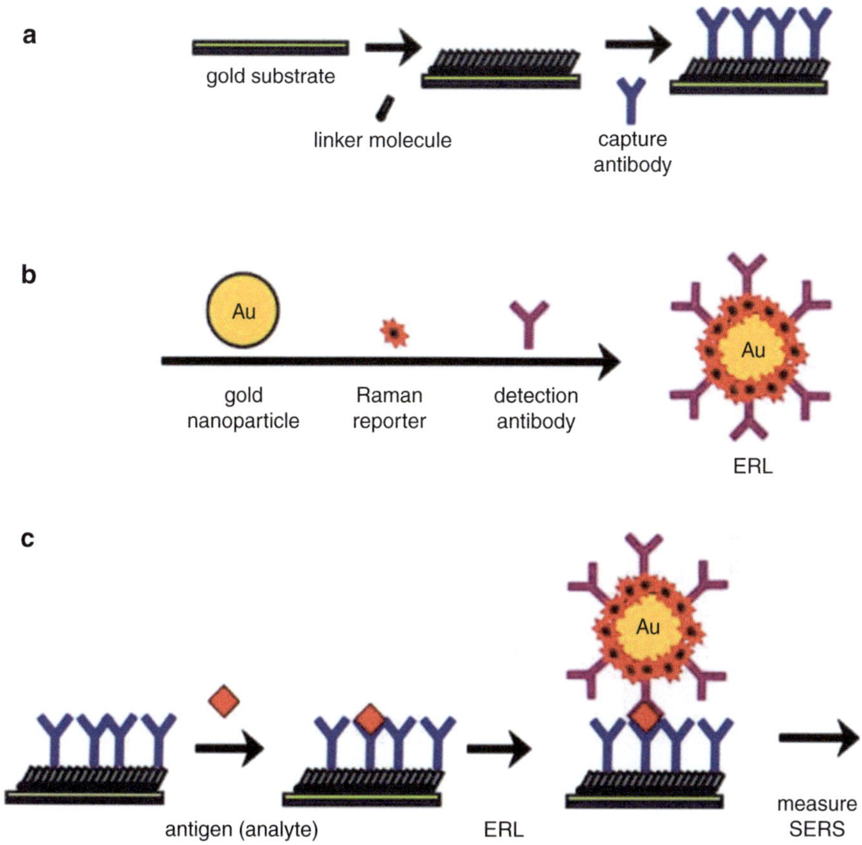

Fig. 8.1 A schematic showing the outline of a general SERS-based sandwich immunoassay platform: (**a**) a capture substrate to specifically extract and concentrate antigens from solution; (**b**) surface-functionalized gold nanoparticles (ERLs) to bind to captured antigens selectively and generate intense SERS signals; and (**c**) sandwich immunoassay with SERS readout. Reproduced with permission from Wang et al. (2011a, b). Copyright 2011 American Chemical Society

antibody specific for the same target antigen, thereby forming sandwich immunocomplexes on the solid plasmonic substrate, which can then be interrogated with a laser for SERS measurements (Porter et al. 2008).

8.4 SERS-Based Immunoassays for Detection of Different Cancer Biomarkers

SERS-based immunoassay works on the principle of binding of antibody-tagged SERS probe to specific biomarkers and detection using Raman spectroscopy. The several biomarkers associated with different cancer types and the SERS platform used for individual biomarkers are summarized in Table 8.1. The detailed description for individual SERS-based biosensors is discussed below.

Table 8.1 SERS-based immunoassays for rapid and sensitive detection of cancer biomarkers and CTCs

Types of cancer	Biomarkers/target analyte	Detection platform	Biosensor design	Limit of detection (LOD)	References
Breast cancer	VEGF	SERS-based sandwich immunoassay	AuNanostar@Si@Ab@RR as SERS nanotag with capture Ab-immobilized Au triangle nanoarray as solid substrate	–	Wang et al. (2011a, b)
	HER2	SERS-based imaging of receptors on cell surface of MCF-7, MDA-MB-468, KPL4 and SK-BR-3 human breast cancer cell lines	SERS nanotags: AuNR@Ab@RR AuNS@Ab@RR AgNP@Ab@RR	–	Krishnamurti and Silverman (2014), Park et al. (2009), Lee et al. (2009)
	EGF, ErbB2, IGF-1 receptors		HGNS@Si@Ab@RR	–	Yang et al. (2011)
Lung cancer	CEA	SERS-based colloidal immunoassay	HGNS@mAb@RR as SERS nanotag with MB@pAb as supporting substrate	1–10 pg/mL	Thomson et al. (1969)
		Magnetic focussing-coupled SERS-based microfluidic platform	NiFe@Au@Ab as capture probe AuNP@Ab@RR as SERS nanotag	0.1pM	Chon et al. (2011)
	CEA, AFP	Multiplexed SERS-based colloidal immunoassay	HGNS@mAb@RR as SERS nanotag with MB@pAb as supporting substrate	5 ng/mL (CEA) 20 ng/mL (AFP)	Chon et al. (2009)
Liver cancer	AFP	SERS-based sandwich immunoassay	AuNP@Ab@RR as SERS nanotag with capture Ab-immobilized gold-coated glass slide as solid substrate	100 pg/mL	Zhao et al. (2013)

	AFP, A1AT	SERS-based sandwich immunoassay combined with gradient microfluidic device	HGNS@Ab@RR as SERS nanotags with gold microarray wells embedded on a glass substrate as detection platform	0–1 ng/mL	Wang et al. 2013a, b)
		Multiplexed SERS-based sandwich immunoassay combined with hollow core photonic crystal fibre	AuNP-PEG@Ab@RR as SERS nanotags with target biomarker-immobilized hollow photonic fibre as solid substrate	-	Lee et al. (2012)
Prostate cancer	PSA	SERS-based sandwich immunoassay	AuNP@Ab@RR as SERS nanotag with capture Ab-immobilized gold-coated glass slide as solid substrate	1 pg/mL	Lilja et al. (1991)
			Peptide-plasmonic nanocrescent hybrid coupled with RR as both SERS nanotag and solid substrate	420 nM	Grubisha et al. (2003)
Pancreatic cancer	MUC-4	SERS-based sandwich immunoassay	AuNP@Ab@RR as SERS nanotag with capture Ab-immobilized gold-coated glass slide as solid substrate	33 ng/mL (total cell protein lysate in PBS)	Moniaux et al. (2007)
Circulating tumour cells:					
Breast cancer cells	SKBR3 cell line spiked into whole blood	SERS-based colloidal immunoassay	AuNP@Ab@RR as SERS nanotag with MNP@Ab as capture particles	50 cells/mL	Sha et al. (2008)
	EpCAM, CD44, keratin 18 and IGF-I receptor β on single-breast cancer cells in unprocessed human blood	SERS-based colloidal immunoassay	AuNR@Ag@Ab@RR as SERS nanotags	–	Nima et al. (2014)
Leukaemia	Chronic lymphocytic leukaemia cells	SERS-based colloidal immunoassay	AuNP@Ab@RR as SERS nanotag	–	Nguyen et al. (2010)

8.4.1 SERS-Based Immunosensor for VEGF Biomarker Detection

Li et al. demonstrated the detection of vascular endothelial growth factor (VEGF) biomarker from the clinical samples of breast cancer patients using an ultrasensitive SERS-based plasmonic immunosensor (Li et al. 2013a, b). VEGF is a very important biomarker for tumour-associated angiogenesis and upregulation of the expression of VEGF or its receptors has been reported in different types of human cancer (Pradeep 2005). The immunosensor consisted of Au triangle nanoarray chip as the SERS-active substrate to which the capture antibodies (anti-VEGF) were immobilized. SERS-active sandwich nanoparticle probes were constructed by conjugating a Raman reporter malachite green isothiocyanate (MGITC) to a gold nanostar followed by SiO_2 coating. The detection antibodies were then conjugated onto these Au@Raman reporter@SiO2 sandwich nanoparticle probes. In the presence of the target biomarker protein in clinical sample, the biomarker (VEGF) gets sandwiched between the capture antibody on the nanoarray platform and the detection antibody on the SERS probes, thereby creating a 3D hierarchical structure with a confined plasmonic field (Fig. 8.2). This confined plasmonic field leads to the generation of

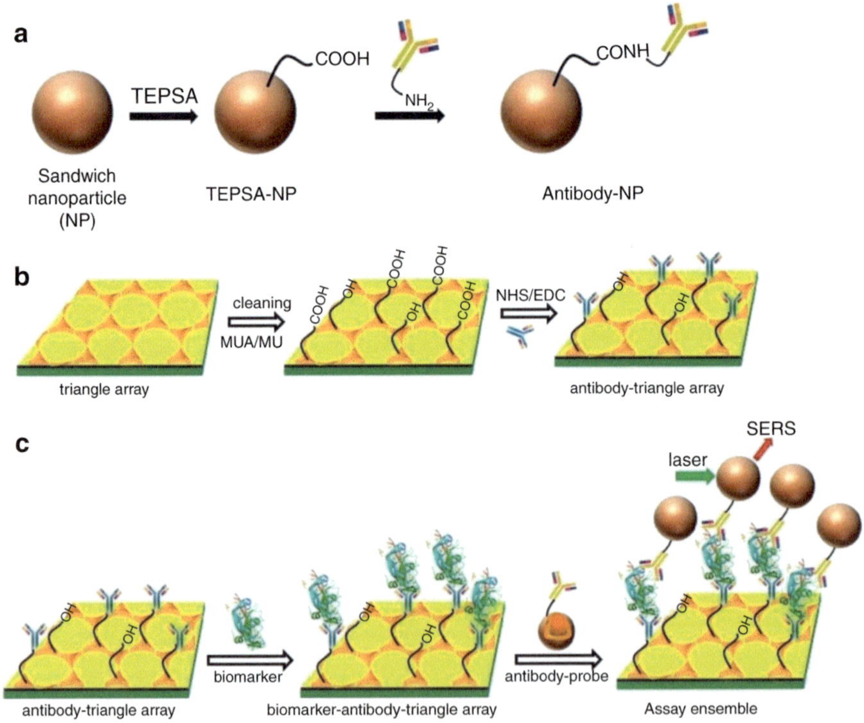

Fig. 8.2 Schematic illustration of conjugation of the (**a**) SERS probe (sandwich nanoparticle) to the detection antibody and (**b**) Au triangle nanoarray chip to the capture antibody. (**c**) Schematic illustration of the operating principle of SERS immunosensor for biomarker detection. The structure of the VEGF biomarker is created by PyMOL with a four-digit code: 1VPF. Reproduced with permission from Li et al. (2013a, b). Copyright 2013 American Chemical Society

high-density electromagnetic "hot spots" between the Au triangle platform and sharp-tipped Au nanostars upon light excitation. As a result, the Raman reporter molecules that are sandwiched between these structures experience the enhanced electromagnetic field and ultimately lead to significant enhancement in the Raman intensity facilitating biomarker detection.

8.4.2 SERS-Based Immunosensor for Simultaneous Detection of p53 and p21

Simultaneous detection of p53 and p21 is of great clinical importance for early diagnosis and monitoring of several types of cancer. The tumour-suppressor gene p53 is mutated in almost 50% of human cancers (Sherr 2004) and plays a very crucial role in regulating cell cycle progression. In normal circumstances, when the p53 gene is not mutated, it activates the expression of p21 gene which is a cyclin-dependant kinase inhibitor that too is associated with normal regulation and control of cell cycle (El-Deiry 1993; Anttila et al. 1999). However, during pathological conditions such as cancer when the p53 gene is mutated, it fails to activate p21 as a result of which the expression of p21 is lowered and the cell cycle becomes deregulated. Hence, the levels of expression of both of these genes are very crucial in keeping the cell cycle in check and can act as indicators or markers for cancer. Studying the expression levels of p21 gene has gained a lot of attention from the scientific community because it directly sheds light on the status of the p53 gene, whether it's mutated or not. Studies have shown that p53 overexpression in the absence of p21 expression (i.e., p53 positive/p21 negative) could be used as a useful marker in the diagnoses of epithelial ovarian cancer (Werness et al. 1999), colorectal cancer (Viale et al. 1999) and hepatocellular carcinoma (Shi et al. 2000). So, from a diagnostic point of view, detection platforms capable of multiplexed detection of both p53 and p21 protein in a single assay from a clinical sample are readily sought after for early cancer diagnosis. Wu et al. have reported a highly reproducible and sensitive SERS-based immunoassay platform for simultaneous detection of p53 and p21 protein in human serum (Wu et al. 2013). They developed two sets of SERS nanotags using gold@silver core-shell nanorods (Au@Ag NRs) as the plasmonic nanostructure conjugated to two different Raman reporter molecules 4-mercaptobenzoic acid (4MBA) and 5,5-dithiobis (2-nitrobenzoic acid) (DTNB). The 4MBA-Au@Ag NRs were conjugated with rabbit anti-p53 antibody and DTNB-Au@Ag NRs were conjugated with rabbit anti-p21 antibody for the detection of p53 and p21, respectively (Fig. 8.3). When mixture of p53 and p21 proteins in different combinations was used, the sensor was able to achieve a sensitivity of 1 pg/mL.

8.4.3 Breast Cancer

Human epidermal growth factor receptor 2, also known as HER2, is known to be over-expressed on the surface membrane of breast cancer cells, thus making this receptor protein a powerful biomarker for breast cancer detection and diagnosis

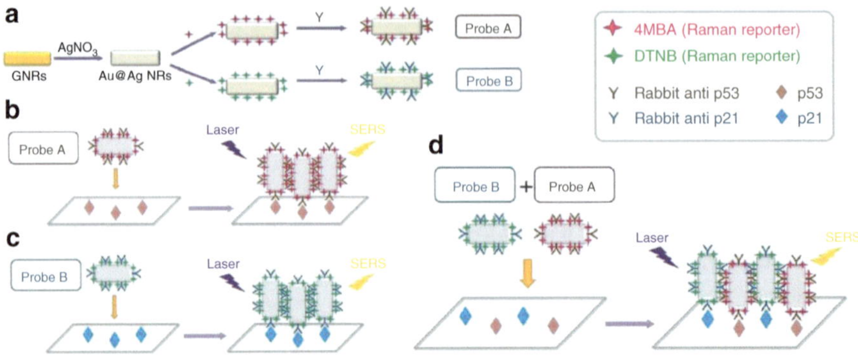

Fig. 8.3 (**a**) Simplified schematic procedure for the fabrication of Au@Ag NRs SERS probes. Schematic procedure for (**b**) separate detection of p53, (**c**) separate detection of p21 and (**d**) simultaneous detection of p53 and p21. Reproduced with permission from Wu et al. (2013). Copyright 2013 The Royal Society of Chemistry

(Krishnamurti and Silverman 2014). SERS-based detection of HER2 on breast cancer cells has been reported where antibody-conjugated gold nanorods (Park et al. 2009) and nanospheres (Lee et al. 2009) coupled with Raman reporter were used as the SERS-active nanoprobes for imaging and detection of HER2 on the surface of MCF7 cells (human breast adenocarcinoma). In another study, Yang et al. reported the use of anti-HER2 antibody-conjugated silver nanoparticles coupled with a Raman reporter p-mercaptobenzoic acid for detection of HER2 on breast cancer cells (Yang et al. 2011). Lee et al. reported a SERS-based multiplexed cellular imaging method for simultaneous imaging, detection and quantification of different breast cancer phenotypic markers (Lee et al. 2014). They developed SERS nanotags using antibody-conjugated silica-encapsulated hollow gold nanospheres (SEHGNs) coupled with different Raman reporters. They used MGITC, RBITC and RuITC for detection and quantification of three different breast cancer biomarkers namely epidermal growth factor (EGF), ErbB2 and insulin-like growth factor-1 (IGF-1) receptors, on the surface of MDA-MB-468, KPL4 and SK-BR-3 human breast cancer cell lines, respectively. The SERS-tagged SEHGNs were able to label the different cancer cells specifically based on the distribution of the individual biomarker proteins enabling quantification of biomarkers and identification of cancer cell phenotype.

8.4.4 Lung Cancer

Carcinoembryonic antigen (CEA) is a well-known biomarker for lung cancer, whose levels are found to be elevated in the blood of chain smokers (Begent 1984). In addition to lung cancer, high levels of CEA have also been reported in the serum of individuals suffering from colorectal, gastric, pancreatic, breast and medullary thyroid carcinomas (Thomson et al. 1969). So, rapid and sensitive detection of this

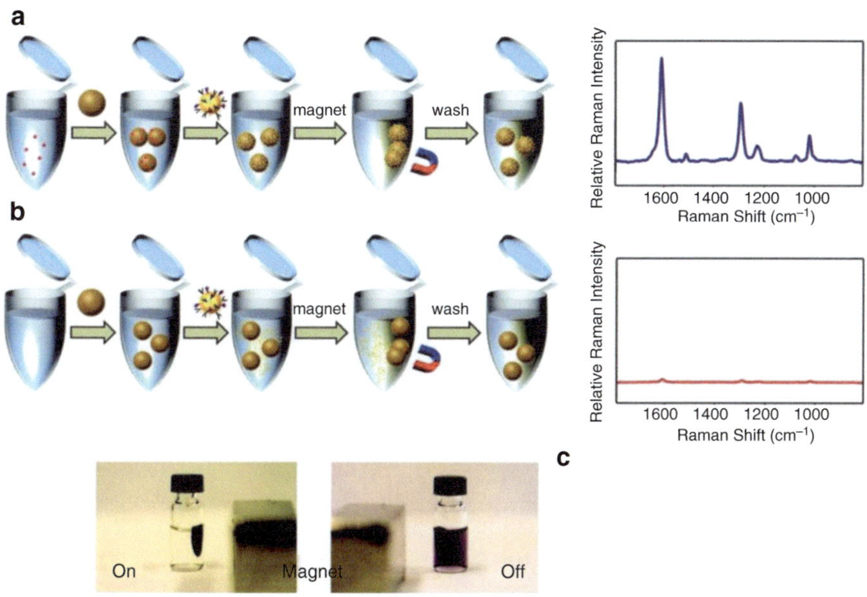

Fig. 8.4 Schematic illustration of immunoassay processes and corresponding Raman spectra: (**a**) with CEA antigens and (**b**) without CEA antigens. (**c**) Photographs showing suspended magnetic beads attracted to the wall of a microtube by a bar magnet: with CEA (*left*) and without CEA (*right*). Reproduced with permission from Chon et al. (2009). Copyright 2009 American Chemical Society

biomarker is very crucial for accurate diagnosis and management of a number of cancer types. H Chon et al. have reported the development of a SERS-based immunosensor using antibody-conjugated hollow gold nanospheres and magnetic beads for rapid, specific and sensitive detection of CEA from clinical samples (Chon et al. 2009). Instead of a solid substrate platform for immobilization of the immunocomplex, the authors developed a colloidal assay platform without any immobilization. Magnetic beads conjugated to polyclonal anti-CEA antibodies were used as supporting substrates and Raman reporter-coupled hollow gold nanospheres conjugated to monoclonal anti-CEA antibodies were used as SERS-active probes. Simple mixing of these magnetic beads and SERS probes with the clinical sample containing the target biomarker resulted in the formation of sandwich immunocomplexes in solution, which were then separated using a magnet for SERS measurements (Fig. 8.4). Formation of immunocomplex resulted in the enhancement of the Raman signal of the reporter due to SERS by multiple closely placed hollow gold nanospheres on the surface of magnetic beads, thereby leading to the specific detection of the target biomarker. Absence of any biomarker did not let any reporter nanoparticles to bind with the magnetic beads and thus no SERS was observed. This colloidal based SERS immunoassay platform reported a LOD of about 1–10 pg/mL for CEA, which is 100–1000 times more sensitive than ELISA. In a follow-up study by the same group, the authors reported multiplexed detection of two lung cancer biomarkers, CEA (LOD of 5 ng/mL) and AFP (LOD of 20 ng/mL), simultaneously in patient serum using the same colloidal based SERS immunoassay platform (Chon et al. 2011). The quick assay time of less than 1 h, ease of performance and high specificity and accuracy make this SERS-based colloidal immunoassay platform very suitable and ideal for being developed into a reliable point-of-care diagnostic tool for early cancer detection.

In a recent study, Li et al. reported a SERS-based microfluidic platform with a LOD of 0.1pM for rapid and sensitive detection of CEA by employing bifunctional nanocomposite probes for magnetic focussing-coupled SERS detection (Li et al. 2015). Gold shell-coated nickel-iron alloy magnetic nanoparticle (NiFe@Au) conjugated to anti-CEA capture antibody was used as the capture probe and Raman label-coupled Au nanoparticle conjugated to anti-CEA detection antibody was used as the SERS-active probe. In the presence of target biomarker CEA, formation of sandwich immunocomplexes between capture and detection probes took place which was then magnetically focussed to a particular spot on a microfluidic platform for further SERS measurement (Fig. 8.5). Magnetic focussing of the plasmonic immunocomplexes to a specific spot caused aggregation of a number of AuNPs resulting in formation of "hot spots" that ultimately enhanced the Raman signal of the reporter for successful detection of the biomarker up to 0.1 pM level.

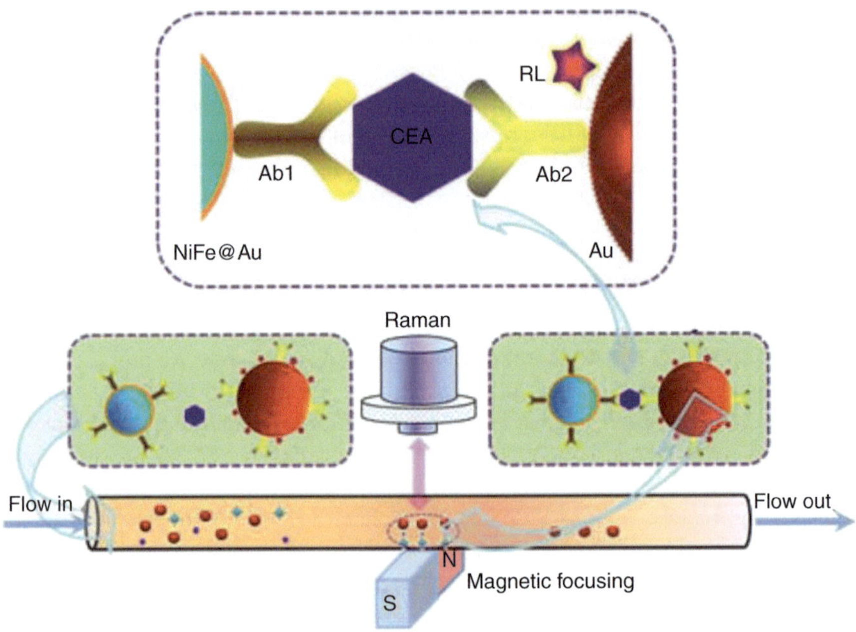

Fig. 8.5 Illustration of SERS detection of cancer biomarker CEA using functional nanoprobes consisting of Au-coated NiFe magnetic nanoparticle (NiFe@Au), Ab1, capture antibody; Ab2, detection antibody; and RL, Raman label. Reproduced with permission from Li et al. (2015). Copyright 2015 American Chemical Society

8.4.5 Liver Cancer

Alpha-fetoprotein (AFP) is considered as an important biomarker for hepatocellular carcinoma (Zhao et al. 2013). An elevated level of AFP in blood is an indicator of liver cancer. So, accurate and early detection of AFP in clinical samples is of utmost importance for determining the correct stage of the cancer as well as for monitoring the progression of the disease upon treatment. In a study, Wang et al. reported a solid substrate-based SERS immunoassay for rapid and specific detection of AFP at very low concentrations (Wang et al. 2013a, b). They utilized Raman reporter-labelled AuNPs conjugated to anti-AFP antibody as a SERS-active probe (antiAFP-MBA-AuNP) that formed a sandwich immunocomplex upon incubation with AFP-precoated glass slide modified with gold colloid. The assay could detect AFP at a concentration as low as 100 pg/mL, thus showing the high sensitivity of this SERS-based method for biosensing applications. Combining SERS-based immunoassay platforms with other detection technologies have also been studied and reported for detection of cancer. One such example is the development of a diagnostic chip by combining the features of gradient microfluidic device with SERS-based immunoassay. Lee et al. developed a programmable, highly reproducible and fully automated gold array-embedded gradient microfluidic chip for fast and highly sensitive detection of AFP using SERS-based immunoassay (Lee et al. 2012). The gradient microfluidic device was developed using a glass substrate embedded with gold microarray wells, which served as the detection platform for the SERS-based immunoassay. The surfaces of the gold wells were immobilized with anti-AFP monoclonal antibody for capture of target biomarker AFP. Hollow gold nanospheres (HGNs) labelled with a Raman reporter MGITC and conjugated with anti-AFP polyclonal antibody were used as the SERS nanoprobes for sandwich immunocomplex formation. The target antigen AFP and SERS nanoprobes were injected sequentially into the microfluidic channels through two separate inlets and were allowed to incubate for specific periods of time. Upon incubation, Raman spectra of the formed immunocomplexes were measured using Raman spectrometer for the purpose of detection and quantitation (Fig. 8.6). It was observed that when the target biomarker protein AFP was not present, there was a very weak SERS signal indicating the presence of few HGNs in the solution due to nonspecific binding. However, with sequential increase in AFP concentration (0–10 ng/mL), the intensity of the Raman peaks in the 1560–1650 cm^{-1} range increased concomitantly, thereby demonstrating highly specific detection as well as excellent sensitive quantitation of AFP with a LOD of 0–1 ng/mL. This SERS-based opto-fluidic sensor with high sensitivity and quick assay time of less than 1 h can be coupled with a handheld portable Raman spectrometer for rapid multiplexed detection of different cancer biomarkers in a point-of-care diagnostic setup.

Similarly, Dinish et al. developed a highly sensitive biosensing platform for multiplexed detection of cancer biomarkers by combining SERS with hollow core photonic crystal fibre (HCPCF) (Dinish et al. 2013). HCPCFs, due to its improved light confinement properties, allows for much better interaction between the guided laser and the target analyte. This results in improved sensitivity of this platform when

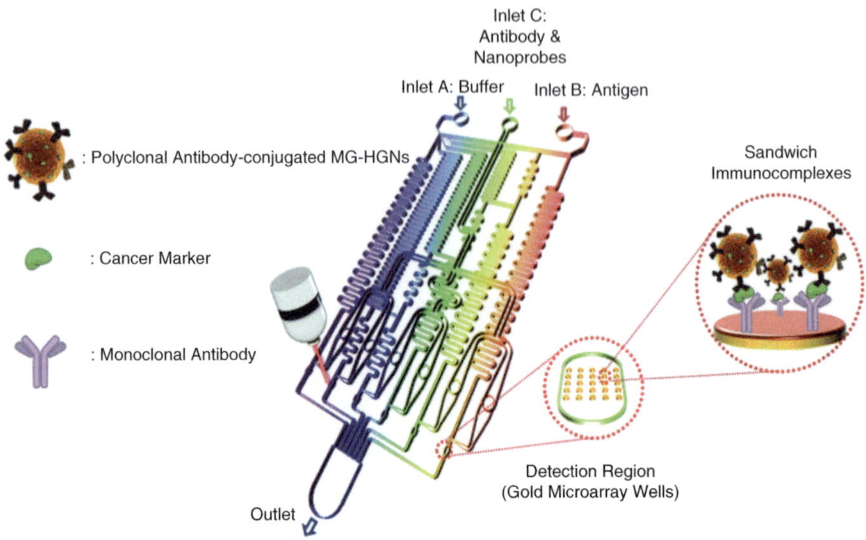

Fig. 8.6 Layout of a gold array-embedded gradient chip for the SERS-based immunoassay. The illustrations in the enlarged circles represent the formation of immunocomplexes on the surface of 5 × 5 round gold wells embedded in the gradient channel. Reproduced with permission from Lee et al. (2012). Copyright 2012 The Royal Society of Chemistry

coupled with SERS for detecting bioanalyte even at very low concentration. Using this platform, the authors reported the multiplexed detection of two hepatocellular carcinoma biomarkers, alpha-fetoprotein (AFP) and alpha-1-anti-trypsin (A1AT), in a very low sample volume of 20 nL. The biomarkers were immobilized on the inner surface of the hollow photonic fibre and were then detected using biomarker-specific antibody-conjugated SERS nanotags. Two different Raman reporters were used in the SERS nanotags for simultaneous detection of two different biomarkers. Multiplexed detection of biomarkers provides great help in accurate diagnosis and monitoring of complex diseases like cancer, where stage of detection is very crucial in determining the fate of the patient.

8.4.6 Prostate Cancer

Prostate-specific antigen (PSA) has been extensively used as a biomarker for the diagnosis and monitoring of prostate cancer in men (Lilja et al. 2008). PSA is a 33 kDa serine protease produced by the prostate gland whose normal level in blood lies between 4 and 10 ng/mL. It has been found to exist either in a free form (fPSA) or in a complex with alpha 1-anti-chymotrypsin (PSA-ACT complex) in the serum (Lilja et al. 2008). The ratio of the concentration of these two forms in serum is clinically very important because the chances of getting prostate cancer increase with decrease in fPSA levels (Stenman et al. 1991; Lilja et al. 1991). Thus, monitoring the level of different forms of PSA specifically would enable early diagnosis of prostate cancer. Working towards this goal, Grubisha et al. developed a solid substrate-based SERS immunoassay platform for rapid and highly sensitive detection of PSA at concentrations as low as 1 pg/mL in human serum (Grubisha et al. 2003). A gold colloid-coated glass slide immobilized with anti-PSA capture antibodies served as the solid assay substrate onto which SERS probes consisting of Raman reporter-coupled AuNP conjugated to anti-PSA detection antibodies formed immunocomplexes in the presence of PSA in test sample. Increased intensity of Raman signal of the reporter molecule as measured by a Raman spectrophotometer upon immunocomplex formation directly correlated with the concentration of PSA present in the sample. In another interesting study, Liu et al. demonstrated a SERS-based approach for detection of PSA by exploiting the serine protease activity of PSA (Liu et al. 2006). The authors developed a peptide-nanoparticle hybrid SERS probe by combining plasmonic nanocrescent particles with peptides and Raman reporters, in which the peptides used had a high sensitivity for PSA and the plasmonic nanocrescents acted both as substrate and SERS enhancer for the Raman reporter. Proteolysis of the peptide by PSA upon enzyme-substrate complex formation was detected in the form of enhanced Raman signal of the reporter molecule due to SERS, which indirectly reported the presence of PSA in the test sample. This assay technique was capable of detecting PSA in nanomolar concentration range (0–420 nM) in very small sample volumes (femtoliter).

8.4.7 Pancreatic Cancer

Pancreatic cancer is another deadliest cancer and a leading cause of cancer-related deaths worldwide (Siegel et al. 2013). Detection of pancreatic cancer at an early stage is very crucial for accurate diagnosis and for determining the disease outcome. However, due to lack of reliable disease biomarkers, most cancers, especially pancreatic cancer, remain undetected at early stages which ultimately lead to rapid progression of the disease to a malignant stage resulting in mortality (Hidalgo 2010). Several studies have reported aberrant expression of MUC4, a mucin protein in pancreatic adenocarcinoma cell lines and tissues (Chaturvedi et al. 2008; Moniaux et al. 2007). This protein is not detected in normal pancreas and chronic pancreatitis, thus making this protein a potential biomarker for early detection and prognosis of pancreatic cancer. Detection assays based on conventional approaches such as ELISA and RIA have been unsuccessful in detecting MUC4 in the sera of clinical samples (Wang et al. 2011a, b). However, Wang et al. reported for the first time sensitive detection of MUC4 (LOD of 33 ng/mL) in the serum of pancreatic cancer patients using a solid substrate SERS-based immunoassay format (Wang et al. 2011a, b). A gold substrate and Raman reporter (4-nitrobenzene thiol)-tagged gold nanoparticles were conjugated with anti-MUC4 monoclonal antibody 8G7 (reactive against the tandem repeat of MUC4) independently. MUC4 protein, if present, attaches to the antibody-conjugated substrate further leading to the binding of the

SERS-tagged nanoparticles to it, which was subsequently detected using Raman spectroscopy. This SERS-based detection method was also successful in showing the importance of MUC4 as a reliable biomarker for pancreatic cancer detection and monitoring because the Raman measurements clearly indicated a significantly higher SERS response for MUC4 in the sera of pancreatic cancer patients in comparison to sera from healthy individuals and patients suffering from benign pancreatitis.

8.5 Label-Free SERS Assays of Biofluids for Cancer Detection

In addition to SERS-based sandwich immunoassays incorporating extrinsic Raman labels for cancer detection, label-free SERS assays have also been developed for studying and analyzing the biochemical composition of different human biofluids in diseased conditions, especially cancer. Biofluids like whole blood, plasma, serum, saliva, urine, semen and even tears are rich sources of clinically important biomolecules such as nucleic acids, lipids, carbohydrates, amino acids and proteins, whose levels can be important biomarkers for disease detection and monitoring. Studies have already shown that pathologic conditions are often associated with abnormal levels of many of such biomolecules in the biofuids (Bauca et al. 2013). So, a systematic study and analysis of these complex physiological fluids from a therapeutic or diagnostic point of view can be very helpful in understanding and mapping the differences in expression of different biomolecules in response to specific disease conditions, which can ultimately be translated into a diagnostic setup for detection, monitoring and management of different diseases, most importantly cancers. In this regard, researchers have utilized SERS for developing a non-invasive, easy-to-perform, "label-free" diagnostic platform for detecting potential biomarkers in the biofluids of patients suffering from different types of cancers (Bonifacio et al. 2015). The basic principle on which a label-free SERS assay works is that it utilizes a plasmonic metal nanostructure as the SERS-active substrate to detect and measure the different biomolecules that are present in the sample by simply enhancing the intensity of the Raman signals of these biomolecules that adsorb on the surface of the plasmonic nanostructure to a detectable range. Since each biomolecule has its own unique Raman signal known as the Raman fingerprint, the spectroscopic measurements result in the generation of a Raman spectral map of the specimen biofluid where individual peaks can be mapped to a specific biomolecule and its intensity to its concentration, thereby giving a detailed picture about the biochemistry of that fluid sample (Bonifacio et al. 2015). These Raman spectral maps are then analyzed using highly specialized and advanced diagnostic algorithms such as principal component analysis (PCA) and linear discriminate analysis (LDA) to determine the specific biochemical differences between normal and pathological conditions with more than 80–90% sensitivity and specificity (Bonifacio et al. 2015). The comparative analysis of this SERS results from a normal healthy person and a diseased patient will enable distinction between the presence and absence of cancer as well as staging of the disease. Such, label-free SERS assay has been used for early-stage detection and screening of different types of cancers, such as colorectal cancer (Lin et al. 2011; Feng et al. 2015), gastric cancer (Feng et al. 2011; Ito et al. 2014), nasopharyngeal cancer (Feng et al. 2010; Lin et al. 2014a, b), cervical cancer (Feng et al. 2013), esophageal cancer (Lin et al. 2014a, b; Li et al. 2013a, b; Huang et al. 2014), oral cancer (Kah et al. 2007), lung cancer (Yang et al. 2014; Li 2012) and prostate cancer (Li et al. 2014; Del Mistro et al. 2015) using serum, plasma, urine and saliva.

8.6 Detection of Circulating Tumour Cells in Whole Blood Using SERS

According to the "seed and soil" theory of circulating tumour cells, cancer cells are often shed from the solid tumours and they circulate in the bloodstream of the patient (Fidler 2003). Detection of these very-low-titre circulating tumour cells (CTCs) directly from blood has gained a lot of attention in cancer diagnostics for easy, simple and rapid detection and monitoring of cancer. CTCs are believed to be a rich source for new prognostic markers and thus are very crucial for determining the correct stage of the cancer (Mathot and Stenninger 2012). Use of SERS-based techniques for detection of CTCs is currently being researched for developing next-generation cancer diagnostic platforms. Michael et al. have developed a simple and sensitive no-wash assay for direct detection of circulating tumour cells in whole blood by using a combination of magnetic particles and Nanoplex SERS biotags (Sha et al. 2008). As a proof-of-concept experiment, they showed the rapid detection of breast cancer cell line SKBR3 spiked into whole blood by using magnetic particles conjugated to anti-EpCAM antibody (epithelial cell adhesion molecule antibody), and SERS biotags conjugated to anti-her2 antibody (human epidermal growth factor receptor-2). In the absence of SKBR3 cells, there was no significant Raman signal from the whole blood. However, in the presence of SKBR3 cells, a strong Raman signal was detected showing that the magnetic bead–EpCAM and SERS tag–her2 conjugates bound specifically to the target breast cancer cells in whole blood and thus helped in its detection. A LOD of 50 cells/mL with a 99.7% confidence level was reported. In another study (Wang et al. 2011a, b), CTCs were detected in the peripheral blood of patients suffering from squamous cell carcinoma of the head and neck using SERS probes conjugated to epidermal growth factor ligands (EGF). Identification and detection of very low litres of circulating chronic lymphocytic leukaemia cells in minimally processed blood samples of cancer patients have also been reported using SERS-based colloidal assay platforms (Nguyen et al. 2010). Nima et al. demonstrated SERS-based detection of single-breast cancer cells in unprocessed human blood using silver-coated gold nanorods

(AuNR@Ag) (Nima et al. 2014). They developed four sets of AuNR@Ag, each conjugated to a unique Raman reporter molecule and a specific antibody to identify four different breast cancer markers (EpCAM, CD44, keratin 18 and IGF-I receptor β) (Fig. 8.7). This SERS-based CTC detection assays have the potential to be developed into future point-of-care diagnostic platforms for successful on-site cancer detection and diagnosis because of their operational simplicity, high specificity and rapid detection ability.

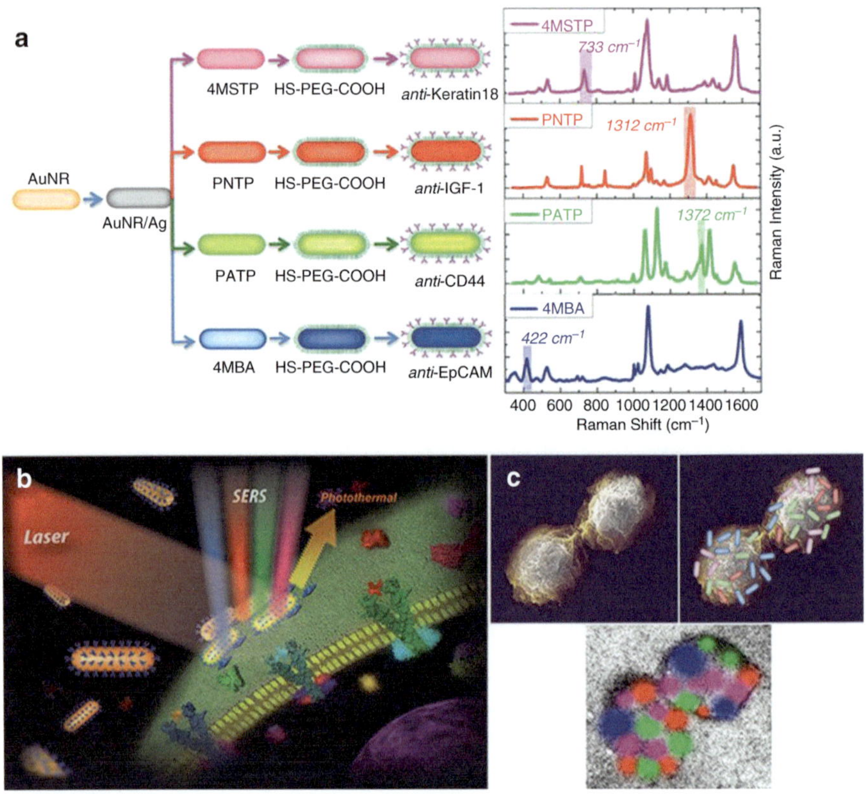

Fig. 8.7 (**a**) Schematic diagram (preparation steps) and Raman spectra (acquisition time 50 s) for the four families of SERS nano-agents. A colour was assigned to a non-overlapping peak from each SERS spectrum as follows: Blue: AuNR/Ag/4MBA/anti-EpCAM, red: AuNR/Ag/PNTP/anti-IGF-1 Receptor b, green: AuNR/Ag/PATP/anti-CD44, magenta: AuNR/Ag/4MSTP/anti-keratin18. Abbreviations are as follows: 4MBA54-mercaptobenzoic acid, PNTP5p-nitrobenzoic acid, PATP5p-aminobenzoic acid, 4MSTP54-(methylsulphanyl) thiophenol. (**b**) Schematics of breast cancer cell surface targeting by four nano-agents and SERS/PT detection technique. (**c**) Schematics of 2D multi-colour SERS data correlation with nano-agents' distribution on cell surface. Reproduced with permission from Nima et al. (2014). Copyright 2014 Nature

8.7 Detection of Cancer-Specific Gene Sequences and Single-Nucleotide Polymorphisms Using SERS

Studies have revealed that cancers are associated with mutations or changes in sequences of specific genes that lead to the development and progression of the disease (Bertram 2000). These mutations can either be in a single base of a gene referred to as single-nucleotide polymorphism (SNP) or in multiple bases (Bertram 2000). Each cancer type has its own set of specific SNPs and gene mutations that are very helpful in detection and diagnosis of a particular type of cancer. Hence, development of diagnostic platforms for detection of cancer specific genes and SNPs, in particular, has received attention from the scientific community and a number of reports have come up with different types of assays for such purpose. Among these, SERS-based detection assays employing nucleic acid hybridization as the basic tool for detection have gained popularity as an emerging alternative for rapid, specific, highly sensitive and multiplexed detection of cancer-specific genes and SNPs. Mustafa et al. reported the development of a SERS substrate for detection of the breast cancer susceptibility gene, BRCA1, by immobilizing a monolayer of 5′-mercaptohexane-labelled ssDNA (BRCA1) probe along with a MCH spacer (SH-(CH2)6-ssDNA/MCH) on a silver-coated glass slide (Culha et al. 2003). Rhodamine B-labelled BRCA1 ssDNA was used as the SERS-active target probe which upon incubation with the SERS substrate hybridized specifically with the immobilized capture probes, thereby causing a significant enhancement in the Raman intensity of the reporter dye rhodamine B, as measured through Raman spectroscopy. Detection of alternative splice variants of the BRCA1 gene has also been reported using a multiplex SERS-based sandwich assay platform (Sun et al. 2008). A SERS-active substrate was developed using a gold-coated glass slide onto which thiolated capture strands (CS) for four different splice variants of BRCA1 were immobilized. The four alternative splice variants of BRCA1 were used as target strands (TS) that hybridized with their complementary capture strands forming a CS-TS complex on the gold substrate. Next, SERS-active probes (DNA-AuNP-RTag) consisting of AuNPs conjugated with specific probe strands and four different Raman reporter molecules were used that specifically hybridized with corresponding CS-TS complex through overhanging complementary sequences resulting in the formation of a sandwich complex (Fig. 8.8). This complexation ultimately led to the enhancement of the Raman intensity of the reporter molecules, thereby facilitating multiplexed detection of four different splice variants of BRCA1 simultaneously in a single-assay platform with a detection sensitivity of up to 1 fM.

SERS has also been used for the detection of SNPs associated with different cancer types. Detection of point mutations in K-RAS oncogene associated with colorectal cancer using ligase detection reaction-coupled SERS (LDR-SERS) technique has also been reported (Huh et al. 2009, Lowe et al. 2010).

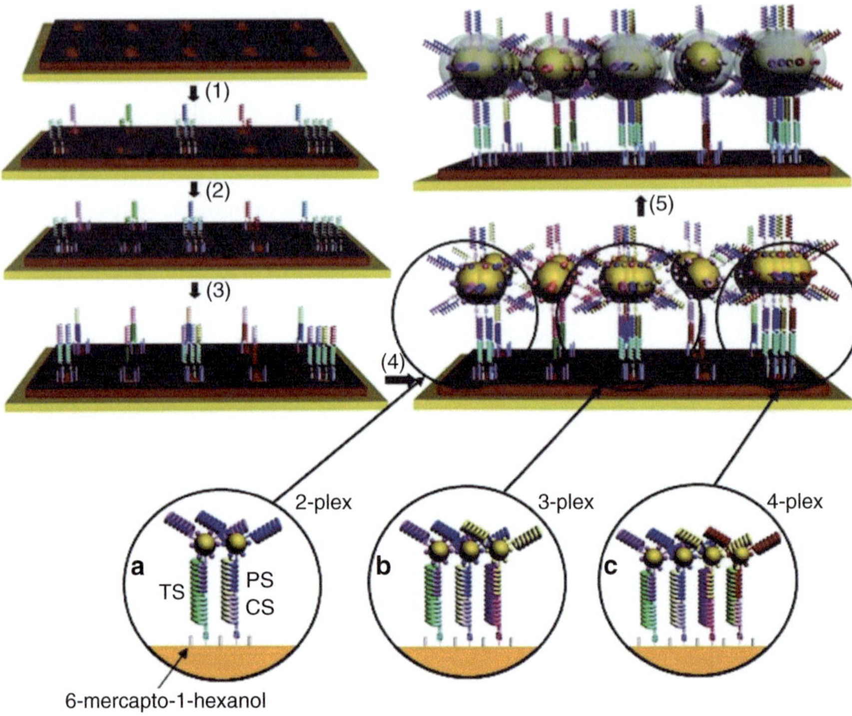

Fig. 8.8 DNA detection with a sandwich complex for BRCA1 alternative splice variants. (1) Immobilization of capturing strands (CS); (2) immobilization of 6-mercapto-1-hexanol to reduce nonspecific binding; (3) hybridize target strands (TS) to CS; (4) hybridize DNA-AuP-RTag probes to the overhanging region of TS; (5) silver enhancement. (a), (b) and (c) represent multiplex detection using DNA sequences specific to BRCA1 alternative splice variants. Reproduced with permission from Sun et al. (2008). Copyright 2008 American Chemical Society

8.8 SERS-Based Molecular Sentinel (MS) Technology

Another novel technology that has emerged as a powerful diagnostic tool in recent years is the molecular sentinel (MS) probe for detection of nucleic acid targets of interest (Wang and Vo-Dinh 2009; Ngo et al. 2013). Vo-Dinh's group at Duke University has pioneered the technique of molecular sentinel technology by designing oligonucleotide-based Raman-active probes for SERS detection of target nucleic acids. A molecular sentinel (MS) nanoprobe is made up of two components, an oligonucleotide sequence made up of 30–45 nucleotides which is arranged in a hairpin loop structure and a plasmonic nanoparticle such as silver. One end of the hairpin probe is labelled with a Raman reporter and the other end of the probe is modified with a thiol group so that it can be conjugated to the metal nanoparticle surface. The loop region of the probe is designed in such a way that the sequence of the loop is complementary to a target nucleic acid sequence. The MS nanoprobes exist in a 'closed state' when the target gene sequence is not present. The Raman reporter remains close to the plasmonic metal surface and as a result produces a strong SERS

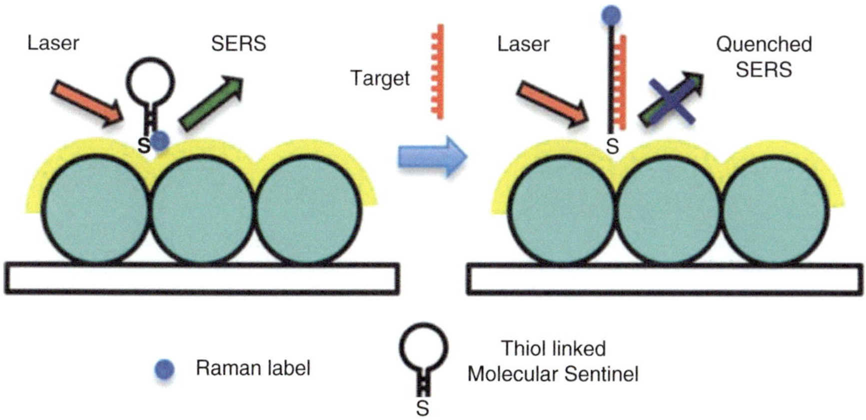

Fig. 8.9 Schematic showing the operating principle of SERS-based molecular sentinel (MS) nanoprobes. Reproduced with permission from Ngo et al. (2013). Copyright 2013 American Chemical Society

signal upon excitation with a laser. However, in the presence of the target gene sequence, the MS nanoprobes shift to an 'open-state' configuration due to complementary hybridization between target sequence and probe sequence. This results in the displacement of the Raman label from the surface of the metal nanoparticle leading to significant quenching of the SERS signal. This reduction in the SERS signal indicates the presence of the target DNA sequence in the sample (Fig. 8.9). Using these SERS-based MS nanoprobes as detection tools, Vo-Dinh's group demonstrated for the first time multiplexed detection of two breast cancer biomarkers erbB-2 and ki-67 in a homogenous solution, thus showing the high specificity and selectivity of the MS nanoprobes (Wang and Vo-Dinh 2009). This method also eliminated the washing steps usually required after hybridization reactions, thus reducing the time of the assay and also making it simple and suitable for diagnostic purposes.

Taking this technology further, Vo-Dinh's group reported the development of another novel technique known as 'molecular sentinel-on-chip' (MSC) for SERS-based DNA detection. A molecular sentinel-on-chip comprises a well-designed and fabricated plasmonic substrate which acts as the SERS enhancer, onto which MS hairpin probes are immobilized. The group fabricated triangular shaped nanowire (TSNW) arrays of silicon coated with a thin layer of gold on a 6-inch wafer as the plasmonic substrate. The TSNW were arranged on the wafer surface with a controlled spacing of around 10 nm between each nanowire in order to create 'SERS hot spots'. The MS probes were immobilized near or inside these gaps (Fig. 8.10a). Using such a molecular sentinel-on-chip platform, Vo-Dinh and group showed the SERS-based detection of the Ki-67 gene sequence (Fig. 8.10a) which is a critical breast cancer biomarker (Wang et al. 2013a, b). In another study, the same group demonstrated SERS-based multiplexed detection of two nucleic acid transcripts—interferon alpha-inducible protein 27 (IFI27) and interferon-induced protein 44-like (IFI44L) using MSC technology (Ngo et al. 2014)—thus establishing the multiplexing abilities of this technique that can be applied for simultaneous detection of multiple cancer biomarkers for diagnostic purposes.

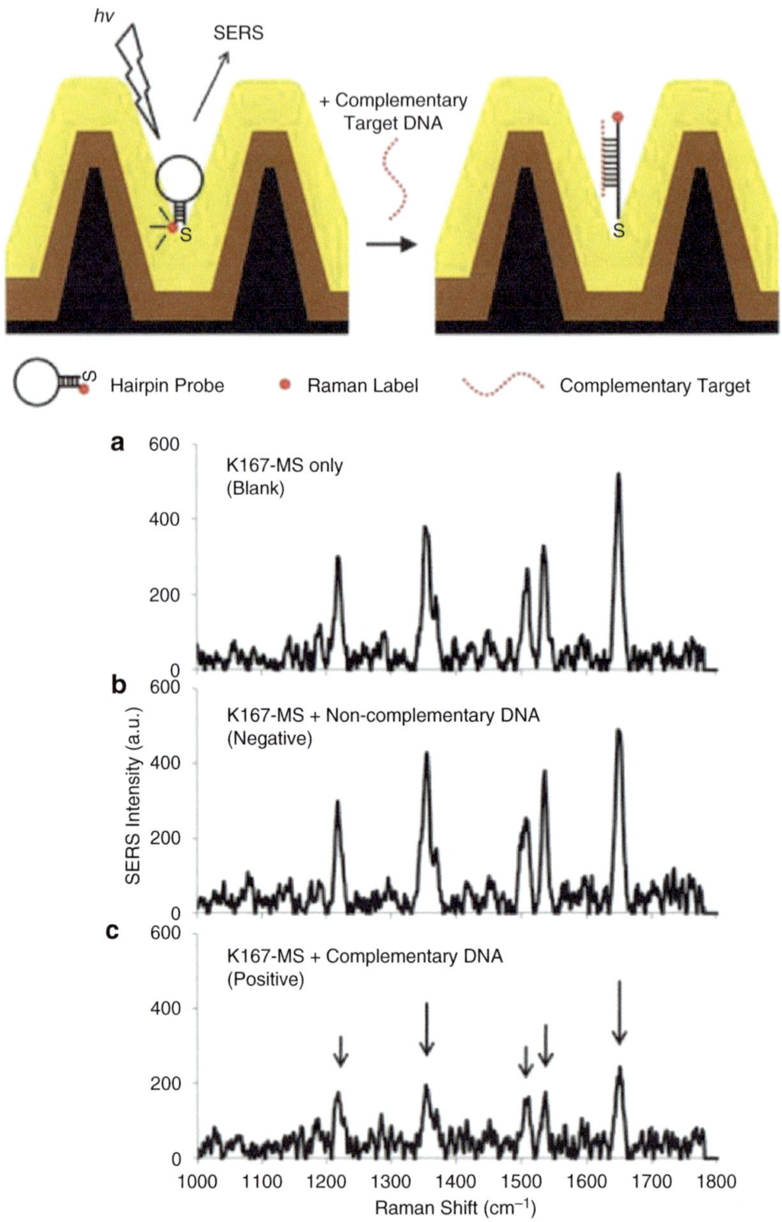

Fig. 8.10 (**a**) The operating principle of the MS-based DNA detection on a TSNW substrate. (**b**) SERS spectra of the immobilized KI67-MSC nanoprobes in the presence or absence of complementary DNA targets. (*a*) Upper spectrum: blank (no target DNA present). (*b*) Middle spectrum: in the presence of 1 μM non-complementary DNA (negative control). (*c*) Lower spectrum: in the presence of 1 μM complementary target DNA (positive diagnostic). The arrow signs illustrate the decreased SERS intensity of the major Raman bands in the presence of complementary target DNA. Reproduced with permission from Wang et al. (2013a, b). Copyright 2013 The Royal Society of Chemistry

In addition to MS nanoprobes and MSC technology for SERS-based nucleic acid detection, Vo Dinh and his group have recently reported the development of another plasmonic based nanobiosensing platform known as the "inverse molecular sentinel" (iMS) nanoprobe technology for nucleic acid detection (Wang et al. 2015). The iMS technology is based on a 'OFF-to-ON' SERS signal switch mechanism which is just the opposite of the 'ON-to-OFF' signal switch mechanism of MS technology. The group developed a iMS nanoprobe consisting of a silver-coated gold nanostar (AuNS@Ag) as a SERS enhancer linked to a DNA probe. Similar to the MS nanoprobe, the DNA probe consisted of a hairpin loop structure whose one end is attached to the plasmonic nanostar through a thiol-metal linkage and the other end is labelled with a Raman reporter. In the absence of the target sequence, a single-stranded 'placeholder sequence' binds to the hairpin probe and disrupts the hairpin loop structure, thereby moving the Raman label away from the nanostar surface. As a result, in this 'open' configuration there is very low SERS intensity (OFF state). However, when the target sequence is present, it first binds to an overhanging region of the probe-placeholder complex and starts to displace the placeholder sequence through a branch migration process ultimately releasing the placeholder from the probe. This allows the probe sequence to return back to its hairpin loop conformation ('close' configuration), thereby bringing the Raman label closer to the surface of the plasmonic nanostar, resulting in a strong SERS signal (ON state) (Fig. 8.11). Using this

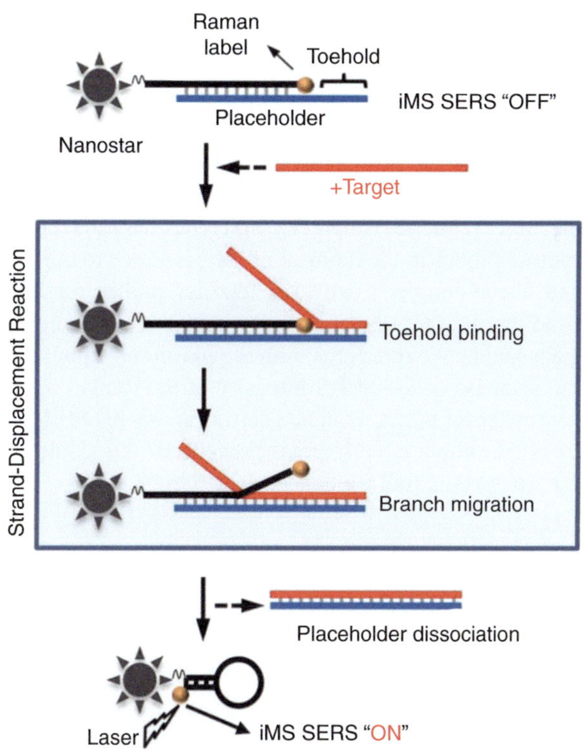

Fig. 8.11 Detection scheme of the SERS iMS nanoprobes. Reproduced with permission from Wang et al. (2016). Copyright 2016 American Chemical Society

iMS nanobiosensor platform, the group reported multiplexed detection of two important breast cancer biomarkers—miR-21 and miR-34a miRNA—from total small RNA extracted from breast cancer cell lines in a single sensing platform with high sensitivity. Different amounts of total small RNA (250 ng, 500 ng and 1 μg) extracted from MCF-7 breast cancer cell line were probed with a 10 μL nanoprobe mixture containing 5 pM of miR-21 nanoprobes and 10 pM of miR-34a nanoprobes, for simultaneous multiplexed detection of these two biomarkers (Wang et al. 2016).

8.9 Plasmonic Coupling Interference-Based SERS Nanoprobes (PCI-SERS) for Detection of miRNA and Other Nucleic Acids

Nucleic acid (DNA/RNA) detection using SERS-based nanosensors employing a technique referred to as plasmonic coupling interference (PCI) has been developed by Wang et al. (Wang and Vo-Dinh 2011). Plasmonic coupling refers to the combination of the plasmons of closely spaced nanoparticles resulting in an intense enhancement of the SERS signal of a reporter molecule. In this study, a novel target label-free platform has been developed for detection of human miR-21 sequence, which is known to be associated with a number of cancers and other diseases (Krichevsky and Gabriely 2008). Plasmonic silver nanoparticles (AgNPs) functionalized with thiolated oligonucleotides were used as the capture probes (Capture-NPs) and complementary Raman-labelled reporter probes (Reporter-NPs). In the absence of target sequence (miR-21), Capture-NPs and Reporter-NPs could easily assemble into aggregates in solution through complementary nucleic acid hybridization. This brings the plasmonic nanoparticles in close proximity to cause a significant enhancement of the SERS intensity of the Raman reporter label located within inter-particle hot-spot domains through plasmonic coupling. However, when the target sequence (miR-21), which has the same complementary sequence for the capture probes as the reporter probes, is added to the solution, it acts as a competitor and hence competes with the reporter probes for binding to the capture probes. Binding of target sequence to the capture probes prevents formation of nanoaggregates and hence interferes with the plasmonic coupling effect (PCI). This results in the decrease of the SERS intensity of the Raman reporter, which can be used as a parameter for target sequence detection. As this SERS-PCI-based sensing technique involves simple experimental procedures this could potentially act as a point-of-care diagnostic tool for cancer detection.

8.10 Limitations

Over the years, SERS-based biosensing has emerged as a powerful tool for specific and sensitive detection of disease biomarkers from a diverse range of biological specimens at very low concentrations, which have opened up the possibility of using this technique for point-of-care diagnostic purposes. However, there are some limitations that have slowed down the translation of SERS-based biosensors from research laboratories to on-site clinics for point-of-care diagnosis. One such limitation is interference of fluorescence emissions from laser-excited complex target biological specimens as well as Raman-active targets that has the possibility to mask the potential SERS signals of Raman reporters, thereby making it very difficult to discern and identify the correct Raman peaks of target analytes and markers. In addition to fluorescence, elastic scattering from different components of biological fluids and intracellular structures also contribute to background signals that impact the resolution of the Raman spectra. Finally, as SERS-based detection is dependent on the use of a Raman spectrometer for laser excitation and subsequent detection, it makes the overall cost high. However, nowadays, handheld Raman spectrophotometer is developed which has the potential to incorporate assay, sensing, detection and read-out components in a single platform, and thus can significantly reduce the cost and also make operation simpler that can then be easily used at on-site locations for point-of-care diagnosis.

8.11 Summary and Outlook

This chapter summarizes the application of SERS-based biosensors for cancer detection and diagnosis. We have presented the different intrinsic and extrinsic SERS-based biosensors reported in last few years for sensitive detection of cancer cells and related biomarkers. The high sensitivity combined with excellent multiplexing ability has made SERS a very powerful spectroscopic based detection technique, especially for developing novel biosensors and diagnostic platforms for detection of biomolecules and biomarkers of clinical significance that can ultimately help in the early diagnosis, treatment, monitoring and management of diseases such as cancer. The high accuracy rate, specificity, ease of performance and non-invasiveness of SERS-based detection technique and the development of portable handheld Raman spectrometer as the detection device make it a very promising tool for potential next-generation point-of-care diagnostic applications. Future research should therefore be devoted towards the translation of SERS-based diagnostic techniques for on-site disease diagnosis using portable handheld Raman spectrophotometer.

Acknowledgements The authors gratefully acknowledge the financial support from Indian Institute of Technology, Mandi, Department of Science and Technology (DST), under project number, SERB/F/5627/2015-16, and Department of Biotechnology (DBT), Government of India, under project number, BT/PR14749/NNT/28/954/2015.

References

Albrecht MG, Creighton JA (1977) Anomalously intense Raman spectra of pyridine at a silver electrode. J Am Chem Soc 99(15):5215–5217

Altintas Z, Tothill IE (2015) Molecular biosensors: promising new tools for early detection of cancer. Nanobiosensors in Disease Diagnosis 4:1–10

Amer MS (2010) Raman spectroscopy, fullerenes and nanotechnology. Royal Society of Chemistry, Cambridge, pp 43–44

American Cancer Society. Cancer Facts & Figures 2015. American Cancer Society, Atlanta

Anttila MA, Kosma V-M, Hongxiu J, Puolakka J, Juhola M, Saarikoski S, Syrjänen K (1999) P21/WAF1 expression as related to p53, cell proliferation and prognosis in epithelial ovarian cancer. Br J Cancer 79(11–12):1870–1878

Bauca JM, Martinez-Morillo E, Diamandis EP (2013) Peptidomics of urine and other Biofluids for cancer diagnostics. Clin Chem 60(8):1052–1061

Begent RHJ (1984) The value of Carcinoembryonic antigen measurement in clinical practice. Ann Clin Biochem 21(4):231–238

Bertram JS (2000) The molecular biology of cancer. Mol Asp Med 21(6):167–223

Bonifacio A, Cervo S, Sergo V (2015) Label-free surface-enhanced Raman spectroscopy of biofluids: fundamental aspects and diagnostic applications. Anal Bioanal Chem 407(27):8265–8277

Brian B, Shaker AM (2011) Biosensors: the new wave in cancer diagnosis. Nanotechnol Sci Appl 4:1–10

Brown CR, Higgins KW, Frazer K, Schoelz LK, Dyminski JW, Marinkovich VA, Miller SP, Burd JF (1985) Simultaneous determination of total IgE and allergen-specific IgE in serum by the MAST chemiluminescent assay system. Clin Chem 31:1500–1505

Butler JE (2000) Solid supports in enzyme-linked Immunosorbent assay and other solid-phase immunoassays. Methods 22(1):4–23

Chaturvedi P, Singh AP, Chakraborty S, Chauhan SC, Bafna S, Meza JL, Singh PK, Hollingsworth MA, Mehta PP, Batra SK (2008) MUC4 Mucin interacts with and stabilizes the HER2 Oncoprotein in human pancreatic cancer cells. Cancer Res 68(7):2065–2070

Cho EJ, Collett JR, Szafranska AE, Ellington AD (2006) Optimization of aptamer microarray technology for multiple protein targets. Anal Chim Acta 564(1):82–90

Chon H, Lee S, Son SW, Oh CH, Choo J (2009) Highly sensitive immunoassay of lung cancer marker Carcinoembryonic antigen using surface-enhanced Raman scattering of hollow gold Nanospheres. Anal Chem 81(8):3029–3034

Chon H, Lee S, Yoon SY, Chang SI, Lim DW, Choo J (2011) Simultaneous immunoassay for the detection of two lung cancer markers using functionalized SERS nanoprobes. Chem Commun 47(46):12515

Cree IA (2015) Liquid biopsy for cancer patients: principles and practice. Pathogenesis 2(1–2):1–4

Culha M, Stokes D, Allain LR, Vo-Dinh T (2003) Surface-enhanced Raman scattering substrate based on a self-assembled monolayer for use in Gene diagnostics. Anal Chem 75(22):6196–6201

Del Mistro G, Cervo S, Mansutti E, Spizzo R, Colombatti A, Belmonte P, Zucconelli R, Steffan A, Sergo V, Bonifacio A (2015) Surface-enhanced Raman spectroscopy of urine for prostate cancer detection: a preliminary study. Anal Bioanal Chem 407(12):3271–3275

Dinish US, Balasundaram G, Chang YT, Olivo M (2013) Sensitive multiplex detection of serological liver cancer biomarkers using SERS-active photonic crystal fiber probe. J Biophotonics 7(11–12):956–965

Doering WE, Piotti ME, Natan MJ, Freeman RG (2007) SERS as a foundation for Nanoscale, optically detected biological labels. Adv Mater 19(20):3100–3108

Driscoll AJ, Harpster MH, Johnson PA (2013) The development of surface-enhanced Raman scattering as a detection modality for portable in vitro diagnostics: progress and challenges. Phys Chem Chem Phys 15(47):20415

El-Deiry W (1993) WAF1, a potential mediator of p53 tumor suppression. Cell 75(4):817–825

Etzioni R, Urban N, Ramsey S, McIntosh M, Schwartz S, Reid B, Radich J, Anderson G, Hartwell L (2003) The case for early detection. Nat Rev Cancer 3(4):243–252

Feng S, Chen R, Lin J, Pan J, Chen G, Li Y, Cheng M, Huang Z, Chen J, Zeng H (2010) Nasopharyngeal cancer detection based on blood plasma surface-enhanced Raman spectroscopy and multivariate analysis. Biosens Bioelectron 25(11):2414–2419

Feng S, Chen R, Lin J, Pan J, Wu Y, Li Y, Chen J, Zeng H (2011) Gastric cancer detection based on blood plasma surface-enhanced Raman spectroscopy excited by polarized laser light. Biosens Bioelectron 26(7):3167–3174

Feng S, Lin D, Lin J, Li B, Huang Z, Chen G, Zhang W, Wang L, Pan J, Chen R, Zeng H (2013) Blood plasma surface-enhanced Raman spectroscopy for non-invasive optical detection of cervical cancer. Analyst 138(14):3967

Feng S, Wang W, Tai IT, Chen G, Chen R, Zeng H (2015) Label-free surface-enhanced Raman spectroscopy for detection of colorectal cancer and precursor lesions using blood plasma. Biomed Optics Exp 6(9):3494

Fidler IJ (2003) Timeline: the pathogenesis of cancer metastasis: the 'seed and soil' hypothesis revisited. Nat Rev Cancer 3(6):453–458

Fleischmann M, Hendra PJ, McQuillan AJ (1974) Raman spectra of pyridine adsorbed at a silver electrode. Chem Phys Lett 26(2):163–166

Gremlich HU, Yan B (2001) Infrared and Raman spectroscopy of biological materials. Marcel Dekker Inc., New York

Grubisha DS, Lipert RJ, Park H-Y, Driskell J, Porter MD (2003) Femtomolar detection of prostate-specific antigen: an immunoassay based on surface-enhanced Raman scattering and Immunogold labels. Anal Chem 75(21):5936–5943

Hayes FJ, Halsall HB, Heineman WR (1994) Simultaneous immunoassay using electrochemical detection of metal ion labels. Anal Chem 66:1860–1865

Haynes CL, McFarland AD, Duyne RPV (2005) Surface-enhanced Raman spectroscopy. Analyt Chem 77(17):338a–346a

Henry LN, Hayes DF (2012) Cancer biomarker. Mol Oncol 6(2):140–146

Hidalgo M (2010) Pancreatic cancer. N Engl J Med 362:1605–1617

Huang S, Wang L, Chen W, Feng S, Lin J, Huang Z, Chen G, Li B, Chen R (2014) Potential of non-invasive esophagus cancer detection based on urine surface-enhanced Raman spectroscopy. Laser Phys Lett 11(11):115604

Huh YS, Lowe AJ, Strickland AD, Batt CA, Erickson D (2009) Surface-enhanced Raman scattering based ligase detection reaction. J Am Chem Soc 131(6):2208–2213

Ito H, Inoue H, Hasegawa K, Hasegawa Y, Shimizu T, Kimura S, Onimaru M, Ikeda H, Kudo S (2014) Use of surface-enhanced Raman scattering for detection of cancer-related serum-constituents in gastrointestinal cancer patients. Nanomedicine 10(3):599–608

Jaiswal A, Tian L, Tadepalli S, Liu K-K, Fei M, Farrell ME, Pellegrino PM, Singamaneni S (2014) Plasmonic nanorattles with intrinsic electromagnetic hot-spots for surface enhanced Raman scattering. Small 10:4287–4292

Jeanmaire D (1977) Surface Raman spectroelectrochemistry part I. Heterocyclic, aromatic, and aliphatic amines adsorbed on the anodized silver electrode. J Electroanal Chem 84(1): 1–20

Jensen L, Aikens CM, Schatz GC (2008) Electronic structure methods for studying surface-enhanced Raman scattering. Chem Society Rev 37(5):1061

Kah JCY, Kho KW, Lee CGL, James C, Sheppard R, Shen ZX, Soo KC, Olivo MC (2007) Early diagnosis of oral cancer based on the surface plasmon resonance of gold nanoparticles. Int J Nanomedicine 2:785–798

Ko H, Singamaneni S, Tsukruk VV (2008) Nanostructured surfaces and assemblies as SERS media. Small 4(10):1576–1599

Krichevsky AM, Gabriely G (2008) MiR-21: a small multi-faceted RNA. J Cell Mol Med 13(1):39–53

Krishnamurti U, Silverman JF (2014) HER2 in breast cancer. Adv Anat Pathol 21(2):100–107

Kyle CB, Audrey FM, Wittenberg NJ, Hyungsoon I, Özge K, Si HL, Nathan CL, Sang-Hyun O, Christy LH (2011) Recent progress in SERS Biosensing. Phys Chem Chem Phys 3(24):11551–11567

Lee S, Chon H, Lee M, Choo J, Shin SY, Lee YH, Rhyu IJ, Son SW, Oh CH (2009) Surface-enhanced Raman scattering imaging of HER2 cancer markers overexpressed in single MCF7 cells using antibody conjugated hollow gold nanospheres. Biosens Bioelectron 24(7):2260–2263

Lee M, Lee K, Kim KH, Oh KW, Choo J (2012) SERS-based immunoassay using a gold array-embedded gradient microfluidic Chip. Lab Chip 12(19):3720

Lee S, Chon H, Lee J, Ko J, Chung BH, Lim DW, Choo J (2014) Rapid and sensitive phenotypic marker detection on breast cancer cells using surface-enhanced Raman scattering (SERS) imaging. Biosens Bioelectron 51:238–243

Li X (2012) Spectral analysis of human saliva for detection of lung cancer using surface-enhanced Raman spectroscopy. J Biomed Opt 17(3):037003

Li M, Cushing SK, Zhang J, Suri S, Evans R, Petros WP, Gibson LF, Ma D, Liu Y, Wu N (2013a) Three-dimensional hierarchical Plasmonic Nano-architecture enhanced surface-enhanced Raman scattering Immunosensor for cancer biomarker detection in blood plasma. ACS Nano 7(6):4967–4976

Li SX, Zeng QY, Li LF, Zhang YJ, Wan MM, Liu ZM, Xiong HL, Guo ZY, Liu SH (2013b) Study of support vector machine and serum surface-enhanced Raman spectroscopy for noninvasive esophageal cancer detection. J Biomed Opt 18(2):027008

Li S, Zhang Y, Xu J, Li L, Zeng Q, Lin L, Guo Z, Liu Z, Xiong H, Liu S (2014) Noninvasive prostate cancer screening based on serum surface-enhanced Raman spectroscopy and support vector machine. Appl Phys Lett 105(9):091104

Li J, Skeete Z, Shan S, Yan S, Kurzatkowska K, Zhao W, Ngo QM, Holubovska P, Luo J, Hepel M, Zhong C-J (2015) Surface enhanced Raman scattering detection of cancer biomarkers with Bifunctional Nanocomposite probes. Anal Chem 87(21):10698–10702

Lilja H, Christensson A, Dahlen U, Matikainen MT, Nilsson O, Pettersson K, Lovgren T (1991) Prostate-specific antigen in serum occurs predominantly in complex with alpha 1 antichymotrypsin. Clin Chem 1991(37):1618–1625

Lilja H, Ulmert D, Vickers AJ (2008) Prostate-specific antigen and prostate cancer: prediction, detection and monitoring. Nat Rev Cancer 8(4):268–278

Lin D, Feng S, Pan J, Chen Y, Lin J, Chen G, Xie S, Zeng H, Chen R (2011) Colorectal cancer detection by gold nanoparticle based surface-enhanced Raman spectroscopy of blood serum and statistical analysis. Opt Exp 19(14):13565

Lin D, Feng S, Huang H, Chen W, Shi H, Liu N, Chen L, Yu Y, Chen R (2014a) Label-free detection of blood plasma using silver nanoparticle based surface-enhanced Raman spectroscopy for esophageal cancer screening. J Biomed Nanotechnol 10(3):478–484

Lin D, Pan J, Huang H, Chen G, Qiu S, Shi H, Chen W, Yu Y, Feng S, Chen R (2014b) Label-free blood plasma test based on surface-enhanced Raman scattering for tumor stages detection in nasopharyngeal cancer. Sci Rep 4:4751

Liu GL, Chen FF, Ellman JA, Lee LP (2006) Peptide-nanoparticle hybrid SERS probe for dynamic detection of active cancer biomarker enzymes. Conf Proc IEEE Eng Med Biol Soc 1:795–798

Lowe AJ, Huh YS, Strickland AD, Erickson D, Batt CA (2010) Multiplex single nucleotide polymorphism genotyping utilizing ligase detection reaction coupled surface enhanced Raman spectroscopy. Anal Chem 82(13):5810–5814

Mahadevan JA, Richards KR (1996) Raman spectroscopy for the detection of cancers and precancers. J Biomed Optics 1:31–70

Marc V, Kaustabh KM, Kevin D, Young-Tae C (2013) Surface-enhanced Raman scattering in cancer detection and imaging. Trends Biotechnol 31(4):249–257

Mathot L, Stenninger J (2012) Behavior of seeds and soil in the mechanism of metastasis: a deeper understanding. Cancer Sci 103(4):626–631

Mayeux R (2004) Biomarkers: potential uses and limitations. Neurotherapeutics 1(2):182–188

Moniaux N, Chaturvedi P, Varshney GC, Meza JL, Rodriguez-Sierra JF, Aubert JP, Batra SK (2007) Human MUC4 mucin induces ultra-structural changes and tumorigenicity in pancreatic cancer cells. Br J Cancer 97(3):345–357

Ngo HT, Wang HN, Fales AM, Vo-Dinh T (2013) Label-free DNA biosensor based on SERS molecular sentinel on Nanowave chip. Anal Chem 85(13):6378–6383

Ngo HT, Wang HN, Burke T, Ginsburg GS, Vo-Dinh T (2014) Multiplex detection of disease biomarkers using SERS molecular sentinel-on-chip. Anal Bioanal Chem 406(14):3335–3344

Nguyen CT, Nguyen JT, Rutledge S, Zhang J, Wang C, Walker GC (2010) Detection of chronic lymphocytic leukemia cell surface markers using surface enhanced Raman scattering gold nanoparticles. Cancer Lett 292(1):91–97

Nie S (1997) Probing single molecules and single nanoparticles by surface-enhanced Raman scattering. Science 275(5303):1102–1106

Nima ZA, Mahmood M, Xu Y, Mustafa T, Watanabe F, Nedosekin DA, Juratli MA, Fahmi T, Galanzha EI, Nolan JP, Basnakian AG, Zharov VP, Biris AS (2014) Circulating tumor cell identification by functionalized silver-gold nanorods with multicolor, super-enhanced SERS and photothermal resonances. Sci Rep 4:4752

Pallela R, Chandra P, Noh HB, Shim YB (2016) An amperometricnanobiosensor using a biocompatible conjugate for early detection of metastatic cancer cells in biological fluid. Biosens Bioelectron 85:883–890

Park H, Lee S, Chen L, Lee EK, Shin SY, Lee YH, Son SW, Oh CH, Song JM, Kang SH, Choo J (2009) SERS imaging of HER2-overexpressed MCF7 cells using antibody-conjugated gold nanorods. Phys Chem Chem Phys 11(34):7444

Porter MD, Lipert RJ, Siperko LM, Wang G, Narayanan R (2008) SERS as a bioassay platform: fundamentals, design, and applications. Chem Soc Rev 37(5):1001

Pradeep CR (2005) Expression of vascular endothelial growth factor (VEGF) and VEGF receptors in tumor angiogenesis and malignancies. Integr Cancer Ther 4(4):315–321

Satish BN, Mukesh DS, Keum-Soo S, Taisun K (2016) Biomarker detection technologies and future directions. Analyst 141:740

Schatz GC, Young MA, Van Duyne RP (2006) Electromagnetic mechanism of SERS. Topics in Applied Physics Springer-Verlag, Berlin, Heidelberg 103:19–46

Sha MY, Xu H, Natan MJ, Cromer R (2008) Surface-enhanced Raman scattering tags for rapid and homogeneous detection of circulating tumor cells in the presence of human whole blood. J Am Chem Soc 130(51):17214–17215

Sharma B, Frontiera RR, Henry A-I, Ringe E, Van DRP (2012) SERS: materials, applications, and the future. Mater Today 15(1–2):16–25

Sherr CJ (2004) Principles of tumor suppression. Cell 116(2):235–246

Shi YZ, Hui AM, Takayama T, Li X, Cui X, Makuuchi M (2000) Reduced p21$^{WAF1/CIP1}$ protein expression is predominantly related to altered p53 in hepatocellular carcinomas. Br J Cancer 83:50–55

Siegel R, Naishadham D, Jemal A (2013) Cancer statistics. CA Cancer J Clin 63(1):11–30

Stenman UH, Leinonen J, Alfthan H, Rannikko S, Tuhkanen K, Alfthan O (1991) A complex between prostate-specific antigen and alpha 1-antichymotrypsin is the major form of prostate-specific antigen in serum of patients with prostatic cancer: assay of the complex improves clinical sensitivity for cancer. Cancer Res 1991(51):222–226

Stephen DH, George C (2009) Bioanalytical applications of SERS (surface enhanced Raman spectroscopy). Anal Bioanal Chem 394:679–686

Strianese M, Staiano M, Ruggiero G, Labella T, Pellecchia C, D'Auria S (2012) Fluorescence based biosensors. Methods Mol Biol 875:193–216

Sun L, Yu C, Irudayaraj J (2008) Raman multiplexers for alternative gene splicing. Anal Chem 80(9):3342–3349

Thomson DMP, Krupey J, Freedman SO, Gold P (1969) The radioimmunoassay of circulating CARCINOEMBRYONIC antigen of the human digestive system. Proc Natl Acad Sci 64(1):161–167

Viale G, Pellegrini C, Mazzarol G, Maisonneuve P, Silverman ML, Bosari S (1999) P21WAF1/CIP1 expression in colorectal carcinoma correlates with advanced disease stage and p53 mutations. J Pathol 187(3):302–307

Vuori J, Rasi S, Takala T, Vaananen K (1991) Dual-label time-resolved fluoroimmunoassay for simultaneous detection of myoglobin and carbonic anhydrase III in serum. Clin Chem 37:2087–2092

Wang HN, Vo-Dinh T (2009) Multiplex detection of breast cancer biomarkers using plasmonic molecular sentinel nanoprobes. Nanotechnology 20(6):065101

Wang HN, Vo-Dinh T (2011) Plasmonic coupling interference (PCI) Nanoprobes for nucleic acid detection. Small 7(21):3067–3074

Wang G, Lipert RJ, Jain M, Kaur S, Chakraboty S, Torres MP, Batra SK, Brand RE, Porter MD (2011a) Detection of the potential pancreatic cancer marker MUC4 in serum using surface-enhanced Raman scattering. Anal Chem 83(7):2554–2561

Wang X, Qian X, Beitler JJ, Chen ZG, Khuri FR, Lewis MM, Shin HJC, Nie S, Shin DM (2011b) Detection of circulating tumor cells in human peripheral blood using surface-enhanced Raman scattering nanoparticles. Cancer Res 71(5):1526–1532

Wang HN, Dhawan A, Du Y, Batchelor D, Leonard DN, Misra V, Vo-Dinh T (2013a) Molecular sentinel-on-chip for SERS-based biosensing. Phys Chem Chem Phys 15(16):6008

Wang A, Ruan W, Song W, Chen L, Zhao B, Jung YM, Wang X (2013b) Detection of the potential tumor marker of AFP using surface-enhanced Raman scattering-based immunoassay. J Raman Spectrosc 44(12):1649–1653

Wang HN, Fales AM, Vo-Dinh T (2015) Plasmonics-based SERS nanobiosensor for homogeneous nucleic acid detection. Nanomedicine 11(4):811–814

Wang HN, Crawford BM, Fales AM, Bowie ML, Seewaldt VL, Vo-Dinh T (2016) Multiplexed detection of MicroRNA biomarkers using SERS-based inverse molecular sentinel (iMS) Nanoprobes. J Phys Chem C 120(37):21047–21055

Werness BA, Freedman AN, Piver MS, Romero-Gutierrez M, Petrow E (1999) Prognostic significance of p53 and p21waf1/cip1 Immunoreactivity in epithelial cancers of the ovary. Gynecol Oncol 75(3):413–418

Willard HH, Merritt LL, Dean JA (1974) Raman spectroscopy. Instrumental methods of analysis, 5th edn. D. Van Nostrand Co., New York, pp 189–197

Willets KA, Van Duyne RP (2007) Localized surface Plasmon resonance spectroscopy and sensing. Annu Rev Phys Chem 58(1):267–297

Wu L, Wang Z, Zong S, Chen H, Wang C, Xu S, Cui Y (2013) Simultaneous evaluation of p53 and p21 expression level for early cancer diagnosis using SERS technique. Analyst 138(12):3450

Xu YY, Pettersson K, Blomberg K, Hemmila I, Mikola H, Lovgren T (1992) Simultaneous quadruple-label fluorometric immunoassay of thyroid-stimulating hormone, 17 alpha-hydroxyprogesterone, immunoreactive trypsin, and creatine kinase MM isoenzyme in dried blood spots. Clin Chem 38:2038–2043

Yang J, Wang Z, Zong S, Song C, Zhang R, Cui Y (2011) Distinguishing breast cancer cells using surface-enhanced Raman scattering. Anal Bioanal Chem 402(3):1093–1100

Yang T, Guo X, Wu Y, Wang H, Fu S, Wen Y, Yang H (2014) Facile and label-free detection of lung cancer biomarker in urine by magnetically assisted surface-enhanced Raman scattering. ACS Appl Mater Interfaces 6(23):20985–20993

Zhao YJ, Ju Q, Li GC (2013) Tumor markers for hepatocellular carcinoma. Mol Clin Oncol 1(4):593–598

Nucleic Acid-Based Aptasensors for Cancer Diagnostics: An Insight into Immobilisation Strategies

9

Pawan Jolly, Marina R. Batistuti, Serife Ustuner, Marcelo Mulato, Sunil K. Arya, and Pedro Estrela

9.1 Introduction

Whilst antibodies remain the most popular and trusted choice for molecular recognition, they may still pose challenges for biosensing applications due to their high cost, low reproducibility and large size. One long championed alternative to antibodies are nucleic acid aptamers. Nucleic acid aptamers are single-stranded DNA or RNA sequences that can bind to a target with high affinity and specificity. Nucleic acid aptamers, due to their varied advantages, have been gaining significant importance in both diagnostic and theranostic applications. Among various diseases, early diagnosis of cancer is one of the biggest concerns for patients and healthcare professionals worldwide. For the case of cancer, it is crucial to be able to deliver treatment whilst monitoring therapy response in real time. This is required in order to prevent over- or under-dosing the patients whilst treatment occurs. One of the most commonly used techniques for cancer diagnosis is to detect biomarkers (cancer-related proteins, small molecules and cancer cells) found in body fluids specific for a particular cancer type. In this chapter, we present a discussion on the use of aptamer-based biosensors (termed as "aptasensors") for cancer diagnosis. The development of aptamer-based biosensor devices is an interdisciplinary field and relies on very distinct aspects such as characterisation of bio-recognition probes with their respective analytes, immobilisation onto electrode surfaces, development of anti-fouling surface chemistries, sensor design and fabrication and microfluidics. Special attention is given to different types of surface chemistry used for the development of

P. Jolly • S. Ustuner • S.K. Arya • P. Estrela (✉)
Department of Electronic and Electrical Engineering, University of Bath, Bath BA2 7AY, UK
e-mail: pawan.jolly@hotmail.com, P.Estrela@bath.ac.uk

M.R. Batistuti • M. Mulato
Department of Physics, Faculty of Philosophy, Sciences and Letters, University of São Paulo, Ribeirão Preto, SP 14040-901, Brazil

P. Chandra et al. (eds.), *Next Generation Point-of-care Biomedical Sensors Technologies for Cancer Diagnosis*, DOI 10.1007/978-981-10-4726-8_9

simple, sensitive and cost-effective aptasensors. Utilisation of aptamers is an encouraging tool for the development of point-of-care (PoC) biosensors for the detection of different types of cancer. In the view of unparalleled merits of aptamers, recent achievements and future perspectives of the applications of aptamers are also discussed.

9.1.1 Nucleic Acid Aptamers

Nucleic acid aptamers are short single-stranded DNA or RNA oligonucleotides, which have gained massive attention as bioreceptors in biosensing applications (aptasensors) or medical therapy (Hianik and Wang 2009; Iliuk et al. 2011). Their specificity is similar or higher than that of antibodies, with dissociation constants (K_d) in the range of nanomolar (nM) to picomolar (pM) levels. An important advantage of their use in the development of biosensor devices is their high affinity and specificity to proteins and other molecules with low molecular weight, for example, toxins (Castillo et al. 2012; Wang et al. 2011). When compared with their biological counterparts, aptamers have several advantages; for instance, aptamers are more stable than antibodies, making them suitable for applications in special conditions such as high temperature or extreme pH.

Furthermore, aptasensors can be regenerated without loss of integrity and selectivity (Mairal et al. 2008; Tombelli et al. 2005). One very interesting property of an aptamer is the conformational change that it undergoes once bound to its target, a property which has been widely utilised within biosensing applications (Jolly et al. 2015a; Radi et al. 2006). The aptamers reported to date are known to form loops, stems, hairpins, triplexes or quadruplex structures. For instance, the DNA aptamer specific to prostate-specific antigen (PSA), a protein biomarker for prostate cancer, forms a stable single-loop configuration. On the other hand, the DNA aptamer specific to alpha-methylacyl-CoA racemase (AMACR), another protein biomarker for prostate cancer, forms multiple-loop structures (Savory et al. 2010; Yang et al. 2014). The formation of loops is due to the specific interactions between nucleotides, adenine and thymine or guanine and cytosine present in DNA aptamer chains. Quadruplexes are nucleic acid sequences that are rich in guanine and are able to form a four-stranded structure. The DNA aptamer specific to thrombin forms a quadruplex structure, which is further stabilised by the presence of a cation, especially potassium, which sits in a central channel between each pair of tetrads (Macaya et al. 1993).

Aptamers can be easily chemically modified with various functional groups, such as thiol, amine or azide as well as biotin groups. This modification allows the immobilisation of aptamers to various solid supports. For example, modification of aptamers with thiols allows their association on the gold surface using Au–S interactions (Jolly et al. 2015a) or modification with azido groups via click chemistry (Hayat et al. 2013). In another case, one end of DNA or RNA aptamers can be modified with biotin and binding of these biomolecules through solid support is realised via avidin, streptavidin or neutravidin bridges (Cavic and Thompson 2002; Centi et al. 2007; Liss et al. 2002; Ostatná et al. 2008).

Aptamers were introduced in 1990 by two independent research groups: Tuerk et al. used the term SELEX (generalised scheme of systematic evolution of ligands by exponential enrichment) for selecting RNA ligands against T4 DNA polymerase; and Ellington et al. coined the term in vitro selection (Ellington and Szostak 1990; Robertson and Joyce 1990; Tuerk and Gold 1990). In contrast to antibodies that are obtained by molecular biology techniques, aptamers are prepared by a synthetic method using an in vitro selection procedure. Once an aptamer sequence is identified, it can be synthesised with high purity, reproducibility and relatively low cost. Although aptamers have a greater advantage over antibodies, they still possess a number of limitations, for example, degradation by nucleases (DNAse and RNAse) or protein fouling in serum due to DNA-binding proteins (Keum and Bermudez 2009; Sylvia et al. 1975).

A generalised SELEX method is presented in Fig. 9.1. The SELEX approach starts with a random library containing 10^{13}–10^{16} ssDNA or RNA sequences (Iliuk et al. 2011; Song et al. 2008). The library is incubated with a target to initiate the first cycle of selection. This is typically followed by iterative cycles of absorption, recovery of bound DNA/RNA and amplification. Isolation of the bound DNA/RNA

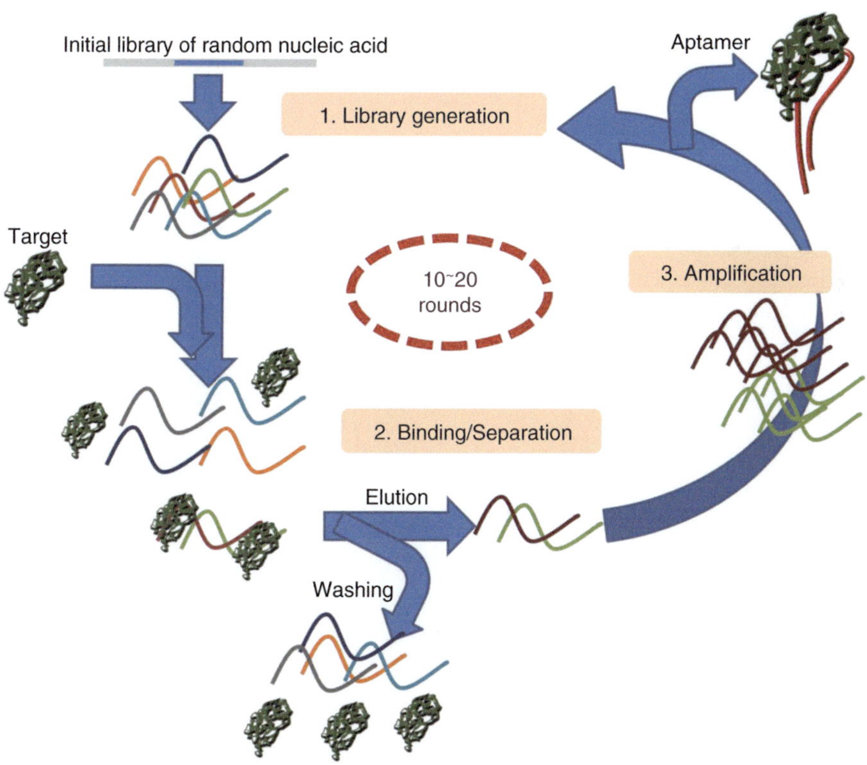

Fig. 9.1 Generalised scheme of systematic evolution of ligands by exponential enrichment (SELEX)

is the most critical step to ensure purity and selectivity. For example, the aptamer-target complex can be separated by filtration through nitrocellulose or by affinity chromatography from the unbound DNA/RNA sequences. The bound aptamers are then eluted and amplified by RT-PCR (for RNA libraries) or PCR (for DNA libraries) to generate new pools for the next selection cycle. An ideal aptamer selection procedure requires around 10–15 cycles.

9.1.2 Aptamer Assay Configuration

The first application of aptamers for biosensing was reported in 1996 where an optical aptasensor based on fluorescence-labelled aptamers was developed, for the detection of immunoglobulin G (IgG) (Davis et al. 1996). Since then, several aptasensors have been developed using different transduction techniques. Replacing the antibodies in a classical ELISA configuration has enabled the development of sophisticated assays, which are more robust, reproducible and economical. The high specificity of aptamers and the possibility of developing aptamers against different binding sites of target analyte offer high variability of assay configuration (Fig. 9.2) (Hianik and Wang 2009; Song et al. 2008). Like an antibody-based assay, aptamer-based assays can have different configurations to design the biorecognition events.

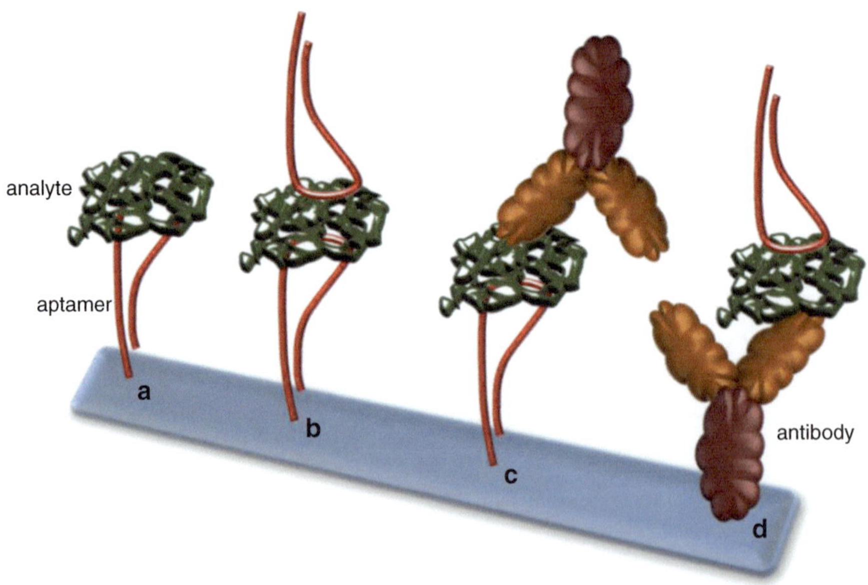

Fig. 9.2 Examples of different assays based on aptamers. (**a**) Capture of the analyte by immobilised aptamers. (**b**) Sandwich-type assay with aptamers using two aptamers specific to two different sites of the analyte. (**c**) Capture of the analyte by the immobilised aptamers whilst a secondary antibody is used to detect in a modified ELISA format. (**d**) Capture of analyte by an immobilised antibody whilst aptamer is used as a secondary probe in a modified ELISA format

Various assay configurations for aptamers have been reported. Nonetheless, the assay can be generalised into two categories: single-site binding and dual-site binding. The first simple, single-site binding assay consists of the attachment of aptamers onto a support and the interactions of aptamer/target can be directly monitored (Fig. 9.2a).

Although the single-site binding format is simple, fast and cheap, it may still impose problems with sensitivity and/or selectivity. Whereas, sandwich assay or dual-site binding assays consist of capturing the target by specific aptamer method followed by interactions with another aptamer or antibody (Fig. 9.2b–d). Again, a dual-site binding assay may enable detection of the target with high sensitivity and selectivity. However, such an assay involves several incubation steps, making it time consuming and expensive (Song et al. 2008).

9.2 Immobilisation Techniques

In order to fabricate a successful biosensor, surface engineering of the sensor transducer plays a key role. Surface chemistry has been demonstrated as a tool to engineer biosensor surfaces and is one of the most crucial aspects of biosensor construction. A proper control of the immobilisation step is essential in order to have good sensitivity, selectivity and stability of the biosensor (Campuzano et al. 2006). Broadly, immobilisation techniques could be divided into physical adsorption, covalent binding, affinity and entrapment. These techniques pose advantages and disadvantages, some of which are listed in Table 9.1.

A presentation on aptamer immobilisation strategy is presented based on the type of surface immobilisation coupled with the type of electrode surface. Briefly, we report the different types of aptasensors for the detection of cancer, based on physical adsorption, self-assembled monolayers (SAM) and polymer-based approach. Furthermore, with the advancements in the field of nanotechnology, exciting and powerful tools for the development of aptasensors have been developed. And therefore, this chapter further describes the different types of nanoparticles used for the fabrication of sensitive aptasensors for cancer diagnosis.

9.2.1 Physical Adsorption of Aptamers

Physical adsorption is a type of direct immobilisation technique of aptamers on a substrate via weak, liable bonds. The interactions involved are non-covalent such as van der Waals, hydrogen bonding, electrostatic and hydrophobic interactions. Although adsorption is one of the simplest and cost-effective methods to immobilise aptamers onto a surface of interest, it may result in a random orientation of the aptamer on the surface. Random orientation may lead to reduced activity along with reduced surface density. Furthermore, the interaction of aptamers via weak bonds could be easily broken, resulting in the loss of aptamers, limiting the immobilisation strategy. Nevertheless, Su et al. in 2007 reported the adsorption of

Table 9.1 General characterisation of immobilisation techniques

Immobilisation strategy	Type of interaction	Advantages	Disadvantages	References
Physical	Non-covalent interactions such as van der Waals, hydrophobic, electrostatic interactions	It is the simplest method which is rapid, simple and of low cost	It may result in random orientation of probe followed by instability of layer in different conditions like change in ionic strength of buffer, pH or other reagents	Du et al. (2012), Gulbakan et al. (2010)
Covalent, e.g. self-assembled monolayer (SAM)	Involves chemical bond between the probe and the electrode surface	It ensures good stability (bond can be broken under extreme conditions), results into well-ordered layer, high degree of control and thickness of electrode surface, high sensitivity and orientation	It may need prior modification of probes and linker molecules. It can be a slow, irreversible process and expensive	Li et al. (2012), Zhu et al. (2012)
Affinity	Specific interactions such as those between biotin and avidin or streptavidin	Appreciable orientation, ensures high functionalization through specificity and well controlled	It employs the use of biocompatible linkers and ends up being expensive	Ma et al. (2013), Zhang et al. (2013)
Entrapment or encapsulation	Involves trapping of probes within a polymer like pyrrole, chitosan, dopamine and acrylamide	Results in high entrapment of probes, high thermal stability, stability against nucleases, no covalent modification needed, well-controlled polymer growth	Could result in leaching of probes and reduced binding efficiency	Prabhakar et al. (2007), Jolly et al. (2016b)

low-molecular-weight ATP-binding DNA aptamers onto a microcrystalline cellulose membrane (Su et al. 2007). In 2015, Li et al. developed an optical aptasensor for the detection of multiple cancer in vitro and in vivo (Li et al. 2015). The study reports an aptasensor that binds specifically to the cell surface mucin 1 (MUC1) marker, which is overexpressed in a number of malignant tumours including prostate cancer and breast cancer (Gaidzik et al. 2013; Kimoto et al. 2013; Li et al. 2015). In 2013, Choi et al. developed a chemiluminescent-based aptasensor for the detection of PSA by immobilising DNA aptamers conjugated with fluorescent dye on magnetic Fe_3O_4 graphene oxide nanoparticles via π–π stacking. The aptamer labelled with a dye (Cy3) was immobilised on the surface of oxidised mesoporous carbon nanospheres via π–π stacking for an optical detection technique using fluorescence (Choi and Lee 2013).

To address the drawback of the random orientation of aptamers, Tzouvadaki et al. in 2016 reported the development of ultrasensitive memristive aptasensor based on a physical adsorption method for electrochemical detection of PSA. The immobilisation strategy was a combination of physical adsorption and affinity method where the silicon nanowires were first activated with hydroxyl groups by exposing the nanowires to piranha solution. First, a physically absorbed layer of biotin was prepared on the silicon nanowire. Later, the nanowire was incubated with streptavidin where it specifically binds to biotin on the surface. Finally, biotinylated aptamers were used to occupy the free spaces of streptavidin leading to the controlled orientation of DNA aptamers. The combined effect of memristor and immobilisation strategy led to detection down to attomolar levels (Tzouvadaki et al. 2016). On the other hand, Liu et al. reported a PSA aptasensor based on gold nanoparticles encapsulated by graphitized mesoporous carbon and bovine serum albumin (BSA) was used as a blocking molecule to reduce non-specific binding. Physical adsorption of nanoparticles on glassy carbon electrodes followed by affinity-based DNA aptamer immobilisation was used for the fabrication process. Differential pulse voltammetry (DPV) was used as a measurement technique with a limit of detection (LOD) of 7.58 pM (Liu et al. 2012).

Feng et al. in 2011 demonstrated a reusable graphene-based electrochemical aptasensor for the detection of nucleon on the cell surfaces of cancer cells using electrochemical impedance spectroscopy (EIS). The study reports the use of 3,4,9,10-perylene tetracarboxylic acid (PTCA), a water-soluble perylene derivative that can strongly bind on the graphene surfaces via hydrophobic and π–π interactions. Thereafter, amine terminated AS1411 aptamer was conjugated with free carboxylate groups of PTCA via ethyl(dimethylaminopropyl) carbodiimide (EDC) and *N*-hydroxysuccinimide (NHS) chemistry. The aptasensor developed successfully differentiated cancer cells from non-cancerous cells (Feng et al. 2011).

9.2.2 Aptamers and Self-Assembled Monolayers

There are many techniques reported for aptamer immobilisation depending on the electrode surface and application. Specifically for gold electrode surfaces, self-assembled monolayer (SAM) is one of the most commonly reported techniques because it results in highly controlled density and thickness of the transducer surface (Hong et al. 1999). The key aspect to take into consideration for the immobilisation of oligonucleotide-based probes on gold surfaces is to have an optimum density (Keighley et al. 2008). Thiols, sulphides and disulphides have demonstrated very high affinity towards gold by forming a covalent bond between the sulphur and gold atoms, making alkane thiol-based SAMs very popular (Bain et al. 1989; Bain and Whitesides 1989; Love et al. 2005). For instance Su et al. reported a lab-on-paper electrochemical cyto-device to demonstrate detection of cancer cells together with multi-glycan profiling on living cancer cells. In such an approach thiolated DNA aptamers specific to K562 cancer cells were directly immobilized on a three-dimensional macroporous Au-paper electrode (Su et al. 2015). More recently, Rahi et al. have reported a relatively simple aptasensor for the detection of prostate cancer based on electrodeposited gold nanospheres using an arginine template to achieve a 50 pg/mL detection limit using differential pulse voltammetry technique. Again a thiolated anti-PSA DNA aptamer was used to bind to the gold surface followed by incubation with 6-mercapto hexanol (MCH) to cover the free gold spaces and to obtain a well-aligned monolayer (Rahi et al. 2016).

Sulphur in the proximity of gold undergoes the following reaction:

$$RSH + Au \Leftrightarrow RS - Au + H^{+} + e^{-}$$

The sulphur–gold reaction is spontaneous with 80–90% coverage within the first few minutes, but a well-organised layer is formed in no less than 12–16 h (Schreiber 2000). The formation of well-organised SAMs depends not only on factors such as presence of contaminants and surface quality (roughness and purity), but also on the length of alkanethiols (Finklea 1996). The higher kinetics for longer alkyl thiols could be attributed to amplified van der Waals interactions (Darling et al. 2002). In 2005, Love et al. determined the maximum density of alkane thiol on gold surfaces to be 4.64×10^{14} molecules/cm^2 with a gap of 4.99 Å between two adjacent molecules. The time-dependent well-organised SAM formation undergoes two main steps where there is spontaneous assembly within the first few minutes. In the early stages, alkane thiols lie flat on the gold electrode surface through physisorption, called the 'lying-down' phase (Camillone et al. 1994). Thereafter follows a chemisorption process where a crystalline or a semi-crystalline structure is formed due to van der Waals forces resulting in lateral movement until a tilt angle of about 30° between the hydrocarbon chains is achieved (Love et al. 2005). It is worth mentioning that the ability of the thiols to move laterally along the gold surface results in the formation of well-ordered layering and healing the defects (Ulman 1996).

In such a process, the terminal groups of the alkanethiol affect the SAM properties that define its interaction with biological molecules. Chun et al. reported an electrochemical aptasensor for the detection of HER2, which is a breast cancer biomarker. The group fabricated the aptasensor by using first a SAM layer of 3-mercaptopropanoic acid (MPA). Thereafter, the carboxylate group of MPA was used to covalently attach amine-terminated anti-HER2 DNA aptamer using EDC/NHS activation step (Chun et al. 2013). Another anti-HER2 aptasensor based on gold micro capacitor electrodes was reported to develop a label-free capacitive aptasensor for breast cancer biomarker (Qureshi et al. 2015). A similar surface chemistry was also reported to develop an aptasensor for the detection of human liver hepatocellular carcinoma cells using electrochemical methods (Kashefi-Kheyrabadi et al. 2014). In recent years, various nanomechanical systems have been developed (Eom et al. 2011). For instance, by monitoring the changing frequencies or deflection of microcantilevers, cantilever-based biosensors can be fabricated (Boisen et al. 2011). Such an approach was utilized to develop label-free microcantilever array sensor in static mode to detect lung cancer cells with a detection limit down to 300 cells/mL. Thiol chemistry was utilized to form a SAM on eight identical microcantilever gold surfaces, by using thiolated DNA aptamer specific to lung cancer cells, followed by backfilling with MCH (Chen et al. 2016).

Furthermore, the tailoring of the SAM layers has led to the establishment of mixed SAMs. Exploring different SAM layers, Jolly et al. demonstrated an optimised SAM of co-immobilised thiol-terminated sulfo-betaine and 11-mercaptoundecanoic acid (MUA) that can be used to create enhanced anti-fouling properties. Thereafter, carboxylate groups of MUA were used to bind an amine-terminated anti-PSA DNA aptamer using EDC/NHS activation step (Jolly et al. 2015b). In 2010 Wu et al. reported a ternary SAM on screen-printed gold electrodes as a potential anti-fouling SAM for amperometric detection of oligonucleotides. Ternary SAM here comprised of an optimised ratio of thiolated oligonucleotide probe and 1,6-hexanedithiol (HDT) and further passivated with 1-mercapto-6-hexanol (MCH). Based on similar surface chemistry with screen-printed electrodes, Miodek et al. demonstrated the development of an electrochemical aptasensor for the detection of thrombin (Miodek et al. 2015). Thrombin is a protein biomarker that plays an important role in cardiovascular diseases and inflammatory process and can indicate many pathological conditions, including cancer (Yeh et al. 2014). Furthermore, by tuning the composition of mixed SAMs, the amount of functional groups can be monitored for the efficient efficacy of biosensors. In fact, it is worth mentioning that by carefully selecting the spacer molecules, desired hydrophobicity or hydrophilicity as well as significant chemical reactivity could be achieved. Such a strategy could impede non-specific binding and therein improve the electrochemical signals (Herne and Tarlov 1997).

9.2.2.1 Silane-Based SAM and Aptamers

Deviating from metallic electrode surfaces for the development of aptasensors, non-metallic surfaces have also gained a lot of interest for the development of point-of-care devices. Different surface chemistries, mostly based on silane layers, have been demonstrated for the conjugation of bioreceptors. Sagiv (1980) demonstrated for the first time a chemisorbed monolayer using siloxane chemistry on a silicon surface. Since then, many reports have been published on bioconjugation techniques using different types of silane (Bañuls et al. 2013; Haensch et al. 2010). For instance, Sharma et al. demonstrated the development of electrochemical aptasensor by covalent attachment of amine-terminated DNA aptamers on 3-(2-aminoethylamino) propyl trimethoxysilane on indium tin oxide (ITO) surfaces via a glutaraldehyde linker. The developed aptasensor was used to detect lung cancer cells in the concentration of 103–107 cells/mL with a detection limit of 103 cells/mL within 60 s (Sharma et al. 2012). More recently, Pasquardini et al. developed an innovative single-photon avalanche diode (SPAD) system for the detection of vascular endothelial growth factor (VEGF), which is a circulating protein biomarker for cancer detection. The system employs the immobilisation of thiol-terminated anti-VEGF DNA aptamer via covalent bond on the SAM of 3-mercaptopropyltrimethoxysilane on silicon dioxide wafers. A secondary detector antibody labelled with horseradish peroxidase

(HRP) was used to complete the sandwich assay with a demonstrated stability of the SAM up to 42 days (Pasquardini et al. 2015). A simple microfluidic assay was reported by Jolly et al. for an aptamer-based ELISA for both quantification and glycoprofiling of PSA for prostate cancer diagnosis. The group utilised an amine-terminated anti-PSA DNA aptamer immobilised in a microfluidic channel on a SAM of 3-glycidyloxypropyl) trimethoxysilane on glass surfaces. A secondary antibody or a lectin is used to quantify, by chemiluminescence, both the amount of PSA and its glycosylation levels (Jolly et al. 2016a).

9.2.3 Nanomaterials and Aptasensors

In the past decades, there has been a considerable interest in nanomaterials for their application in medicine and biology. Nanomaterials have been used as a physical approach to improving the pharmacokinetic and pharmacodynamic properties of different drug molecules, increasing their therapeutic benefit and, at the same time, minimizing side effects. Since then, the applications of these structures increased and nowadays have become one of the biggest research fields. Nanoparticles are particulate dispersions or solid particles with a size range from 1 to 100 nm. They are the fundamental components to fabricate nanostructures and were well described as smaller than the world of everyday objects described by Newton's laws of motion, but bigger than an atom or a simple molecule that is governed by quantum mechanics (Horikoshi and Serpone 2013).

The major goal in designing nanoparticles is their controlled size and surface properties for specific applications. Once control parameters are achieved, nanoparticles can be introduced into medicine and biology in different ways, for example as drug delivery, fluorescent biological labels, probing DNA structure, detection of pathogens and proteins, tissue engineering, imaging contrast and, one of the newest applications, aptasensors. Due to all these possibilities, nanomaterial has been applied on biosensor development. The different structures and properties already improve the sensitivity lowering the LOD down to femtomolar levels and opening new possibilities for biosensing applications (Li et al. 2010). Using these nanomaterials, different signal transducers are employed for detection.

Given that a biosensor's signal is generally proportional to the surface coverage, most methods of increasing the sensitivity of label-free biosensors revolve around surface modification to increase probe loading. An obvious method is simply multiplex detection with an array of sensors, but this often has the disadvantage of increasing the required sample volume and electronic complexity. Alternatively forming meso- and microporous surfaces with methods such as electrodeposition can provide increased surface area whilst still maintaining low sample volumes. One of the simplest ways to now increase surface area is to anchor nanoparticles to the surface. These nanoparticles may be formed from metals, oxides, semiconductors and conducting polymers, but it is the use of gold nanoparticles (AuNPs) which has attracted most attention for biosensing applications, in particular for biosensors based on optical and electrochemical transduction.

Nanomaterials are classified into different approaches, as based on the dimension or type of material that is produced (Hett 2004): in the first case, one-dimensional system such as thin film or monolayer, two-dimensional nanoparticles as carbon nanotubes (CNTs) single- or multi-walled carbon and three-dimensional nanoparticles as dendrimers and quantum dots (QDs). However, classifying them as a function of the material will be a better approach for discussing their application on biosensors.

Intentionally produced nanomaterials are divided into metallic (iron oxide, gold, silver), carbon structures (fullerene, carbon nanotubes), ceramic (silica, alumina), semiconductor (QDs), organic (protein based, DNA based, liposomes, polymers, dendrimers) and hybrid (magneto liposomes) (Estelrich et al. 2014). From all the possibilities, this topic will focus on nanomaterials that are commonly applied to aptasensors.

9.2.3.1 Carbon-Based Materials

These nanomaterials commonly take the form of hollow spheres or tubes with many potential applications, including improved films, surface coatings and applications in electronics (Wang et al. 2016). The biosensing applications are very wide, including aptasensors. For instance, Zhang et al. (2014) reported an electrochemical aptasensor for thrombin. Thrombin plays an important role in cardiovascular diseases and inflammatory process and can indicate many pathological conditions, including cancer (Yeh et al. 2014). In this study a glassy carbon electrode modified with a graphene and porphyrin nanocomposite was used to immobilise thrombin aptamer via aptamer/graphene π–π stacking interactions and aptamer/porphyrin π–π stacking simultaneously. The result displays a linear response with a LOD of 0.2 nM. This aptasensor benefits from the synergetic effects of graphene, a nanomaterial with high conductivity and high surface area, its ability to interact with porphyrin and the aptamer specificity (Zhang et al. 2014).

Recently, Nawaz et al. functionalized carbon nanotubes (CNTs) resulting in CNTs bearing benzoic acid and subsequently fabricated films on carbon. This assembly method offers an efficient protocol by using water as a solvent and one simple step for fabrication using a very small amount of CNTs. These modified electrodes were used to develop a DNA aptamer-based biosensor to detect mucin 1 (MUC1), a prevalent gene associated with breast cancer with a LOD of 0.02 U/mL (Nawaz et al. 2016).

Dendrimers are three-dimensional nano-sized polymers synthesized as spherical structures and the number of terminal groups increases exponentially as a function of the number of layers. Polyamidoamine dendrimers (PAMAM) are some of the most used and commercialised structures (Baker 2009). For instance Zhang et al. used gold electrode modified with PAMAM dendrimer to immobilise thrombin aptamer for the development of an EIS-based aptasensor (Zhang et al. 2009).

9.2.3.2 Metal Nanoparticles

Metal nanoparticles include noble metals, heavy metals, iron and metal oxides, such as titanium dioxide, that have unique physicochemical properties depending on their size and material, an easy and simple functionalisation, conductance and a high surface-to-volume ratio (Borghei et al. 2016). These properties provide an enhanced application for the commonly used biosensing techniques, particularly optical techniques.

Gold nanoparticles (AuNPs) should be highlighted as they are the most extensively studied nanomaterial and led to the development of numerous methods for molecular diagnostics, imaging, drug delivery and therapeutics because of their unique properties (Doria et al. 2012). For biosensing, AuNPs present excellent biocompatibility that allows using them for interfacing biological events. They can attach to biological probes, as described before, such as antibodies, enzyme, lectins, nucleic acids and glycans. AuNPs are already used as imaging and therapeutic agents (Borghei et al. 2016). In addition, because of their versatility in biological and medical applications, AuNPs have been investigated as signal enhancement probes for biosensors. Furthermore, AuNP conductivity permits direct electron transfer between many electroactive species and electrodes, which enables to use them for signal amplification, enhancing the analytical performance compared to other biosensor designs. Another important property is the high surface area. The diameter of AuNPs varies between 1 and 100 nm, offering a structure that increases the amount of biomolecules anchored on the surface, maintaining their bioactivity and increasing the sensitivity (Cao et al. 2011; Javier et al. 2008).

The application of AuNPs in medicine started almost two decades ago. In 2008, Javier et al. reported a platform for molecular specific using aptamer-based gold nanoparticles as contrast agents. They demonstrated an approach for prostate-specific membrane antigen (PSMA) detection obtaining reflectance images of cell lines treated with the anti-PSMA aptamer-gold conjugates (Javier et al. 2008). More recently, Borghei et al. reported a simple but highly sensitive colorimetric method based on aptamer/cell interaction for the detection of cancer cells. Cancer cells were able to uptake specific aptamers having affinity with receptors that are overexpressed in cancer cells. Such a process resulted in the removal of the aptamers from the solution, and leaving no free aptamers that can hybridise with complementary ssDNA/AuNP probes, leaving the solution red. Whereas in a negative control, with the absence of target cells or presence of normal cells, ssDNA/AuNP probes were able to hybridise with free aptamers and produced a purple solution. A linear response for MCF-7 cells was obtained with a LOD of 10 cells. This strategy can be extended to detect other receptors from different cancer cells (Borghei et al. 2016).

Quantum dots (QD) and magnetic beads (MB) are other nanoparticles very frequently reported for specific bio-application for molecular diagnostics and cancer applications. A QD is a semiconductor nanostructure that confines the motion of conduction in three spatial directions. QD has a discrete quantized energy spectrum; changing the size of QDs changes their optical properties and this property is the highlight of QDs. Due to these reasons, most of the transducers involved with QDs are optical. An example of QD application for biosensing is conjugate MUC1 aptamers with QDs to recognise MUC1 peptide. The strand includes additional bases capable of hybridisation with the complementary ssDNA sequence that carries the label. QD irradiation and FRET were observed for these MUC1-targeted probes down to 1 pM/μL (Singh et al. 2016).

MBs comprise a ferromagnetic elemental, alloy, oxide or composite structure. According to their material, they are divided into paramagnetic, antiferromagnetic and ferromagnetic (Xiao et al. 2016). As with QDs, optical biosensing is often developed for these nanomaterials due to their properties. Mostly, affinity-based method like avidin/biotin or streptavidin/biotin is used for the immobilisation of aptamers on magnetic beads. For instance, in 2006, Herr et al. developed an aptamer-conjugated MB for selective collection and detection of leukaemia cells. The study reports the conjugation of two types of aptamer-modified nanoparticles: aptamer-modified MB for target cell extraction, while aptamer-modified fluorescent nanoparticles for the detection using fluorescent imaging. By doing so, leukaemia cells were extracted from complex samples including whole-blood samples (Herr et al. 2006). In 2014, Hu et al. proposed a simple optical aptasensor for cancer biomarker AGR2 detection using UV–vis spectrometry. In this case, the aptasensor is sandwich-typed AuNPs/DNA/MBs where the aptamers/MBs target proteins and DNA probes on the AuNPs compete with proteins to hybridise with catchers. As a result, the increasing number of target proteins reduces the possibility of the sandwich structure formation with a picomolar range of detection limit (Hu et al. 2015). The sum of different properties allows an extraordinary performance of the aptasensor.

Researchers have also demonstrated the use of bivalent metal ions (M^{2+}) for biosensing applications. Immobilisation of aptamers to the electrode surface can also be performed via histidine-tag/M^{2+}/Nα,Nα-bis(carboxymethyl)-L-lysine hydrate (ANTA) chemistry. During this process, bivalent metal ions such as Cu^{2+} bound to ANTA anchored on the surface by a tetravalent chelation, leaving two available coordination sites for linking of histidine-modified aptamers. ANTA/Cu^{2+} form a stable complex and the binding of Cu^{2+} to the ANTA-modified self-assembled monolayer was studied by Stora et al., showing a dissociation constant of 5 nM obtained with impedimetric measurements (Stora et al. 1997). Because of the high affinity to the histidine sequence, the complex of ANTA with Cu^{2+} but also with other metal cations such as Ni^{2+}, Zn^{2+} and Co^{2+} is widely used in protein purification methods and is known as 'immobilised metal-ion affinity chromatography' (IMAC) (Kronina et al. 1999; Ueda et al. 2003). The binding of histidine-modified molecules can be reversible; however, it needs a presence of highly concentrated (~0.2 μM) imidazole solution (Haddour et al. 2005).

Such strategy based on His-tag/M^{2+}/ANTA chemistry has already been described in the case of immunosensors (Chebil et al. 2010) and aptasensors (Xu et al. 2013) and in both cases a wide linear response range was obtained. For example, the aptasensor based on ANTA/Cu^{2+} complex allowed to detect thrombin protein with a detection limit of 4.4 pM (Xu et al. 2013).

9.2.3.3 Assay Designs with Nanoparticles and Aptamers

The integration of nanoparticles with biosensors is also divided into immobilisation, amplification or both at the same time, as illustrated in Fig. 9.3. For example, AuNPs can be immobilised on different carbon-based matrices such as a carbon nanotube, graphene or graphene oxide, gold surfaces or other polymers (Sardar et al. 2009). This is an effective method to increase the amount of probes immobilised,

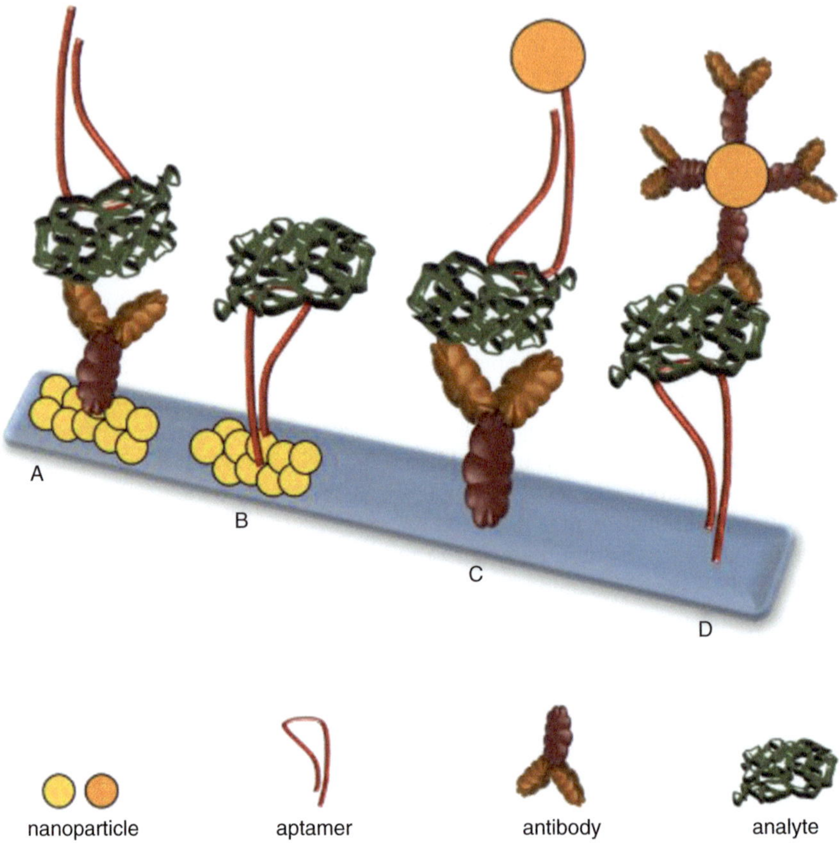

Fig. 9.3 Examples of different assays based on the use of nanoparticles, where nanoparticles can be either used to modify the surface of electrode or used as a label for amplification

especially thiol molecules which present high affinity to bind AuNPs. Moreover, due to the availability of high surface area of nanoparticles, the amount of biomolecules anchored is increased, consequently amplifying the signal.

Furthermore, metal NPs can be used to mediate reactions amplifying the signal. For example, capture probes can be first immobilised on a substrate. Thereafter, AuNPs modified with target probes are introduced for specific recognition, resulting in signal amplification. Using AuNPs as a well-established model, with the properties previously described and, additionally, due to their high absorption coefficient as well as the ability to enhance electromagnetic fields and fluorescence, allows exploring these particles in many different ways for biosensing purposes, especially in the development of optical, electrochemical and piezoelectric biosensors (Sardar et al. 2009). The different transducers can be applied to many biological probes, including aptamers.

For instance, Chai et al. in 2011 reported the use of SAMs to deposit AuNPs on electrode surfaces. The study reports a simple strategy to immobilise streptavidin-coated AuNPs on the surface of gold by employing a SAM of 1,3-propanedithiol. Thereafter, biotinylated DNA aptamer specific to PDGF-BB was immobilised using the conventional affinity interaction between streptavidin and biotin. Such a platform was used to capture the target PDGF-BB and finally a secondary label was used as a detector. The label comprised of AuNPs modified with *N*-(aminobutyl)-*N*-ethylisoluminol (ABEI) and DNA aptamers. The binding event of the sandwich assay was monitored by electrochemiluminescence method. The signal amplification by the AuNP was specific, simple and stable, with a detection limit as low as 2.7×10^{-14} M being achieved (Chai et al. 2011).

Other proteins, such as thrombin, have been extensively studied. Fang et al. proposed an electrochemiluminescence (ECL) aptasensor where aptamers labelled with AuNPs were first immobilised onto ITO electrode surface and, after catching the thrombin, signal aptamers tagged with ECL labels were attached to the assembled electrode surface. As a result, a sandwich-type assay was formed achieving 10 nM as LOD (Fang et al. 2008). Thrombin can also be detected using other approaches. Screen-printed electrodes (SPE) are economical electrochemical substrates with advantageous properties, such as disposability and simplicity and can be used for the rapid in situ analysis. Yeh et al. developed a system with a carbon SPE capture thrombin and an amplifier for recognising thrombin which is the multiple molecules of anti-thrombin antibody-modifying AuNPs. The electrochemical response presented a LOD of 1.5 pM (Yeh et al. 2014).

The whole cell can be captured by the aptasensor with good selectivity once they usually have specific proteins on the surface. Mucin 1 is a tumour maker protein in human breast carcinoma MCF-7 cells. An assay used aptamers to bind tumour markers on the surface of cancer cells, in this case, MCF-7 cells. Nanoporous materials modified with mucin 1 aptamers attach on the cell improving the biosensor performance and exhibiting a detection limit of 38 cells/mL (Yan et al. 2013).

Optical transducers are also applied in cancer diagnosis using aptasensors. Colorimetric detection is a signal transducer based on absorbance curves with low cost, fastness and a point-of-care compatible testing technique for cancer cells. Wang et al. recently designed a specific detection for MCF-7 cells with mucin 1 aptamer, the same samples previously described. But, in this case, the aptamer and PtAu nanoparticle with high catalytic activity were able to differentiate cancer cells from non-cancer cells and different cancer cell types with a LOD of 10 cells/mL in phosphate buffer solution and in the serum samples (Wang et al. 2015). Different nanoparticles can be used based on their properties. In some cases, nanoparticles can be combined. For example, Ye et al. published an interesting study with bimetallic nanoparticle (Cu–AuNP) combined with an iodide-catalysed system. The nanoparticle was modified with aptamers to human leukaemia CCRF-CEM cells, thus indirectly inducing the colorimetric signal variation of the system. The LOD of 5 cells in 100 μL shows a strategy that can be extended to other cancer cell assays (Ye et al. 2015).

Similar strategies have been developed for other optical transducers. Surface-enhanced Raman scattering (SERS) was applied to detect vascular endothelial growth factor (VEGF). A different strategy involves silver nanoparticle-ornamented gold nanoparticle pyramids (Ag–AuNPs) using an aptamer-based sensor that was ultrasensitive with a LOD of 22.6 aM. Limits lower than this have already been achieved with optical techniques (Zhao et al. 2015). Vance and Sandros presented an application of surface plasmon resonance imaging (SPRi) system to detect 7 zM (or 5 fg/mL) of C-reactive protein (CRP), in this case, combining aptamer-modified quantum dots (QDs) and microwave-assisted surface functionalisation (Vance and Sandros 2014).

Deviating from conventional nanoparticles, graphene oxide (GO) has also been used to develop aptasensors based on a fluorescence technique. By using the inherent capability of the ring in DNA guanine residues to absorb on the surface of graphene oxide by π–π interactions, He et al. demonstrated the development of an aptasensor that characterises epithelial malignancy by targeting MUC1. In such an approach, an anti-MUC1 labelled with a fluorescent dye (Cy5) was adsorbed on the GO resulting in close proximity of the dye to the surface. Consequently, a quenching effect is observed via energy transfer from dyes to GO. However in the presence of the target, the aptamer changes conformation resulting into increased distance between the dye and the GO, inducing the fluorescence restoration. The aptasensor was successfully tested in both buffer and blood serum with a detection limit of 28 nM (He et al. 2012). A similar strategy of using GO and the quenching effect was used for the detection of hepatocellular carcinoma, where the aptasensor was able to detect human liver cancer cell lines SMMC-7721 with a detection limit of 200 cells in 200 μL buffer (Xie et al. 2014).

9.2.4 Polymer-Based Aptasensors

The cost has always been a crucial factor in the development of novel biosensor devices for medical purposes. In order to reduce the cost of biosensor fabrication, the adoption of noble metals and their cleanroom processing are required to be kept at a minimum (Kiilerich-Pedersen et al. 2011). These factors have led to a shift from the use of gold and platinum to degradable polymer materials. As a result, the application of polymers is experiencing an increasing importance over traditional systems, especially within the area of immobilisation of aptamers for the detection of damages within target DNA (Liao et al. 2008; Radi 2011). Moreover, the development of novel polymeric materials led to a new trend within the area of biosensors for cancer diagnostics. The necessity for introducing the adoption of polymers aroused with the requirement of individualised and tailored methods of treatment for a more heterogeneous disease such as cancer (Luk and Zhang 2014). Polymer-based biosensors offer the ability of theranostic applications (emergence of therapy and diagnostics imaging into a single package), with their additional advantage of possessing excellent biocompatibility (Luk and Zhang 2014). These polymer-based nanomaterials demonstrate desirable biodegradability and structural versatility. Except at high concentrations, biopolymers are typically non-toxic and naturally degrade into safe materials (Clawson et al. 2011; Hu et al. 2010).

It is worth mentioning that depending on the type of polymer, different immobilisation strategies for aptamers have been reported. For example, a reagent-less electrochemical transduction-based aptasensor for the detection of PSA was developed: the study demonstrated the elimination of any redox labels by adopting a quinone-containing conducting polymer (Souada et al. 2015) on glassy carbon electrode surface. Short amine-terminated DNA aptamers have been first immobilised on the quinone-based conducting copolymer via EDC/NHS chemistry. When subjected to PSA, a strong current decrease ('signal-off') was generated due to heavier molecules of PSA compared to aptamer strands on the probe surface. This was next switched to a current drop ('signal-on') by the hybridisation of probe aptamer with its complementary strand DNA which breaks PSA–aptamer interactions. As a result, the developed switch signal system was able to detect PSA in ng/mL range and also evaluated the PSA–aptamer dissociation constant (K_d), of ca. 2.6 nM. This dual-check system provided a full assurance of a perfectly specific recognition event (Souada et al. 2015).

There are many conducting polymers like polypyrrole (PPy), polythiophene and polyaniline that have been used for biosensor construction. Among them, PPy is one of the most extensively used conducting polymers in the design of bioanalytical sensors apart from polythiophene and polyaniline (Peng et al. 2009; Ramanavičius et al. 2006). This is due to its copious properties such as redox activity (Han et al. 2005), ion exchange and ion discrimination capacities (Johanson et al. 2005; Weidlich et al. 2005), strong absorptive properties (Azioune et al. 2005; Chehimi et al. 1999), catalytic activity (Khomenko et al. 2005) as well as biocompatibility (Wang et al. 2004). PPy as a polymer can be further characterised by its high electrical conductivity, hydrophilic character and high stability in water (Andrade 1985). In fact, its low oxidation potential enables a pyrrole polymer film to be grown from aqueous solutions which is compatible with most of the biological elements (Asavapiriyanont et al. 1984). Researchers have also worked on the modification of pyrrole monomers in order to provide enhanced anti-fouling properties. For example, in 2002 Rodrigez et al. synthesised pyrrole modified with biotin set apart by polyethylene glycol chain which prevented biosensor against non-specific interactions. More recently, Jolly et al. (2016c) reported the development of an aptasensor for AMACR detection using a voltammetry detection technique. In the study, a simple and efficient electrochemical patterning of amine-bearing PEG derivatives on PPy films has been developed. Such a method paved way to simple pyrrole monomer modification and demonstrated enhanced anti-fouling properties of PPy with PEG. His-tagged DNA aptamers specific to AMACR were then immobilised via coordination chemistry with copper ions. A very low detection limit of 1.4 fM was established in human plasma samples. The progression of a tumour and its transformation to different stages can be related to platelet-derived growth factor (PDGF), especially PDGF-B chain. Recently in 2016, Lee et al. fabricated an aptasensor based on multidimensional hybrid conductive nanoplate for the detection of the PDGF. PDGF has emerged as a critical cancer biomarker as it is associated with diverse cancers and other diseases (Lee et al. 2016; Shih et al. 2004). The sensor was fabricated by using multidimensional hybrid carboxylated polypyrrole plates (MHCPPs) that were functionalised with the PDGF-B-specific DNA aptamer. The vertical decoration of the polypyrrole nanosheets has managed to maximise the active surface area of the MHCPPs. The strategy demonstrated a dramatic increase in the interaction between the developed MHCPP-based sensor and the PDGF-BB analyte. Detection limit as low as 1.78 fM has been achieved for PDGF-B, by using field-effect transistor (FET)-type aptamer sensor (Lee et al. 2016).

Polymer-based approaches have also penetrated graphene-based FETs and are witnessing a rapid development growth and even considered as an alternative for post-silicon electronics. Kwon et al. demonstrated how to grow polypyrrole-converted nitrogen-doped few-layer graphene (PPy-NDFLG) on copper substrates which provided the recognition of the target molecules at really low concentrations of 100 fM. Such a process was carried out using a combined technique of chemical vapour deposition and vapour deposition polymerisation. The developed platform was used to immobilise amino-terminated anti-VEGF RNA aptamer using Schiff-base reaction via glutaraldehyde-conjugated 1,5-diaminonaphthalene (DAN) (Kwon et al. 2012).

Development of aptasensor based on single-polymer-based nanowire has also been reported. Huang et al. in 2011 reported a single-step electrochemical deposition of single-PPy nanowire between two gold electrode junctions in a patterned polymethylmethacrylate (PMMA) nanochannel. A different immobilisation strategy was utilised where a mixture of pyrrole monomer and aptamer was used, with the aptamers encapsulated within the polymer without any covalent bonding. By using microfluidic systems the group demonstrated the detection of immunoglobulin E (IgE) and mucin 1 (MUC1) with their specific aptamer. Mucin 1 is a protein biomarker which has been reported to be over-expressed in almost all human epithelial cell carcinomas like breast cancer, ovarian cancer and lung cancer (Croce et al. 2003; Hough et al. 2000; Maeshima et al. 1997). The detection of the protein immunoglobulin E (IgE) was achieved within a range from 0.01 to 100 nM and the aptasensor performed excellent sensitivity with a fast response and rapid stabilisation time (~20 s). A detection limit of 2.66 nM was obtained for MUC1 using conductance measurements, which is significantly sensitive compared to commercially available MUC1 diagnosis assay (800 nM) (Huang et al. 2011).

Although conducting polymers have always been in the interest for the development of electrochemical biosensors, there are also reports on the use of non-conducting polymers such as chitosan. For instance, Tahmasebi et al. adopted carbon nanotube (CNT)-based polymer materials in order to provide a sensitive electrochemical aptasensor for the detection of PSA using EIS technique. The study demonstrates the benefits of nanomaterials composed of carboxylic acid-functionalised CNTs and chitosan for the immobilisation of PSA aptamer on probe surface with a LOD of 22 pM. The experimental results also proved the higher aptamer immobilising capability of chitosan-CNT composite when compared to CNTs or chitosan alone (Tahmasebi and Noorbakhsh 2016).

9.2.4.1 Molecular Imprinting and Aptamers

In response to the challenges faced by the conventional methods of protein imprinting for biosensor studies, molecular imprinting has adopted new approaches to overcome its limitations (Menger et al. 2016). One of the approaches is to incorporate aptamers within the molecularly imprinted polymers (MIPs). A study carried out on an aptamer-MIP hybrid receptor for the detection of prostate-specific antigen in 2016 reported much more sensitive recognition characteristics compared to that of the aptamer alone (Jolly et al. 2016b). The MIP cavity that had been developed was observed to act synergistically with the enclosed aptamer forming a hybrid receptor which provided much sensitive recognition characteristics (termed as 'apta-MIP'). A simple strategy was demonstrated by using dopamine monomers and PSA-specific thiolated DNA aptamers. A thiolated anti-PSA DNA aptamer was complexed with PSA prior to being immobilised on the gold electrode surface. Thereafter, a controlled electropolymerisation of dopamine around the complex enabled the entrapment of the complex. The PSA was then removed from the template and the fabricated sensor was used to study the subsequent rebinding of PSA. The study has adopted EIS as a method of evaluation by looking into capacitance changes, where the apta-MIP sensor showed a detection limit of 1 pg/mL of PSA (Jolly et al. 2016b). Within the same year, another novel method of luminescent 'double recognition' for the detection of ENR (enrofloxacin) was developed. This method also had two stages of recognition. Firstly, ENR-specific biotinylated aptamers were immobilised on a surface of up-conversion nanoparticles (UCNPs) in order to capture and concentrate ENR as the initial imprinting recognition safeguard. This was then followed by the polymerisation of methacrylic acid monomers neighbouring the aptamers of ENR, which interacted with the residual groups of ENR by using MIP techniques as the second recognition safeguard. The sensor demonstrated detection and quantification limits of 0.04 and 0.12 ng/mL, indicating the feasibility of the method for the detection of ENRs in real samples (Liu et al. 2017).

The studies have shown that MIPs offer an exchange rate of the target aptamer that is significantly higher than that of the antibodies. However, the recent methods of using MIPs have been only developed for the detection of a limited range of proteins. As a result, the research in MIPs still requires improvement in terms of sensitivity and the application on real samples (Menger et al. 2016).

9.3 Outlook

Nucleic acid aptamers represent a challenging and fascinating venue and a possible replacement of antibodies for both therapeutics and diagnostics. Indeed early technology for the application of aptamers for sensing applications has its own limitations but it has continued for more than two decades with the integration of many modifications and advancements in the field. Aptamers, although they have been raised against a number of targets over the last two decades, are expected to see further improvement in their variety, affinity, diversity and half-lives. Therapeutic use of aptamers is a well-established field; for example, aptamer compound named Pegaptanib was approved as a drug for clinical use for endovascular age-related macular degeneration (Gragoudas et al. 2004). Conversely, in the field of diagnostics, especially diagnostic systems based on biomolecular binding events, are still under the dominancy of immunoassays. However, the diagnostic field is now showing that with the use of aptamers, some of the limits of current diagnostic tools can be circumvented, such as flexibility for signal transduction and detection (Liu et al. 2009). For instance, the ability of the aptamers to distinguish small differences between proteins sharing similar surface homology may allow aptamers to differentiate cells based on cancerous and non-cancerous. The use of natural and synthetic nucleotides is still developing and paving the way towards advanced biosensor development. Developments in biochemistry and molecular biology have led to a deeper understanding of the role of nucleic acids and showed that the functions they play are far greater than originally expected. This leads to a new world of biosensing applications, where nucleic acid-based biosensing approaches can have an unparalleled impact on clinical diagnosis, prognosis and monitoring. It can be seen from the plethora of reported literature how researchers from different fields are coming together to realise high-throughput aptamer-based biosensors for use with complex matrix samples such as clinical or environmental. The ease of manipulation of oligonucleotides, controlled surface chemistry approaches and 'straightforward' charge distribution make them optimal bioreceptors for biosensing applications. Moreover, the commercialisation of biosensors has been fuelled right after the first glucose test in a PoC format.

However, the process of commercialisation of aptamer-based biosensors for the detection of cancer biomarkers is still at an early stage (proof of concept) and requires further developments. There is still not a self-contained answer to why nucleic acid aptamers have not yet penetrated into the clinical laboratories (Baird 2010). Although there are several challenges that need to be addressed for real clinical application, one of the most important is related to the conformational change of aptamers. Most of the aptamers generated via SELEX is known to have well-known interaction between the aptamer and its target which is dependent on the conformation of aptamers. However, such a conformation is largely affected by chemical and physical environment. As a result, the aptamers selected through in vitro SELEX procedures might have decreased or completely lose their binding efficiency. Overall, the development of aptamer-based biosensors for biomarker detection is expected to attract increasing interest because of its ease of synthesis and the possibilities of multiple modifications. Furthermore, there are many successful reports that have been published for cancer diagnosis; however, most of them have been carried out either in ideal buffer solutions or in vitro cultures. There are very few reports that have shown the tests with real cultures or in animals.

These studies not only demonstrate the enormous potential, but also prove how they can be used for a wide range of other biomarkers for various diseases that exploit target/probe features similar to those of the systems that have been reported previously. It has become apparent that the aptasensor field has reached a new level of maturity where it has been employed to detect multiple biomarkers. Aptasensors have already shown their specificity and versatility to be applied to different surfaces and systems with different transducers. The use of nanoparticles combined to aptamers addressing the increase of sensibility opens the door to new detection levels not previously conceived. The future of these approaches aims to develop a multiplexing platform with the capacity to distinguish different biomarkers with ultra-low levels.

Nevertheless, the sensitivity of an aptasensor is affected not only by the surface chemistry used but also the analytical method used for the detection. Furthermore, the development of a multiplexed platform for parallel sensing of different biomarkers of cancer would help assist clinicians with deeper information on the pathological and physiological state of the patient (in particular the disease). Such a device could be represented as a point-of-use device that can provide a first assessment of the patient's state which would accelerate the diagnosis and or prognosis speed. Not only limited to diagnosis, the device can be used for surveillance purposes in order to monitor patients at risk or those being treated (either post-surgery or during medication). In the near future, it is likely that cancer detection using aptamers will undoubtedly benefit from the integration of novel aptamers with miniaturised transducer platforms, and therefore in some regard revolutionise the cancer diagnostics on a global platform.

Acknowledgements PJ was funded by the European Commission FP7 Programme through the Marie Curie Initial Training Network PROSENSE (Grant No. 317420, 2012–2016). MRB was funded by FAPESP (process number 2013/26133-7). SU was funded by the UK EPSRC Centre for Doctoral Training in Sustainable Chemical Technologies. SKA was funded by the European Commission's Horizon 2020 Programme through a Marie Skłodowska-Curie Individual Fellowship (Grant No. 655176). MM and PE acknowledge support from FAPESP and the University of Bath through the SPRINT program.

References

Andrade JD (ed) (1985) Surface and interfacial aspects of biomedical polymers: volume 1—surface chemistry and physics. Plenum Press, New York

Asavapiriyanont S, Chandler G, Gunawardena G, Pletcher D (1984) The electrodeposition of polypyrrole films from aqueous solutions. J Electroanal Chem Interfacial Electrochem 1:229–244

Azioune A, Siroti F, Tanguy J, Jouini M, Chehimi MM, Miksa B, Slomkowski S (2005) Interactions and conformational changes of human serum albumin at the surface of electrochemically synthesized thin polypyrrole films. Electrochim Acta 50:1661–1667

Bain CD, Whitesides GM (1989) Formation of monolayers by the coadsorption of thiols on gold: variation in the length of the alkyl chain. J Am Chem Soc 111:7164–7175

Bain CD, Troughton EB, Tao YT, Evall J, Whitesides GM, Nuzzo RG (1989) Formation of monolayer films by the spontaneous assembly of organic thiols from solution onto gold. J Am Chem Soc 111:321–335

Baird GS (2010) Where are all the aptamers? Am J Clin Pathol 134:529–531

Baker JR (2009) Dendrimer-based nanoparticles for cancer therapy. Hematology Am Soc Hematol Educ Program 2009:708–719

Bañuls MJ, Puchades R, Maquieira Á (2013) Chemical surface modifications for the development of silicon-based label-free integrated optical (IO) biosensors: a review. Anal Chim Acta 777:1–16

Boisen A, Dohn S, Keller SS, Schmid S, Tenje M (2011) Cantilever-like micromechanical sensors. Rep Prog Phys 74:036101

Borghei YS, Hosseini M, Dadmehr M, Hosseinkhani S, Ganjali MR, Sheikhnejad R (2016) Visual detection of cancer cells by colorimetric aptasensor based on aggregation of gold nanoparticles induced by DNA hybridization. Anal Chim Acta 904:92–97

Camillone N III, Eisenberger P, Leung T, Schwartz P, Scoles G, Poirier G, Tarlov M (1994) New monolayer phases of n-alkane thiols self-assembled on au (111): preparation, surface characterization, and imaging. J Chem Phys 101:11031–11036

Campuzano S, Pedrero M, Montemayor C, Fatás E, Pingarrón JM (2006) Characterization of alkanethiol-self-assembled monolayers-modified gold electrodes by electrochemical impedance spectroscopy. J Electroanal Chem 586:112–121

Cao X, Ye Y, Liu S (2011) Gold nanoparticle-based signal amplification for biosensing. Anal Biochem 417:1–16

Castillo G, Trnkova L, Hrdy R, Hianik T (2012) Impedimetric aptasensor for thrombin recognition based on CD support. Electroanalysis 24:1079–1087

Cavic BA, Thompson M (2002) Interfacial nucleic acid chemistry studied by acoustic shear wave propagation. Anal Chim Acta 469:101–113

Centi S, Tombelli S, Minunni M, Mascini M (2007) Aptamer-based detection of plasma proteins by an electrochemical assay coupled to magnetic beads. Anal Chem 79:1466–1473

Chai Y, Tian D, Gu J, Cui H (2011) A novel electrochemiluminescence aptasensor for protein based on a sensitive N-(aminobutyl)-N-ethylisoluminol-functionalized gold nanoprobe. Analyst 136:3244–3251

Chebil S, Hafaiedh I, Sauriat-Dorizon H, Jaffrezic-Renault N, Errachid A, Ali Z, Korri-Youssoufi H (2010) Electrochemical detection of d-dimer as deep vein thrombosis marker using single-chain d-dimer antibody immobilized on functionalized polypyrrole. Biosens Bioelectron 26:736–742

Chehimi MM, Abel ML, Perruchot C, Delamar M, Lascelles SF, Armes SP (1999) The determination of the surface energy of conducting polymers by inverse gas chromatography at infinite dilution. Synth Met 104:51–59

Chen X, Pan Y, Liu H, Bai X, Wang N, Zhang B (2016) Label-free detection of liver cancer cells by aptamer-based microcantilever biosensor. Biosens Bioelectron 79:353–358

Choi HK, Lee JH (2013) Role of magnetic Fe_3O_4 graphene oxide in chemiluminescent aptasensors capable of sensing tumor markers in human serum. Anal Methods 5:6964–6968

Chun L, Kim SE, Cho M, Choe WS, Nam J, Lee DW, Lee Y (2013) Electrochemical detection of HER2 using single stranded DNA aptamer modified gold nanoparticles electrode. Sensors Actuators B Chem 186:446–450

Clawson C, Ton L, Aryal S, Fu V, Esener S, Zhang L (2011) Synthesis and characterization of lipid–polymer hybrid nanoparticles with pH-triggered poly (ethylene glycol) shedding. Langmuir 27:10556–10561

Croce MV, Isla-Larrain MT, Demichelis SO, Segal-Eiras A, Gori JR, Price MR (2003) Tissue and serum MUC1 mucin detection in breast cancer patients. Breast Cancer Res Treat 81: 195–207

Darling S, Rosenbaum A, Wang Y, Sibener S (2002) Coexistence of the $(23\times\sqrt{3})$ au (111) reconstruction and a striped phase self-assembled monolayer. Langmuir 18:7462–7468

Davis KA, Abrams B, Lin Y, Jayasena SD (1996) Use of a high affinity DNA ligand in flow cytometry. Nucleic Acids Res 24:702–706

Doria G, Conde J, Veigas B, Giestas L, Almeida C, Assunção M, Rosa J, Baptista PV (2012) Noble metal nanoparticles for biosensing applications. Sensors 12:1657–1687

Du M, Yang T, Zhao C, Jiao K (2012) Electrochemical logic aptasensor based on graphene. Sensors Actuators B Chem 169:255–260

Ellington AD, Szostak JW (1990) In vitro selection of RNA molecules that bind specific ligands. Nature 346:818–822

Eom K, Park HS, Yoon DS, Kwon T (2011) Nanomechanical resonators and their applications in biological/chemical detection: nanomechanics principles. Phys Rep 503:115–163

Estelrich J, Quesada-Pérez M, Forcada J, Callejas-Fernández J (2014) Introductory aspects of soft nanoparticles. In: Callejas-Fernández J, Estelrich J, Quesada-Pérez M, Forcada J (eds) Soft nanoparticles for biomedical applications. Royal Society of Chemistry, Cambridge, pp 1–18

Fang L, Lü Z, Wei H, Wang E (2008) An electrochemiluminescence aptasensor for detection of thrombin incorporating the capture aptamer labeled with gold nanoparticles immobilized onto the thio-silanized ITO electrode. Anal Chim Acta 628:80–86

Feng L, Chen Y, Ren J, Qu X (2011) A graphene functionalized electrochemical aptasensor for selective label-free detection of cancer cells. Biomaterials 32:2930–2937

Finklea H (1996) Electrochemistry of organized monolayers of thiols and related molecules on electrodes. Electroanal Chem 19:110–335

Gaidzik N, Westerlind U, Kunz H (2013) The development of synthetic antitumour vaccines from mucin glycopeptide antigens. Chem Soc Rev 42:4421–4442

Gragoudas ES, Adamis AP, Cunningham ET Jr, Feinsod M, Guyer DR (2004) Pegaptanib for neovascular age-related macular degeneration. N Engl J Med 351:2805–2816

Gulbakan B, Yasun E, Shukoor MI, Zhu Z, You M, Tan X, Sanchez H, Powell DH, Dai H, Tan W (2010) A dual platform for selective analyte enrichment and ionization in mass spectrometry using aptamer-conjugated graphene oxide. J Am Chem Soc 132:17408–17410

Haddour N, Cosnier S, Gondran C (2005) Electrogeneration of a poly (pyrrole)-NTA chelator film for a reversible oriented immobilization of histidine-tagged proteins. J Am Chem Soc 127:5752–5753

Haensch C, Hoeppener S, Schubert US (2010) Chemical modification of self-assembled silane based monolayers by surface reactions. Chem Soc Rev 39:2323–2334

Han DH, Lee HJ, Park SM (2005) Electrochemistry of conductive polymers XXXV: electrical and morphological characteristics of polypyrrole films prepared in aqueous media studied by current sensing atomic force microscopy. Electrochim Acta 50:3085–3092

Hayat A, Sassolas A, Marty JL, Radi AE (2013) Highly sensitive ochratoxin A impedimetric aptasensor based on the immobilization of azido-aptamer onto electrografted binary film via click chemistry. Talanta 103:14–19

He Y, Lin Y, Tang H, Pang D (2012) A graphene oxide-based fluorescent aptasensor for the turn-on detection of epithelial tumor marker mucin 1. Nanoscale 4:2054–2059

Herne TM, Tarlov MJ (1997) Characterization of DNA probes immobilized on gold surfaces. J Am Chem Soc 119:8916–8920

Herr JK, Smith JE, Medley CD, Shangguan D, Tan W (2006) Aptamer-conjugated nanoparticles for selective collection and detection of cancer cells. Anal Chem 78:2918–2924

Hett A (2004) Nanotechnology: small matter, many unknowns, Risk perception series. Swiss Reinsurance Company, Zürich

Hianik T, Wang J (2009) Electrochemical aptasensors–recent achievements and perspectives. Electroanalysis 21:1223–1235

Hong HS, Kim SJ, Lee KS (1999) Long-term oxidation characteristics of oxygen-added modified Zircaloy-4 in 360° C water. J Nucl Mater 273:177–181

Horikoshi S, Serpone N (2013) Microwaves in nanoparticle synthesis: fundamentals and applications. Wiley, New York

Hough CD, Sherman-Baust CA, Pizer ES, Montz F, Im DD, Rosenshein NB, Cho KR, Riggins GJ, Morin PJ (2000) Large-scale serial analysis of gene expression reveals genes differentially expressed in ovarian cancer. Cancer Res 60:6281–6287

Hu CMJ, Kaushal S, Cao HST, Aryal S, Sartor M, Esener S, Bouvet M, Zhang L (2010) Half-antibody functionalized lipid–polymer hybrid nanoparticles for targeted drug delivery to carcinoembryonic antigen presenting pancreatic cancer cells. Mol Pharm 7:914–920

Hu Y, Li L, Guo L (2015) The sandwich-type aptasensor based on gold nanoparticles/DNA/magnetic beads for detection of cancer biomarker protein AGR2. Sensors Actuators B Chem 209:846–852

Huang J, Luo X, Lee I, Hu Y, Cui XT, Yun M (2011) Rapid real-time electrical detection of proteins using single conducting polymer nanowire-based microfluidic aptasensor. Biosens Bioelectron 30:306–309

Iliuk AB, Hu L, Tao WA (2011) Aptamer in bioanalytical applications. Anal Chem 83:4440–4452
Javier DJ, Nitin N, Levy M, Ellington A, Richards-Kortum R (2008) Aptamer-targeted gold nanoparticles as molecular-specific contrast agents for reflectance imaging. Bioconjug Chem 19:1309–1312
Johanson U, Marandi M, Tamm T, Tamm J (2005) Comparative study of the behavior of anions in polypyrrole films. Electrochim Acta 50:1523–1528
Jolly P, Formisano N, Estrela P (2015a) DNA aptamer-based detection of prostate cancer. Chem Pap 69:77–89
Jolly P, Formisano N, Tkáč J, Kasák P, Frost CG, Estrela P (2015b) Label-free impedimetric aptasensor with antifouling surface chemistry: a prostate specific antigen case study. Sensors Actuators B Chem 209:306–312
Jolly P, Damborsky P, Madaboosi N, Soares RR, Chu V, Conde JP, Katrlik J, Estrela P (2016a) DNA aptamer-based sandwich microfluidic assays for dual quantification and multi-glycan profiling of cancer biomarkers. Biosens Bioelectron 79:313–319
Jolly P, Tamboli V, Harniman RL, Estrela P, Allender CJ, Bowen JL (2016b) Aptamer–MIP hybrid receptor for highly sensitive electrochemical detection of prostate specific antigen. Biosens Bioelectron 75:188–195
Jolly P, Miodek A, Yang DK, Chen LC, Lloyd MD, Estrela P (2016c) Electro-engineered polymeric films for the development of sensitive aptasensors for prostate cancer marker detection. ACS Sens 1:1308–1314
Kashefi-Kheyrabadi L, Mehrgardi MA, Wiechec E, Turner AP, Tiwari A (2014) Ultrasensitive detection of human liver hepatocellular carcinoma cells using a label-free aptasensor. Anal Chem 86:4956–4960
Keighley SD, Li P, Estrela P, Migliorato P (2008) Optimization of DNA immobilization on gold electrodes for label-free detection by electrochemical impedance spectroscopy. Biosens Bioelectron 23:1291–1297
Keum JW, Bermudez H (2009) Enhanced resistance of DNA nanostructures to enzymatic digestion. Chem Commun (45):7036–7038.
Khomenko V, Frackowiak E, Beguin F (2005) Determination of the specific capacitance of conducting polymer/nanotubes composite electrodes using different cell configurations. Electrochim Acta 50:2499–2506
Kiilerich-Pedersen K, Poulsen CR, Daprà J, Christiansen NO, Rozlosnik N (2011) Polymer based biosensors for pathogen diagnostics. In: Somerset V (ed) Environmental biosensors. INTECH, Rijeka, pp 193–212
Kimoto M, Yamashige R, Matsunaga KI, Yokoyama S, Hirao I (2013) Generation of high-affinity DNA aptamers using an expanded genetic alphabet. Nat Biotechnol 31:453–457
Kronina VV, Wirth HJ, Hearn MT (1999) Characterisation by immobilised metal ion affinity chromatographic procedures of the binding behaviour of several synthetic peptides designed to have high affinity for Cu (II) ions. J Chromatogr A 852:261–272
Kwon OS, Park SJ, Hong JY, Han AR, Lee JS, Lee JS, JH O, Jang J (2012) Flexible FET-type VEGF aptasensor based on nitrogen-doped graphene converted from conducting polymer. ACS Nano 6:1486–1493
Lee J, Kim W, Cho S, Jun J, Cho KH, Jang J (2016) Multidimensional hybrid conductive nanoplate-based aptasensor for platelet-derived growth factor detection. J Mater Chem B 4:4447–4454
Li Y, Schluesener HJ, Xu S (2010) Gold nanoparticle-based biosensors. Gold Bull 43:29–41
Li Y, Deng L, Deng C, Nie Z, Yang M, Si S (2012) Simple and sensitive aptasensor based on quantum dot-coated silica nanospheres and the gold screen-printed electrode. Talanta 99: 637–642
Li C, Meng Y, Wang S, Qian M, Wang J, Lu W, Huang R (2015) Mesoporous carbon nanospheres featured fluorescent aptasensor for multiple diagnosis of cancer in vitro and in vivo. ACS Nano 9:12096–12103
Liao W, Randall BA, Alba NA, Cui XT (2008) Conducting polymer-based impedimetric aptamer biosensor for in situ detection. Anal Bioanal Chem 392:861–864

Liss M, Petersen B, Wolf H, Prohaska E (2002) An aptamer-based quartz crystal protein biosensor. Anal Chem 74:4488–4495

Liu J, Cao Z, Lu Y (2009) Functional nucleic acid sensors. Chem Rev 109:1948–1998

Liu B, Lu L, Hua E, Jiang S, Xie G (2012) Detection of the human prostate-specific antigen using an aptasensor with gold nanoparticles encapsulated by graphitized mesoporous carbon. Microchim Acta 178:163–170

Liu X, Ren J, Su L, Gao X, Tang Y, Ma T, Zhu L, Li J (2017) Novel hybrid probe based on double recognition of aptamer-molecularly imprinted polymer grafted on upconversion nanoparticles for enrofloxacin sensing. Biosens Bioelectron 87:203–208

Love JC, Estroff LA, Kriebel JK, Nuzzo RG, Whitesides GM (2005) Self-assembled monolayers of thiolates on metals as a form of nanotechnology. Chem Rev 105:1103–1170

Luk BT, Zhang L (2014) Current advances in polymer-based nanotheranostics for cancer treatment and diagnosis. ACS Appl Mater Interfaces 6:21859–21873

Ma W, Yin H, Xu L, Xu Z, Kuang H, Wang L, Xu C (2013) Femtogram ultrasensitive aptasensor for the detection of OchratoxinA. Biosens Bioelectron 42:545–549

Macaya RF, Schultze P, Smith FW, Roe JA, Feigon J (1993) Thrombin-binding DNA aptamer forms a unimolecular quadruplex structure in solution. Proc Natl Acad Sci 90:3745–3749

Maeshima A, Miyagi A, Hirai T, Nakajima T (1997) Mucin-producing adenocarcinoma of the lung, with special reference to goblet cell type adenocarcinoma: Immunohistochemical observation and Ki-ras gene mutation. Pathol Int 47:454–460

Mairal T, Özalp VC, Sánchez PL, Mir M, Katakis I, O'Sullivan CK (2008) Aptamers: molecular tools for analytical applications. Anal Bioanal Chem 390:989–1007

Menger M, Yarman A, Erdőssy J, Yildiz HB, Gyurcsányi RE, Scheller FW (2016) MIPs and aptamers for recognition of proteins in biomimetic sensing. Biosensors 6:35

Miodek A, Regan EM, Bhalla N, Hopkins NA, Goodchild SA, Estrela P (2015) Optimisation and characterisation of anti-fouling ternary SAM layers for impedance-based aptasensors. Sensors 15:25015–25032

Nawaz MAH, Rauf S, Catanante G, Nawaz MH, Nunes G, Marty JL, Hayat A (2016) One step assembly of thin films of carbon nanotubes on screen printed interface for electrochemical aptasensing of breast cancer biomarker. Sensors 16:1651

Ostatná V, Vaisocherová H, Homola J, Hianik T (2008) Effect of the immobilisation of DNA aptamers on the detection of thrombin by means of surface plasmon resonance. Anal Bioanal Chem 391:1861–1869

Pasquardini L, Pancheri L, Potrich C, Ferri A, Piemonte C, Lunelli L, Napione L, Comunanza V, Alvaro M, Vanzetti L (2015) SPAD aptasensor for the detection of circulating protein biomarkers. Biosens Bioelectron 68:500–507

Peng H, Zhang L, Soeller C, Travas-Sejdic J (2009) Conducting polymers for electrochemical DNA sensing. Biomaterials 30:2132–2148

Prabhakar N, Arora K, Singh SP, Singh H, Malhotra BD (2007) DNA entrapped polypyrrole–polyvinyl sulfonate film for application to electrochemical biosensor. Anal Biochem 366:71–79

Qureshi A, Gurbuz Y, Niazi JH (2015) Label-free capacitance based aptasensor platform for the detection of HER2/ErbB2 cancer biomarker in serum. Sensors Actuators B Chem 220:1145–1151

Radi AE (2011) Electrochemical aptamer-based biosensors: recent advances and perspectives. Int J Electrochem 2011:863196

Radi AE, Acero Sánchez JL, Baldrich E, O'Sullivan CK (2006) Reagentless, reusable, ultrasensitive electrochemical molecular beacon aptasensor. J Am Chem Soc 128:117–124

Rahi A, Sattarahmady N, Heli H (2016) Label-free electrochemical aptasensing of the human prostate-specific antigen using gold nanospears. Talanta 156:218–224

Ramanavičius A, Ramanavičienė A, Malinauskas A (2006) Electrochemical sensors based on conducting polymer—polypyrrole. Electrochim Acta 51:6025–6037

Robertson DL, Joyce GF (1990) Selection in vitro of an RNA enzyme that specifically cleaves single-stranded DNA. Nature 344:467–468

Rodrígez LMT, Billon M, Roget A, Bidan G (2002) Electrosynthesis of a biotinylated polypyrrole film and study of the avidin recognition by QCM. J Electroanal Chem 523:70–78
Sagiv J (1980) Organized monolayers by adsorption 1: formation and structure of oleophobic mixed monolayers on solid surfaces. J Am Chem Soc 102:92–98
Sardar R, Funston AM, Mulvaney P, Murray RW (2009) Gold nanoparticles: past, present, and future. Langmuir 25:13840–13851
Savory N, Abe K, Sode K, Ikebukuro K (2010) Selection of DNA aptamer against prostate specific antigen using a genetic algorithm and application to sensing. Biosens Bioelectron 26:1386–1391
Schreiber F (2000) Structure and growth of self-assembling monolayers. Prog Surf Sci 65:151–257
Sharma R, Agrawal VV, Sharma P, Varshney R, Sinha R, Malhotra B (2012) Aptamer based electrochemical sensor for detection of human lung adenocarcinoma A549 cells. J Phys Conf Ser 358:012001
Shih AH, Dai C, Hu X, Rosenblum MK, Koutcher JA, Holland EC (2004) Dose-dependent effects of platelet-derived growth factor-B on glial tumorigenesis. Cancer Res 64:4783–4789
Singh S, Jha P, Singh V, Sinha K, Hussain S, Singh MK, Das P (2016) A quantum dot–MUC1 aptamer conjugate for targeted delivery of protoporphyrin IX and specific photokilling of cancer cells through ROS generation. Integr Biol 8:1040–1048
Song S, Wang L, Li J, Fan C, Zhao J (2008) Aptamer-based biosensors. Trends Anal Chem 27:108–117
Souada M, Piro B, Reisberg S, Anquetin G, Noël V, Pham M (2015) Label-free electrochemical detection of prostate-specific antigen based on nucleic acid aptamer. Biosens Bioelectron 68:49–54
Stora T, Hovius R, Dienes Z, Pachoud M, Vogel H (1997) Metal ion trace detection by a chelator-modified gold electrode: a comparison of surface to bulk affinity. Langmuir 13:5211–5214
Su S, Nutiu R, Filipe CD, Li Y, Pelton R (2007) Adsorption and covalent coupling of ATP-binding DNA aptamers onto cellulose. Langmuir 23:1300–1302
Su M, Ge L, Kong Q, Zheng X, Ge S, Li N, Yu J, Yan M (2015) Cyto-sensing in electrochemical lab-on-paper cyto-device for in-situ evaluation of multi-glycan expressions on cancer cells. Biosens Bioelectron 63:232–239
Sylvia P, Brehm SO, Hoch HJA (1975) DNA-binding proteins in human serum. Biochem Biophys Res Commun 63:24–31
Tahmasebi F, Noorbakhsh A (2016) Sensitive electrochemical prostate specific antigen aptasensor: effect of carboxylic acid functionalized carbon nanotube and glutaraldehyde linker. Electroanalysis 28:1134–1145
Tombelli S, Minunni M, Mascini M (2005) Analytical applications of aptamers. Biosens Bioelectron 20:2424–2434
Tuerk C, Gold L (1990) Systematic evolution of ligands by exponential enrichment: RNA ligands to bacteriophage T4 DNA polymerase. Science 249:505–510
Tzouvadaki I, Jolly P, Lu X, Ingebrandt S, De Micheli G, Estrela P, Carrara S (2016) Label-free ultrasensitive memristive aptasensor. Nano Lett 16:4472–4476
Ueda E, Gout P, Morganti L (2003) Current and prospective applications of metal ion–protein binding. J Chromatogr A 988:1–23
Ulman A (1996) Formation and structure of self-assembled monolayers. Chem Rev 96:1533–1554
Vance SA, Sandros MG (2014) Zeptomole detection of C-reactive protein in serum by a nanoparticle amplified surface plasmon resonance imaging aptasensor. Sci Rep 4:5129
Wang X, Gu X, Yuan C, Chen S, Zhang P, Zhang T, Yao J, Chen F, Chen G (2004) Evaluation of biocompatibility of polypyrrole in vitro and in vivo. J Biomed Mater Res A 68:411–422
Wang L, Ma W, Chen W, Liu L, Ma W, Zhu Y, Xu L, Kuang H, Xu C (2011) An aptamer-based chromatographic strip assay for sensitive toxin semi-quantitative detection. Biosens Bioelectron 26:3059–3062
Wang K, Fan D, Liu Y, Wang E (2015) Highly sensitive and specific colorimetric detection of cancer cells via dual-aptamer target binding strategy. Biosens Bioelectron 73:1–6

Wang Z, Yu J, Gui R, Jin H, Xia Y (2016) Carbon nanomaterials-based electrochemical aptasensors. Biosens Bioelectron 79:136–149

Weidlich C, Mangold KM, Jüttner K (2005) EQCM study of the ion exchange behaviour of polypyrrole with different counterions in different electrolytes. Electrochim Acta 50:1547–1552

Wu J, Campuzano S, Halford C, Haake DA, Wang J (2010) Ternary surface monolayers for ultrasensitive (zeptomole) amperometric detection of nucleic acid hybridization without signal amplification. Anal Chem 82:8830–8837

Xiao D, Lu T, Zeng R, Bi Y (2016) Preparation and highlighted applications of magnetic microparticles and nanoparticles: a review on recent advances. Microchim Acta 183:2655–2675

Xie Q, Tan Y, Guo Q, Wang K, Yuan B, Wan J, Zhao X (2014) A fluorescent aptasensor for sensitive detection of human hepatocellular carcinoma SMMC-7721 cells based on graphene oxide. Anal Methods 6:6809–6814

Xu H, Gorgy K, Gondran C, Le Goff A, Spinelli N, Lopez C, Defrancq E, Cosnier S (2013) Label-free impedimetric thrombin sensor based on poly (pyrrole-nitrilotriacetic acid)-aptamer film. Biosens Bioelectron 41:90–95

Yan M, Sun G, Liu F, Lu J, Yu J, Song X (2013) An aptasensor for sensitive detection of human breast cancer cells by using porous GO/Au composites and porous PtFe alloy as effective sensing platform and signal amplification labels. Anal Chim Acta 798:33–39

Yang DK, Chen LC, Lee MY, Hsu CH, Chen CS (2014) Selection of aptamers for fluorescent detection of alpha-methylacyl-CoA racemase by single-bead SELEX. Biosens Bioelectron 62:106–112

Ye X, Shi H, He X, Wang K, He D, Yan LA, Xu F, Lei Y, Tang J, Yu Y (2015) Iodide-responsive Cu–Au nanoparticle-based colorimetric platform for ultrasensitive detection of target cancer cells. Anal Chem 87:7141–7147

Yeh FY, Liu TY, Tseng IH, Yang CW, LC L, Lin CS (2014) Gold nanoparticles conjugates-amplified aptamer immunosensing screen-printed carbon electrode strips for thrombin detection. Biosens Bioelectron 61:336–343

Zhang Z, Yang W, Wang J, Yang C, Yang F, Yang X (2009) A sensitive impedimetric thrombin aptasensor based on polyamidoamine dendrimer. Talanta 78:1240–1245

Zhang K, Xie M, Zhou B, Hua Y, Yan Z, Liu H, Guo LN, Wu B, Huang B (2013) A new strategy based on aptasensor to time-resolved fluorescence assay for adenosine deaminase activity. Biosens Bioelectron 41:123–128

Zhang H, Shuang S, Sun L, Chen A, Qin Y, Dong C (2014) Label-free aptasensor for thrombin using a glassy carbon electrode modified with a graphene-porphyrin composite. Microchim Acta 181:189–196

Zhao S, Ma W, Xu L, Wu X, Kuang H, Wang L, Xu C (2015) Ultrasensitive SERS detection of VEGF based on a self-assembled Ag ornamented–AU pyramid superstructure. Biosens Bioelectron 68:593–597

Zhu Y, Chandra P, Shim YB (2012) Ultrasensitive and selective electrochemical diagnosis of breast cancer based on a hydrazine–Au nanoparticle–aptamer bioconjugate. Anal Chem 85:1058–1064

10 Nanobiosensing Technologies for Prostate Cancer Diagnostics/Prognostics: Tiny Smart Medicine

Renu Singh and Chandini C. Mohan

10.1 Introduction

The widespread incidence of disease, high death rate, and relapse has significantly increased the concern for development of better cancer diagnosis and treatment. Breast cancer being the most prevalent form of cancer among women, prostate cancer (PCa) has turned out to be one of the most common lethal cancer type among men aged 50–80 years old (Oesterling et al. 1993). Tumors develop in the prostate gland of the male reproductive system causing significant pain during urinating and sexual intercourse. Other than lung and skin cancer, prostate cancer is the most common cancer prevalent among American men. As per American Cancer Society's estimate for prostate cancer, there are ~180,890 new cases and ~26,120 deaths in the United States for year 2016. According to recent statistics, ~1 out of 7 men will be diagnosed with prostate cancer during his lifetime (Siegel et al. 2016). PCa is very difficult to detect in its early stage which is the cause for increased mortality rates every year (Schröder et al. 2009). Hence, in such a condition where there is an alarming increase of PCa cases, it is important that the disease is accurately detected at an early stage to improve patient outcomes in terms of morbidity, mortality, and relapse (Zieglschmid et al. 2005). This demands for an effective diagnosis and prognosis tools with improved sensitivity and specificity towards PCa.

R. Singh (✉)
Department of Bioproducts and Biosystems Engineering, University of Minnesota, Twin Cities, 2004 Folwell Ave, Saint Paul, MN 55108, USA
e-mail: raina1785@gmail.com

C.C. Mohan
Center for Soft and Living Matter, Institute for Basic Science (IBS), UNIST-gil, 50, Ulsan, Republic of Korea
e-mail: singhr@umn.edu

P. Chandra et al. (eds.), *Next Generation Point-of-care Biomedical Sensors Technologies for Cancer Diagnosis*, DOI 10.1007/978-981-10-4726-8_10

DNA mutation induces development of a malignant tumor inside prostate gland. Currently used clinical diagnosis tools for PCa includes magnetic resonance imaging (MRI), transrectal ultrasonography, digital rectal inspection, Prostate biopsies, computed tomography (CT) scan, and biochemical assays (Brock et al. 2015; Crawford and De Antoni 1993; Kang et al. 2015; Haese et al. 2008; Elabbady and Khedr 2006; Ghai and Toi 2012; Najeeb et al. 2014). Among these, biochemical assays are used for initial disease screening (Kang et al. 2015). Prostate-specific antigen (PSA) is the most widely used serum marker for the clinical diagnosis of PCa as their serum levels will be elevated during malignant conditions (>4.0 ng mL^{-1}) (Crawford and De Antoni 1993; Kang et al. 2015). Many of these traditional diagnostic methods are less effective to detect cancer at a very early stage. The quest for primary diagnosis of cancer with better sensitivity and precision has led to an increased work on the detection of numerous biomarkers including molecular DNA and RNA markers (Ploussard et al. 2011; Stuopelyte et al. 2016; Velonas et al. 2013).

The past few years witnessed a rapid growth in nanotechnology and its application in various sectors (Kang et al. 2015; Huang et al. 2011; Bellan et al. 2011). Tremendous attempts have been made to synthesize and study nanomaterials of desirable properties. Nanomaterials have been widely used to construct biosensing devices with improved performance and cost-effectiveness due to their inimitable physical, chemical, optical, magnetic, and mechanical properties (Huang et al. 2011; Bellan et al. 2011). Several nanoscale structures, such as single-walled carbon nanotubes (SWCNTs), silicon nanowires (SiNWs), or metal oxide nanowires (e.g., indium oxide (In_2O_3) NWs), have been fabricated for their use in more sensitive and precise prostate cancer detection (Zheng et al. 2005; Chikkaveeraiah et al. 2009; Li et al. 2005). Nanomaterials in the form of nanowires, nanorods, and nanotubes offer high surface-to-volume ratios and are highly sensitive and can be further surface modified with well-known methods, making it a propitious platform for sensitive detection of biomarkers (Huang et al. 2011). These nanomaterials can contribute to design transducers that help in biosensing by converting the biochemical recognition events to electrical/optical/piezoelectric signals (Huang et al. 2011; Bellan et al. 2011). Based on the presence of various transducing systems present, several types of nanobiosensors have been reported (Guardia and Garrigues 2012).

10.2 Conventional vs. Advanced Cancer Diagnosis

Conventionally, cancer diagnosis begins with a detailed clinical history of the person. It involves a complete physical examination, combined with laboratory testing of various body fluids i.e. blood, urine, saliva, and semen etc. Prostate biopsy followed by histological examination remain standing as gold standard for the diagnosis of prostate cancer (Brock et al. 2015). However, high false negatives are associated with more than 50% of cases, remains a current limitation in existing prostate cancer screening. In the current clinical setting, patients with clinical symptoms and negative prostate biopsy are advised to repeat the screening test by 5 years (Welch et al. 2007). Repeated biopsies impart significant morbidity, anxiety, and pain to the patient and moreover adds to the unavoidable cost. An elevated level of PSA in patient serum is widely used as a protein marker for PCa screening. However, the specificity and sensitivity of PSA test is very low resulting in high false-negative (15%) and false-positive (66%) diagnosis (Catalona et al. 1991, 1994; Bretton 1994; Thompson et al. 2004; Schroder et al. 2000). These limitations have initiated an extensive investigation for alternate markers with superior selectivity such as PCA3 for prostate cancer that might help to reduce the number of unnecessary biopsies (Loeb and Partin 2011). Imaging techniques using X-rays, ultrasound, CT scans, or MRI are often combined with biochemical assays and biopsies for better screening as well as disease diagnosis (Sanz et al. 2004; Reske et al. 2006; Mintz 2014; Jadvar 2015). In addition to the traditional diagnostic methods, technological advancements have led to the introduction of newer technologies that uses radioisotopes for cancer detection in situ, i.e., ProstaScint-scan, positron emission tomography (PET) scanning, and CT with PET (Sanz et al. 2004; Reske et al. 2006; Mintz 2014; Jadvar 2015). One of the highly advanced technologies like 68Ga-prostate-specific membrane antigen (PSMA) PET/CT which was recently approved by Food and Drug Administration for human trial are capable of molecular imaging of PSMA as PCa biomarker, with improved sensitivity for early cancer detection (Jadvar 2015). Among the various detection methods, biochemical assays offer a more rapid, easy, and cost-effective strategy for disease diagnostic platforms. Several researches around the globe are involved in the discovery of new biomarkers for precise detection of cancer. Studies on molecular diagnostics have been increased exponentially over recent times, wherein specific DNA and RNA signatures have been identified to detect specific cancer types with great precision from body fluids (Alhasan et al. 2016; Fabrisa et al. 2016; Geybels et al. 2016). This is often referred to as "liquid biopsies," as they are collected via a minimal invasive route, like a needle prick. Newer biosensing platforms with the incorporation of reliable biomarkers could improve diagnostic accuracy to maximize the efficiency of subsequent therapies for the diagnosed patients.

10.3 Nanobiosensing Technologies for Prostate Cancer Diagnosis

Nanobased biosensing technologies are highly promising and guaranteed to exhibit improvements in their selectivity for cancer biomarkers (Mouli et al. 2010; Ma et al. 2009). Several studies have demonstrated their effectiveness for the development of next-generation diagnostics, contributing towards more accurate diagnosis (Park et al. 2013a, b; Hwang et al. 2013; Singh et al. 2014a). The present chapter discusses mainly electrochemical, mass-based, microfluidics, field effect transistors, and optical-based nanobiosensing technologies for prostate cancer detection (Fig. 10.1).

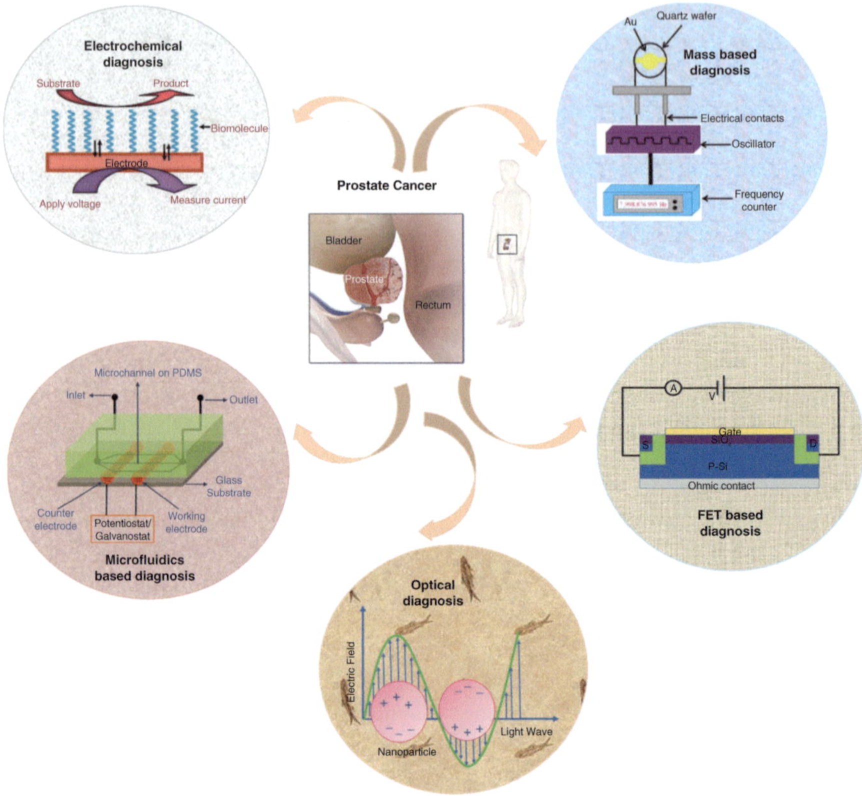

Fig. 10.1 Schematic showing the various transducers-based nanobiosensing technologies for prostate cancer diagnostics/prognostics

10.3.1 Electrochemical-Based Nanobiosensing Technologies

Electrochemical biosensors are the most widely used systems for development of rapid, easy, and cost-effective point-of-care diagnostics (Hammond et al. 2016). Ample amount of investigations have been focused on the development of miniaturized easy-to-use portable electrochemical systems that can easily incorporate inexpensive electrodes with simple electronics to perform rapid measurements. Recent progress in nano-biotechnology enabled the development of a variety of reproducible coatings of the active biomolecules such as enzymes on the miniature transducer element suitable for sensing application. Nanomaterials are often advantageous to provide high surface area for immobilization of biomolecules to electrode surface for the fabrication of ultrasensitive electrochemical biosensors with better electronic and electrocatalytic properties (Singh et al. 2009, 2010, 2011, 2012a, b). Electrochemical biosensor works due to easy electron transfer that occurs between the active redox centers of biomolecules and the working electrode. The use of nanostructured coatings on working electrodes greatly increases the electrical conductivity because of thin-layer phenomena with abundant no. of redox reaction sites (Streeter et al. 2008).

The detection efficiency of the sensor is greatly influenced by the electrode material, the surface modification/properties, or its dimensions. Electrochemical sensor design usually involves a reference electrode, a counter electrode, and a working/redox electrode. Ag/AgCl electrodes are commonly used in electrochemical sensors for reference. The working electrode serves as transducer though counter electrode establishes a connection with the electrolytic solution. Nobel metals (viz., platinum, gold, silver), carbon (e.g., graphite), and silicon compounds which are chemically stable and conductive are commonly used as electrodes depending on the analyte. Biosensing devices require a variety of biomolecules as recognition elements, e.g., enzymes, antibodies, nucleic acids, aptamers, cells, and microorganisms (Eggins 2002; Chaubey and Malhotra 2002). Many of the earlier studies focused predominantly on enzymes for electrochemical detection techniques owing to their specific binding proficiencies and biocatalytic activity (Eggins 2002; D'Orazio 2003; Schoning and Poghossian 2002). However, rigorous attempts are being made to develop ultrasensitive devices using specific antibodies and DNA/RNA probes, aptamers. Surface density, stability, and specificity of the recognizing elements are extremely important for the improved sensitivity of device. Appropriate arrangement of molecule could be achieved through different molecular ways, i.e., langmuir-blodgett (Cabaj et al. 2010), self-assembled monolayers or layer-by-layer (LbL) as well as electrolytic deposition (Cabaj et al. 2012). However, the retention of molecular activity is often challenged in the process of deposition.

The researches by *Giaver, and Kronick* and Little began the introduction of antibody-based biosensors into being from 1970s (Giaever 1973; Kronick and Little 1975). Among several applications, an electrochemical immunosensor fabricated using Graphene sheet–methylene blue nanocomposite was employed for detection of PCa by analyzing PSA levels. Silver-hybridized mesoporous silica nanoparticles (NPs) were used as electrode and hydroquinone as a mediator with an observed limit of detection (LOD) of 13 pg mL^{-1} (Wang et al. 2013). *Sarkar* et al. performed the immunobased detection of free PSA (f-PSA) using immunolabeled magnetic beads premixed with horseradish peroxidase (HRP)-labelled secondary antibodies on a 3-electrode screen-printed sensor (Sarkar et al. 2008). The amperometric response of the HRP labels was used to measure the target concentrations on the beads upon addition of hydrogen peroxide substrates, with LOD of <0.1 ng mL^{-1} fPSA. *Chen* et al. stated that the nanoscale-spacing of immobilized antibodies affects the sensitivity of the sensor critically. They used gold nanoparticles (AuNPs) to further amplify the sensing signal which resulted into extremely high sensitivity along with LOD of 1 pg mL^{-1} (Xiaoqing et al. 2014). Nanoparticle probes, i.e., carbon nanotubes (CNTs), can be used to load large number of biomolecules/multi-enzymes for target protein signal amplification. Signal amplification can also be achieved by labelling multienzymes with magnetic beads, polymer beads, and streptavidin-biotin complexes; however, the sensitivity of the detection system can be limited by nonspecific binding of bioconjugates other than the antigenic sites. *Wang* et al. used multi-enzyme-labelled CNTs for ultrasensitive detection of DNA and protein (IgG) at fM level with the immobilized antibodies and thousands of alkaline phosphatase enzymes on the surface (Wang et al. 2004). In subsequent work, they used an LbL approach in conjunction with square wave voltammetry to achieve LOD of 67 aM (tenfold increase) for IgG in buffer (Munge et al. 2005). *Topkaya* et al. developed a carbon graphite-based electrochemical DNA biosensor for detection of hypermethylation of the glutathione S-transferase P1 gene, prostate cancer-specific marker (Topkaya et al. 2012). In a study reported by *Jolly* et al., thiolated DNA aptamer was complexed with PSA prior to being immobilized on the surface of a gold electrode. The aptamer-molecularly imprinted polymer-based sensor showed high sensitivity with linear response from 100 pg mL^{-1} to 100 ng mL^{-1} of PSA and LOD of 1 pg mL^{-1}, which was threefold higher than alone aptamer sensor for PSA (Jolly et al. 2016). *Tang* et al. developed an electrochemical sensor to offer triple sensitivity amplification for detection of PSA using conductive reduced graphene oxide-Au nanocomposites coating on the Au-poly modified glassy carbon electrode. Specific peptides (CEHSSKLQLAK-NH_2) immobilized on modified electrodes underwent enzymatic cleavage upon reaction with PSA and resulted in a flux. The designed biosensor presented a wide linear range from 1.0 fg mL^{-1} to 100 ng mL^{-1} with an ultralow LOD of 0.11 fg mL^{-1} (Tang et al. 2016a, b).

10.3.2 Mass-Based Nanobiosensing Technologies

Mass-sensing biosensors; quartz crystal microbalance (QCM) and surface acoustic wave (SAW) are highly promising candidates as they can detect transductions based on small changes in mass with high sensitivity, fast response, and low cost (Velusamy et al. 2010). QCM-based detection is widely used as they are relatively facile, cost-effective method and provides direct label-free detection with enhanced sensing characteristics. Mass analysis generally depends on the use of piezoelectric crystals which vibrates at specific frequencies and change in this frequency upon binding of the analyte is electrically measured. In this technology, device uses the piezoelectric effect to measure physical changes in mass due to bio-recognition events and convert it to electrical signals by piezoelectric crystal transducer elements of the sensor. However, this area of biosensing still needs more exploration to demonstrate their successful application in high background samples, i.e., serum (Prakrankamanant 2014).

Basically, a mass sensor disc is covered by a metallic deposit on the top and bottom sides referred to as front and back electrodes, respectively. The disc is coupled to an oscillating circuit that applies an alternating electrical field to the crystals, inducing an oscillation at their center. The application of electrical potential causes a displacement of the crystal atoms parallel to the surface, and any change in these oscillation frequencies could be directly monitored by a frequency counter. Sauerbrey was the first who recognized the potential usefulness of the piezoelectric technology and demonstrated the relationship between mass and resonant frequency (Sauerbrey 1959). The capture of analyte in turn results in an increase of crystal mass ((Δm)) and a proportionate decrease in frequency (frequency change, Δf) from initial frequency oscillation (f_0). Biosensor transducer crystal surfaces are coated with biomolecules such as a DNA probe or antibody that determines the sensitivity and specificity of the device.

Quartz crystal microbalances have emerged as versatile biosensors offering high sensitivity and label-free detection in former few years. The use of NPs viz., AuNPs in QCM sensors have been stated to enhance the sensitivity of electromechanical assays in the detection of human IgG and aflatoxin B1 at clinical levels as amplification agents in a dynamic mode quartz crystal microbalance biosensor (Chu et al. 2006; Jin et al. 2009). *Uludag and Tothill* demonstrated high sensitivity for PSA with LOD of 0.29 ng mL^{-1} by performing a sandwich assay using antibody-modified NPs in QCM platforms (Uludag and Tothill 2010). This platform exhibited higher sensitivity compared to 4.0 ng mL^{-1} of the gold standard method and can be applied clinically revealing the high potential for prostate cancer diagnosis and prognosis. *Wang* et al. successfully developed a mass-sensing BioCD Protein Array for clinical detection of PSA PCa Patient Serum in a 96-well anti-PSA microarray. They could demonstrate LOD of 4 ng mL^{-1} PSA concentrations in full serum as tested in three patient samples (Wang et al. 2011). *Schlensog* et al. reported a special type of SAW sensors named Love-wave biosensor to decrease energy dissipation of acoustic wave into the liquid by using a surface layer of shear horizontal waves (Schlensog et al. 2004). SAW sensor was fabricated by immobilizing a single strand DNA aptamer as ligand (Zhang et al. 2015) to detect Human thrombin with high sensitivity and specificity. In another work, a microfluidic Love-wave biosensor was developed to specifically detect PSA in real time using DNA aptamer-beacon bearing a stem-loop structure which changes the selective target-induced assembly. A novel hybrid sandwich immunoassay which is the combination of mass and optoplasmonic transduction was reported to detect PSA biomarkers in serum at clinically relevant levels. Antibody-labelled AuNPs that act as a mass and plasmonic label; a silicon cantilever that works as a mechanical resonator and plasmonic signal enhancer from the nanoparticles. LOD of 1×10^{-16} g mL^{-1} in serum was attained with both biomarkers, at a significantly higher sensitivity by detecting ultra-low magnitudes at the order of seven in comparison with routine clinical practice with extremely low, $\sim 10^{-4}$ false-positive and false-negative rates (Kosaka et al. 2014). Another study utilized signal transduction biosensor as a novel electrical sensor for identifying PSA where surface stress changes occurred due to antigen–antibody-specific binding using micro-fabricated self-sensing nanomechanical membrane (Omidi et al. 2014).

10.3.3 Microfluidics-Based Nanobiosensing Technologies

Microfluidic-based technologies are generally called as lab-on-a-chip systems or micro-total-analysis system (μTAS) and deals with the flow of micrometer-size channels on miniaturized forms of traditional laboratory systems (Manz et al. 1990). In recent years, biosensor nanotechnology is rapidly developing and has become widespread in the biomedical engineering. A prodigious number of studies have been developed for advanced nanobiosensing devices that can replace the conventional laboratory processes performed by technicians and experts in large-scale integration laboratories (Haeberle and Zengerle 2007; Samiei et al. 2016). Among them, microfluidic systems provide high throughput, controlled flow carrier, enhance the blending of various samples, sensitivity, and require less volume of reagents. Microfluidics technique is known to be interdisciplinary technology as it comes by the amalgamation of various streams, i.e., physics chemistry, biochemistry, biotechnology, bioelectronics, engineering, and nano-technology (Wang et al. 2009).

Scheming and construction of microfluidics biosensing devices require careful considerations to improve the fabrication procedures, especially the wettability and compatibility of the devices. Nanolithography and nanostructured patterning techniques are being used to fabricate microfluidic devices by utilizing silicon as substrate as silicon microfabrication is very well established technique but has been ignored in microfluidic fabrication for biosensing applications. Various materials, i.e., silicon, glass, and PDMS polymer hybrids, can be used to fabricate microfluidic devices but polymeric materials have the advantages due to excellent conducting and thermal properties, biological compatibility, and adequate transparency for biosensing applications. Schematic diagram of microchannels using polydimethylsiloxane (PDMS) wall is shown in Fig. 10.2.

Microfluidic actuators with inlet–outlet valves and pumps can be fabricated on the chip by attaching multilayers of PDMS which is an elastomeric material (Unger et al. 2000). Microchannel architecture is used to build basic elements of the microfluidic sensor systems, which is shown in Fig. 10.3.

Augustsson et al. established a microfluidic acoustophoresis method for detection as well as separation of prostate cancer cells from white blood cells via forces generated in microfluidic channels, and the recovery ranged from 93.6 to 97.9% with purity ranging from 97.4 to 98.4% for cells fixed with paraformaldehyde, and for nonfixed cells recovery ranged from 72.5 to 93.9% with purity ranging from 79.6

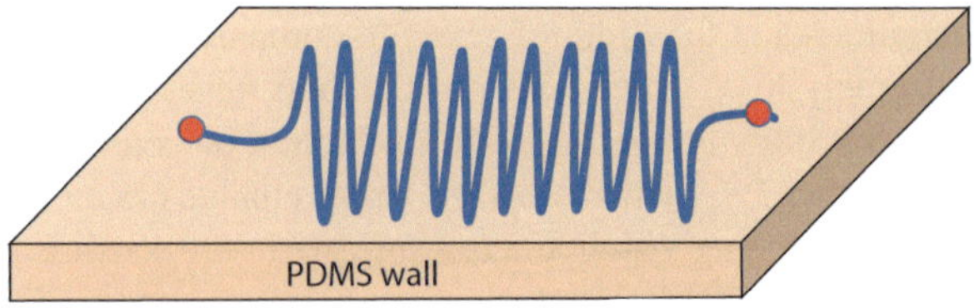

Fig. 10.2 Schematic diagram of microchannels

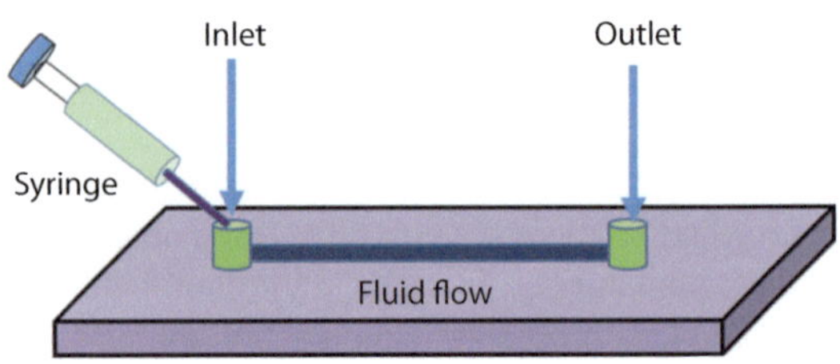

Fig. 10.3 Schematic diagram of microfluidic systems

to 99.7% (Augustsson et al. 2012). *Madaboosi* et al. presented a system that allows the integrated detection of fPSA in PDMS-based microfluidic immunoassay system (Madaboosi et al. 2014). Automated microfluidic device was fabricated using lithium tantalate substrate with SiO_2 film, two set of inter-digital transducers, Au for immobilization of the biorecognition layer, aptamer beacon and PDMS channels for real-time detection of PSA. The fabricated device showed the detection range between 10 ng mL^{-1} and 1 μg mL^{-1}, with LOD of 10 ng mL^{-1} (Zhang et al. 2015). *Islam* et al. implemented a simple, reusable, and efficient microfluidic channel to detect cancer cells directly from blood without any fluorescent tags, surface functionalization, or pre-processing of the blood except dilution for the detection of tumor cells using the electrical measurement of each single cell from the samples. As per the authors the surface nanotexture of the microfluidic channel retarded the metastatic renal cancer cells' translocation by 50%. This interesting phenomenon increased tumor cell detection efficiency by 14% in nanotextured microchannel compared to plain channel (Islam et al. 2015). *Tang* et al. proposed a cost-effective electrochemical array of 256 sensors with 32 individually addressable microelectrodes by creating hydrophobic well via print and peel method to avoid cross-contamination. This system was used to determine PSA, PSMA, interleukin-6 (IL-6), and platelet factor-4 (PF-4) in serum and clinically relevant LODs (0.05–2 pg mL^{-1}) was achieved. They developed the high-throughput system which could detect multiple cancer biomarker proteins in serum. Protein capture using magnetic NPs from 5 μL samples provided a viable strategy for multiplexed detection by minimizing nonspecific binding (Tang et al. 2016a, b). *Hughes* et al. fabricated a completely integrated microfluidic assay to protein measurements of endogenous PSA isoforms in human prostate cancer cell lysate and crude sera from metastatic prostate cancer patients and achieved LOD 1.1 pg (Hughes et al. 2012). A multiplex RT-PCR assay was developed for rapid screening of alterations in prostate cancer genes at miRNA levels by integrating it with microfluidic-based on-chip electrophoresis (Moltzahn et al. 2011). *Kirby* et al. developed system to capture circulating tumor cells (CTCs) isolated from castrate-resistant prostate cancer patients by integrating immunocapture microfluidic device that combines an anti-PSMA antibody with a 3D geometry that captures CTCs while minimizing nonspecific leukocyte adhesion. This device showed a 2–400-fold higher sensitivity compared with the commercially available CellSearch® (Kirby et al. 2012).

10.3.4 Field Effect Transistor-Based Nanobiosensing Technologies

Integration of semiconducting nanomaterials and biological materials provides an attractive interface for emerging applications, such as chemical/biological sensors, for health monitoring devices. The nanostructure of composites as a channel and a sensing material plays a crucial role in the performance of field effect transistors (FETs) (Cui et al. 2001; Duan et al. 2012; Singh et al. 2014b). Therefore, it is highly desirable to prepare elaborate composite that can allow the fabrication of high

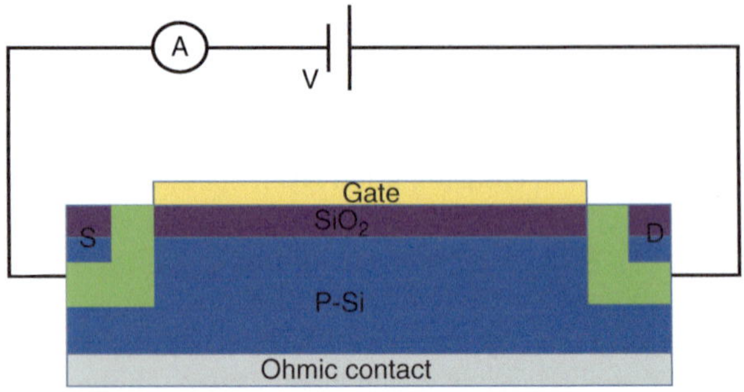

Fig. 10.4 Schematic diagram of field effect transistor systems

performance FETs and provide high sensitivity and selectivity in detecting specific chemical/biological targets (Byeon et al. 2016). Among the silicon-based biosensors, FET based on ion selectivity (ISFET) is known to be the utmost popular electric biosensor because of its excellent properties, i.e., sensitivity, rapidity, miniaturization, and cost-effectiveness.

As innumerable nanostructures, having been utilized to enhance the efficacy and sensitivity of most of the detecting devices, nanobiosensing technologies has become the perfect for the incorporation of biological materials, particularly for the construction of cost-effective, and ultrasensitive FET devices and provide high functionality by binding target in the confined gated region. The typical three-terminal structure containing the drain, source, and gate are shown schematically in Fig. 10.4. Electrical channel of conduction and the conductivity of the charge carriers is controlled by using electric field by FET system. The conductance of the channel between source and drain is generally modulated by voltage at gate terminal. The size and shape of conducting channel can affect the flow of charge carriers between source and drain. In biosensor construction, FET comprises nanomaterials-based channel between the source and drain and nanomaterials surface functionalizes with biomolecule to create an electric field via biomolecular event.

A novel, real-time back-gate Si nanobelt FET device was fabricated for the early diagnosis of PSA using complementary metal oxide semiconductor (CMOS) manufacturing technology and provided high surface-to-volume ratio and back-gate control of the shrank nanobelt structure by the local-oxidation of silicon. Authors also demonstrated the feasibility of nanobelt FET for clinical diagnosis and prostate cancer in the future (Wu et al. 2011). In another report, *Wu* et al. fabricated n-type silicon nanobelt-based immuno FET device for detection of prostate cancer biomarkers by immobilizing PSA-antibody molecules on nanobelt FET utilizing the aldehyde groups of glutaraldehyde linked to the amino groups of 3-aminopropyltriethoxysilane (APTES) and detected PSA concentration up to 5 pg mL^{-1} levels. Arginine molecules between glutaraldehyde and APTES were inserted to enhance the sensitivity of a nanobelt FET immunosensor and improved the LOD by 50 fg mL^{-1} with the detection range of 50 fg mL^{-1} to 500 pg mL^{-1} and suggested that this label-free PSA nanobelt-based FET immuno device might be useful for tool for future prostate cancer screening due to excellent electrical properties (Wu et al. 2012). A dual-channel PDMS microfluidic integrated CMOS-compatible SiNW FET arrays were developed by *Lu* et al. and showed the ultrahigh sensitivity of PSA with LOD of 1 fg mL^{-1} in buffer which showed better performances and high selectivity than other reported techniques. The sensor was applied to the real samples and revealed LOD of 10 fg mL^{-1} in undiluted human serums and due to its outstanding characteristics and miniaturization this sensor provides great prospects for a point-of-care test (POCT) (Lu et al. 2015). Exceedingly responsive n- and p-type CMOS-compatible SiNW arrays were constructed by integrating on a single chip for PSA detection. CMOS compatible anisotropic self-stop etching technique was used to eliminate the requirement for a hybrid method. When n- and p-type nanowires incorporated together it provided an exceptional means of internal control for sensing signal confirmation which revealed complementary electrical response for instantaneous and multiplexed detection of PSA marker at atto-molar levels. The SiNW array was exposed to the artificial and clinical sample of blood serum at different pH and demonstrated LOD as 1 fg mL^{-1} which indicates the realistic advancement of quick, high-performance, and cost-effective diagnostic devices (Fig. 10.5) (Anran et al. 2014).

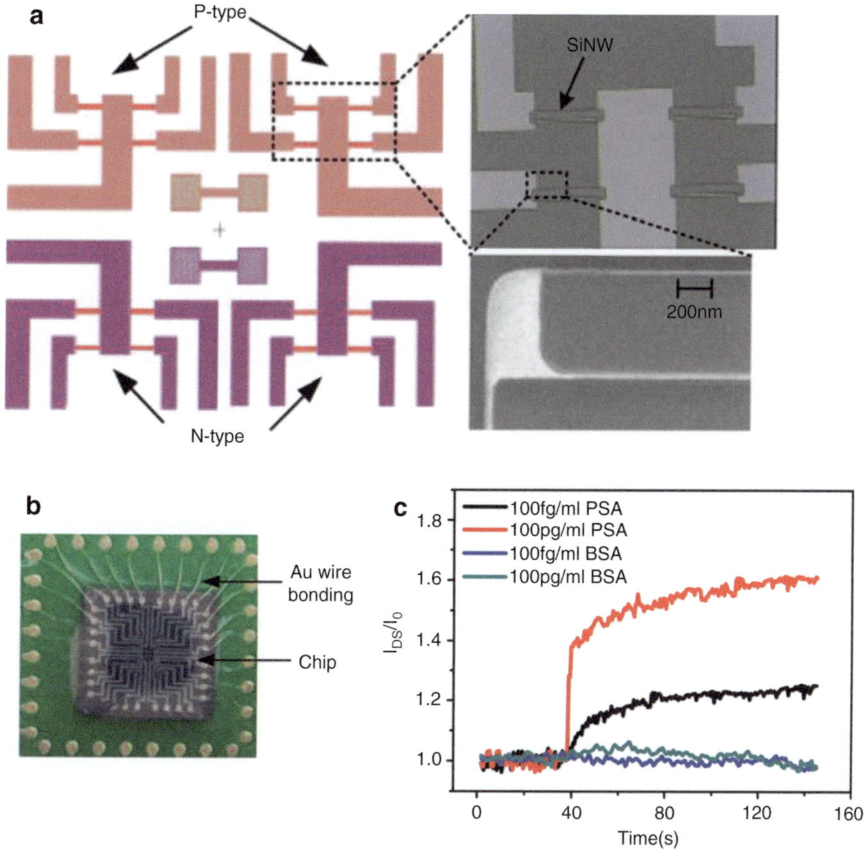

Fig. 10.5 (**a**) Schematic showing the layout of the SiNW device arrays on the chip with the optical image; (**b**) photograph of a SiNW-FET device; (**c**) sensing graph (Anran et al. 2014)

Zheng et al. also fabricated a SiNW-FET array for the detection of multiple cancer markers (Zheng et al. 2005). Reduced graphene oxide (RGO) was self-assembled on an aminated FET device followed by immobilizing PSA antibodies for detection of PSA-antichymotrypsin (PSA-ACT) complex and achieved LOD of 100 fg mL^{-1} (Kim et al. 2013). Molybdenum disulfide (MoS_2) FET-based biosensors were fabricated for PSA detection. Unlike other FET devices, hydrophobicity property of MoS_2 provides simplification in designing of sensor and improves sensitivity as there is no need to add another dielectric layer. Off state current of the device showed a significant decrease as a function of increased PSA concentration and detected 1 pg mL^{-1} below the clinical cutoff level of ~4 ng mL^{-1} (Lee et al. 2014). SWCNT-based FET immunosensor was fabricated to detect the prostate cancer biomarker, osteopontin (OPN) by electrophoretically depositing SWCNT on substrates, and antibodies were immobilized using EDC-NHS chemistry through the binding with NH_2 groups of NHS ester on the carboxy functionalized SWCNTs. SWCNT-FET device showed excellent ohmic contact between SWCNTs and source/drain electrodes with each step of functionalization and exhibited the detection range 1 pg mL^{-1}–1 μg mL^{-1} with LOD of 0.3 pg mL^{-1} in human serum and PBS (Sharma et al. 2015). *Li* et al. reported detection of PSA using n-type In_2O_3 nanowires and p-type CNT by covalently attaching antibodies to In_2O_3 NW surfaces via the onsite surface synthesis of phosphonic acid−succinylimide ester and attained LOD of 5 ng mL^{-1} (Li et al. 2005). Hypermethylation of the glutathione-S-transferase pi gene, which is epigenetic alteration in prostate cancer, is detected using ssDNA biomarker by fabricating dual gate field effect transistor (DGFET). Authors designed a 64-sensor array model arranged in 8 × 8 arrays with reference bias voltage and analyzed five different DGFET sensor structures with ssDNA biomarker and sensitivity was improved by enhancing drain current (Shobha and Muniraj 2014).

10.3.5 Optical-Based Nanobiosensing Technologies

Optical biosensing technologies are very powerful alternative to conventional analytical techniques and the variation in the phase, amplitude, polarization, or frequency of the input light in response to the physical or chemical change produced by the biorecognition process induces the transduction -osch et al. 2007; Sant et al. 2003). There has been an exponential growth in the field of optical biosensors in the last era due to the several advantages, i.e., high-selectivity and specificity, multiplexed sensing, rapidity, miniaturized design, and online measurements (Meharvar et al. 2000).

Light source, optical transmission medium (fiber, waveguide, etc.), immobilized biolmolecules, and optical detection system (Prasad 2003) are the basic components of an optical biosensor. Multiplexed sensors also can be fabricated on to single chip in compact sensing designs by integrating several passive and active optical components in optics technology (Knoll 1998). Optical biosensors are progressively impacting analytical technologies for clinical diagnosis due to the selectivity and sensitivity for the real-time detection of very low levels of clinical samples (Nath and Chilkoti 2002). Interdisciplinary tactic with microelectronics, MEMS, micro/nanotechnologies, molecular biology, nanotechnology, chemistry, and physics are needed for the execution of these analytical devices (Narayanaswamy and Wolfbeis 2004).

To diagnose prostate cancer at attomolar level, *Truong* et al. fabricated an ultra-sensitive method using Rayleigh light scattering spectroscopy of individual gold nanorods (AuNRs) sensors. AuNRs with an aspect ratio of ~3.5 were proved best by measuring the refractive index sensitivity for Localized Surface Plasmon Resonance (LSPR) sensing. AuNRs were immobilized onto the glass substrate by conjugating with the PSA-ACT complex antibody and found the LOD 1 aM and established the promising potential of AuNRs for clinical diagnosis (Truong et al. 2012). Silver nanoparticles when exposed to electromagnetic radiation demonstrated the unique optical properties of LSPR and revealed the quantification of extremely low levels of PSA, a prostate cancer biomarker (No et al. 2008). AuNP-conjugated anti-PSA antibody-based LSPR as a novel approach was developed recently to detect prostatic disease. Stability of particles was increased by pegylation of AuNPs by EDC-NHS chemistry and they found a significant increase in the absorbance and intensity of the particles with extinction peak at 545/2 nm, which was shifted by ~1 nm after conjugation and further shifted 3 nm for a solution of 100 nM unlabeled antigen and improved the sensitivity of PSA in the assessment of prostate disease (Jazayeri et al. 2016). Paralleled LSPR lab-on-a-chip was presented for the first time by the amalgamation of plasmonics, nanofabrication, microfluidics, and surface chemistry and offered parallel, real-time inspection of 32 sensing sites distributed across 8 independent microfluidic channels. They demonstrated LOD of 500 pg mL^{-1} in a complex matrix consisting of 50% human serum (Acimovic et al. 2014). Keeping in the view of the enzymatic activity characteristic of PSA for the specific peptide sequence HSSKLQ, which it recognizes and cleaves, *Cho* et al. fabricated a novel Au-nanoparticle-based biosensor via a sequence-specific peptide cleavage reaction using fluorescein isothiocyanate/peptide-conjugated AuNP complexes. They could detect PSA successfully in the detection range of 10 pM–100 nM and revealed that this approach is far better than immunoassay in terms of the performance, facile nature, and simplicity (Choi et al. 2013). MoS_2 nanosheet (2-D layered nanomaterial) having higher fluorescence-quenching ability than graphene was used to fabricate aptamer-functionalized MoS_2 nanosheet fluorescent biosensor to detect PSA. The authors claimed that binding of the aptamer to the target PSA induced a firm aptamer structure resulting into weak interaction of aptamer with MoS_2 nanosheet very weak and released the aptamer from the nanosheet surface and restored the quenched fluorescence. They revealed the LOD as 0.2 ng mL^{-1} and applied the sensor to detect PSA in human serum as well (Kong et al. 2015). *Barnett* et al. developed lateral flow assays using paramagnetic particles for the measurement of PSA in serum samples and achieved LOD of 0.8 ng mL^{-1} using the resonant coil magnetometer. They compared the data obtained in a pilot study from the analysis of serum samples with commercially available immunoassays and showed good agreement (Barnett et al. 2014). Nanocrescent particles and peptides with artificial tag molecules incorporated hybrid optical probe were used for real-time in situ detection of active proteases present in patients' seminal fluid and serum as it is crucial for early-stage cancer screening and cell signaling pathway study. They observed that the individual nanoparticles react with the specific peptides which results into no cross reaction and minimizes false detection. Hence, high-fidelity and high-signal-to-noise-ratio cancer nanoprobe formed and can be easily integrated with nano/microfluidic devices (Liu et al. 2007). Surface-enhanced Raman scattering (SERS) active nanoparticles were used for molecular imaging and advancement, proof of principle experiments has been carried out over the last years (Zhang et al. 2011). A very interesting study using a novel wash-free magnetic immunoassay technique was conducted recently for PSA detection via SERS-based microdroplet sensor and found the LOD as 0.1 ng mL^{-1} with 174 droplets per minute and stated this approach as fast, very sensitive, and wash free (Gao et al. 2016). Multicellular tumor spheroids (MTS), the best describes 3-D cell cultures are becoming progressively used as tumor models and offer an excellent in vitro screening system that mimics to a great extent the microenvironment prevailing in tumor biology. *Camus* et al. recently used SERS to measure the viability of MTS grown from prostate cancer cells and showed that they could monitor the loss of viability by measuring pH and redox potential in MTS (Camus et al. 2016). *Rodríguez-Lorenzo* et al. developed a method for PSA detection with inverse sensitivity unlike other sensors. Silver ions grow on silver nanostar surface with low concentration of GOx and at high GOx concentrations the nucleation of silver nanoparticles occurred in solution, not on the surface of silver nanostars which resulted into small peak shift and found LOD up to 4×10^{-20} M in whole serum (Rodríguez-Lorenzo et al. 2012). In another report, *Liu* et al. also demonstrated glucose oxidase (GOx)-catalyzed growth of AuNPs for ultrasensitive detection of PSA, and confirmed four orders of magnitude more sensitivity in comparison to ELISA (Liu et al. 2014). Bio-bar-code, the combination of scanometric detection with magnetic microparticles, was demonstrated for detecting PSA with a LOD of 3 aM, more than six orders of magnitude of clinically accepted, conventional assays (Nam et al. 2003). Recently, in a very interesting study researchers stated that there is a dramatic reduction in the zinc content of prostate tissue associated with the inability of cancer cells to accumulate the ion, with the development of malignancy. They developed a fluorescence-based sensor to detect mobile zinc and used a transgenic mouse model of prostate adenocarcinoma and revealed that the progression of prostate cancer can be monitored by watching the decrease in zinc content in the prostates of tumor-bearing mice (Ghosh et al. 2010).

10.3.5.1 Challenges and Future Prospects

Prostate cancer diagnosis in clinics currently faces many problems that need attention. One of the most important concerns is the unavailability of a proper biomarker for diagnostic screening. PSA screening is no longer valid with its reduced sensitivity. Clinical diagnosis of PCa still depends on PSA for initial testing and later confirmation by many imaging and biochemical tests. While most of these tests can effectively detect cancer in a patient it gives minimal information regarding its aggressiveness. A proper stratification of patients based on their death risk should be carried out to prevent patients with low risk from go through unnecessary treatments, pain as well as high cost. Diagnosis of cancer using molecular biosensors is an emerging field for designing highly sensitive, quick, easy-to-use, and reliable procedures for early and precise detection of cancer. Nanomaterials and other technologies such as aptamers combined with micro/nanofluidics are highly promising to evolve as an ideal diagnostic biosensing platform. In conjunction with cost-effectiveness, it is very vital to construct devices that actually can detect early cancer using multiple cancer biomarkers at low concentrations in real biological fluids. Samples such as saliva is easy to obtain yet, the detection of markers are very challenging due to their presence in very low concentrations. Given that early cancer diagnose allows for successful treatment and recovery in patients, stratifying them according to the need is crucial for "Personalized Medicine" and Point-of-Care Testing. Possibility of detecting a cancer at very early stage, before even the development of a tumor, is still a challenge which could be explored in future.

Conclusions

The chapter emphasizes on the significance of nanobased biosensing technologies for detection of prostate cancer and development of future point-of-care devices. Although immense studies based on nanobiosensors and molecular cancer markers have been carried out, their efficacy in clinics still needs to be demonstrated. Further evolution of such nano devices with multiplexing of molecular biomarkers for cancer detection can aid in early and fast diagnosis, less sample volume requirement, and cost-effectiveness. As future medicine shifts towards personalized care, there is an increasing demand for such "tiny" biosensing devices that can be easily carried to any site for a faster and highly sensitive cancer detection at POC or even at home.

Acknowledgements The authors acknowledge Department of Bioproducts and Biosystems Engineering, University of Minnesota.

References

Acimovic SS, Ortega MA et al (2014) LSPR Chip for parallel rapid and sensitive detection of cancer markers in serum. Nano Lett 14:2636–2641

Alhasan AH, Scott AW, Wu JJ et al (2016) Circulating microRNA signature for the diagnosis of very high-risk prostate cancer. PNAS 113:10655–10660

Anran G, Lu N, Dai P et al (2014) Direct ultrasensitive electrical detection of prostate cancer biomarkers with CMOS-compatible n- and p-type silicon nanowire sensor arrays. Nanoscale 6:13036–13042

Augustsson P, Magnusson C, Nordin M, Lilja H, Laurell T (2012) Microfluidic label-free enrichment of prostate cancer cells in blood based on Acoustophoresis. Anal Chem 84:7954–7962

Barnett JM, Wraith P, Kiely J et al (2014) An inexpensive fast and sensitive quantitative lateral flow magneto-immunoassay for total prostate specific antigen. Biosensors 4:204–220

Bellan LM, Wu D, Langer RS (2011) Current trends in nanobiosensor technology. Wiley Interdiscip Rev Nanomed Nanobiotechnol 3:229–246
Bosch ME, Sanchez AJR et al (2007) Review: recent development in optical fiber. Biosensors Sens 7:797–859
Bretton PR (1994) Prostate-specific antigen and digital rectal examination in screening for prostate cancer: a community-based study. South Med J 87:720–723
Brock M, Bodman CV, Palisaar J, Becker W, Martin-Seidel P, Noldus J (2015) Detecting prostate cancer—a prospective comparison of systematic prostate biopsy with targeted biopsy guided by fused MRI and Transrectal Ultrasound. Dtsch Arztebl Int 112:605–611
Byeon HH, Lee SW, Lee EH, Kim W, Yi H (2016) Biologically templated assembly of hybrid semiconducting nanomesh for high performance field effect transistors and sensors. Sci Rep 6:35591
Cabaj J, Soloducho J, Nowakowska-Oleksy A (2010) Langmuir–Blodgett film based biosensor for estimation of phenol derivatives. Sensors Actuators B 143:508–515
Cabaj J, Chyla A, Jedrychowska A, Olech K, Soloducho J (2012) Detecting platform for phenolic compounds-characteristic of enzymatic electrode. Opt Mater 34:1677–1681
Camus VL, Stewart G, Nailon WH, McLaren DB, Campbell CJ (2016) Measuring the effects of fractionated radiation therapy in a 3D prostate cancer model system using SERS nanosensors. Analyst 141:5056–5061
Catalona WJ, Smith DS, Ratliff TL et al (1991) Measurement of prostate-specific antigen in serum as a screening-test for prostate-cancer. N Engl J Med 324:1156–1161
Catalona WJ et al (1994) Comparison of digital rectal examination and serum prostate specific antigen in the early detection of prostate cancer: results of a multicenter clinical trial of 6630 men. J Urol 151:1283–1290
Chaubey A, Malhotra BD (2002) Mediated biosensors. Biosens Bioelectron 17:441–456
Chikkaveeraiah BV, Bhirde A, Malhotra R, Patel V, Gutkind JS, Rusling JF (2009) Single-wall carbon nanotube forest arrays for immunoelectrochemical measurement of four protein biomarkers for prostate cancer. Anal Chem 81:9129–9134
Choi JH, Kim HS, Choi JW, Hong JW, Kim YK, Oh BK (2013) A novel Au-nanoparticle biosensor for the rapid and simple detection of PSA using a sequence-specific peptide cleavage reaction. Biosens Bioelectron 49:415–419
Chu X, Zhao ZL, Shen GL, Yu RQ (2006) Quartz crystal microbalance immunoassay with dendritic amplification using colloidal gold immunocomplex. Sensors Actuators B 114:696–704
Crawford ED, De Antoni EP (1993) PSA as a screening test for prostate cancer. Urol Clin North Am 20:637–646
Cui Y, Wei QQ, Park HK, Lieber CM (2001) Nanowire nanosensors for highly sensitive and selective detection of biological and chemical species. Science 293:1289–1292
D'Orazio P (2003) Biosensors in clinical chemistry. Clinica Chimica Acta 334:41–69
Duan XJ, Gao R, Xie P et al (2012) Intracellular recordings of action potentials by an extracellular nanoscale field-effect transistor. Nat Nanotechnol 7:174–179
Eggins B (2002) Chemical sensors and biosensors. Wiley, New York
Elabbady AA, Khedr MM (2006) Extended 12-core prostate biopsy increases both the detection of prostate cancer and the accuracy of Gleason score. Eur Urol 49:49–53
Fabrisa L, Ceder Y, Chinnaiyan AM et al (2016) The potential of MicroRNAs as prostate cancer biomarkers. Eur Urol 70:312–322
Gao R, Cheng Z, deMello AJ, Choo J (2016) Wash-free magnetic immunoassay of the PSA cancer marker using SERS and droplet microfluidics. Lab Chip 16:1022–1029
Geybels MS, Wright JL, Bibikova M et al (2016) Epigenetic signature of Gleason score and prostate cancer recurrence after radical prostatectomy. Clin Epigenetics 8:97
Ghai S, Toi A (2012) A role of transrectal ultrasonography in prostate cancer. Radiol Clin N Am 50:1061–1073
Ghosh K, Kim P, Zhang XA et al (2010) A novel imaging approach for early detection of prostate cancer based on endogenous zinc sensing. Cancer Res 70:6119–6127
Giaever I (1973) The antibody: antigen interaction: a visual observation. J Immunol 110:1424–1426

Guardia M, Garrigues S (2012) Handbook of green analytical chemistry. Wiley, Chichester
Haeberle S, Zengerle R (2007) Microfluidic platforms for lab-on-a-chip applications. Lab Chip 7:1094–1110
Haese A, de la Taille A, van Poppel H et al (2008) Clinical utility of the PCA3 urine assay in European men scheduled for repeat biopsy. Eur Urol 54:1081–1088
Hammond JL, Formisano N, Estrela P, Carrara S, Tkac J (2016) Electrochemical biosensors and nanobiosensors. Essays Biochem 60:69–80
Huang L, Peng Z, Guo Y, Porter AL (2011) Characterizing a technology development at the stage of early emerging applications: nanomaterial-enhanced biosensors. Tech Anal Strat 23: 527–544
Hughes AJ, Lina RKC, Peehl DM, Herr AE (2012) Microfluidic integration for automated targeted proteomic assays. PNAS 109:5972–5977
Hwang MP, Lee JW, Lee KE, Lee KH (2013) Think modular: a simple apoferritin-based platform for the multifaceted detection of pancreatic cancer. ACS Nano 7:8167–8174
Islam M, Bellah MM, Sajid A et al (2015) Effects of nanotexture on electrical profiling of single tumor cell and detection of cancer from blood in microfluidic channels. Sci Rep 5:13031
Jadvar H (2015) PSMA PET in prostate cancer. J Nucl Med 56:1131–1132
Jazayeri MH, Amani H, Pourfatollah AA, Avan A, Ferns GA, Pazoki-Toroudi H (2016) Enhanced detection sensitivity of prostate-specific antigen via PSA-conjugated gold nanoparticles based on localized surface plasmon resonance: GNP-coated anti-PSA/LSPR as a novel approach for the identification of prostate anomalies. Cancer Gene Ther 23:365–369
Jin X, Chen L, Jiang J, Shen G, Yu R (2009) Piezoelectric immunosensor with gold nanoparticles enhanced competitive immunoreaction technique for quantification of aflatoxin B1. Biosens Bioelectron 24:2580–2585
Jolly P, Tamboli V, Harniman RL, Estrela P, Allender CJ, Bowen JL (2016) Aptamer-MIP hybrid receptor for highly sensitive electrochemical detection of prostate specific antigen. Biosens Bioelectron 75:188–195
Kang BJ, Jeun M, Jang GH, Song SH, Jeong IG, Kim CS, Searson PC, Lee KH (2015) Diagnosis of prostate cancer via nanotechnological approach. Int J Nanomedicine 10:6555–6569
Kim DJ, Sohn IY, Jung JH, Yoon OJ, Lee NE, Park JS (2013) Reduced graphene oxide field-effect transistor for label-free femtomolar protein detection. Biosens Bioelectron 41:621–626
Kirby BJ, Jodari M, Loftus MS et al (2012) Functional characterization of circulating tumor cells with a prostate-cancer-specific microfluidic device. PLoS One 7:e35976
Knoll W (1998) Interfaces and thin films as seen by bound electromagnetic waves. Annu Rev Phys Chem 49:569–638
Kong RM, Ding L, Wang Z, You J, Qu F (2015) A novel aptamer-functionalized MoS_2 nanosheet fluorescent biosensor for sensitive detection of prostate specific antigen. Anal Bioanal Chem 407:369–377
Kosaka PM, Pini V, Ruz JJ et al (2014) Detection of cancer biomarkers in serum using a hybrid mechanical and optoplasmonic nanosensor. Nat Nanotechnol 9:1047–1053
Kronick MN, Little WA (1975) A new immunoassay based on fluorescent excitation by internal reflection spectroscopy. J Immunol Methods 8:235–240
Lee J, Dak P, Lee Y et al (2014) Two-dimensional layered MoS_2 biosensors enable highly sensitive detection of biomolecules. Sci Rep 4:7352
Li C, Curreli M, Lin H, Lei B et al (2005) Complementary detection of prostate-specific antigen using In_2O_3 nanowires and carbon nanotubes. J Am Chem Soc 127:12484–12485
Liu GL, Rosa-Bauza YT et al (2007) Peptide-nanoparticle hybrid SERS probes for optical detection of protease activity. J Nanosci Nanotechnol 7(7):2323–2330
Liu D, Yang J, Wang HF et al (2014) Glucose oxidase-catalyzed growth of gold nanoparticles enables quantitative detection of attomolar cancer biomarkers. Anal Chem 86:5800–5806
Loeb S, Partin AW (2011) Review of the literature: PCA3 for prostate cancer risk assessment and prognostication. Rev Urol 13:e191–e195
Lu N, Gao A, Dai P et al (2015) Ultrasensitive detection of dual cancer biomarkers with integrated CMOS-compatible nanowire arrays. Anal Chem 87:11203–11208

Ma L, He S, Huang J, Cao L, Yang F, Li L (2009) Maximizing specificity and yield of PCR by the quantum dot itself rather than property of the quantum dot surface. Biochimie 91:969–973
Madaboosi N, Pedrosa CR, Reis MF (2014) Microfluidic ELISA for sensing of prostate cancer biomarkers using integrated a–Si:H p–i–n photodiodes. In: IEEE Sensors Proceedings Valencia, pp 881–884
Manz A, Graber N, Widmer HM (1990) Miniaturized total chemical analysis systems: a novel concept for chemical sensing. Sensors Actuators B Chem 1:244–248
Meharvar M, Bis C et al (2000) Fiber optic biosensors-trends and applications. Anal Sci 16
Mintz A (2014) PET/CT in prostate cancer: an unmet clinical need. Oncology (Williston Park) 28:1065–1066
Moltzahn F, Olshen AB, Baehner L et al (2011) Microfluidic-based multiplex qRT-PCR identifies diagnostic and prognostic microRNA signatures in the sera of prostate cancer patients. Cancer Res 71:550–560
Mouli SK, Zhao LC, Omary RA, Thaxton CS (2010) Lymphotropic nanoparticle enhanced MRI for the staging of genitourinary tumors. Nat Rev Urol 7:84–93
Munge B, Liu G, Collins G, Wang J (2005) Multiple enzyme layers on carbon nanotubes for electrochemical detection down to 80 DNA copies. Anal Chem 77:4662–4666
Najeeb MA, Jasmine SH, Chavali M (2014) Recent advancements in nano-based biosensor for early detection of prostate cancer. IJRTE 3:112–120
Nam JM, Thaxton CS, Mirkin CA (2003) Nanoparticle-based bio-bar codes for the ultrasensitive detection of proteins. Science 301:1884–1886
Narayanaswamy R, Wolfbeis O (2004) Optical sensors. Springer, Berlin
Nath N, Chilkoti A (2002) A colorimetric gold nanoparticle sensor to interrogate biomolecular interactions in real time on a surface. Anal Chem 74:504–509
No D, Duyne RV, Bingham J (2008) Localized surface plasmon resonance nanobiosensors for the detection of a prostate cancer. Biomarker Nanoscape 5:15
Oesterling JE, Jacobsen SJ, Chute CG, Guess HA, Girman CJ, Panser LA, Lieber MM (1993) Serum prostate-specific antigen in a community-based population of healthy men establishment of age-specific reference ranges. JAMA 270:860–864
Omidi M, Choolaei M, Asjodi F, Haghiralsadat F, Yazdian F (2014) Measurement of prostate specific antigen using self-sensing nanomechanical membrane. Proc Eng 87:660–663
Park H, Hwang MP, Lee JW, Choi J, Lee KH (2013a) Harnessing immunomagnetic separation and quantum dot-based quantification capacities for the enumeration of absolute levels of biomarker. Nanotechnology 24:285103
Park H, Hwang MP, Lee KH (2013b) Immunomagnetic nanoparticle-based assays for detection of biomarkers. Int J Nanomedicine 8:4543–4552
Ploussard G, Masson-Lecomtec A, Beauval JB et al (2011) Radical prostatectomy for high-risk prostate cancer defined by preoperative criteria: oncologic follow-up in national multicenter study in 813 patients and assessment of easy-to-use prognostic substratification. Urology 78:607–613
Prakrankamanant P (2014) Quartz crystal microbalance biosensors: prospects for point-of-care diagnostics. J Med Assoc Thail 97:S56–S64
Prasad PN (2003) Introduction to biophotonics. Wiley, Hoboken
Reske SN, Blumstein NM, Neumaier B et al (2006) Imaging prostate cancer with 11C-choline PET/CT. J Nucl Med 47:1249–1254
Rodríguez-Lorenzo L, de la Rica R, Álvarez-Puebla RA, Liz-Marzán LM, Stevens MM (2012) Plasmonic nanosensors with inverse sensitivity by means of enzyme-guided crystal growth. Nat Mater 11:604
Samiei E, Tabrizian M, Hoorfar M (2016) A review of digital microfluidics as portable platforms for lab-on a-chip applications. Lab Chip 16:2376
Sant W, Pourciel ML et al (2003) Development of chemical field effect transistors for the detection of urea. Sensors Actuators B 95:309–314
Sanz G, Rioja J, Zudaire JJ, Berian JM, Richter JA (2004) PET and prostate cancer. World J Urol 22:351–352

Sarkar P, Ghosh D, Bhattacharyay D, Setford SJ, Turner APF (2008) Electrochemical immunoassay for free prostate specific antigen (f-PSA) using magnetic beads. Electroanalysis 20:1414–1420
Sauerbrey G (1959) The use of quartz oscillators for weighing thin layers and for microweighing. Zeitschrift Fuer Physik 115:206–222
Schlensog MD, Gronewold TM, Tewe M, Famulok M, Quandt E (2004) A Love-wave biosensor using nucleic acids as ligands. Sensors Actuators B 101:308–315
Schoning MJ, Poghossian A (2002) Recent advances in biologically sensitive field-effect transistors (biofets). Analyst 12:1137–1151
Schroder FH, van der Cruijsen-Koeter I, de Koning HJ et al (2000) Prostate cancer detection at low prostate specific antigen. J Urol 163:806–811
Schröder FH, Hugosson J, Roobol MJ et al (2009) Screening and prostate-cancer mortality in a randomized European study. N Engl J Med 360:1320–1328
Sharma A, Hong S, Singh R, Jang J (2015) Single-walled carbon nanotube based transparent immunosensor for detection of a prostate cancer biomarker osteopontin. Anal Chim Acta 869:68–73
Shobha BN, Muniraj NJR (2014) Design modeling and simulation of prostate cancer biosensor with ssDNA biomarker and DGFET biosensor. IJCSIT 5:2612–2262
Siegel RL, Miller KD, Jemal A (2016) Cancer statistics 2016. CA Cancer J Clin 66:7–30
Singh R, Prasad R, Sumana G, Arora K et al (2009) STD sensor based on nucleic acid functionalized nanostructured polyaniline. Biosens Bioelectron 24:2232–2238
Singh R, Dhand C, Sumana G, Verma R, Sood S, Gupta RK, Malhotra BD (2010) Polyaniline/carbon nanotubes platform for sexually transmitted disease detection. J Mol Recognit 23:472–479
Singh R, Verma R, Kaushik A et al (2011) Chitosan-Iron Oxide Nano-composite Platform for Mismatch-Discriminating DNA Hybridization for detection of *Neisseria gonorrhoeae* causing Sexually Transmitted Disease. Biosensens Bioelectron 26:2967–2974
Singh R, Matharu Z, Srivastava AK, Sood S, Gupta RK, Malhotra BD (2012a) Nanostructured platform for the detection of *Neisseria gonorrhoeae* using electrochemical impedance spectroscopy and differential pulse voltammetry. Microchim Acta 177:201–210
Singh R, Verma R, Sumana G et al (2012b) Nanobiocomposite platform based on polyaniline-iron oxide-carbon nanotubes for bacterial detection. Bioelectrochemistry 86:30–37
Singh R, Mukherjee MD, Sumana D, Gupta RK, Sood S, Malhotra BD (2014a) Biosensors for pathogen detection: a smart approach towards clinical diagnosis. Sensors Actuators B 197:385–404
Singh R, Sharma A, Hong S, Jang J (2014b) Electrical immunosensor based on dielectrophoretically-deposited carbon nanotubes for detection of influenza virus H1N1. Analyst 139:5415–5421
Streeter I, Wildgoose GG, Shao L, Compton RG (2008) Cyclic voltammetry on electrode surfaces covered with porous layers: an analysis of electron transfer kinetics at single-walled carbon nanotube modified electrodes. Sensors Actuators B Chem 133:462–466
Stuopelyte K, Daniunaite K, Bakavicius A et al (2016) The utility of urine-circulating miRNAs for detection of prostate cancer. Br J Cancer 115:707–715
Tang CK, Vaze A, Shen M, Rusling JF (2016a) High-throughput electrochemical microfluidic immunoarray for multiplexed detection of cancer biomarker proteins. ACS Sensors 1:1036–1043
Tang Z, Wang L, Ma Z (2016b) Triple sensitivity amplification for ultrasensitive electrochemical detection of prostate specific antigen. Biosens Bioelectron 92:577–582
Thompson IM, Pauler DK, Goodman PJ et al (2004) Prevalence of prostate cancer among men with a prostate-specific antigen level 40 ng per milliliter. N Engl J Med 351:2239–2246
Topkaya SN, Ozkan-Ariksoysal D, Kosova B, Ozel R, Ozsoz M (2012) Electrochemical DNA biosensor for detecting cancer biomarker related to glutathione S-transferase P1 (GSTP1) hypermethylation in real samples. Biosens Bioelectron 31:516–522
Truong PL, Kim BW, Sim SJ (2012) Rational aspect ratio and suitable antibody coverage of gold nanorod for ultra-sensitive detection of a cancer biomarker. Lab Chip 12:1102–1109
Uludag Y, Tothill IE (2010) Development of a sensitive detection method of cancer biomarkers in human serum (75%) using a quartz crystal microbalance sensor and nanoparticles amplification system. Talanta 82:277–282

Unger MA, Chou HP, Thorsen T, Scherer A, Quake SR (2000) Monolithic microfabricated valves and pumps by multilayer soft lithography. Science 288:113–116

Velonas VM, Woo HH, dos Remedios CG, Assinder SJ (2013) Current status of biomarkers for prostate cancer. Int J Mol Sci 14:11034–11060

Velusamy V, Arshak K, Korostynska O, Oliwa K, Adley C (2010) An overview of foodborne pathogen detection: in the perspective of biosensors. Biotechnol Adv 28:232–254

Wang J, Liu G, Jan MR (2004) Ultrasensitive electrical biosensing of proteins and DNA: carbon-nanotube derived amplification of the recognition and transduction events. J Am Chem Soc 126:3010–3011

Wang J, Renu L, Liu W et al (2009) Microfluidics: a new cosset for neurobiology. Lab Chip 9:644–652

Wang X, Zhao M, Nolte DD, Ratliff TL (2011) Prostate specific antigen detection in patient sera by fluorescence-free BioCD protein array. Biosens Bioelectron 26:1871–1875

Wang H, Zhang Y, Yu H et al (2013) Label-free electrochemical immunosensor for prostate-specific antigen based on silver hybridized mesoporous silica nanoparticles. Anal Biochem 434:123–127

Welch HG, Fisher ES, Gottlieb DJ, Barry MJ (2007) Detection of prostate cancer via biopsy in the Medicare-SEER population during the PSA era. J Natl Cancer Inst 99:1395–1400

Wu C, Ko F, Wu C, Pan T, Chuang C (2011) Prostate cancer marker sensing under nanostructural biochip technique. In: 5th European IFMBE conference IFMBE proceedings, vol 37, pp 983–986

Wu CC, Pan TM, Wu CS et al (2012) Label-free detection of prostate specific antigen using a silicon nanobelt field-effect transistor. Int J Electrochem Sci 7:4432–4442

Xiaoqing C, Zhou G, Song P et al (2014) Ultrasensitive electrochemical detection of prostate-specific antigen by using antibodies anchored on a DNA nanostructural scaffold. Anal Chem 86:7337–7342

Zhang Y, Hong H, Myklejord DV, Cai W (2011) Molecular imaging with SERS-active nanoparticle. Small 7:3261–3269

Zhang F, Li S, Cao K et al (2015) A microfluidic Love-wave biosensing device for PSA detection based on an aptamer beacon probe. Sensors 15:13839–13850

Zheng G, Patolsky F, Cui Y, Wang WU, Lieber CM (2005) Multiplexed electrical detection of cancer markers with nanowire sensor arrays. Nat Biotechnol 23:1294–1301

Zieglschmid V, Hollmann C, Gutierrez B, Albert W, Strothoff D, Gross E, Böcher O (2005) Combination of immunomagnetic enrichment with multiplex rt-pcr analysis for the detection of disseminated tumor cells. Anticancer Res 25:1803–1810

Developments in the Electrochemical Bionanosensors for the Predictive Diagnosis of Prostate and Breast Cancer

11

Suman Singh, Akash Deep, Girish Mohanta, and Vijay Kumar Meena

11.1 Introduction

The abnormal and uncontrolled cell growth due to an accumulation of specific genetic and epigenetic defects is referred as "Cancer." The genetic defects can originate from both environment and hereditary. This unregulated cell growth leads to the formation of a tumor. As the cancer progresses, the tumor begins to spread to other body organs and systems and this spreading is sometimes so fast that by the time of cancer detection, it becomes incurable. It is reported that, in 2004, about 7.4 million people lost their lives and this number is expected to increase to 12 million by 2030. As per the World Health Organisation (WHO) documentation, 30% of people could have been saved if their cancer could be detected earlier (www.who.int/, 2010). Among various cancers, prostate and breast cancer are most common type of cancers occurring in men and women, respectively. Prostate cancer is the second largest cause among all cancer-related deaths in male population and breast cancer contributes to about 23% of the cancer cases, worldwide (Mahfoud et al. 2014).

The detection of cancer biomarkers is considered as a most valuable tool for early cancer detection. The National Cancer Institute (NCI) defines a biomarker as "a biological molecule found in blood, other body fluids, or tissues that is a sign of a normal or abnormal process or of a condition or disease." Biomarkers are typically detected in human fluids such as blood, serum, urine, or cerebral spinal fluid, but they can also be present in or on tumor cells. Biomarkers are further classified as prognostic markers and predictive markers. The prognostic markers are used for the useful selection of patients for treatment; however the response to a treatment can't

S. Singh (✉) • A. Deep • G. Mohanta • V.K. Meena
CSIR-Central Scientific Instruments Organisation (CSIR-CSIO), Sector 30-C, Chandigarh, India
e-mail: sumansingh01@gmail.com, ssingh@csio.res.in

P. Chandra et al. (eds.), *Next Generation Point-of-care Biomedical Sensors Technologies for Cancer Diagnosis*, DOI 10.1007/978-981-10-4726-8_11

be predicted. On the other hand, the predictive markers are useful for the evaluation of the benefits of specific clinical intervention (Mehta et al. 2010). Thus, the biomarker detection can help in staging the accurate pretreatment, determining the response of cancer to chemotherapy treatment, and monitoring the disease progression (Basil et al. 2006).

Presently, clinicians are dependent on conventional techniques like chromatography, enzyme-linked immunoassay (ELISA), magnetic resonance imaging (MRI), biopsy, thermography, radioimmunoassay (RIA), immunohistochemistry (IHC), and many more, depending upon the requirement (Michaelson et al. 2002). Though these techniques are effective but at the same time suffer limitations like false-positive or -negative results which can give wrong interpretation, often require sophisticated, bulky, and costly equipments, which make them unsuitable for routine screening applications or point-of-care diagnostics (Johnson and Kotowski 1993; Acevedo et al. 2002; Kanyong et al. 2016). As a solution to these problems, the researchers are now looking for better detection methods which can be sensitive and fulfil the need of point-of-care diagnosis. Biosensors since their inception have revolutionized the research domain in the area of detection of important analytes, ranging from agricultural to medical applications. A biosensor consists of three components: analyte, bioreceptor, and transducer. All these three components work in coherence to give significant signal. An analyte is any molecule of interest that needs to be detected. The bioreceptor is the molecule of biological origin, specific to an analyte, which interacts with it and undergoes biological reaction to produce a signal. The bioreceptors could be any biological entity like enzymes, antibodies, aptamers, DNA, whole cells, etc., that has specificity towards a particular analyte. The transducer converts the biological signal produced by the interaction of target analyte and bioreceptor into a measurable signal. Thus, the principle of a biosensor is recognition of a target molecule or a specific event by a biological molecule, and the extent to which the target is recognized is detected by a transducer. The biosensors exhibit attractive features like sensitive, specificity, selectivity, quick response, and cost-effectiveness. Moreover, these can be used for all kinds of reagent matrices like blood, serum, urine, milk, saliva, etc. Among various types of transducers, electrochemical transducers are more frequently being used due to their easy handling, simplicity, possibility of miniaturization, fast response, and relatively lower device costs (Chandra 2013). Owing to these inherent properties, the electrochemical transducers are considered as most suitable for construction of on-site detection devices. Their application has diversification in many fields, including food safety, green energy, biomedical and environmental monitoring (Truong et al. 2011; Cesarino et al. 2012; Chen and Chatterjee 2013; Singh et al. 2013, 2016a).

Over the last decade, integration of nanotechnology with biosensors has had a great impact and significant advances have been made in this area due to their high surface area, advantageous electronic properties and electrocatalytic activity as well as proper biocompatibility of nanomaterials. However, all these properties are size and shape dependent, which can be tuned depending upon the type of application to be explored. Nanomaterials and their composites offer the prospects which help in interfacing of biological recognition events with electronic signal transduction and that might result in the design of a new generation bioelectronic devices. The different kinds of nanoparticles play different roles in different sensing systems. These can be used for "electronic wiring" of biorecognition element to an electrode surface, as "catalysts" to promote electrochemical reaction, and as "labels and amplifiers". The electrochemical nanobiosensors are emerging field of biosensors, where the advantages of electrochemical biosensing in combination of nanotechnology has resulted in the generation of low cost, robust, reliable, easy-to-use, and ultrasensitive diagnostics.

In this book chapter, authors have addressed the advancements made in the area of electrochemical sensing of prognostic diagnosis of prostate and breast cancers and role of nanotechnology in enhancing the sensing capabilities of sensors.

11.2 Prostate Cancer

Prostate cancer accounts for about 10% of all deaths from cancers and is the sixth leading cause of cancer deaths among men worldwide (Grönberg 2003). Serum-based prostate-specific antigen (PSA) is the most widely used biomarker of prostate cancer (Kingsmore 2006). PSA is a serine protease, produced by the prostate epithelium to maintain liquefaction of seminal fluid (Nadler et al. 1995; Webber et al. 1995). The detection and monitoring of PSA assumes a great significance for early prognosis of cancer (Garnick 1993). The occurrence of prostate tumor leads to the release of relatively high concentrations of PSA into the circulatory system (Sarkar et al. 2002). On an average, an above cutoff value of 4.0 ng/mL (and more recently 2.5 ng/mL) is generally categorized as PSA positive and indicates the need of biopsy for confirmation (Catalona et al. 1999). PSA testing is also used to evaluate the patient's response to ablation therapy (e.g., radical prostatectomy) and to monitor the chances of disease recurrence (Healy et al. 2007). Due to the above reasons, it is important to develop ultrasensitive biosensing platforms with capability of detecting very low concentrations of serum PSA.

Currently, most PSA testing takes place at dedicated, centralized laboratories on large, automated high-throughput systems. Numerous analyzer-run PSA assays are currently available in the marketplace (Healy et al. 2007; Shi and Yeh 2012). Immunosensor platforms are also used for the sensitive detection of PSA, like ELISA, electrochemiluminescence, fluorescence, electrochemical and surface Plasmon resonance spectrometry (Triroj et al. 2011; Uludag and Tothill 2012; Liu et al. 2013; Lang et al. 2014). Due to the experimental complexity of their operation, the above techniques can only be executed by trained personals for which long analysis times are invariably required. In comparison, the biosensors can be seen as an ideal (point-of-care) POC solution as they can offer rapid, sensitive, and bedside analysis, particularly for the risk stratification and assessment of cancer markers (Ivnitski et al. 1999).

The quantitative biosensing of PSA can also be carried out using optical methods including fluorescent labeling (Song et al. 2010), surface-enhanced Raman scattering (SERS) (Ma et al. 2014), and surface Plasmon resonance (SPR) (Jang et al. 2009).

The electrical and electrochemical signal transduction methods have also been developed for the above said purpose, which employ silicon nanowire field-effect transistors (Chen et al. 2011), gold nanorods (Choi et al. 2010), two-dimensional molybdenum disulfide nanosheets (Kukkar et al. 2016), and single-walled carbon nanotubes (SWNTs) (Okuno et al. 2007). Despite many efforts that have been made in developing the sensors for PSA detection, certain challenges still need to be addressed such as their miniaturization, simplification of the fabrication steps, and cost reduction. Majority of these issues are related to the bulk component and high cost of optical detection methods. FETs and related electronic device architecture are possibly the best solutions for designing of portable, stable, and precise biosensing devices. Some of the important developments made in the area of electrochemical biosensing of PSA are depicted in Figs. 11.1, 11.2, 11.3, 11.4, 11.5, 11.6, 11.7, 11.8, 11.9, 11.10, 11.11, 11.12, 11.13, 11.14, 11.15, and 11.16.

Immunosensors employing highly specific molecular interaction between antigen and antibodies are most widely used approaches. A wide variety of immunosensors have been developed for PSA detection using different immobilization substrates and transducers. An electrochemical immunosensor has been developed by covalently immobilizing the anti-PSA antibodies and redox mediator (thionine) onto AuNPs (gold nanoparticles) incorporated polyamidoamine dendrimer which was then deposited on multiwalled carbon nanotubes/ionic liquid/chitosan nanocomposite (Kavosi et al. 2014). The schematic of this immunosensors is shown in Fig. 11.1.

The amplification of immunoassay was facilitated due to the electron transfer process, accelerated by the AuNPs along with the sandwiching of the antigen between the immobilized anti-PSA and horseradish peroxidase (HRP)-labeled anti-PSA secondary antibody. A detection limit of 1 pg/mL for PSA was achieved through the enhanced electrocatalytic reduction of H_2O_2 by HRP in the presence of antigen. The signal was measured by differential pulse voltammetry technique. The promising results of biosensor as compared to ELISA signified its practicability towards developing other electrochemical immunosensors for clinical investigations.

Another label-free and highly selective three-dimensional electrochemical immunosensor using a highly conducting graphene/Au nanocomposite electrode has been reported for PSA detection (Jang et al. 2015). The stepwise schematic for the development of biosensor is portrayed in Fig. 11.2. The nanocomposite was prepared using aerosol spray technique, followed by the drop casting of antibody solution onto its surface to form the modified electrode. The immunosensor showed a linear activity in the range of 0–10 ng/mL with detection limit of 0.59 ng/mL. The sensor also exhibited high sensitivity, selectivity, stability, and reproducibility. In a recent study, the use of antibody-conjugated thin films of tetracyanoquinodimethane (TCNQ)-doped copper-MOF ($Cu_3(BTC)_2$) over a screen-printed gold electrode has been demonstrated for electrochemical sensing of PSA (Bhardwaj et al. 2017). The TCNQ doping improved the conductance of $Cu_3(BTC)_2$ thin films resulting in a highly sensitive sensing format which could deliver a dynamic linear range of

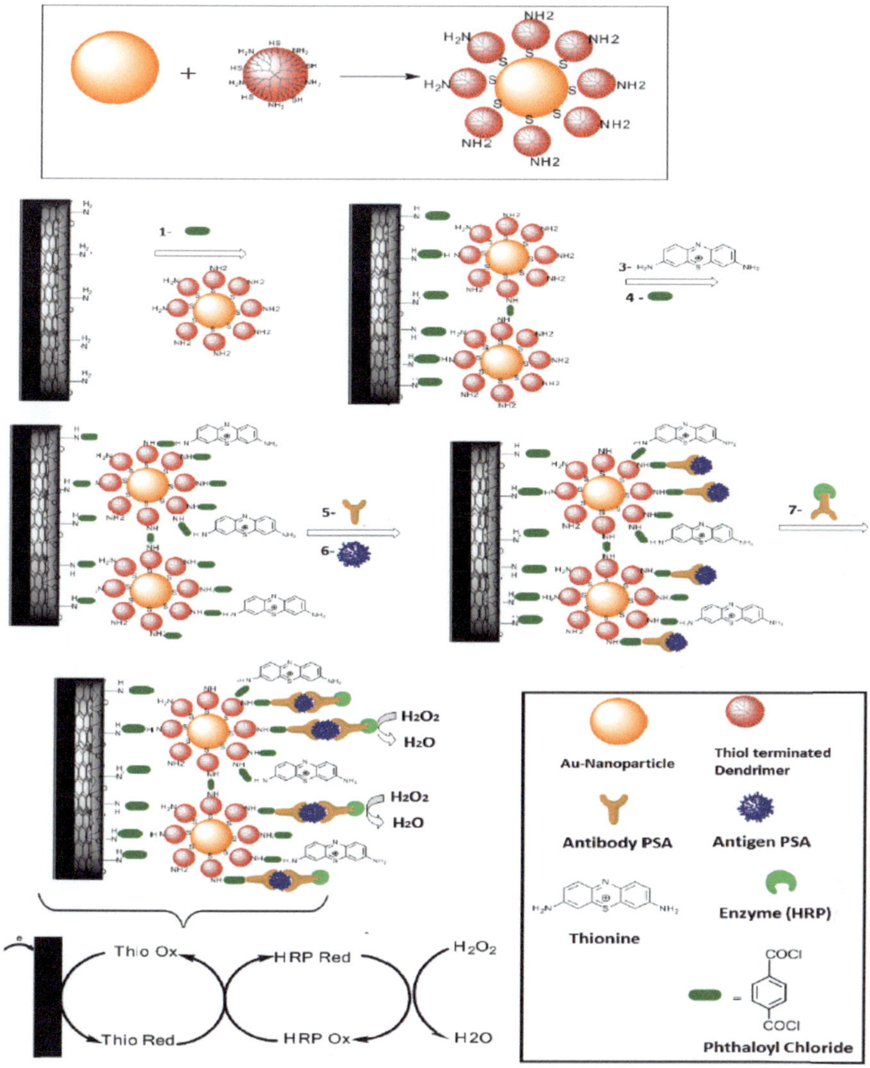

Fig. 11.1 Illustration of the stepwise process for PSA immunosensor fabrication (adapted from reference Kavosi et al. 2014)

detection 0.1–100 ng/mL with a low detection limit of 0.06 ng/mL even in the presence of other proteins.

For achieving real-time ultrasensitive detection of PSA to aid in early diagnosis of prostate cancer along with assessing the chances of disease recurrence after treatment, microcontact-PSA imprinted capacitive biosensor has been developed (Ertürk et al. 2015). Figure 11.3 shows the schematic and preparation of this microcontact imprinted biosensor. The imprinted antigen-based biosensor chips were compared

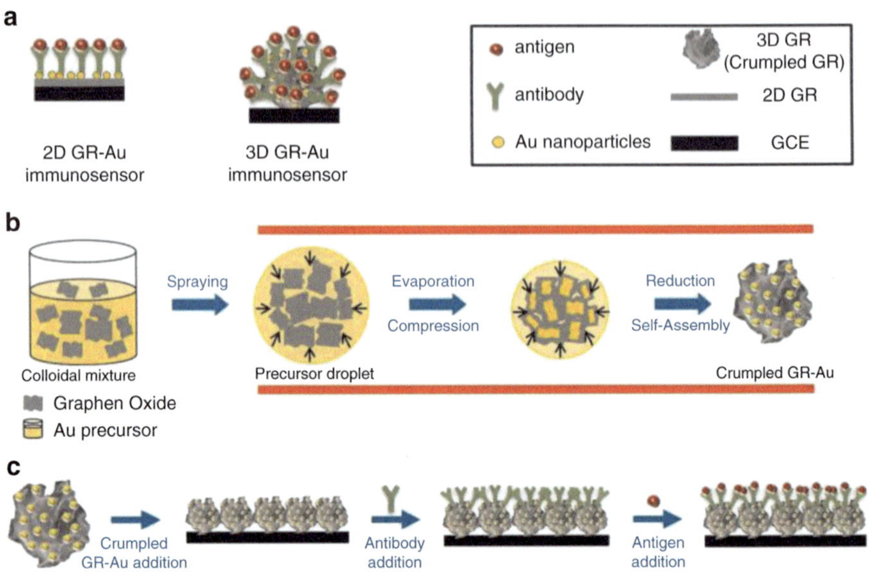

Fig. 11.2 Schematic illustration of (**a**) 2D GR–Au and 3D GR–Au electrodes, (**b**) the formation of crumpled GR–Au composites via aerosol spray pyrolysis, and (**c**) fabricating step of 3D label-free PSA immunosensor using crumpled GR–Au composites (adapted from reference Jang et al. 2015)

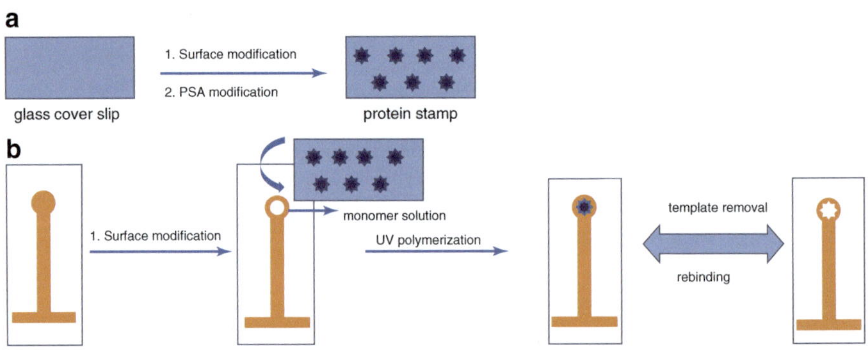

Fig. 11.3 Schematic representation of the microcontact imprinting of PSA onto the capacitive biosensor. (**a**) Preparation of the glass cover slips (protein stamps), (**b**) preparation of the capacitive gold electrodes and micro-contact imprinting via UV polymerization (adapted from reference Ertürk et al. 2015)

with anti-PSA antibody-immobilized electrode for capacitance-based detection performance. A detection limit of 8.0×10^5 ng/mL was observed in case of microcontact imprinted sensor compared to 6.0×10^4 ng/mL for anti-PSA electrodes. Thus, the attainment of reproducible, sensitive, selective, and real sample compatible results with the developed biosensing chips emphasized its potential applications for the real-time detection of analyte even at very low concentration.

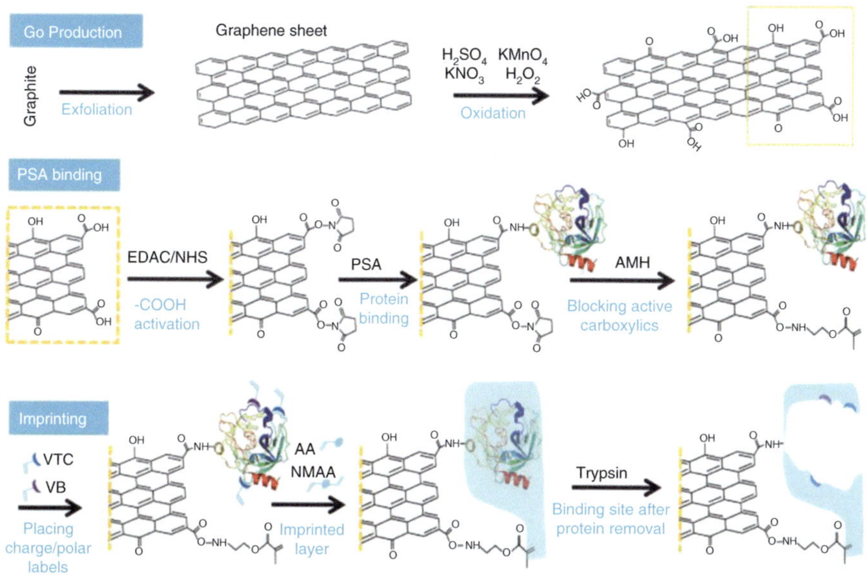

Fig. 11.4 Scheme for synthesis of the C/PIM materials for application in the development of sensor for PSA (adapted from reference Rebelo et al. 2014)

As an alternate to environment-sensitive natural antibodies, the chemically and thermally stable artificial plastic antibodies produced using Protein Imprinted Materials (PIM) in conformation controlled manner are being currently investigated in biosensor development. Such devices for PSA quantification can be coupled with low cost potentiometric electrochemical measurement (Rebelo et al. 2014). Using the concept, it is possible to detect PSA in serum samples with the antigens level of >10 ng/mL. The Protein Imprinted Materials with charged binding sites (C/PIM) can be produced by surface imprinting of covalently attached protein over graphene layers as shown in Fig. 11.4. The binding sites were labeled using charged monomers, followed by their self-organization around the protein particles through polymerization. These imprinted materials were then used as ionophores in PVC membranes for detection of PSA up to a sensitive level of 2 ng/mL. Thus, this system was suggested for the development of simple and inexpensive potentiometric immunosensors for PSA.

Last few decades have witnessed the miniaturization of biomedical devices. In this development, the semiconductor nanowire field-effect transistors (NWFETs) have attracted significant interest due to their nano-dimensions, intrinsic electronic properties, and the label-free real-time detection mode. In an example of AuNPs-modified silicon nanowire NWFETs, a new approach was used based on covalent immobilization of the fragmented antibodies on the thiol-modified silicon (Fig. 11.5) (Presnova et al. 2017). An ultralow detection limit of 23 fg/mL was achieved for PSA with an analysis time of 1 min.

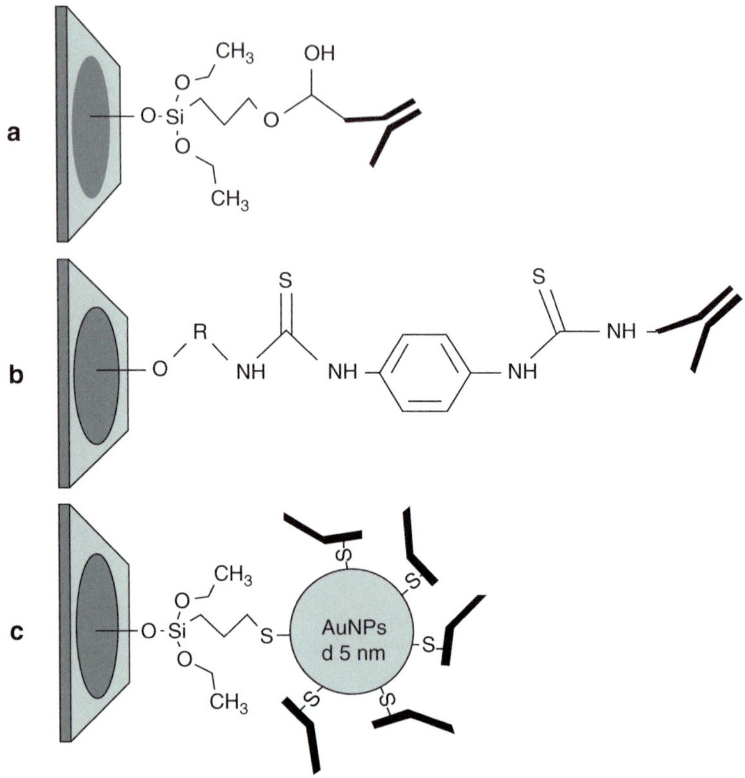

Fig. 11.5 Schematic illustration of stepwise functionalization of silicon surfaces with AuNPs (adapted from reference Presnova et al. 2017)

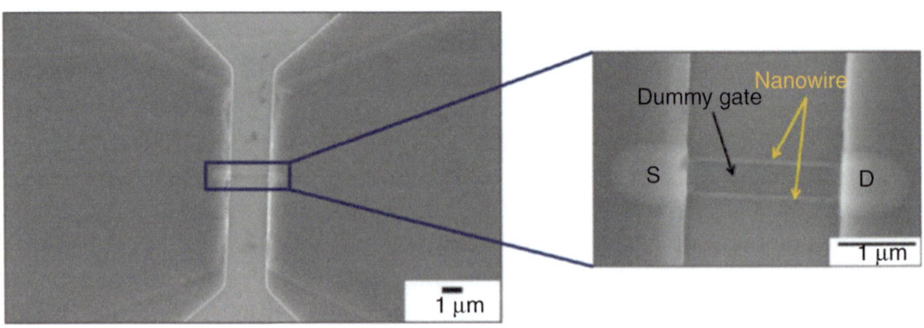

Fig. 11.6 X SEM image of the poly-Si NWFET device (adapted from reference Huang et al. 2013)

In another work exploring a similar approach, a polycrystalline silicon nanowire field-effect transistor (poly-Si NWFET) was developed as a biosensor. The involved strategy employed the use of a sidewall spacer to create nanoscale patterns, thereby avoiding the requirement of costly lithography tools (Fig. 11.6). This technique with advantages of simplicity and low cost could allow the prediction of cancer

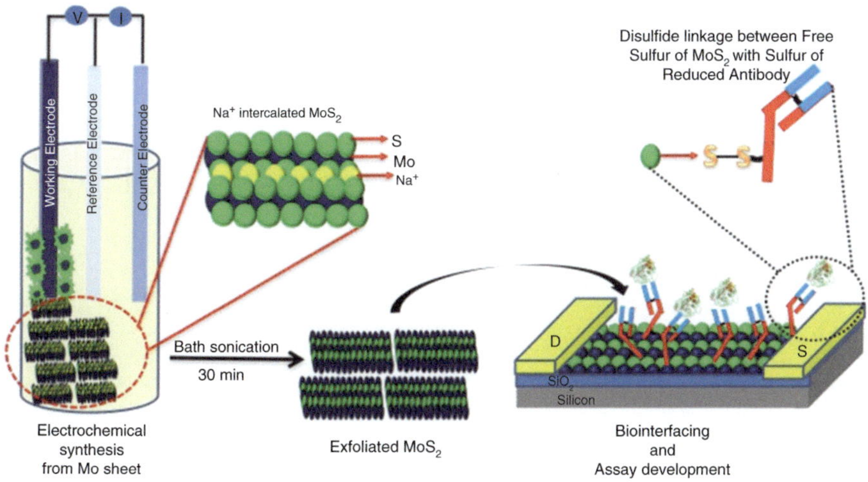

Fig. 11.7 Schematic of the few-layered MoS_2 nanosheets-based FET microdevice for PSA (adapted from reference Kukkar et al. 2016)

recurrence and the poly-Si NWFET device offers real-time, label-free, and ultrahigh-sensitivity detection of PSA in human serum. The detection limit of the above microfluidic channel incorporated sensor for PSA was better than 5 fg/mL and the detections were possible only in buffer but also in human serum samples (Huang et al. 2013).

Recently, a research has been published to demonstrate a FET-based immunosensor for highly sensitive detection of PSA. In the process, the few-layered MoS_2 nanosheets were channelled in the FET microdevice (Fig. 11.7). Subsequently, the chemically reduced anti-PSA antibodies were immobilized on the MoS_2 channel. The antibodies were deliberately reduced to expose their hinge-region disulfide bonds. This approach was suggested to offer realization of a robust and site-directed immunosensing device through bio-interfacing of the sulfhydryl groups (−SH) in the reduced antibody with surface S atoms of MoS_2. The device was validated as an effective immunosensor with a low detection limit (10^{-5} ng/mL) over a wide linear detection range (10^{-5}–75 ng/mL).

Electrochemiluminescence (ECL)-based immunosensors have also been explored for the visual quantification of free PSA in serum. In one of the examples, the ferrocene carboxylate nanocrystals were entrapped in functionalized graphene on the cathodic ECL of peroxodisulfate (Zhang et al. 2014). The peroxodisulfate casted on the glassy carbon electrode (GCE) shows enhanced ECL which decreased in the presence of antigen PSA due to the formation of antigen–antibody complex. Figure 11.8 shows the schematic of the above developed immunosensor. This ECL-based immunosensor could exhibit high detection limit of 1.7 pg/mL along with selectivity and stability.

An ECL immunosensor based on the use of potassium niobate-Au NPs@bismuthsulfide nanosheets-modified GCE (KNbO3–Au NPs@Bi2S3/GCE) has been proposed in the recent literature (Li et al. 2015). The authors of this work explored

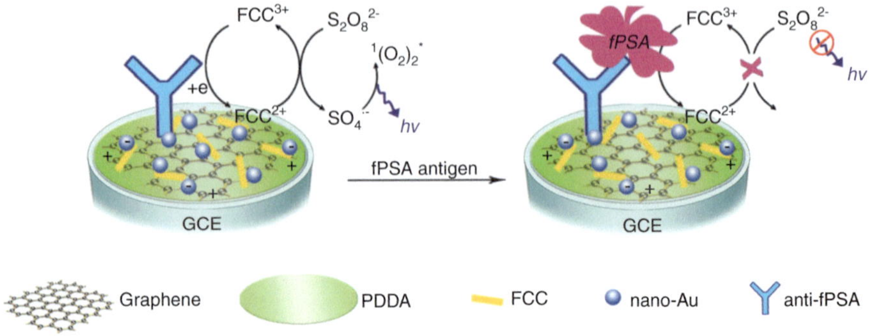

Fig. 11.8 Schematic procedure for the ECL immunoassay of PSA using graphene-modified glass electrode (adapted from reference Zhang et al. 2014)

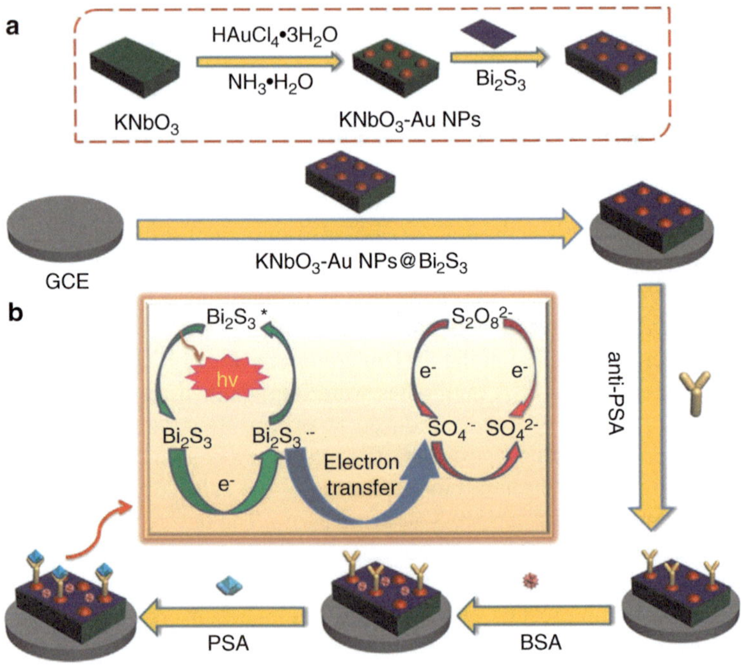

Fig. 11.9 Schematic for fabrication of ECL-based KNbO3–Au NPs@Bi2S3/GCE immunosensor (adapted from reference Li et al. 2015)

first time the bismuthsulfide nanosheets as luminophores in ECL sensors. The use of cross-linked antibodies provides the enhanced specificity, sensitivity, and stability of the immunosensor. The inverse correlation between the ECL signal and the PSA concentration was observed which displayed a detection limit of 3 pg/mL in real samples. The fabrication process and sensing protocol of the above example are depicted in Fig. 11.9.

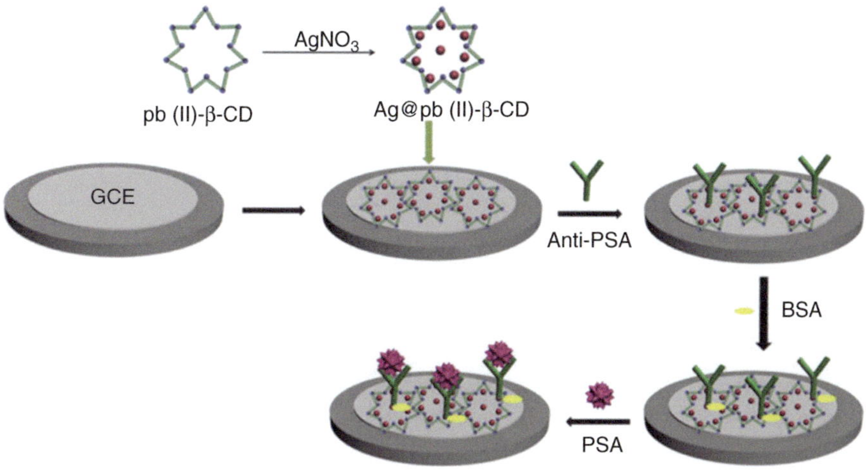

Fig. 11.10 Schematic of ECL-based immunosensor developed using Ag@Pb(II)-β-CD-modified GCE (adapted from reference Ma et al. 2016a, b)

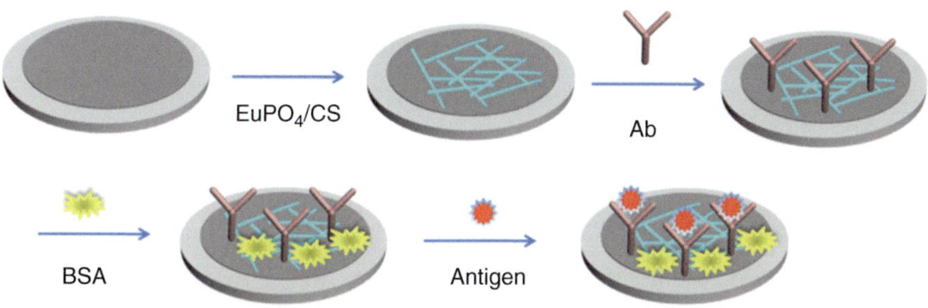

Fig. 11.11 Schematic illustration of the ECL immunosensor fabrication based on $EuPO_4$/CS film (adapted from reference Ma et al. 2016a, b)

The application of silver nanoparticles-doped Pb (II) metal-organic framework (MOF) has also been documented for the ECL immunosensing of PSA (Ma et al. 2016a, b). The above β-cyclodextrin-based MOF (Pb(II)-β-CD) showed excellent ECL behavior and exceptional reducing capacity towards silver ions. As specific recognition moieties, the anti-PSA antibodies were immobilized onto Ag@Pb(II)-β-CD-modified GCE (Fig. 11.10). The specific binding of PSA with the immobilized antibodies resulted in a decreased ECL signal within a linear PSA concentration range of 0.001–50 ng/mL, showing a low detection limit of 0.34 pg/mL.

The use of $EuPO_4$ nanowire has also been suggested for the development of ECL immunosensors for the PSA. The $EuPO_4$ nanowire possesses strong cathodic ECL activity, high quantum yield, and low toxicity (Ma et al. 2016a, b). In the process, the chitosan mediated dispersion of $EuPO_4$ nanowires and the anti-PSA antibodies were covalently bonded via the exploitation of amine groups of chitosan (Fig. 11.11). The ECL emission of $EuPO_4$ nanowires inhibited upon specific binding of PSA

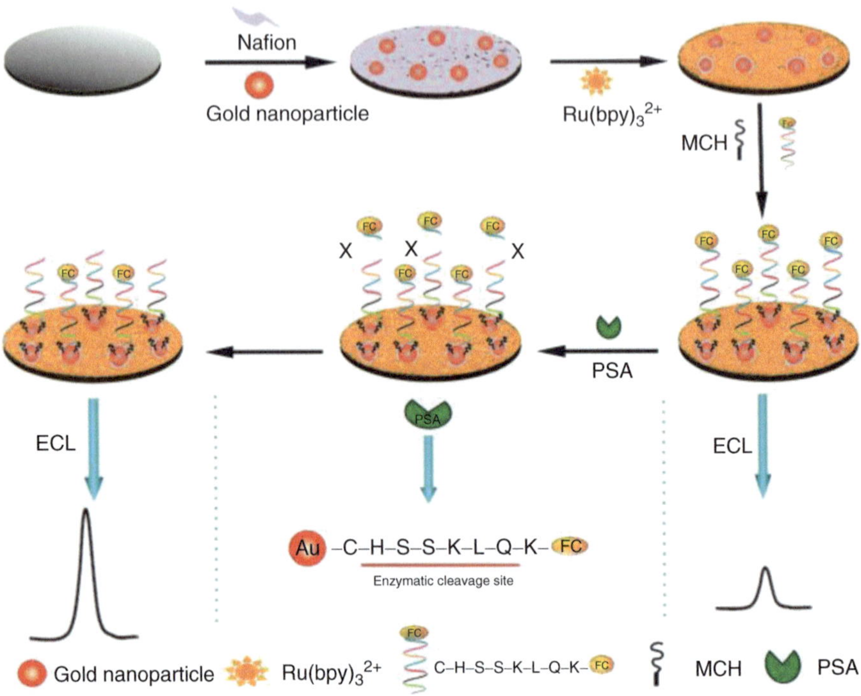

Fig. 11.12 Schematic diagram of the fabrication process of ECL-PB biosensor for the determination of PSA (adapted from reference Qi et al. 2014)

molecules due to steric hindrance effect. The proposed ECL immunosensor exhibited remarkable stability, selectivity with a low detection limit of 177.3 fg/mL.

Even though the immunoassays exhibit high sensitivity and selectivity, they require the use of monoclonal or polyclonal antibodies. The generation of antibodies is itself costly and labor intensive. Lately, the researchers have been exploring short peptides as optional recognition molecules for PSA. The short peptides are small sized, easier to synthesize at lower cost, and resistance to harsh operating conditions. In this category, a novel peptide-based electrogenerated chemiluminescence biosensor (ECL-PB) has been developed (Qi et al. 2014). The schematic of the developed biosensor and the related sensing mechanism have been depicted in Fig. 11.12.

This glassy carbon electrode-based sensing platform was fabricated by facilitating self-assembly of the ferrocene carboxylic acid tagged peptide (Peptide-Fc) along with ECL emitting species tris(2,2′-ripyridine) dichlororuthenium(II) (Ru(bpy)3 2+) onto Nafion incorporated gold nanoparticles. The presence of the PSA molecules induced the cleavage of the peptide leading to the release of the quencher and subsequently increasing the ECL intensity. An extremely low limit of detection 8×10^{-13} g/mL was achieved with the above biosensor due to signal amplification attributes of

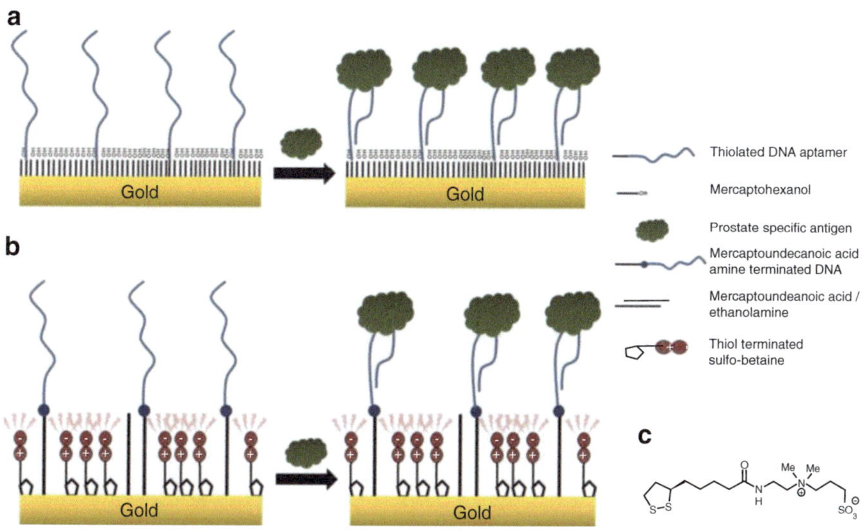

Fig. 11.13 Self-assembled monolayers with DNA aptamers on gold surface. (**a**) Thiolated aptamer with MCH; (**b**) amine terminated aptamer with sulfo-betaine; (**c**) structure of the thiol-modified sulfo-betaine (adapted from reference Jolly et al. 2015)

the AuNPs and background suppression with the use of Fc as ECL quencher. The work demonstrated a promising strategy that can be employed for developing enzymatic cleavage-based highly selective and sensitive biosensing platforms.

In another peptide-based biosensing approach for PSA, a "signal-on" assay was reported employing ferrocene (Fc)-labeled peptide, immobilized onto the Fe_3O_4@Au magnetic beads (Xie et al. 2015). The presence of PSA cleaved and released the Fc region which led to the formation of a host–guest complex of Fc with b-cyclodextrin (b-CD). The subsequent containment of the above complex onto the electrode surface produced electrochemical signal in proportion to the PSA concentrations; the detection limit was 1 ng/mL.

Aptamers have also been investigated in the development of electrochemical sensors for the PSA. These types of sensors are rather new entry to the list of immunosensing techniques for the PSA. One example of the aptamer sensor is shown in Fig. 11.13 (Jolly et al. 2015).

The sensing platform was composed of self-assembled monolayers of 6-mercaptohexanol (MCH) and thiolated-DNA aptamer on the gold surface. The identification of the PSA by the above sensor was recorded by monitoring the changes in the charge transfer resistance (R_{ct}) across the interface. A partial screening of aptamer charge by the analyte PSA was reasoned for such type of signal variations. This was probably the first of its kind DNA-aptamer biosensor for PSA that was developed using thiol-based simple surface chemistry. It offered highly sensitive detection of PSA up to 1 ng/mL in very cost-effective manner.

11.2.1 Other Markers for Prostate Cancer

Though PSA is a widely used biomarker to screen prostate cancer in man, there are some limitations regarding its cancer specificity and sensitivity and specificity. Therefore, need has been realized to develop new generation of prostate cancer biomarkers based upon serum, urine, and tissue-based assays. These new biomarkers have been suggested to supplement the PSA testing. Several of such biomarkers are now commercially available and helping in genomic assays. The following text summarized the important information on such new molecules:

11.2.1.1 Prostate Health Index (phi) and Prostate Cancer Antigen 3 (PCA3)

A PSA subtype "phi" has been recognized as a diagnostic biomarker in men who have a serum PSA level of 2–10 ng/mL (Crawford et al. 2014). PCA3, a noncoding messenger RNA, can reach to an elevated level in more than 90% of men with prostate cancer. This marker is under normal levels in normal prostatic glands. PCA3 can be determined in urine and therefore presents valuable diagnostic information. PCA3 diagnostic test was approved by the FDA in 2012 and now complements the PSA test. This combined diagnosis strategy is useful to decide the requirement of re-biopsy and follow-up of the patients as active surveillance (Salami et al. 2013).

11.2.1.2 ConfirmMDx

As prostate biopsy may also lead to false-negative results to as high as 25%, there are chances of undue risks and associated complications, such as bleeding, sepsis/bacteremia, infection, urinary symptoms/retention, endocarditis, and sexual dysfunction. Therefore, there is a need for precise tests to screen the subjects which really needs to be re-biopsied. In such cases, ConfirmMDx provides an epigenetic assay to help distinguish patients with true-negative biopsy from the ones who may have occult cancer. This test is helpful to detect an epigenetic field effect to diagnose the "cancerization" process at DNA level. Cancer lesions are generally surrounded with the field effect which can show their presence despite the normal appearance of cells. The detection of such a field effect helps in to determine the rule in, or rule out of the cancers. The test also provides useful data and information that can help the physicians and surgeons to advice upon the requirements of repeated biopsies (Wojno et al. 2014).

11.2.1.3 Prostate Core Mitomic Test (PCMT) and TMPRSS2-ERG

This test is useful to diagnose a large-scale depletion in mitochondrial DNA. Such a diagnosis is helpful to determine the cellular change in case of any undiagnosed prostate cancer. The test extends to the study of presence of malignant cells in normal-appearing prostate tissue (Yu et al. 2010; Sartori and Chan 2014). Another valuable test of TMPRSS2-ERG gene (a fusion between the transmembrane protease serine 2 (TMPRSS2) gene and the v-ets erythroblastosis virus E26 oncogene homolog (avian) (ERG) gene) offers the determination of predominant variant in approximately 40–80% of prostate cancers. The quantitative levels of urine TMPRSS2-ERG have been suggested to be associated with the clinically significant prostate cancer. A check of its levels may help to stratify the disease aggressiveness using PSA density (Tomlins et al. 2011).

In case of prostate cancer diagnosis, it might be of great help to combine different biomarker assays. It may improve the predictive accuracy than with the study of individual markers. Therefore, it is considered to enhance the prediction probability of prostate cancer risk if the post-DRE urine TMPRSS2-ERG, urine PCA3, and serum PSA level can be carried out in tandem. This kind of a clinical methodology can help in cancer management in a more effective way (Willard and Koochekpour 2012).

11.2.1.4 PTEN Gene and ProMark

PTEN is a tumor suppressor gene which is involved in cell cycle regulation. Its dysregulation is consistently associated with poor prognosis in prostate cancer. The determination of PTEN can be helpful to assess the risk of cancer progression and its recurrence after therapy. The PTEN is also associated with advanced localized or metastatic disease and death. PTEN is generally tested with fluorescence in situ hybridization (FISH) test, in conjunction with prostate biopsy tests. The quantification of its partial (hemizygous) or complete (homozygous) deletions indicates the rate of disease progression (Krohn et al. 2012).

The testing of ProMark offers a chance to determine the requirement of biopsy. Its quantification with immunofluorescent imaging could classify the tumors while differentiating indolent from aggressive disease (Gaudreau et al. 2016).

11.2.1.5 Urinary Biomarkers

There is an increasing need to discover accurate noninvasive tests for prostate cancer diagnosis to allow the stratification of patients. Urine is an ideal sample to develop noninvasive testing protocols because of the easy collection. Prostate cells are directly released into the urethra through prostatic ducts; therefore there is a definite feasibility of using urine for the noninvasive detection of PCa. Lately, the detection of some RNA biomarkers in urinary exosomes has also been associated with a great potential for noninvasive detections. The important urinary biomarkers include PCA3 and TMPRSS2-ERG. The recent FDA approval of PCA3 has led to its use in clinical practice. A combination of both of the markers has also been suggested.

11.3 Breast Cancer

Cancer is one of the deadliest and costliest (in terms of treatment) diseases, afflicting all communities worldwide. Though most of the cancers are life threatening, but among various types of cancers, breast cancer is the most common invasive type of cancer occurring in women globally. It is the second most common type of cancer worldwide and is second leading cause of cancer death in women (DeSantis et al.

2014). Early detection of breast cancer is therefore key to the most effective and timely therapy. Presently used techniques like MRI, mammography, biopsy, molecular breast imaging, radioimmunoassay (RIA), and immunohistochemistry (IHC) are 80–90% efficient in breast cancer diagnosis; however, these techniques are associated with some limitations like false-positive or false-negative results, biased or wrong interpretations, etc. As a result, there is a need for low cost, effective, and selective biosensors which have inherent properties like rapid response, selectivity, and high sensitivity. The use of biosensor technology can play instrumental role in early cancer detection which will result in effective treatment and improvement in overall chance of survival. In biosensing, though various types of transducers can be used, but electrochemical transduction is generally a choice due to their ease of operation and their selectivity.

11.3.1 Biosensors-Based Detection of Cancer Biomarkers

The carboxylic acid-functionalized single-walled carbon nanotube-modified screen-printed electrodes (SWCNT-SPEs) have been explored for the electrochemical monitoring of direct DNA hybridization related to the specific sequences on breast cancer 1 (BRCA1) (Li et al. 2012). Breast Cancer 1 (BRCA1) gene is a human caretaker gene which is expressed in the cells of breast and other tissues. In genetic research and clinical prospect, the breast cancer 1 (BRCA1) DNA mutant detection in disease-related gene fragments is considered as a critical parameter. Approximately 80% of inherited breast and ovarian cancers occur due to mutations in BRCA1 gene (Tutt and Ashworth 2002; Sifri et al. 2004). The carbon-based structures are considered very often for the modification of electrode/chip surfaces due to their large surface area, fast heterogeneous electron transfer, and long-range electron transfer. These features make these carbon-based nano/materials suitable for electroanalytical applications. For signal monitoring, the differential pulse voltammetry (DPV) was used. The signal got generated from the oxidation of guanine in the presence of DNA hybridization between BRCA1 probe and its complementary target. A graphene-based electrochemical DNA biosensor has also been reported for the detection of BRCA 1 gene using sandwich configuration in which a capture probe (DNA-c) and reporter probe (DNA-r) DNAs got hybridized to target probe DNA (DNA-t) (Rasheed and Sandhyarani 2014). The DNA-r was conjugated to gold nanoparticle and the oxidation of gold nanoparticle was used for the electrochemical detection of DNA-t using cyclic voltammetry (CV) and chronoamperometry. The genomagnetic electrochemical assay has also been used for label-free detection of BRCA1 using pencil graphite electrode (PGE), carbon paste electrode (CPE), and also magnetic CPE (m-CPE) (Wang et al. 2001). Apart from carbon-based structures, other materials have also been reported to be effective in detection of BRCA biomarker. A recyclable electrochemical sensing platform has been reported by Yang et al. (2016a, b) for the detection of BRCA. The schematic illustration of recyclable electrochemical sensing platform is shown in Fig. 11.14. The sensing is based on the hybridization of target sequence with ferrocene-labeled DNA-gold

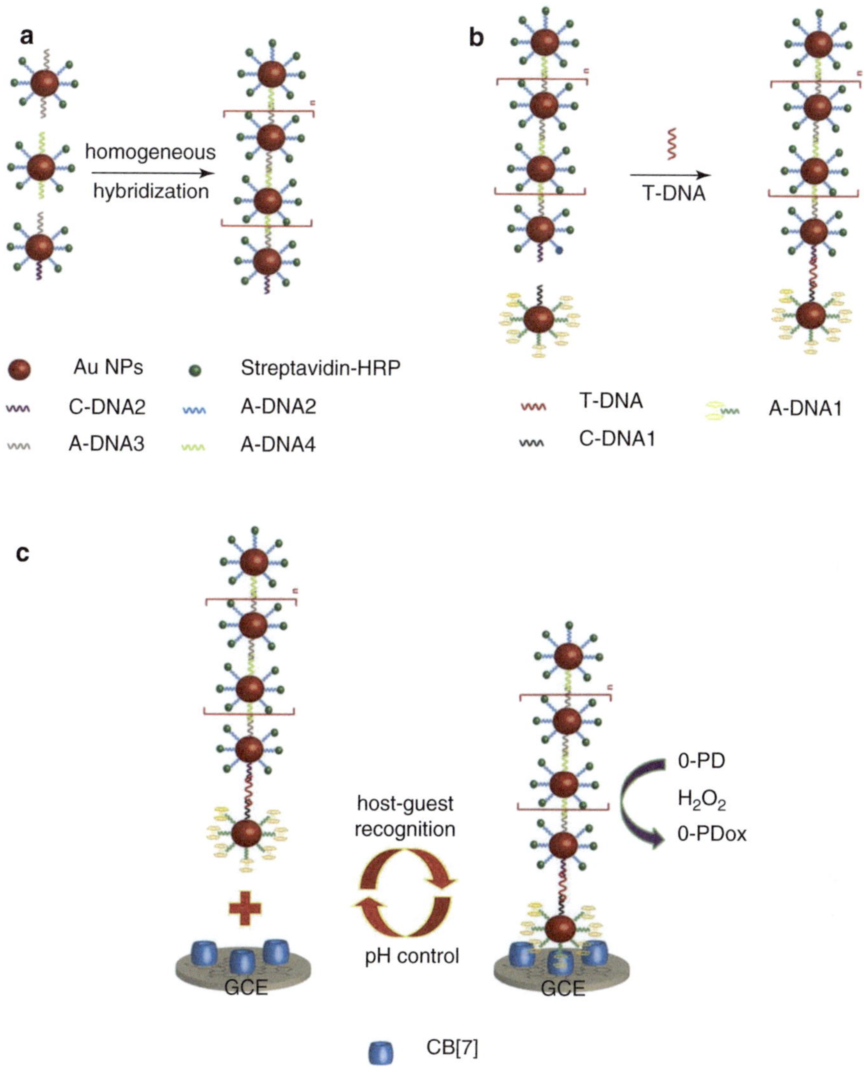

Fig. 11.14 Schematic illustration of recyclable electrochemical sensing platform for BRCA DNA detection (adapted from reference Yang et al. 2016a, b)

nanospheres (FcNS) and horseradish peroxidase-labeled DNA/Au nanospheres (HRPNS) concatamers, and the host–guest interaction between cucurbit [7] uril (CB [7]) adsorbed on the electrode surface. The same group has used host–guest interaction in the development of a recyclable DNA detection system (Yang et al. 2016a, b). Cucurbit[n]urils, a cage-like macromolecule with high symmetry, is composed of glycoluril units and has a hydrophobic cavity and ferrocene (Fc) has strong affinity towards cucurbit [7]uril (CB [7]). The use of FcNS and HRPNS concatamers served as dual signal amplifiers. Since the complex between Fc and CB [7] can get easily dissociated with change in pH (Kaifer et al. 2012), this strategy has been exploited to make the sensing platform recyclable.

The gold nanoparticles/multiwall carbon nanotube-ionic liquid electrode (AuNPs/MW-CILE) was fabricated for the detection of human epidermal growth factor receptor 2 (HER2) (Arkan et al. 2015). HER2 is a tyrosine kinase receptor and belongs to the epidermal growth factor receptor (EGFR) family and its over expression is an indicator for the breast cancer. About 15–20% breast carcinomas have an overexpression of human epidermal growth factor receptor-2 (HER2), and in case of aggressive breast tumor with increased levels of HER2, the survival chances of the patient is very less (Ludovini et al. 2008; Reix et al. 2016). The immunosensing in this study has been performed using change in impedance on the interactions of antibody with antigen, as the signal. Impedance spectroscopy is simple and is based on sampling of the impedance when reactions take place at the electrode/matrix surface. Depending upon the type of complex formation and matrix used, the impedance can increase or decrease when interfaces take place. In this reported study, the MW-CILE electrode was modified using self-assembly of carboxylated AuNPs via 1,6-hexanedithiol (HDT) as a cross linker. The fabricated AuNPs/MW-CILE-based impedimetric immunosensor was used for the determination of HER2 in the sera of several breast cancer patients.

A copolymer of chitosan and polyaniline (PANI), (CHIT-*co*-PANI) was used for fabrication of an electrochemical breast cancer biosensor (Tiwari and Gong 2009). The copolymer was coated on indium–tin-oxide (ITO) and a complementary DNA (cDNA) probe (42 bases long) associated with the breast cancer susceptible gene BRCA1 was immobilized. The increase in the logarithm of molar concentration of the single-stranded target DNA (ssDNA) resulted in decreased amperometric current due to formation of hybridized complex on the electrode surface. The electrochemical technique is one of the promising approaches for the early detection of breast cancer susceptible gene-1 (BRCA1; 5592 bp) specific cDNA probe (Tilley et al. 1989). However, the performance of electrochemical biosensors is dependent on physicochemical properties of the electrode materials and sensing element immobilized over the electrode surface.

Sezginturk reported an impedometric biosensor using vascular endothelial growth factor receptor-1 (VEGFR1) for the determination of vascular endothelial growth factor (VEGF) (Sezgintürk 2011). The VEGF-R1 is a disulfide-linked dimeric protein with two 905 amino acid residue polypeptides and it binds VEGF with high affinity. Since VEGF is associated with recurrence, it is used as a marker for unfavorable prognosis in case of breast cancer. The patients with metastasis have higher levels of serum VEGF. For the study, the VEGF-R1 was immobilized using covalent linkage with self-assembled monolayers of mercapto propionic acid on gold electrodes. This impedance-based biosensor exhibited very high affinity to VEGF, and showed good reproducibility and repeatability. The Kramers–Kronig transforms showed impedance data very stable and linear.

The protein G-functionalized magnetic beads (MBs) were used for the detection of MUC1 using sandwich immunoassay (Taleat et al. 2013), Fig. 11.15. The determination of serum MUC1 level is very important in the early detection of breast and

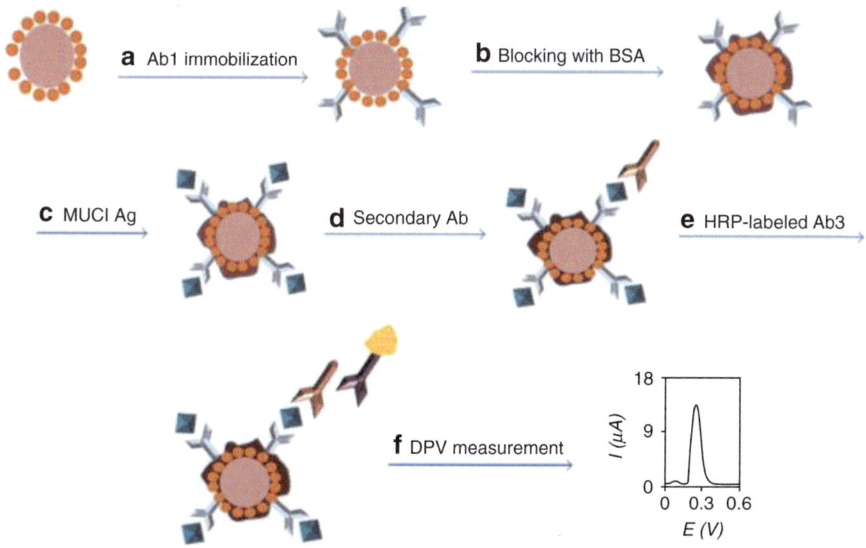

Fig. 11.15 Sandwich immunosensor preparation for MUC1 detection: (**a**) primary antibody attachment on MBs surface; (**b**) MBs free binding sites blocking using BSA; (**c**) incubation with MUC1 antigen solutions and Ab1/MUC1 Ag complex formation on the MBs; (**d**) reaction with Ab2, (**e**) incubation with Ab3 labeled with HRP; (**f**) DPV measurements of MBs-bound MUC1 in the presence of acetaminophen/H_2O_2 (adapted from Taleat et al. 2013)

ovarian cancer. The magnetic beads are employed as the platforms for the immobilization and immunoreaction process. The MUC1 protein was captured by a pair of primary and secondary antibodies. Then it was labeled with horseradish peroxidase (HRP)-conjugated third antibody and was trapped at graphite-based screen-printed electrodes. The electrochemical sensing was done by differential pulse voltammetry (DPV) using acetaminophen (APAP) as the redox mediator.

The two co-expressing tumor markers, human mucin-1 (MUC1) and carcinoembryonic antigen (CEA) have been detected by Li et al. using electrochemical immunoassay (Li et al. 2010). These two biomarkers are expressed on the surface of the cancer cells at the same time and are most commonly used for monitoring the metastatic breast tumor. The co-determination of these biomarkers can result in more precise prognostic diagnosis (Molina and Gion 1998). For the electrochemical detection, anodic stripping voltammetry was employed and electrode surface was modified with cadmium nanoparticles. The electrochemical signal originating from the stripping of cadmium ions was recorded as sensing signal for the detection of MUC1 and CEA. Authors affirmed that since the breast cancer cell MCF-7 can be easily distinguished from other kinds of cells such as acute leukemia cells CCRF-CEM, etc., this cytosensor is capable of monitoring the breast cancer cell MCF-7 in a wide range (10^4–10^7 cell mL^{-1}).

A genosensor array was fabricated using PEGylated DNA probe for the detection of breast cancer biomarker estrogen receptor-α in a PCR product prepared from the genetic material extracted from 20 MCF7 breast cancer cells (Henry et al. 2010).

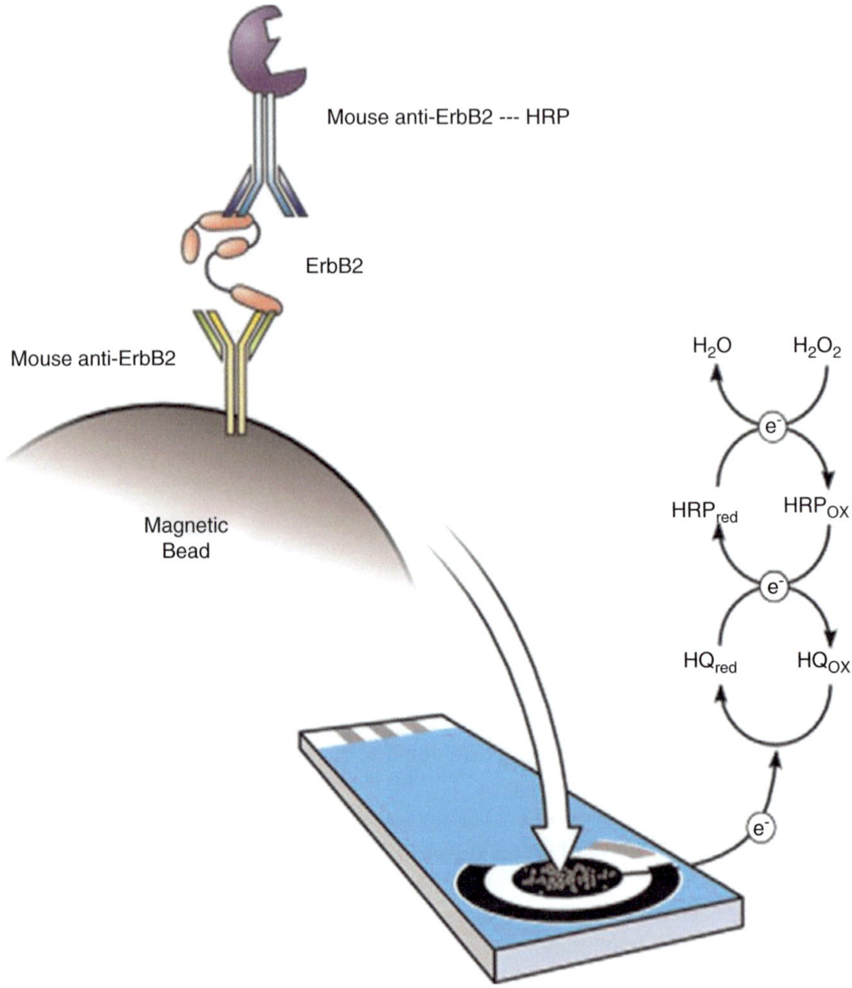

Fig. 11.16 Schematic representation of the ErbB2 sandwich magnetoimmunosensor

The estrogen receptor-α capture probes and bipodal aromatic polyethylene glycol (PEG) alkanethiol were co-immobilized in a ratio of 1:100 and detection was probed using electrochemical technique and surface plasmon resonance. The use of alkanethiol helps in the improvement of organization of the DNA monolayer and provides protection against nonspecific binding and electrode fouling (Peeters et al. 2008). The detection of ErbB2protein, which is known to be associated with disease like breast cancer has been reported using magnetic immunosensor (Eletxigerra et al. 2015). The immunosensing was performed in sandwich format in which a specific capture antibody was immobilized covalently on the magnetic beads, schematic presented in Fig. 11.16. These modified magnetic beads were incubated with a solution consisting of antigen and horse radish peroxidase (HRP)-labeled detector antibody. The magnetic beads were then captured on the surface of screen-printed electrodes for amperometric studies. The developed magnetosensor showed efficiency in detecting the ErbB2 in intact breast cancer cells also.

A sandwich-type electrochemical immunosensor has been fabricated for the determination of Sloan–Kettering breast cancer (SKBR-3) breast cancer cell using reduced graphene oxide (rGO) as an immobilization platform for primary Herceptin antibody (Anti-HCT) (Amouzadeh Tabrizi et al. 2017). SKBR-3 breast cancer cell is one of the important breast cancer cells that has Her2 antigen on its plasma membrane (Serena Cecchetti et al. 2015; Jeong et al. 2016). Various composites of reduced graphene oxide were used as electrochemical labels for secondary herceptin antibody. In recent years, rGO has emerged as one of the potential materials that can be used for biomedical applications due to its biocompatibility, high surface area, low processing cost, and ease in its functionalization (Singh et al. 2016a, b). The sensing was done using differential pulse voltammetry (DPV) technique and the immunosensor exhibited high selectivity, liner range responsibility, and good stability.

A composite of graphene oxide and gold nanorod was used for fabrication of electrochemical nanobiosensor for plasma miRNA-155, which is an emerging novel biomarker for cancer (Azimzadeh et al. 2016). Similar to HER2, overexpression of miR-155 is an indication of breast cancer. An early detection and prognosis of the breast cancer strongly relies on the exact and selective quantification of the miR-155 in the serum/plasma. Sun et al. studied the miR-155 level in patients suffering from breast cancer (Sun et al. 2012). Their study demonstrated increased levels of serum miR-155 in breast cancer patients (n = 103) compared with healthy subjects (n = 55) ($p < 0.001$). Authors also studied change in its concentration after surgery and after four cycles of chemotherapy to evaluate the effects of clinical treatment on serum levels of candidate miRNAs. The serum miR-155 concentration decreased after treatment which proved its importance in diagnosis of breast cancer. The electrochemical nanobiosensor showed dynamic linearity in the range 2.0 fM–13 8.0 pM with detection limit of 0.6 fM. The nanobiosensor could easily discriminate between complementary target miRNA, single-, three-base mismatch, and non-complementary miRNA.

11.4 Concluding Remarks

As cancer is not a single disease but a composite of multiple diseases, a viable strategy for its early detection may not depend solely upon the plasma tumor markers. There is a requirement of detecting a batch of biomarkers instead of a single one. These biomarker may include tumor antigens as well as the antibodies against those tumor antigens. The immune responses of the human body against the tumor antigens could be an efficient way for disease detection. The application of antigens and autoantibodies for multiple marker detection strategy offers the advantages of individual markers and results in both improved sensitivity and specificity over a single marker. Ironically, the recent research in this area has focussed more on the sensitivity at the expense of specificity. However, there are some vital signs of significant development in multiplexed sensing to improve the overall accuracy of the prostate cancer screening process.

There are several techniques available nowadays for the clinical analysis of prostate-specific biomarkers, including ELISA, electrochemiluminescence, fluorescence, colorimetric, surface plasmon resonance spectrometry, and mass spectrometry. The biosensors have also come up as a potential tool for the easy and point-of-care type of testing of various cancers. There seems an exciting market in the area of such biosensors to be employed for the detection of more biomarkers and also for the multiplexed analysis of several markers within a single lab on chip.

The electrochemical biosensors could play an important role in the development of viable point-of-care diagnostic devices for the cancer biomarkers. These devices are extremely useful for obtaining results in a fast, simple, and low cost fashion with the aid of compact (handheld) analyzers. Different nanoplatforms can be conjugated with biorecognition elements, such as antigens, autoantibodies, and DNA, to yield remarkable sensitivity essential for early cancer detection. There exists a potential opportunity for the electrochemical devices with nanoscale materials to offers multiplexing capability for simultaneous measurements of multiple cancer markers.

Acknowledgement Authors acknowledge the support from Director, CSIR-Central Scientific Instruments Organisation (CSIR-CSIO), Chandigarh, India, for motivating us to work on our research ideas.

References

Acevedo B, Perera Y, Ruiz M, Rojas G, Benítez J, Ayala M, Gavilondo J (2002) Development and validation of a quantitative ELISA for the measurement of PSA concentration. Clin Chim Acta 317(1–2):55–63

Amouzadeh Tabrizi M, Shamsipur M, Saber R, Sarkar S, Zolfaghari N (2017) An ultrasensitive sandwich-type electrochemical immunosensor for the determination of SKBR-3 breast cancer cell using rGO-TPA/FeHCFnano labeled anti-HCT as a signal tag. Sensors Actuators B Chem 243:823–830

Arkan E, Saber R, Karimi Z, Shamsipur M (2015) A novel antibody–antigen based impedimetric immunosensor for low level detection of HER2 in serum samples of breast cancer patients via modification of a gold nanoparticles decorated multiwall carbon nanotube-ionic liquid electrode. Anal Chim Acta 874:66–74

Azimzadeh M, Rahaie M, Nasirizadeh N, Ashtari K, Naderi-Manesh H (2016) An electrochemical nanobiosensor for plasma miRNA-155, based on graphene oxide and gold nanorod, for early detection of breast cancer. Biosens Bioelectron 77:99–106

Basil CF, Zhao Y, Zavaglia K, Jin P, Panelli MC, Voiculescu S, Mandruzzato S, Lee HM, Seliger B, Freedman RS, Taylor PR, Hu N, Zanovello P, Marincola FM, Wang E (2006) Common cancer biomarkers. Cancer Res 66(6):2953

Bhardwaj SK, Sharma AL, Bhardwaj N, Kukkar M, Gill AA, Kim K-H, Deep A (2017) TCNQ-doped Cu-metal organic framework as a novel conductometric immunosensing platform for the quantification of prostate cancer antigen. Sensors Actuators B Chem 240:10–17

Catalona WJ, Partin AW, Finlay JA, Chan DW, Rittenhouse HG, Wolfert RL, Woodrum DL (1999) Use of percentage of free prostate-specific antigen to identify men at high risk of prostate cancer when PSA levels are 2.51 to 4 ng/mL and digital rectal examination is not suspicious for prostate cancer: an alternative model. Urology 54(2):220–224

Cesarino I, Moraes FC, Lanza MRV, Machado SAS (2012) Electrochemical detection of carbamate pesticides in fruit and vegetables with a biosensor based on acetylcholinesterase immobilised on a composite of polyaniline–carbon nanotubes. Food Chem 135(3):873–879

Chandra P (2013) Miniaturized multiplex electrochemical biosensor in clinical bioanalysis. J Bioanal Biomed 5(5):e122

Chen A, Chatterjee S (2013) Nanomaterials based electrochemical sensors for biomedical applications. Chem Soc Rev 42(12):5425–5438

Chen K-I, Li B-R, Chen Y-T (2011) Silicon nanowire field-effect transistor-based biosensors for biomedical diagnosis and cellular recording investigation. Nano Today 6(2):131–154

Choi Y-E, Kwak J-W, Park JW (2010) Nanotechnology for early cancer detection. Sensors 10(1):428–455

Crawford ED, Ventii K, Shore ND (2014) New biomarkers in prostate cancer. Oncology (Williston Park) 28(2):135–142

DeSantis C, Ma J, Bryan L, Jemal A (2014) Breast cancer statistics, 2013. CA Cancer J Clin 64(1):52–62

Eletxigerra U, Martinez-Perdiguero J, Merino S, Barderas R, Torrente-Rodríguez RM, Villalonga R, Pingarrón JM, Campuzano S (2015) Amperometric magnetoimmunosensor for ErbB2 breast cancer biomarker determination in human serum, cell lysates and intact breast cancer cells. Biosens Bioelectron 70:34–41

Ertürk G, Hedström M, Tümer MA, Denizli A, Mattiasson B (2015) Real-time prostate-specific antigen detection with prostate-specific antigen imprinted capacitive biosensors. Anal Chim Acta 891:120–129

Garnick MB (1993) Prostate cancer: screening, diagnosis, and management. Ann Intern Med 118(10):804–818

Gaudreau P-O, Stagg J, Soulières D, Saad F (2016) The present and future of biomarkers in prostate cancer: proteomics, genomics, and immunology advancements. Biomark Cancer 8(Suppl 2):15

Grönberg H (2003) Prostate cancer epidemiology. Lancet 361(9360):859–864

Healy DA, Hayes CJ, Leonard P, McKenna L, O'Kennedy R (2007) Biosensor developments: application to prostate-specific antigen detection. Trends Biotechnol 25(3):125–131

Henry OYF, Sanchez JLA, O'Sullivan CK (2010) Bipodal PEGylated alkanethiol for the enhanced electrochemical detection of genetic markers involved in breast cancer. Biosens Bioelectron 26(4):1500–1506

Huang Y-W, Wu C-S, Chuang C-K, Pang S-T, Pan T-M, Yang Y-S, Ko F-H (2013) Real-time and label-free detection of the prostate-specific antigen in human serum by a polycrystalline silicon nanowire field-effect transistor biosensor. Anal Chem 85(16):7912–7918

Ivnitski D, Abdel-Hamid I, Atanasov P, Wilkins E (1999) Biosensors for detection of pathogenic bacteria. Biosens Bioelectron 14(7):599–624

Jang HS, Park KN, Kang CD, Kim JP, Sim SJ, Lee KS (2009) Optical fiber SPR biosensor with sandwich assay for the detection of prostate specific antigen. Opt Commun 282(14):2827–2830

Jang HD, Kim SK, Chang H, Choi J-W (2015) 3D label-free prostate specific antigen (PSA) immunosensor based on graphene–gold composites. Biosens Bioelectron 63:546–551

Jeong J, VanHouten JN, Dann P, Kim W, Sullivan C, Yu H, Liotta L, Espina V, Stern DF, Friedman PA, Wysolmerski JJ (2016) PMCA2 regulates HER2 protein kinase localization and signaling and promotes HER2-mediated breast cancer. Proc Natl Acad Sci 113(3):E282–E290

Johnson ED, Kotowski TM (1993) Detection of prostate specific antigen by ELISA. J Forensic Sci 38(2):250–258

Jolly P, Formisano N, Tkáč J, Kasák P, Frost CG, Estrela P (2015) Label-free impedimetric aptasensor with antifouling surface chemistry: a prostate specific antigen case study. Sensors Actuators B Chem 209:306–312

Kaifer AE, Li W, Silvi S, Sindelar V (2012) Pronounced pH effects on the kinetics of cucurbit[7]uril-based pseudorotaxane formation and dissociation. Chem Commun 48(53):6693–6695

Kanyong P, Rawlinson S, Davis J (2016) Immunochemical assays and nucleic-acid detection techniques for clinical diagnosis of prostate cancer. J Cancer 7(5):523–531

Kavosi B, Salimi A, Hallaj R, Amani K (2014) A highly sensitive prostate-specific antigen immunosensor based on gold nanoparticles/PAMAM dendrimer loaded on MWCNTS/chitosan/ionic liquid nanocomposite. Biosens Bioelectron 52:20–28

Kingsmore SF (2006) Multiplexed protein measurement: technologies and applications of protein and antibody arrays. Nat Rev Drug Discov 5(4):310–321

Krohn A, Diedler T, Burkhardt L, Mayer P-S, De Silva C, Meyer-Kornblum M, Kötschau D, Tennstedt P, Huang J, Gerhäuser C (2012) Genomic deletion of PTEN is associated with tumor progression and early PSA recurrence in ERG fusion-positive and fusion-negative prostate cancer. Am J Pathol 181(2):401–412

Kukkar M, Tuteja SK, Sharma AL, Kumar V, Paul AK, Kim K-H, Sabherwal P, Deep A (2016) A new electrolytic synthesis method for few-layered MoS2 nanosheets and their robust biointerfacing with reduced antibodies. ACS Appl Mater Interfaces 8(26):16555–16563

Lang Q, Wang F, Yin L, Liu M, Petrenko VA, Liu A (2014) Specific probe selection from landscape phage display library and its application in enzyme-linked immunosorbent assay of free prostate-specific antigen. Anal Chem 86(5):2767–2774

Li T, Fan Q, Liu T, Zhu X, Zhao J, Li G (2010) Detection of breast cancer cells specially and accurately by an electrochemical method. Biosens Bioelectron 25(12):2686–2689

Li C-z, Karadeniz H, Canavar E, Erdem A (2012) Electrochemical sensing of label free DNA hybridization related to breast cancer 1 gene at disposable sensor platforms modified with single walled carbon nanotubes. Electrochim Acta 82:137–142

Li J, Ma H, Wu D, Li X, Zhao Y, Zhang Y, Du B, Wei Q (2015) A label-free electrochemiluminescence immunosensor based on KNbO3–Au nanoparticles@Bi2S3 for the detection of prostate specific antigen. Biosens Bioelectron 74:104–112

Liu D, Huang X, Wang Z, Jin A, Sun X, Zhu L, Wang F, Ma Y, Niu G, Hight Walker AR (2013) Gold nanoparticle-based activatable probe for sensing ultralow levels of prostate-specific antigen. ACS Nano 7(6):5568–5576

Ludovini V, Gori S, Colozza M, Pistola L, Rulli E, Floriani I, Pacifico E, Tofanetti FR, Sidoni A, Basurto C, Rulli A, Crinò L (2008) Evaluation of serum HER2 extracellular domain in early breast cancer patients: correlation with clinicopathological parameters and survival. Ann Oncol 19(5):883–890

Ma W, Yin H, Xu L, Wu X, Kuang H, Wang L, Xu C (2014) Ultrasensitive aptamer-based SERS detection of PSAs by heterogeneous satellite nanoassemblies. Chem Commun 50(68):9737–9740

Ma H, Li X, Yan T, Li Y, Zhang Y, Wu D, Wei Q, Du B (2016a) Electrochemiluminescent immunosensing of prostate-specific antigen based on silver nanoparticles-doped Pb (II) metal-organic framework. Biosens Bioelectron 79:379–385

Ma H, Zhou J, Li Y, Han T, Zhang Y, Hu L, Du B, Wei Q (2016b) A label-free electrochemiluminescence immunosensor based on EuPO4 nanowire for the ultrasensitive detection of Prostate specific antigen. Biosens Bioelectron 80:352–358

Mahfoud OK, Rakovich TY, Prina-Mello A, Movia D, Alves F, Volkov Y (2014) Detection of ErbB2: nanotechnological solutions for clinical diagnostics. RSC Adv 4(7):3422–3442

Mehta S, Shelling A, Muthukaruppan A, Lasham A, Blenkiron C, Laking G, Print C (2010) Predictive and prognostic molecular markers for cancer medicine. Ther Adv Med Oncol 2(2):125–148

Michaelson J, Satija S, Moore R, Weber G, Halpern E, Garland A, Puri D, Kopans DB (2002) The pattern of breast cancer screening utilization and its consequences. Cancer 94(1):37–43

Molina R, Gion M (1998) Use of blood tumour markers in the detection of recurrent breast cancer. Breast 7(4):187–189

Nadler RB, Humphrey PA, Smith DS, Catalona WJ, Timothy L (1995) Effect of inflammation and benign prostatic hyperplasia on elevated serum prostate specific antigen levels. J Urol 154(2):407–413

Okuno J, Maehashi K, Kerman K, Takamura Y, Matsumoto K, Tamiya E (2007) Label-free immunosensor for prostate-specific antigen based on single-walled carbon nanotube array-modified microelectrodes. Biosens Bioelectron 22(9):2377–2381

Peeters S, Stakenborg T, Reekmans G, Laureyn W, Lagae L, Van Aerschot A, Van Ranst M (2008) Impact of spacers on the hybridization efficiency of mixed self-assembled DNA/alkanethiol films. Biosens Bioelectron 24(1):72–77

Presnova G, Presnov D, Krupenin V, Grigorenko V, Trifonov A, Andreeva I, Ignatenko O, Egorov A, Rubtsova M (2017) Biosensor based on a silicon nanowire field-effect transistor functionalized by gold nanoparticles for the highly sensitive determination of prostate specific antigen. Biosens Bioelectron 88:283–289

Qi H, Li M, Dong M, Ruan S, Gao Q, Zhang C (2014) Electrogenerated chemiluminescence peptide-based biosensor for the determination of prostate-specific antigen based on target-induced cleavage of peptide. Anal Chem 86(3):1372–1379

Rasheed PA, Sandhyarani N (2014) Graphene-DNA electrochemical sensor for the sensitive detection of BRCA1 gene. Sensors Actuators B Chem 204:777–782

Rebelo TS, Santos C, Costa-Rodrigues J, Fernandes M, Noronha JP, Sales MGF (2014) Novel Prostate Specific Antigen plastic antibody designed with charged binding sites for an improved protein binding and its application in a biosensor of potentiometric transduction. Electrochim Acta 132:142–150

Reix N, Malina C, Chenard M-P, Bellocq J-P, Delpous S, Molière S, Sevrin A, Neuberger K, Tomasetto C, Mathelin C (2016) A prospective study to assess the clinical utility of serum HER2 extracellular domain in breast cancer with HER2 overexpression. Breast Cancer Res Treat 160(2):249–259

Salami SS, Schmidt F, Laxman B, Regan MM, Rickman DS, Scherr D, Bueti G, Siddiqui J, Tomlins SA, Wei JT (2013) Combining urinary detection of TMPRSS2:ERG and PCA3 with serum PSA to predict diagnosis of prostate cancer. Urol Oncol 31:566–571. Elsevier

Sarkar P, Pal PS, Ghosh D, Setford SJ, Tothill IE (2002) Amperometric biosensors for detection of the prostate cancer marker (PSA). Int J Pharm 238(1):1–9

Sartori DA, Chan DW (2014) Biomarkers in prostate cancer: what's new? Curr Opin Oncol 26(3):259

Serena Cecchetti IB, Ferri R, Mercurio L, Canevari S, Podo F, Miotti S, Iorio E (2015) Inhibition of phosphatidylcholine-specific phospholipase C interferes with proliferation and survival of tumor initiating cells in squamous cell carcinoma. PLoS One 10(9):e0136120

Sezgintürk MK (2011) A new impedimetric biosensor utilizing VEGF receptor-1 (Flt-1): early diagnosis of vascular endothelial growth factor in breast cancer. Biosens Bioelectron 26(10):4032–4039

Shi H, Yeh JI (2012) Biosensors and related methods, Google Patents

Sifri R, Gangadharappa S, Acheson LS (2004) Identifying and testing for hereditary susceptibility to common cancers. CA Cancer J Clin 54(6):309–326

Singh S, Jain DVS, Singla ML (2013) One step electrochemical synthesis of gold-nanoparticles-polypyrrole composite for application in catechin electrochemical biosensor. Anal Methods 5(4):1024–1032

Singh S, Meena VK, Mizaikoff B, Singh SP, Suri CR (2016a) Electrochemical sensing of nitroaromatic explosive compounds using silver nanoparticles modified electrochips. Anal Methods 8(39):7158–7169

Singh S, Tuteja SK, Sillu D, Deep A, Suri CR (2016b) Gold nanoparticles-reduced graphene oxide based electrochemical immunosensor for the cardiac biomarker myoglobin. Microchim Acta 183(5):1729–1738

Song S, Qin Y, He Y, Huang Q, Fan C, Chen H-Y (2010) Functional nanoprobes for ultrasensitive detection of biomolecules. Chem Soc Rev 39(11):4234–4243

Sun Y, Wang M, Lin G, Sun S, Li X, Qi J, Li J (2012) Serum microRNA-155 as a potential biomarker to track disease in breast cancer. PLoS One 7(10):e47003

Taleat Z, Cristea C, Marrazza G, Robert S (2013) Electrochemical sandwich immunoassay for the ultrasensitive detection of human MUC1 cancer biomarker. Int J Electrochem 2013:6

Tilley WD, Marcelli M, Wilson JD, McPhaul MJ (1989) Characterization and expression of a cDNA encoding the human androgen receptor. Proc Natl Acad Sci U S A 86(1):327–331

Tiwari A, Gong S (2009) Electrochemical detection of a breast cancer susceptible gene using cDNA immobilized chitosan-co-polyaniline electrode. Talanta 77(3):1217–1222

Tomlins SA, Aubin SM, Siddiqui J, Lonigro RJ, Sefton-Miller L, Miick S, Williamsen S, Hodge P, Meinke J, Blase A (2011) Urine TMPRSS2: ERG fusion transcript stratifies prostate cancer risk in men with elevated serum PSA. Sci Transl Med 3(94):94ra72–94ra72

Triroj N, Jaroenapibal P, Shi H, Yeh JI, Beresford R (2011) Microfluidic chip-based nanoelectrode array as miniaturized biochemical sensing platform for prostate-specific antigen detection. Biosens Bioelectron 26(6):2927–2933

Truong LTN, Chikae M, Ukita Y, Takamura Y (2011) Labelless impedance immunosensor based on polypyrrole–pyrolecarboxylic acid copolymer for hCG detection. Talanta 85(5):2576–2580

Tutt A, Ashworth A (2002) The relationship between the roles of BRCA genes in DNA repair and cancer predisposition. Trends Mol Med 8(12):571–576

Uludag Y, Tothill IE (2012) Cancer biomarker detection in serum samples using surface plasmon resonance and quartz crystal microbalance sensors with nanoparticle signal amplification. Anal Chem 84(14):5898–5904

Wang J, Kawde A-N, Erdem A, Salazar M (2001) Magnetic bead-based label-free electrochemical detection of DNA hybridization. Analyst 126(11):2020–2024

Webber MM, Waghray A, Bello D (1995) Prostate-specific antigen, a serine protease, facilitates human prostate cancer cell invasion. Clin Cancer Res 1(10):1089–1094

Willard SS, Koochekpour S (2012) Regulators of gene expression as biomarkers for prostate cancer. Am J Cancer Res 2(6):620–657

Wojno KJ, Costa FJ, Cornell RJ, Small JD, Pasin E, Van Criekinge W, Bigley JW, Van Neste L (2014) Reduced rate of repeated prostate biopsies observed in ConfirmMDx clinical utility field study. Am Health Drug Benefits 7(3):129

Xie S, Zhang J, Yuan Y, Chai Y, Yuan R (2015) An electrochemical peptide cleavage-based biosensor for prostate specific antigen detection via host–guest interaction between ferrocene and β-cyclodextrin. Chem Commun 51(16):3387–3390

Yang S, Liu L, You M, Zhang F, Liao X, He P (2016a) The novel pillar[5]arene derivative for recyclable electrochemical sensing platform of homogeneous DNA hybridization. Sensors Actuators B Chem 227:497–503

Yang S, You M, Yang L, Zhang F, Wang Q, He P (2016b) A recyclable electrochemical sensing platform for breast cancer diagnosis based on homogeneous DNA hybridization and host-guest interaction between cucurbit [7]uril and ferrocene-nanosphere with signal amplification. J Electroanal Chem 783:161–166

Yu J, Yu J, Mani R-S, Cao Q, Brenner CJ, Cao X, Wang X, Wu L, Li J, Hu M (2010) An integrated network of androgen receptor, polycomb, and TMPRSS2-ERG gene fusions in prostate cancer progression. Cancer Cell 17(5):443–454

Zhang F, Mao L, Zhu M (2014) Ultrasensitive immunoassay for free prostate-specific antigen based on ferrocenecarboxylate enhanced cathodic electrochemiluminescence of peroxydisulfate. Microchim Acta 181(11):1285–1291

Oligopeptides for Cancer and Other Biomedical Sensing Applications

12

Xiaokang Ding and Kun-Lin Yang

Abbreviations

5-FAM	5-Carboxyfluorescein
AFP	Alpha-fetoprotein
BHQ-1	Black hole quencher-1
BBI	Bowman–Birk inhibitor
CEA	Carcinoembryonic antigen
CT	Computed tomography
CTC	Circulating tumor cells
ctDNA	Circulating tumor DNA
EGFR	Epidermal growth factor receptor
ELISA	Enzyme-linked immunosorbent assay
EphA2	Ephrin type-A receptor 2
EphB4	Ephrin type-B receptor 4
EpCAM	Epithelial cell adhesion molecule
Fmoc	*N*-Fluorenyl-9-methoxycarbonyl
HER2	Human epidermal growth factor receptor 2
HIV-1	Human immunodeficiency virus type 1
HPS	Hexaphenylsilole
hCG	Human chorionic gonadotropin
IL-6	Interleukin 6
MRI	Magnetic resonance imaging
PET	Positron emission tomography
PSA	Prostate-specific antigen

X. Ding
College of Materials Science and Engineering, Beijing University of Chemical Technology, 15 North 3rd Ring Road East, Beijing 100029, China

K.-L. Yang (✉)
Department of Chemical and Biomolecular Engineering, National University of Singapore, 4 Engineering Drive 4, Singapore 117585, Singapore
e-mail: cheyk@nus.edu.sg

P. Chandra et al. (eds.), *Next Generation Point-of-care Biomedical Sensors Technologies for Cancer Diagnosis*, DOI 10.1007/978-981-10-4726-8_12

QCM	Quartz crystal microbalance
RT-PCR	Reverse transcription polymerase chain reaction
RIV	Restriction of intramolecular vibration
RIR	Restriction of intramolecular rotation
SPR	Surface plasmon resonance
TAMRA	Carboxytetramethylrhodamine
TPE	Tetraphenylethene

12.1 Introduction

For decades, protein biomarkers have been employed for early diagnosis and screening of cancers, even before the tumor tissues can be detected by imaging techniques, such as CT imaging, MRI, PET, or endoscopy. Until 2013, 23 protein biomarkers have been approved by the Food and Drug Administration (FDA) for clinical applications (Füzéry et al. 2013). As recommended by FDA, the most reliable detection methods for these protein biomarkers are based on immunoassays such as ELISA, immunohistochemistry, immunomagnetic-capture, or immune-fluorescence, and several standard protocols have been established for a routine examination. In particular, ELISA is prevalently used in regular health screenings for detection of biomarkers such as AFP, CEA, and PSA. Despite their effectiveness, the current immunoassays suffer from some major drawbacks such as the susceptibility to temperature and the batch-to-batch variation in the production of antibodies.

More recently, the range of biomarkers for cancer diagnosis has been expanded to circulating tumor DNAs (ctDNAs), circulating tumor cells (CTCs), and cell-surface receptors, due to the advances in the analytical diagnostic techniques. The above in vitro diagnostics can be used to access the development of cancer or to evaluate the efficacy of therapeutic treatment (Krishnamoorthy 2015; Zhang et al. 2015). However, the above diagnostics of cancer biomarkers are usually performed in a well-equipped laboratory. It is still challenging to achieve point-of-care diagnostics of the cancer biomarkers with adequate sensitivity at remote and resource-limited clinical sites. Motivated by the "Precision Medicine Initiative" announced by the U.S. President Obama in January 2015, many efforts have been devoted to developing next-generation, point-of-care sensing devices for detecting cancer biomarkers to overcome the current limitations of immunoassays.

Oligopeptides are short biopolymers of 5–20 amino acids linked by peptide bonds. Because of their diversity in chemical functionality and structure, oligopeptides are able to bind a wide range of targets with high specificity. Oligopeptides are promising candidates to replace antibodies as molecular receptors in biosensors (Iqbal et al. 2000; Dover et al. 2009). Short oligopeptides have several advantages over antibodies owing to the following features. (1) The oligopeptides are more robust and resistant to harsh environments. (2) The oligopeptides can be synthesized with high purity (Merrifield 1963) and immobilized on solid surfaces with a single anchoring point. As a result, it is much easier to immobilize oligopeptides on solid surfaces with a well-defined orientation, either by a reaction between a thiol group and a gold surface (Jaworski et al. 2008; Cerruti et al. 2009), or through the formation of thiazolidine ring between N-terminal cysteine and an aldehyde decorated surface (Bi et al. 2008). (3) The oligopeptides screened from random libraries (phage library or one-bead-one-compound library) are able to serve as ligands that specifically bind to target biomarkers. (4) The sequences of oligopeptides can be rationally designed to serve as substrates for enzymes (e.g., proteases or kinase). Due to the above advantages, numerous of oligopeptide-based biomedical sensing devices have been developed. In this chapter, we aim to introduce recent advances in oligopeptide-based biomedical sensing devices for detection of protein biomarkers, CTCs, proteases, and kinase. In addition, the oligopeptide chips will also be introduced to achieve high-throughput screening of target molecules. Lastly, to identify the margins of the tumor tissues in a surgery, the application of oligopeptide probes for biomedical imaging of tumor tissues will also be discussed.

The major challenge of using oligopeptide as a receptor is how to identify a specific amino acid sequence that provides the strongest multivalent binding among all possible combination. One way is using known binding sites of proteins that bind to the target. For example, Kuang et al. developed a trinitrotoluene (TNT) sensor by using an oligopeptide HSSYWYAFNNKTGGGGWFVI as a sensitive layer on single-wall carbon nanotubes (Kuang et al. 2010). The role of HSSYWYAFNNKT is to bind to single-wall carbon nanotubes (SWNTs), and the role of GGGG is a spacer. The four amino acids residues WFVI were derived from the binding site of the honeybee odor binding protein ASP1, which binds to TNT. Similarly, Sankaran et al. developed a QCM-based biosensor using an oligopeptide SLMAGTVNKKGEF, which was derived from an odorant binding protein from *Drosophila* (a fruit fly), as a sensitive layer to detect alcohols (Sankaran et al. 2011). However, this method is limited because binding sites of the odorant binding proteins are only known for a few targets. To address this issue, biopanning processes using random libraries will be elaborated below.

12.2 Oligopeptides Screened from Phage Library

Phage library was evolved as a genetic technique to study protein–protein binding interactions (Smith 1985). In this technology, the most commonly employed bacteriophage is filamentous phage (shaped like a rod filament), including M13, f1, fd, etc. Figure 12.1a shows the structure of an M13 phage, which is composed of circular single-stranded DNA (ssDNA) encapsulated in approximately 2700 copies of the major coat protein capsid, called pVIII (Mullen et al. 2006). The surface exposed a coating protein pIII, which is located at one end of the rod-like phage particle. The M13 phage can be engineered with random oligopeptide sequences by shot-gun cloning of random oligonucleotide segments at the 5′ end of the pVIII or pIII genes of filamentous phage (Hopp et al. 1988; Cwirla et al. 1990; Smith and Scott 1993). Phage display screening is a powerful tool to identify a specific oligopeptide sequence that provides the strongest binding to the target.

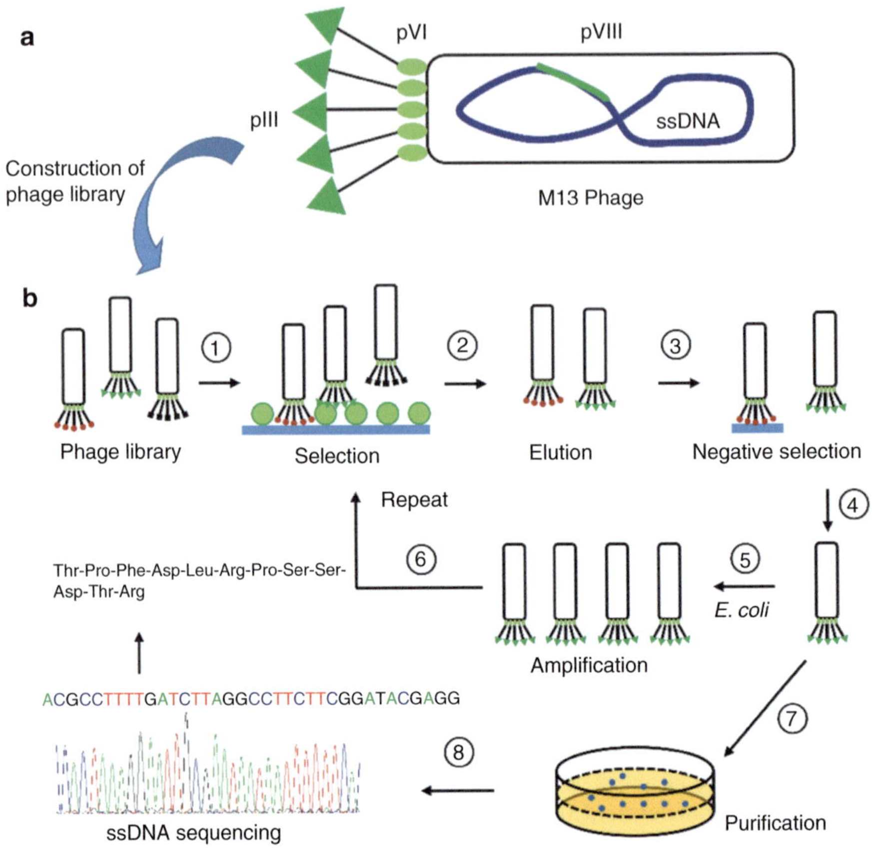

Fig. 12.1 Schematic illustration of (**a**) the structure of an M13 phage and (**b**) phage display screening procedure. Reproduced with permission from reference Ding and Yang (2013a)

Figure 12.1b shows the experimental procedure of phage display screening. Briefly, phage panning is carried out by incubating a phage library containing billions of phages with random peptides with a plate (or bead) coated with the target, washing away the unbound phage, and eluting the specifically bound phage with an excess of a known ligand for the target or by lowering pH. The eluted phage is then amplified by infecting *Escherichia coli* (*E. coli*) and the replicated phage is collected for the next round of panning. Additional binding/amplification cycles are conducted to enrich the phage that expresses the most desirable oligopeptide sequence. After 3–4 rounds, individual clones are characterized by DNA sequencing.

Over decades, the binding capability of oligopeptides screened from phage library has been proven when various targets have been used, such as inorganic nanoparticles (Whaley et al. 2000; Lee et al. 2002), small organic molecules (Goldman et al. 2002;

Takakusagi et al. 2005; Jaworski et al. 2008; Cerruti et al. 2009), and proteins (Parmley and Smith 1988; Scott and Smith 1990; Smith and Scott 1993). In the pioneer works, phage display was used to select oligopeptides that specifically binds to certain targets, such as antibodies, enzymes, and receptors on cell surfaces, by using panning process (Parmley and Smith 1988; Smith and Scott 1993). For example, random peptide phage libraries were used for mapping the epitopes (Cortese et al. 1995). When an antibody is used as the screening target, the consensus sequence of the oligopeptides identified from random phage library are often easily recognized in the sequence of the natural antigen, thus allowing the epitope recognized by the antibody to be mapped (Cwirla et al. 1990; Scott and Smith 1990; Felici et al. 1991). Later, phage library was used to identify oligopeptide sequences to bind specifically to certain protein molecules. For example, Devlin et al. successfully identified oligopeptides containing a His-Pro-Gln (HPQ) motif that selectively binds to streptavidin (Devlin et al. 1990). However, in this study, the binding affinity of the oligopeptide to streptavidin is unknown because phage particles, rather than oligopeptides, were used in the binding experiment. Pillutla et al. identified oligopeptides which can bind to insulin receptor by using phage library (Pillutla et al. 2002). They also studied interactions between insulin and insulin receptor in the presence of oligopeptides by using radioreceptor assays. To study the binding kinetics, an oligopeptide, GAMHLPWHMGTL, was identified from phage library panning against Bowman–Birk inhibitor (BBI) (Fields et al. 2012). Then the oligopeptide was synthesized, and the binding kinetics between BBI and the oligopeptide was studied by using a surface plasmon resonance (SPR) platform. They also used oligopeptide-modified magnetic particles to isolate BBI from soybean extracts. The above studies show that phage library is a powerful tool to identify oligopeptides which bind to target proteins with high affinity and specificity. More recently, to explore the application of oligopeptides in biosensors, one oligopeptide receptor, PPLRINRHILTR, was identified from phage library to capture hCG, a hormone biomarker for pregnancy tests and for some cancer diagnosis (Ding and Yang 2013a, b). Using a phage library, Stevens' group identified an oligopeptide, WSRVGYW, which was utilized for detection of HIV-1 protease together with another peptide ligand LLEYSL (Herpoldt et al. 2015). Zou and Yang identified an oligopeptide SHSLPASADLRR that can bind to penicillinase, an enzyme evolved in bacteria to resist the antibiotic activity of penicillin (Zou and Yang 2016). This oligopeptide can be either utilized for detection of penicillinase in an SPR biosensor or to work as a penicillinase inhibitor to reduce the resistance of bacteria toward β-lactam antibiotics.

12.3 Oligopeptides Screened from One-Bead One-Compound (OBOC) Library

Although phage display has been proven as a powerful tool to screen oligopeptide ligands that specifically bind to target molecules, the nonuniform amplification of eluted phages by infecting the host of *E. coli* is still a concern (Derda et al. 2010,

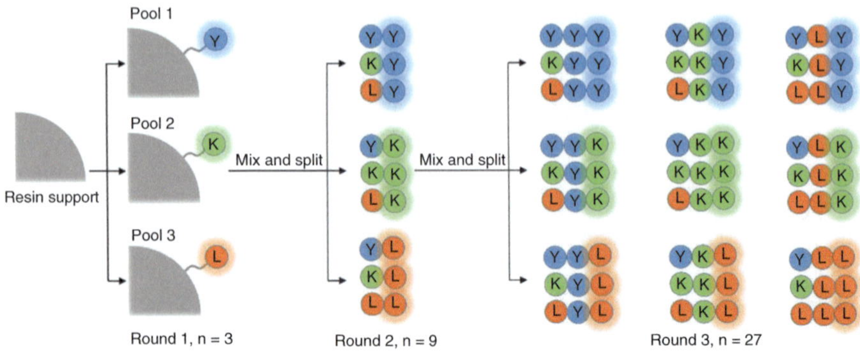

Fig. 12.2 Schematic illustration for constructing an OBOC oligopeptide library via mix-and-split approach. An example of the combinatorial synthesis of tripeptides composed of tyrosine (Y), lysine (K), and leucine (L) is shown, where the *glowing circles* represent the newly synthesized amino acids. The resin support is omitted in Round 2 and Round 3. Reproduced with permission from reference Lam et al. (1997)

2011). The one-bead-one-compound (OBOC) oligopeptide library which is pioneered by Lam and his coworkers provides an alternative approach to discover oligopeptide ligands (Lam et al. 1991, 1997, 2003). Figure 12.2 shows a schematic illustration for constructing an OBOC oligopeptide library based on the standard solid-phase synthesis of oligopeptides and a mix-and-split approach for combinatorial chemistry. For example, the resin beads (~100 μm) were divided into three pools and conjugated with an amino acid of tyrosine (Y), lysine (K), and leucine (L), respectively. Next, the resin beads from all pools were combined and randomly split into three new pools before the second round of amino acid conjugation. This mix-and-split procedure was repeated until a full-length oligopeptide library was obtained. Interestingly, by using this OBOC oligopeptide library, Lam and his coworkers also identified an oligopeptide sequence of HPQ that exhibited high binding affinity toward streptavidin (Lam et al. 1991). This result is similar to that when a phage library is used (Devlin et al. 1990). By expanding this method, the diversity of OBOC oligopeptide library can achieve up to 10^8, which is close to that of phage library. However, most of OBCB libraries are smaller in size (e.g., 10^5–10^6) due to the cost, yield, and time consumption (Gray and Brown 2014).

Next, the OBOC library consisting of millions of beads was subjected to a biological panning against protein molecules or whole cells. For the on-bead assays, a standard enzyme-linked colorimetric assay could be used to identify the beads that exhibit high binding affinity to the target molecules, and the bearing oligopeptides can be sequenced by Edman degradation using an oligopeptide microsequencer (Lam et al. 1991). Because of the Edman degradation proceeds from the N-terminus of the oligopeptide, this sequencing method will not work for the cyclic or branched oligopeptides or the N-terminal amino acid has been

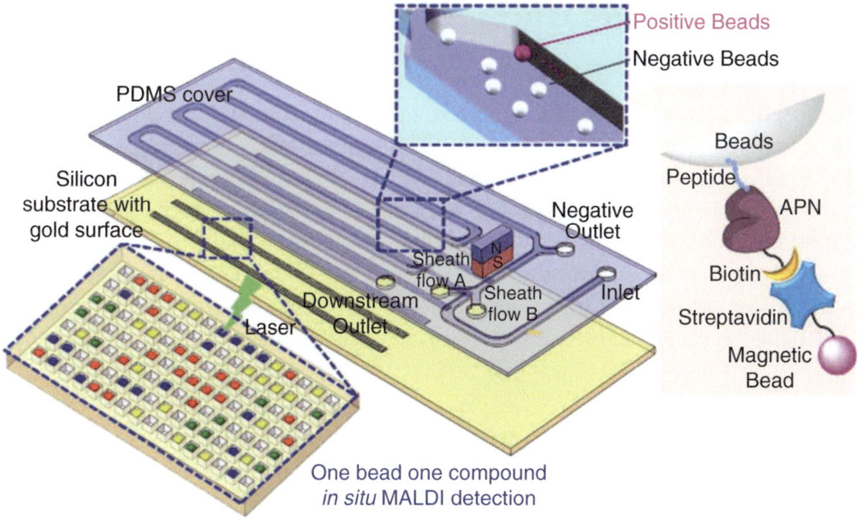

Fig. 12.3 Schematic illustration of the architecture of the integrated microfluidic system for on-bead screening, sorting, and sequencing. Reproduced with permission from reference Wang et al. (2014)

chemically modified. To address this issue, several approaches have been proposed by incorporating interior tags of bilayer beads (Liu et al. 2002; Wang et al. 2005). Although the Edman degradation-based sequencing is still considered as gold standard, it is still time-consuming compared to the DNA sequencing techniques used in phage display. With the advances in oligopeptide sequencing technology, matrix-assisted laser desorption ionization time-of-flight mass spectrometry (MALDI-TOF-MS) has been evolved as a fast and accurate technique (Lee et al. 2010).

More recently, the OBOC approach has been miniaturized and automated by using a microfluidic system (Wang et al. 2014, 2015). In this approach, the OBOC library was constructed to screen an oligopeptide ligand of YVEYHLC which specifically binds to a tumor marker aminopeptidase N (APN), a key protein biomarker in tumor angiogenesis. Figure 12.3 shows the architecture of the integrated microfluidic system, in which the mixture of OBOC library, the biotinylated APN, and the streptavidin-conjugated magnetic beads are introduced. After applying a magnetic field, the target-binding positive beads carrying a magnetic bead can be separated and then subsequently sorted into the micro-wells in a one-well-one-bead manner, and the oligopeptides can be sequenced by in situ MALDI-TOF-MS. The OBOC approach offers more flexibility in designing of the library with unnatural blocking materials, such as the D-amino acids, β-amino acids, or peptoids, which resist the proteolytic cleavage of the oligopeptides in presence of proteases (Aina et al. 2007).

12.4 Oligopeptides as Recognizing Components in Cancer Diagnosis

Oligopeptides are promising candidates to replace antibodies as a molecular recognizing component to capture or bind to a target molecule in a biosensor. In the point-of-care cancer diagnosis, the most common targets are circulating protein biomarkers or the protein receptors on the surface of tumor cells. To this end, two sensing configurations have been reported in the literature. In the first case, the oligopeptides are immobilized on a solid surface to serve as recognition component to capture target molecules or tumor cells. In the second case, the fluorescent labeled oligopeptides bind to the targets and the sensing can be performed in homogeneous solution.

12.4.1 Oligopeptides Recognizing Protein Biomarkers

The sequence-specific oligopeptides-protein recognition has been attracting much attention because the synthetic oligopeptides can be easily modified or immobilized at well-defined sites. The binding affinity between the oligopeptides and protein biomarkers are crucial in developing oligopeptide-based biosensors. To evaluate the binding affinity, a recombinant antibody fragment, Fab57P, which specifically binds to the tobacco mosaic virus (TMV) coat protein, was used as a target molecule (Andersson et al. 2009). An oligopeptide $CH_3CO\text{-}E^{Biotin}$RGTGSYNRSSFESSSGLV-$CONH_2$ (P1), derived from the amino acids 134–151 of TMV coat protein, was immobilized on the sensor chip of imaging surface plasmon resonance (iSPR) among other oligopeptides $CH_3CO\text{-}E^{Biotin}$KSYNRSSFETNSGLT-$CONH_2$ (P2), $CH_3CO\text{-}E^{Biotin}$NKTSFSPPPLSI-$CONH_2$ (P3), and $CH_3CO\text{-}E^{Biotin}$RGSGTRSYSNEGFSSLSV-$CONH_2$ (P4) to capture Fab57P. The kinetic assay shows that P1 exhibits the highest binding affinity with a dissociation constant (K_D) of 0.95 nM, which is 13.7 and 389-fold lower than that of P2 and P3. By combining the above oligopeptides, an affinity array consisting P1, P2, P3, P1 + P2, and P2 + P3 was developed to extend the dynamic range for protein quantification (Fig. 12.4).

In another approach, a ratiometric fluorescent biosensor was developed by using an 18 amino acid oligopeptide RGTGSYNRSSFESSSGLV to detect Fab57P. The oligopeptide was labeled with an environmentally sensitive fluorophore (6-bromom ethyl-2-(2-furanyl)-3-hydroxychromone) with a two-band emission. Since the emission properties of the fluorophore are sensitive to the changes in the local environment induced by the binding event, the change of emission intensity of the fluorophore will be detected when the labeled oligopeptide binds to the target molecule of Fab57P (Fig. 12.5).

More recently, Stevens' group screened out an oligopeptide of WSRVGYW that bound to HIV-1 protease by using a phage library (Herpoldt et al. 2015). To explore the feasibility of detecting HIV-1 protease, this oligopeptide was conjugated with a fluorophore of Alexa Fluor 647 (AF647). Another oligopeptide

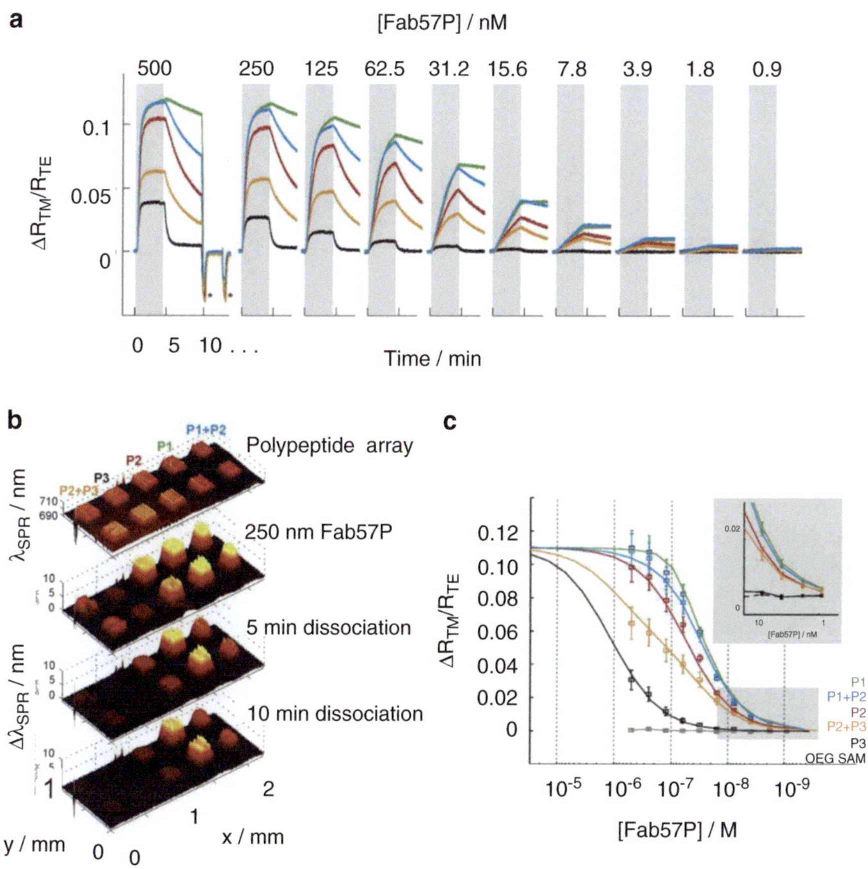

Fig. 12.4 The affinity recognition of the oligopeptide array toward Fab57P using iSPR. (**a**) Sensorgrams showing the shift in reflectivity upon injection of Fab57P. (**b**) SPR wavelength maps showing the shift in λ_{SPR} upon introduction of 250 nM Fab57P to two affinity arrays in parallel comprising P1 (*green*), P2 (*red*), P3 (*black*), P1 + P2 (*blue*), and P2 + P3 (*orange*). (**c**) Calibration curves for each oligopeptide element of the array. Reproduced with permission from reference Andersson et al. (2009)

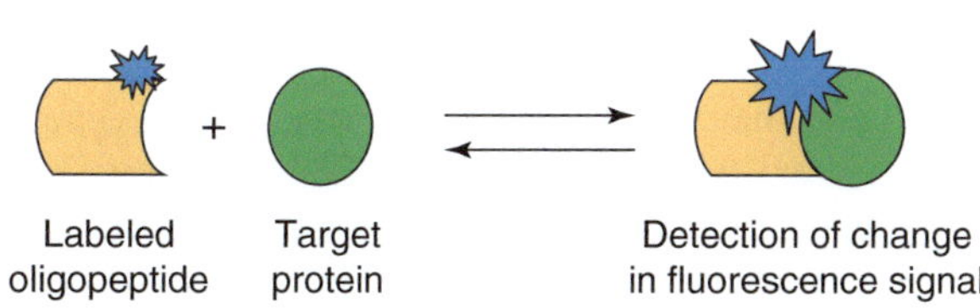

Fig. 12.5 Schematic illustration of ratiometric fluorescent biosensor based on the binding of fluorescently labeled oligopeptide and the target protein. This figure was modified after the reference Enander et al. (2008)

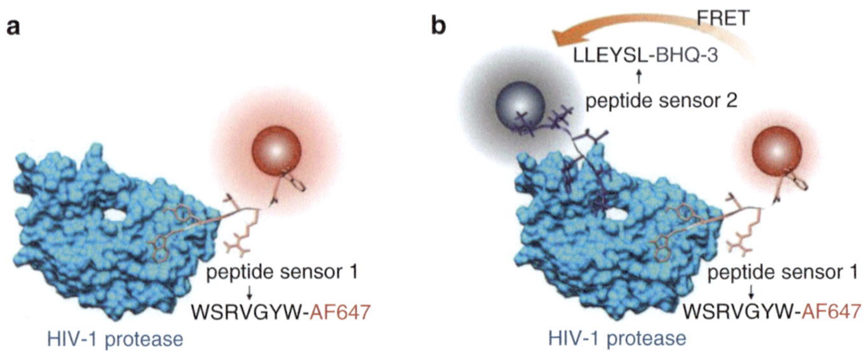

Fig. 12.6 Schematic illustration for the detection of HIV-1 protease by using a pair of oligopeptide probes. The oligopeptide sensor 1 (WSRVGYW) is labeled with a fluorophore of AF647, and the oligopeptide sensor 2 (LLEYSL) is labeled with a quencher of BHQ-3. In the presence of HIV-1 protease, the fluorescence of AF647 is quenched via FRET effect. Reproduced with permission from reference Herpoldt et al. (2015)

LLEYSL, an inhibitory oligopeptide isolated from thermolysin hydrolysate of oyster proteins, was tagged with a quencher of Black Hole Quencher-3 (BHQ-3). Once the oligopeptides bound to HIV-1 protease, the fluorescence of AF647 was significantly quenched due to Förster resonance energy transfer (FRET) effect (Fig. 12.6). This approach employs two independent recognition events of oligopeptides to the target molecule of HIV-1 protease to improve the specificity of the biosensing. Moreover, this approach does not require multiple wash steps in conventional ELISA assays, making it easier to be streamlined for bench-top diagnostic assays.

Since the labeling of fluorophores or quenchers may hamper the binding affinity between the oligopeptide and the protein target, developing label-free biosensors is still appealing. For example, an SPR biosensor has been reported by using oligopeptides to detect human chorionic gonadotropin (hCG), a glycoprotein hormone produced by placental trophoblasts. hCG is a common biomarker for the diagnosis of pregnancy (Tsampalas et al. 2010; Haarburger and Pillay 2011) and several cancers such as prostate cancer, testicular cancer, trophoblastic cancer, and gestational choriocarcinoma (Bagshawe 1992; Acevedo et al. 1995; Birken et al. 2001). Therefore, detection of hCG in serum or urine is often carried out during routine medical screening. The most commonly used hCG diagnostic kit is based on lateral-flow immunoassays that rely on a pair of specific anti-hCG antibodies (Wong and Tse 2009). If hCG exists in a test sample, hCG will bind to a probe antibody labeled with gold nanoparticles (or other dyes) and form an antigen–antibody complex. Subsequently, this complex can be captured by a surface-immobilized antibody to form a sandwich-type structure, and a positive line will appear. This lateral-flow immunoassay is simple to use and it can readily give qualitative results with a limit of detection (LOD) around 20 mIU/mL in urine samples. However, the production of antibodies is expensive and tedious, and the labeling of the probe antibodies would deteriorate the binding affinity between the probe antibodies and the target molecules. To detect hCG in a label-free configuration, an oligopeptide sequence of PPLRINRHILTR with highest binding affinity was identified from phage library to capture hCG, and this oligopeptide is incorporated in an SPR biosensor as the capture component (Ding and Yang 2013a, b). The SPR results reveal the dissociation constant (K_D) is 0.9 nM, showing a relatively high binding affinity between the oligopeptide and the hCG. Although SPR biosensor is a label-free technique, the detection of an optical shift in resonance angle (Lofas 1995) requires delicate instrumentation, which constrains the SPR biosensors for point-of-care applications.

To circumvent these limitations, the authors developed a label-free method for the detection of hCG by using liquid crystals (LCs) (Ding and Yang 2013a, b). LCs are birefringent materials which have been widely used in flat-panel displays because the orientation of LCs can be easily tuned by surface anchoring (Kahn 1973; Jerome 1991) or by applying an electric field (Haas et al. 1970). Abbott and his coworkers discover that LCs are able to transduce the presence of proteins on solid surfaces into optical signals without any labeling procedure (Gupta et al. 1998). Afterward, numerous LC-based immunosensors have been developed using antibodies as a capture component (Lockwood et al. 2008; Carlton et al. 2013). However, the immobilization of antibodies also disrupts the homeotropic LC orientation when its surface density exceeds a critical value (9.6×10^{-3} molecule/nm^2) (Alino and Yang 2011). This is unfavorable because a higher surface density of immobilized antibody (it is often required to achieve a lower LOD) would result in false positive result in a diagnosis. Oligopeptide is a good candidate to replace antibodies because it has a smaller size, and it will not disrupt LC orientation easily before binding to target molecules. Figure 12.7 shows a schematic illustration of an LC biosensor using oligopeptides to capture hCG (Ding and Yang 2013a, b). First, the oligopeptide with a sequence of PPLRINRHILTRGGG–biotin (derived from the phage library) was immobilized on a glass slide via biotin-streptavidin conjugation. When the hCG molecules were captured by the oligopeptide, the presence of hCG easily disrupted the orientation of LCs and a colorful spot was observed under crossed polarizers. In contrast, the optical image of LC biosensor remained dark if no hCG was present. In this oligopeptide/LC-based sensing architecture, only a pair of polarizers is required to report the diagnostic signal. This feature is advantageous for point-of-care biomedical diagnostics in remote clinical sites where resources are limited.

Alternatively, the cancer biomarkers can be detected using electrochemical biosensors, in which oligopeptides are used as sensitive layers. For example, an oligopeptide probe (also called "peptamer") was designed to incorporate a pair of Cu(II) ion binding sites, a pair of cucurbituril (Q8) affinity motifs, and an integrin affinity binding site (Fig. 12.8a) (Li et al. 2016). This oligopeptide probe was first immobilized on the electrode and underwent conformational transition in the presence of Q8 and the target of integrin. This conformational transition led to enhanced binding of Cu(II) ions, which successively catalyzed a signal amplification reaction (Fig. 12.8b, c).

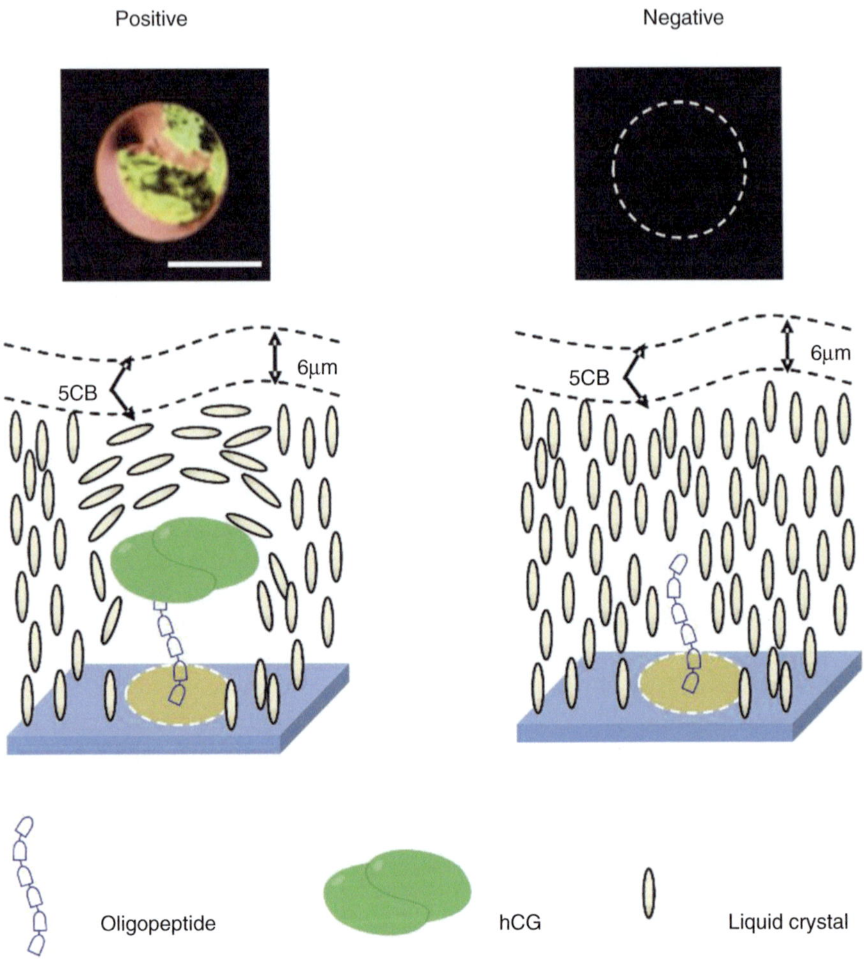

Fig. 12.7 The optical images of positive and negative output when the LC biosensors were observed under crossed polarizers. The schematic illustration of the LC orientation is shown below in the presence or absence of hCG. The concentration of hCG is 10 IU/mL in buffer solution. The scale bar is 1 mm. Reproduced with permission from reference Ding and Yang (2013b)

12.4.2 Oligopeptides Recognizing Tumor Cells

The circulating tumor cells (CTCs) are the cells shed from tumor tissues into the blood stream. As the CTCs in the patient's blood reveal the progression and metastasis of the tumor, the detection of CTCs in a "liquid biopsy" is important for cancer diagnosis, prognosis, and evaluation of therapeutic implications. Compared to conventional tissue biopsies, the detection of CTCs is advantageous because it is non-invasive (or minor-invasive), and it facilitates routine blood tests to monitor the patients' disease status over time, making it possible to evaluate the efficacy of

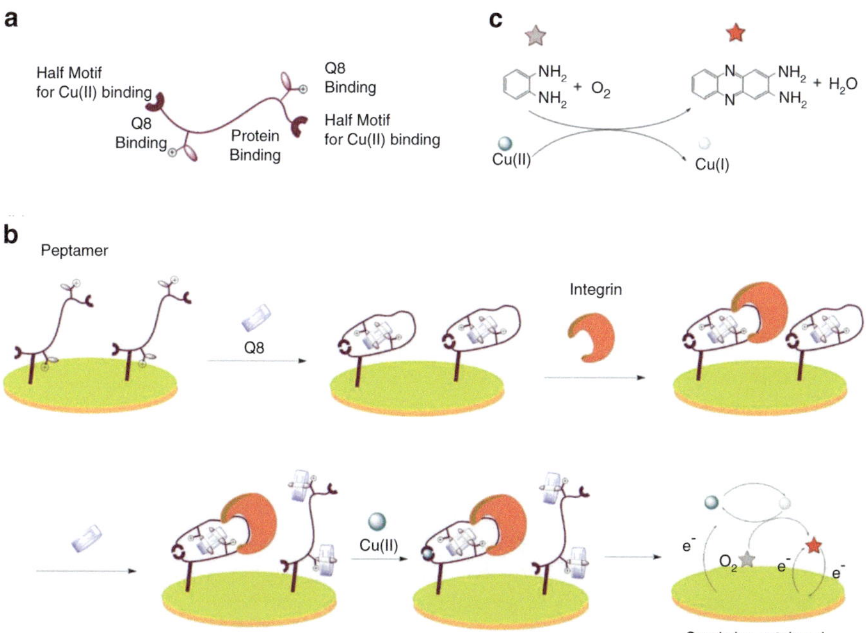

Fig. 12.8 Schematic illustration of the oligopeptide-based electrochemical biosensor for detection of integrin, a biomarker for tumor invasion. (**a**) Design of the oligopeptide probe. (**b**) Assay procedure for the electrochemical biosensor. (**c**) Reaction of the catalytic signal amplification. Reproduced with permission from reference Li et al. (2016)

patient's therapy. However, identification and counting of CTCs from blood are still challenging due to their low abundance in comparison with normal blood cells. For example, the abundance of CTCs is around 1–10 cells per mL of whole blood, which contains a few millions of white blood cells and a billion of red blood cells (Paterlini-Brechot and Benali 2007; Yu et al. 2011). To achieve high sensitivity and specificity, enrichment steps are commonly required in commercially available test kits (Paterlini-Brechot and Benali 2007). For example, the immunomagnetic beads or ferrofluids (colloidal iron) are employed in the magnetic-activated cell sorting (MACS) system or CellSearch system, respectively, to capture epithelial cells expressing antigen proteins such as EpCAM, BerEP4, and cytokeratins (CK). Next, the CTCs can be isolated by immunolabelling in a negative selection by using a fluorescent antibody to CD 45 to rule out leucocytes combining a positive selection by using fluorescent antibodies to CK markers (CK-8, 18, and 19) which are specific to epithelial cells (Mostert et al. 2009). One of the major drawbacks of this method is the interference of epithelial non-tumor cells. Alternatively, the CTCs can be identified through the detection of genetic mutations that are specific for tumor cells by using molecular biotechnologies such as RT-PCR (Mostert et al. 2009).

Over decades, many cancer-targeting oligopeptides have been isolated from combinatorial oligopeptide libraries (e.g., phage libraries, bacterial libraries, and

Table 12.1 Oligopeptides capturing cancer cells

Oligopeptides	Target cancer cells	Method	Reference
cDGWGPNc[a]	Ovarian cancer cell lines	OBOC library	Aina et al. (2005)
pA peptide (cNGQGEQc[a])	A549 non-small-cell lung cancer cells	OBOC library	Lau et al. (2006)
QMARIPKRLARH	LNCaP cells (human prostate cancer cell line)	OBOC library	Aggarwal et al. (2005)
SAKTAVSQRVWLPSHRGGEP	B-cell lymphoma cell line	Phage library	McGuire et al. (2006)
YSAYPDSVPMMS	Ovarian cancer cells	Phage library	Scarberry et al. (2008)

[a]The letters in lower case represent dextrorotatory amino acids

OBOC libraries) for the diagnostic and therapeutic applications (Gray and Brown 2014). The selected oligopeptides specifically bind to the cellular target proteins (e.g., HER2, EGFR, and IL-6 receptor, EphA2, EphB4) that are excessively expressed in tumor cells. At the beginning, the cancer-targeting oligopeptides were used for either in vivo imaging of the tumor tissues (Kumar et al. 2007, 2010) or delivery of therapeutic agents (e.g., drugs or oligonucleotides) (Cheng et al. 2011; Schafer et al. 2011). More recently, the cancer-targeting oligopeptides have been reported to capture CTCs from blood samples in vitro. For example, Lam and his coworkers identified a tripeptide-based peptidomimetic ligand, LLP2A, by using an OBOC peptidomimetic library (Peng et al. 2006). The LLP2A exhibits high binding affinity against α4β1 integrin, which is excessively expressed in cancer cells and plays important roles in cancer metastasis and development. The LLP2A-coated beads show a capability to capture Jurkat cells (an immortalized line of human T lymphocyte) from peripheral blood mononuclear cells (PBMC) at a dilution of 1:100,000 (Peng et al. 2006; Aina et al. 2007). In another study, an oligopeptide of VRRDAPRFSMQGLDACGGNNCNN was identified from de novo designed peptide pool, showing an equilibrium dissociation constant (K_D) of 1.98×10^{-9} mol L^{-1} against EpCAM, which is comparable to that of anti-EpCAM (Bai et al. 2014). This EpCAM-binding oligopeptide was immobilized onto iron oxide magnetic nanoparticles (MNPs) to isolate the CTCs from blood samples for downstream analysis, and the capture efficiency of CTCs can reach above 90% compared to the anti-EpCAM modified MNPs. Up to now, several oligopeptides have been reported in the literature to isolate cancer cells from whole blood, as listed in Table 12.1.

12.5 Oligopeptides for In Vitro Protease Assays

Proteases exist ubiquitously in all living forms and play important roles in regulating numerous biological and physiological processes (e.g., food digestion, blood clotting, cell apoptosis, and disease development) by catalyzing the hydrolysis of peptide bonds in proteins or peptides (Pan et al. 2012; Vickers et al. 2013). Many diseases such as cardiovascular disease (Krizkova et al. 2011; Jin et al. 2015), Alzheimer disease (Vassar et al. 1999), and cancers (Harris et al. 2006; Chen et al. 2013) are correlated with dysfunction or overactivity of proteases. Therefore, monitoring of protease activity is important for both of disease diagnosis and screening of protease inhibitors for developing new drugs. Rather than directly detecting protease molecules by immunoassays, quantification of protease activity often attracts more attention. The quantification of protease activity is more challenging, because the differentiation of protease-cleaved peptide fragments usually requires bulky and expensive instrumentation, such as high performance liquid chromatography (HPLC) (Zhang et al. 2014), gel electrophoresis (Lefkowitz et al. 2010; Zhao et al. 2012), or mass spectrometry (Na et al. 2013; Vosyka et al. 2013). Moreover, the above techniques are end-point analytical methods, which do not provide real-time data of the proteolytic activity of proteases. To overcome these limitations, many efforts have been made to quantify the protease activities by using rationally designed oligopeptides incorporating the cleavage site of the particular protease of interest.

12.5.1 Förster Resonance Energy Transfer (FRET)

Figure 12.9 shows the schematic illustration of FRET-based protease assay. The oligopeptide substrate is labeled with a pair of energy donor and acceptor. In a general scenario, the donor/acceptor pair can be both fluorophores (e.g., fluorescein/TAMRA) (Zauner et al. 2011), or a fluorophore/quencher pair (e.g., 5-FAM/BHQ-1) (Lock et al. 2016). When the distance between the donor and acceptor is less than 10 nm, the energy of an excited state donor can be transferred nonradiatively to a proximal ground state acceptor via resonant dipole–dipole interactions (Kim and Kim 2012). In the presence of proteases, the oligopeptide is cleaved and the donor and acceptor are separated. As a result, the energy transfer is disrupted, resulting in a restoration of the fluorescence. This FRET-based protease assay is successful because it is able to monitor the proteolytic activity in a real-time manner, and numerous of protease activity assay kits based on this principle are commercially available.

However, the use of organic fluorophores suffers from some drawbacks such as fast photo-bleaching, pH sensitivity, photo/chemical susceptibility, and relatively small Stokes shifts. To improve the performance of this FRET-based protease assay, quantum dots (QDs) have been conjugated to the oligopeptide substrates to replace organic fluorophores as energy donors (Medintz et al. 2006). The use of QDs is advantageous because their photoemission spectra can be easily tuned by controlling the size so as to overlap the absorption spectra of the acceptor. Besides, the QDs typically have broad absorption spectra with large molar extinction coefficients ($0.5–5 \times 10^6\ M^{-1}\ cm^{-1}$) and higher quantum yield (65–75%) (Kim and Kim 2012). The design of the oligopeptide linker is crucial in the performance of the FRET-based protease assay. Figure 12.10 shows a schematic illustration of an FRET-based protease assay using QDs as a signal reporter (Medintz et al. 2006). In this configuration,

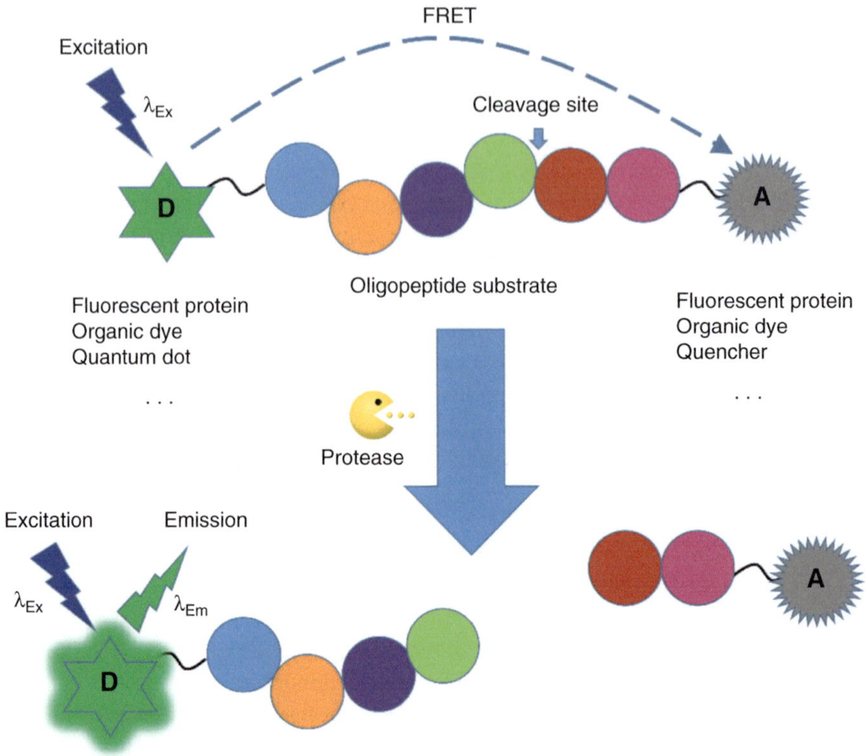

Fig. 12.9 Schematic illustration of FRET-based protease assay

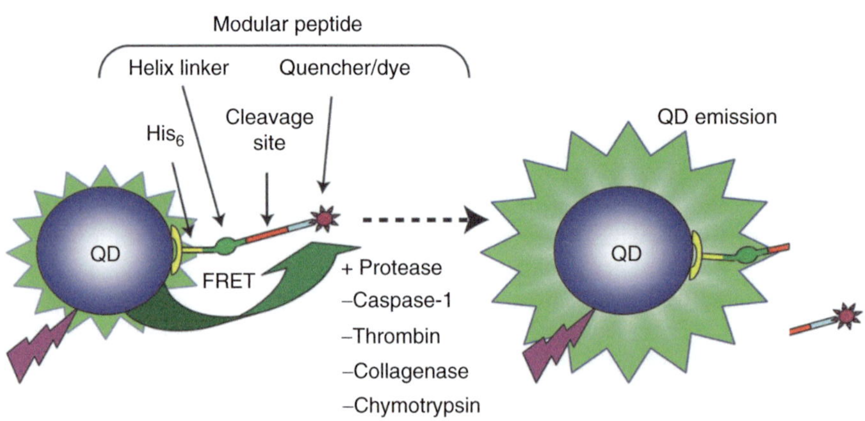

Fig. 12.10 Schematic illustration of an FRET-based protease assay using QDs as a signal reporter. Reproduced with permission from reference Medintz et al. (2006)

the oligopeptide was attached to the QD through the self-assembly of amino-terminal hexahistidine (His_6) on dihydrolipoic acid (DHLA)-modified QD. Moreover, a helix-linker spacer was included to provide rigidity, and push the cleavage site away from the QD surface to facilitate the proteolytic reaction. As an acceptor, organic fluorophores or quenchers (e.g., Cy3 or QXL-520) were attached to the oligopeptide at the carboxyl-terminal cysteine. As shown in these studies, it is essential to control the center-to-center distance between the donor (QD) and the acceptor (dye/quencher) to achieve optimized FRET. As a result, the length of the oligopeptide linker needs to be rationally designed to match the Förster radius (R_0).

12.5.2 Aggregation-Induced Emission (AIE)

Aggregation-induced emission (AIE) is an intriguing phenomenon discovered by Luo et al. (Luo et al. 2001). Unlike conventional fluorophores quenched by aggregation (aggregation-caused quenching, ACQ), the AIEgens (the luminogens exhibiting AIE attributes) are non-emissive if they are dissolved in a good solvent (Fig. 12.11). In contrast, when the AIEgens are aggregated in a poor solvent, the AIEgens become emissive due to the restriction of intramolecular vibration (RIV) and/or restriction of intramolecular rotation (RIR) mechanisms (Mei et al. 2014). Over decades, several AIEgens and their derivatives, including tetraphenylethene (TPE), hexaphenylsilole (HPS), and 10,10′,11,11′-tetrahydro-5,5′-bidibenzo[*a,d*][7]annulenylidene (THBA), have been synthesized and applied to biosensing (Wang et al. 2010; Ding et al. 2013; Kwok et al. 2015).

For instance, an oligopeptide bearing a recognition/cleavage sequence to a particular protease was conjugated to the TPE core to construct an AIE-probe (Fig. 12.12a) for detecting the protease activity (Shi et al. 2012). The oligopeptide with a sequence of DEVDK was conjugated to the hydrophobic TPE core (an AIEgen) via "click" chemistry. The oligopeptide is hydrophilic, rendering a good solubility of this AIE-probe in aqueous solution. In this case, the fluorescence of the AIE-probe is low due to the RIV and RIR mechanisms. In the

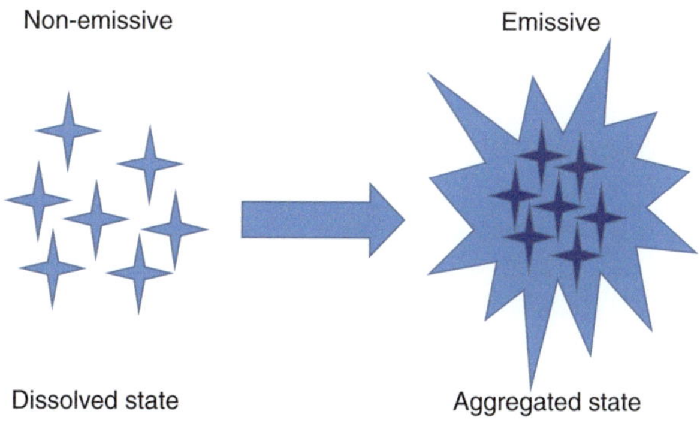

Fig. 12.11 Schematic illustration of an AIE phenomenon

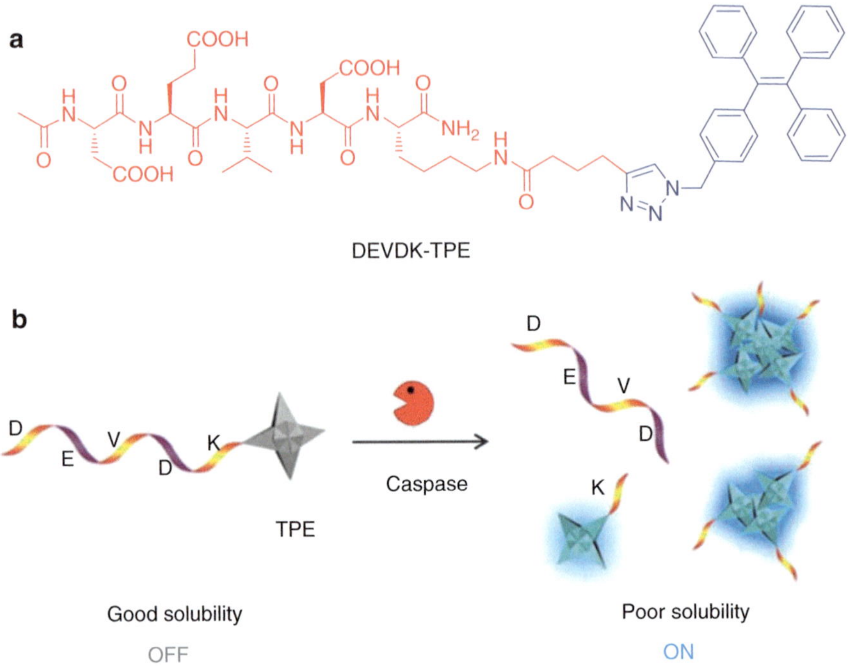

Fig. 12.12 (**a**) Molecular structure of the AIE-probe, DEVDK-TPE, for protease assay. (**b**) Schematic illustration of the AIE probe for protease assay. Reproduced with permission from reference Shi et al. (2012)

presence of caspase-3 (a protease responsible for the apoptosis of cells), this protease recognized the sequence of DEVD and cleaved the oligopeptide at the carboxyl-terminal of aspartic acid (D), and the z of DEVD was released, resulting in the aggregation of the hydrophobic residue of K-TPE due to its poor solubility in aqueous solution (Fig. 12.12b). After aggregation, the fluorescence of the TPE core recovered. In this way, the AIE-probe enables a "turn-on" strategy to monitor the proteolytic activity of caspase-3 either in a buffer solution or in living cells.

12.5.3 Gold Nanoparticles (AuNPs)

In the recent years, gold nanoparticles (AuNPs) have been emerging into promising scaffolds for the fabrication of biosensors (Zhou et al. 2015). The color of the colloidal gold strongly depends on the size, shape, and the refractive index of the surrounding medium due to the effect of localized surface plasmon resonance (LSPR) (Underwood and Mulvaney 1994; Burda et al. 2005). The radiation of electromagnetic field induces the polarization of the electron cloud on the surface of spherical AuNPs, causing intense light absorption at a specific wavelength (Fig. 12.13a)

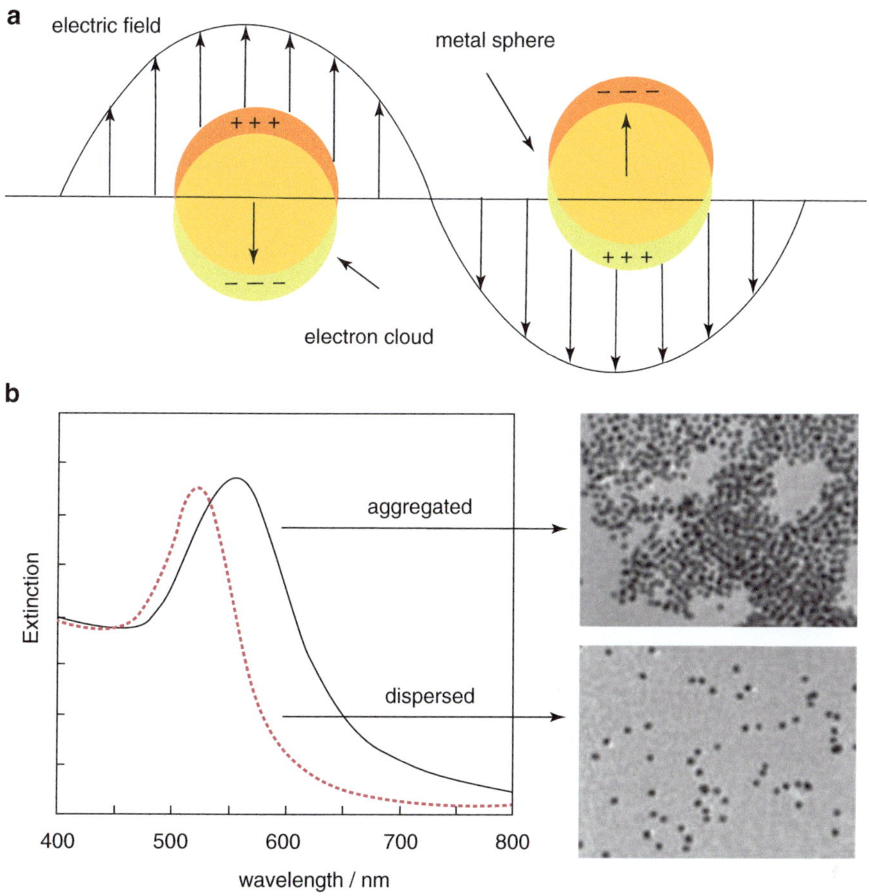

Fig. 12.13 (**a**) Schematic illustration of the localized surface plasmon resonance (LSPR). (**b**) UV-vis spectra of dispersed (*red dashed line*) and aggregated (*blue solid line*) AuNPs. The TEM images of dispersed and aggregated AuNPs are also shown, respectively. Reproduced with permission from reference Aili and Stevens (2010)

(Kelly et al. 2003). In particular, the color of the colloidal gold changes from red to blue when the dispersed AuNPs are aggregated together (Fig. 12.13b) (Aili and Stevens 2010).

Numerous colorimetric biosensors have been reported to detect proteases based on the analyte-induced aggregation or dispersion of the AuNPs (Guarise et al. 2006; Laromaine et al. 2007; Ding et al. 2014). For example, the oligopeptide sequence is rationally designed with a recognition and cleavage site to a particular protease, and a pair of anchoring points (e.g., cysteine and lysine) to induce cross-linking between the AuNPs and thus causes the aggregation of AuNPs (Fig. 12.14a). In this case, a distinct color change (from red to blue) of the colloidal gold can be observed by the naked eye. When the oligopeptide linker was pretreated with a protease (e.g., thrombin, lethal factor, or trypsin), the cleaved oligopeptides could not induce the

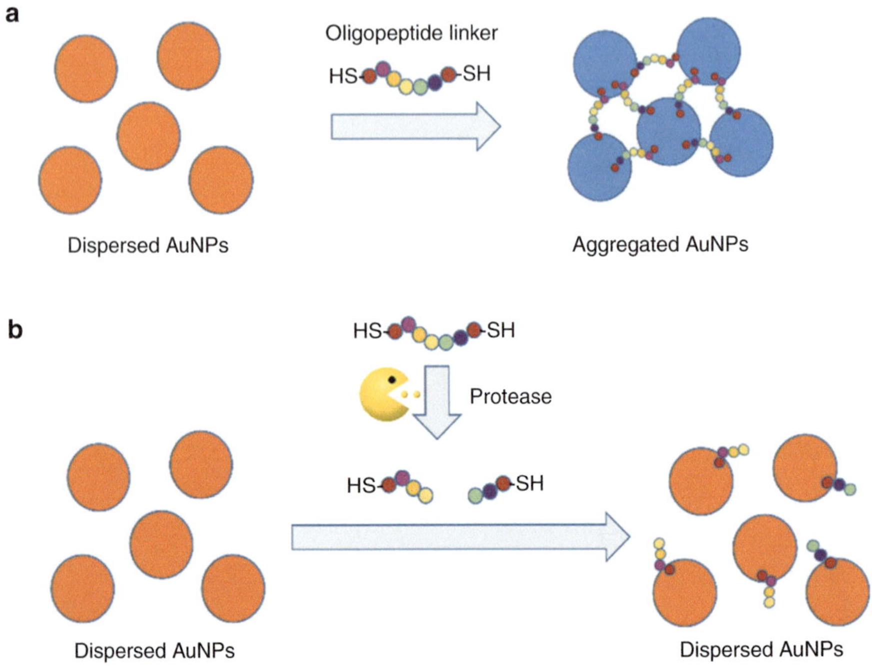

Fig. 12.14 Schematic illustration of the AuNP-based colorimetric protease assay. (**a**) In the absence of protease, the oligopeptide linkers trigger the aggregation of AuNPs and result in blue colour. (**b**) In the presence of protease, the oligopeptide linkers are digested, resulting in non-aggregated AuNPs showing red colour

aggregation of the AuNPs, and the color of the colloidal gold did not change (Fig. 12.14b). This colorimetric assay is advantageous because the presence of the analyte can be easily interpreted by the naked eye. However, this two-step protease assay is not applicable for real-time protease assay. To address this issue, Stevens' group demonstrated a new strategy for designing an oligopeptide with (1) a cysteine anchoring point, (2) a cleavage site, and (3) an assembly directing actuator of Fmoc to promote physical assembly of the AuNPs through π-stacking interactions (Laromaine et al. 2007). When the oligopeptide is cleaved, the Fmoc is removed and the AuNPs are re-dispersed due to the electrostatic repulsion (Fig. 12.15).

More recently, synthetic biomarkers have been designed by conjugating an oligopeptide-based reporter to iron oxide nanoworms (NWs) to achieve long circulation time in vivo (Warren et al. 2014). The oligopeptide contains the cleavage sites of particular proteases. After i.v. administration, the NWs accumulated in the diseased tissues where the proteases (e.g., thrombin or MMP-9) were excessively expressed. Next, the aforementioned oligopeptides were cleaved, and the synthetic reporters were released into the blood stream and finally collected from the urine after renal clearance. Thereafter, the collected reporters were detected by using ELISA or lateral immunoassay, which enables the point-of-care diagnostics of the synthetic reporters.

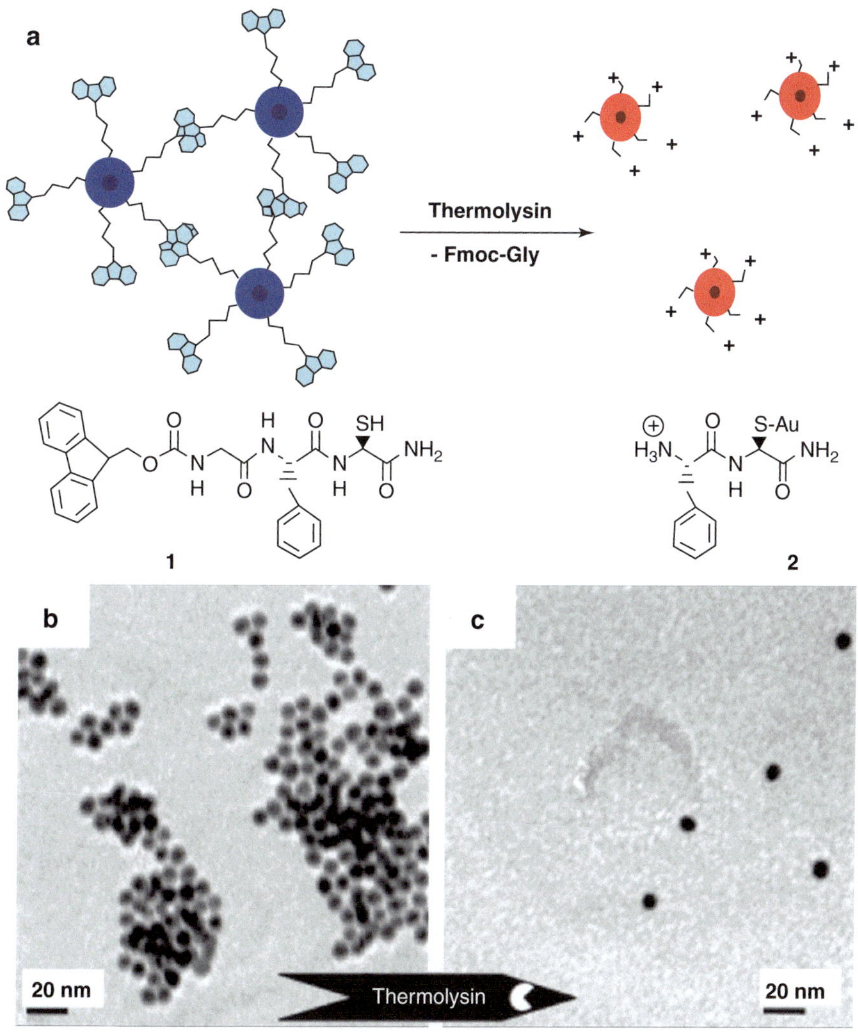

Fig. 12.15 (**a**) Schematic illustration of a real-time protease assay based on the protease-triggered dispersion of AuNPs. (**b**) and (**c**) TEM images of aggregated and dispersed AuNPs, respectively. Reproduced with permission from reference Laromaine et al. (2007)

In conclusion, this chapter aims to summarize recent advances of oligopeptides for biomedical sensing applications, which promote the development of biomedical sensors for "point-of-care" cancer diagnosis. In general, in a biosensing device, the oligopeptides are able to serve as a recognition component that selectively binds to a target molecule, or to serve as a substrate for a protease to recognize and cleave. In the first case, combinatorial libraries (e.g., phage library and OBOC library) are employed to identify the oligopeptide with high binding affinity to the target molecule. In the second case, the oligopeptides are conjugated to other reporting components (e.g., organic dyes, QDs, AuNPs, and NWs) to generate detectable signals after the proteolytic reaction. The development of biomedical sensing devices by using oligopeptides offers new approaches for rapid and sensitive detection of biomarkers which would benefit the disease diagnosis at remote clinical sites where the resources are limited.

Acknowledgments We would like to thank http://muchong.com/ in gathering the relevant literature which are not accessible from the author's affiliation.

References

Acevedo HF, Tong JY, Hartsock RJ (1995) Human chorionic gonadotropin-beta subunit gene expression in cultured human fetal and cancer cells of different types and origins. Cancer 76:1467–1475

Aggarwal S, Janssen S, Wadkins RM, Harden JL, Denmeade SR (2005) A combinatorial approach to the selective capture of circulating malignant epithelial cells by peptide ligands. Biomaterials 26:6077–6086

Aili D, Stevens MM (2010) Bioresponsive peptide-inorganic hybrid nanomaterials. Chem Soc Rev 39:3358–3370

Aina OH, Liu RW, Sutcliffe JL, Marik J, Pan CX, Lam KS (2007) From combinatorial chemistry to cancer-targeting peptides. Mol Pharm 4:631–651

Aina OH, Marik J, Liu R, Lau DH, Lam KS (2005) Identification of novel targeting peptides for human ovarian cancer cells using "one-bead one-compound" combinatorial libraries. Mol Cancer Ther 4:806–813

Alino VJ, Yang KL (2011) Using liquid crystals as a readout system in urinary albumin assays. Analyst 136:3307–3313

Andersson O, Nikkinen H, Kanmert D, Enander K (2009) A multiple-ligand approach to extending the dynamic range of analyte quantification in protein microarrays. Biosens Bioelectron 24:2458–2464

Bagshawe KD (1992) Choriocarcinoma: a model for tumor markers. Acta Oncol 31:99–106

Bai L, Du Y, Peng J, Liu Y, Wang Y, Yang Y, Wang C (2014) Peptide-based isolation of circulating tumor cells by magnetic nanoparticles. J Mater Chem B 2:4080–4088

Bi XY, Hartono D, Yang KL (2008) Controlling orientations of immobilized oligopeptides using N-terminal cysteine labels. Langmuir 24:5238–5240

Birken S, Kovalevskaya G, O'Connor J (2001) Immunochemical measurement of early pregnancy isoforms of hCG: potential applications to fertility research, prenatal diagnosis, and cancer. Arch Med Res 32:635–643

Burda C, Chen X, Narayanan R, El-Sayed MA (2005) Chemistry and properties of nanocrystals of different shapes. Chem Rev 105:1025–1102

Carlton RJ, Hunter JT, Miller DS, Abbasi R, Mushenheim PC, Tan LN, Abbott NL (2013) Chemical and biological sensing using liquid crystals. Liq Cryst Rev 1:29–51

Cerruti M, Jaworski J, Raorane D, Zueger C, Varadarajan J, Carraro C, Lee SW, Maboudian R, Majumdar A (2009) Polymer-oligopeptide composite coating for selective detection of explosives in water. Anal Chem 81:4192–4199

Chen P, Selegard R, Aili D, Liedberg B (2013) Peptide functionalized gold nanoparticles for colorimetric detection of matrilysin (MMP-7) activity. Nanoscale 5:8973–8976

Cheng Y, Meyers JD, Agnes RS, Doane TL, Kenney ME, Broome A-M, Burda C, Basilion JP (2011) Addressing brain tumors with targeted gold nanoparticles: a new gold standard for hydrophobic drug delivery? Small 7:2301–2306

Cortese R, Monaci P, Nicosia A, Luzzago A, Felici F, Galfre G, Pessi A, Tramontano A, Sollazzo M (1995) Identification of biologically active peptides using random libraries displayed on phage. Curr Opin Biotechnol 6:73–80

Cwirla SE, Peters EA, Barrett RW, Dower WJ (1990) Peptides on phage: a vast library of peptides for identifying ligands. Proc Natl Acad Sci U S A 87:6378–6382
Derda R, Tang SKY, Li SC, Ng S, Matochko W, Jafari MR (2011) Diversity of phage-displayed libraries of peptides during panning and amplification. Molecules 16:1776–1803
Derda R, Tang SKY, Whitesides GM (2010) Uniform amplification of phage with different growth characteristics in individual compartments consisting of monodisperse droplets. Angew Chem Int Ed 49:5301–5304
Devlin JJ, Panganiban LC, Devlin PE (1990) Random peptide libraries: a source of specific protein binding molecules. Science 249:404–406
Ding D, Li K, Liu B, Tang BZ (2013) Bioprobes based on AIE fluorogens. Acc Chem Res 46:2441–2453
Ding X, Ge D, Yang K-L (2014) Colorimetric protease assay by using gold nanoparticles and oligopeptides. Sensors Actuators B Chem 201:234–239
Ding XK, Yang KL (2013a) Development of an oligopeptide functionalized surface plasmon resonance biosensor for online detection of glyphosate. Anal Chem 85:5727–5733
Ding XK, Yang KL (2013b) Antibody-free detection of human chorionic gonadotropin by use of liquid crystals. Anal Chem 85:10710–10716
Dover JE, Hwang GM, Mullen EH, Prorok BC, Suh SJ (2009) Recent advances in peptide probe-based biosensors for detection of infectious agents. J Microbiol Methods 78:10–19
Enander K, Choulier L, Olsson AL, Yushchenko DA, Kanmert D, Klymchenko AS, Demchenko AP, Mély Y, Altschuh D (2008) A peptide-based, ratiometric biosensor construct for direct fluorescence detection of a protein analyte. Bioconjug Chem 19:1864–1870
Füzéry AK, Levin J, Chan MM, Chan DW (2013) Translation of proteomic biomarkers into FDA approved cancer diagnostics: issues and challenges. Clin Proteomics 10:13
Felici F, Castagnoli L, Musacchio A, Jappelli R, Cesareni G (1991) Selection of antibody ligands from a large library of oligopeptides expressed on a multivalent exposition vector. J Mol Biol 222:301–310
Fields C, Mallee P, Muzard J, Lee GU (2012) Isolation of Bowman-Birk-inhibitor from soybean extracts using novel peptide probes and high gradient magnetic separation. Food Chem 134:1831–1838
Goldman ER, Pazirandeh MP, Charles PT, Balighian ED, Anderson GP (2002) Selection of phage displayed peptides for the detection of 2,4,6-trinitrotoluene in seawater. Anal Chim Acta 457:13–19
Gray BP, Brown KC (2014) Combinatorial peptide libraries: mining for cell-binding peptides. Chem Rev 114:1020–1081
Guarise C, Pasquato L, De Filippis V, Scrimin P (2006) Gold nanoparticles-based protease assay. Proc Natl Acad Sci U S A 103:3978–3982
Gupta VK, Skaife JJ, Dubrovsky TB, Abbott NL (1998) Optical amplification of ligand-receptor binding using liquid crystals. Science 279:2077–2080
Haarburger D, Pillay TS (2011) Historical perspectives in diagnostic clinical pathology: development of the pregnancy test. J Clin Pathol 64:546–548
Haas W, Adams J, Flannery JB (1970) AC-field-induced Grandjean plane texture in mixtures of room-temperature nematics and cholesterics. Phys Rev Lett 24:577–578
Harris TJ, von Maltzahn G, Derfus AM, Ruoslahti E, Bhatia SN (2006) Proteolytic actuation of nanoparticle self-assembly. Angew Chem Int Ed 45:3161–3165
Herpoldt K-L, Artzy-Schnirman A, Christofferson AJ, Makarucha AJ, de la Rica R, Yarovsky I, Stevens MM (2015) Designing fluorescent peptide sensors with dual specificity for the detection of HIV-1 protease. Chem Mater 27:7187–7195
Hopp TP, Prickett KS, Price VL, Libby RT, March CJ, Cerretti DP, Urdal DL, Conlon PJ (1988) A short polypeptide marker sequence useful for recombinant protein identification and purification. Nat Biotechnol 6:1204–1210
Iqbal SS, Mayo MW, Bruno JG, Bronk BV, Batt CA, Chambers JP (2000) A review of molecular recognition technologies for detection of biological threat agents. Biosens Bioelectron 15:549–578

Jaworski JW, Raorane D, Huh JH, Majumdar A, Lee SW (2008) Evolutionary screening of biomimetic coatings for selective detection of explosives. Langmuir 24:4938–4943

Jerome B (1991) Surface effects and anchoring in liquid crystals. Rep Prog Phys 54:391–451

Jin S, Wan J, Meng L, Huang X, Guo J, Liu L, Wang C (2015) Biodegradation and toxicity of protease/redox/pH stimuli-responsive PEGlated PMAA nanohydrogels for targeting drug delivery. ACS Appl Mater Interfaces 7:19843–19852

Kahn FJ (1973) Orientation of liquid crystals by surface coupling agents. Appl Phys Lett 22:386–388

Kelly KL, Coronado E, Zhao LL, Schatz GC (2003) The optical properties of metal nanoparticles: the influence of size, shape, and dielectric environment. J Phys Chem B 107:668–677

Kim GB, Kim YP (2012) Analysis of protease activity using quantum dots and resonance energy transfer. Theranostics 2:127–138

Krishnamoorthy S (2015) Nanostructured sensors for biomedical applications—a current perspective. Curr Opin Biotechnol 34:118–124

Krizkova S, Zitka O, Masarik M, Adam V, Stiborova M, Eckschlager T, Chavis GJ, Kizek R (2011) Assays for determination of matrix metalloproteinases and their activity. TrAC Trends Anal Chem 30:1819–1832

Kuang ZF, Kim SN, Crookes-Goodson WJ, Farmer BL, Naik RR (2010) Biomimetic chemosensor: designing peptide recognition elements for surface functionalization of carbon nanotube field effect transistors. ACS Nano 4:452–458

Kumar SR, Gallazzi FA, Ferdani R, Anderson CJ, Quinn TP, Deutscher SL (2010) In vitro and in vivo evaluation of cu-64-radiolabeled KCCYSL peptides for targeting epidermal growth factor receptor-2 in breast carcinomas. Cancer Biother Radiopharm 25:693–703

Kumar SR, Quinn TP, Deutscher SL (2007) Evaluation of an in-111-radiolabeled peptide as a targeting and imaging agent for ErbB-2 receptor-expressing breast carcinomas. Clin Cancer Res 13:6070–6079

Kwok RTK, Leung CWT, Lam JWY, Tang BZ (2015) Biosensing by luminogens with aggregation-induced emission characteristics. Chem Soc Rev 44:4228–4238

Lam KS, Lebl M, Krchňák V (1997) The "one-bead-one-compound" combinatorial library method. Chem Rev 97:411–448

Lam KS, Liu RW, Miyamoto S, Lehman AL, Tuscano JM (2003) Applications of one-bead one-compound combinatorial libraries and chemical microarrays in signal transduction research. Acc Chem Res 36:370–377

Lam KS, Salmon SE, Hersh EM, Hruby VJ, Kazmierski WM, Knapp RJ (1991) A new type of synthetic peptide library for identifying ligand-binding activity. Nature 354:82–84

Laromaine A, Koh LL, Murugesan M, Ulijn RV, Stevens MM (2007) Protease-triggered dispersion of nanoparticle assemblies. J Am Chem Soc 129:4156–4157

Lau D, Guo L, Liu R, Marik J, Lam K (2006) Peptide ligands targeting integrin α3β1 in non-small cell lung cancer. Lung Cancer 52:291–297

Lee SS, Lim J, Tan S, Cha J, Yeo SY, Agnew HD, Heath JR (2010) Accurate MALDI-TOF/TOF sequencing of one-bead-one-compound peptide libraries with application to the identification of multiligand protein affinity agents using in situ click chemistry screening. Anal Chem 82:672–679

Lee SW, Mao CB, Flynn CE, Belcher AM (2002) Ordering of quantum dots using genetically engineered viruses. Science 296:892–895

Lefkowitz RB, Marciniak JY, CM H, Schmid-Schonbein GW, Heller MJ (2010) An electrophoretic method for the detection of chymotrypsin and trypsin activity directly in whole blood. Electrophoresis 31:403–410

Li H, Li W, Liu F, Wang Z, Li G, Karamanos Y (2016) Detection of tumor invasive biomarker using a peptamer of signal conversion and signal amplification. Anal Chem 88:3662–3668

Liu R, Marik J, Lam KS (2002) A novel peptide-based encoding system for "one-bead one-compound" peptidomimetic and small molecule combinatorial libraries. J Am Chem Soc 124:7678–7680

Lock LL, Reyes CD, Zhang P, Cui H (2016) Tuning cellular uptake of molecular probes by rational design of their assembly into supramolecular nanoprobes. J Am Chem Soc 138:3533–3540

Lockwood NA, Gupta JK, Abbott NL (2008) Self-assembly of amphiphiles, polymers and proteins at interfaces between thermotropic liquid crystals and aqueous phases. Surf Sci Rep 63:255–293

Lofas S (1995) Dextran modified self-assembled monolayer surfaces for use in biointeraction analysis with surface plasmon resonance. Pure Appl Chem 67:829–834

Luo J, Xie Z, Lam JWY, Cheng L, Chen H, Qiu C, Kwok HS, Zhan X, Liu Y, Zhu D, Tang BZ (2001) Aggregation-induced emission of 1-methyl-1,2,3,4,5-pentaphenylsilole. Chem Commun:1740–1741

McGuire MJ, Samli KN, Chang Y-C, Brown KC (2006) Novel ligands for cancer diagnosis: selection of peptide ligands for identification and isolation of B-cell lymphomas. Exp Hematol 34:443–452

Medintz IL, Clapp AR, Brunel FM, Tiefenbrunn T, Tetsuo Uyeda H, Chang EL, Deschamps JR, Dawson PE, Mattoussi H (2006) Proteolytic activity monitored by fluorescence resonance energy transfer through quantum-dot-peptide conjugates. Nat Mater 5:581–589

Mei J, Hong Y, Lam JWY, Qin A, Tang Y, Tang BZ (2014) Aggregation-induced emission: the whole is more brilliant than the parts. Adv Mater 26:5429–5479

Merrifield RB (1963) Solid phase peptide synthesis. I. The synthesis of a tetrapeptide. J Am Chem Soc 85:2149–2154

Mostert B, Sleijfer S, Foekens JA, Gratama JW (2009) Circulating tumor cells (CTCs): detection methods and their clinical relevance in breast cancer. Cancer Treat Rev 35:463–474

Mullen LM, Nair SP, Ward JM, Rycroft AN, Henderson B (2006) Phage display in the study of infectious diseases. Trends Microbiol 14:141–147

Na YR, Kim SY, Gaublomme JT, Shalek AK, Jorgolli M, Park H, Yang EG (2013) Probing enzymatic activity inside living cells using a nanowire-cell "sandwich" assay. Nano Lett 13:153–158

Pan YL, Guo ML, Nie Z, Huang Y, Peng Y, Liu AF, Qing M, Yao SZ (2012) Colorimetric detection of apoptosis based on caspase-3 activity assay using unmodified gold nanoparticles. Chem Commun 48:997–999

Parmley SF, Smith GP (1988) Antibody-selectable filamentous fd phage vectors: affinity purification of target genes. Gene 73:305–318

Paterlini-Brechot P, Benali NL (2007) Circulating tumor cells (CTC) detection: clinical impact and future directions. Cancer Lett 253:180–204

Peng L, Liu R, Marik J, Wang X, Takada Y, Lam KS (2006) Combinatorial chemistry identifies high-affinity peptidomimetics against [alpha]4[beta]1 integrin for in vivo tumor imaging. Nat Chem Biol 2:381–389

Pillutla RC, Hsiao KC, Beasley JR, Brandt J, Ostergaard S, Hansen PH, Spetzler JC, Danielsen GM, Andersen AS, Brissette RE, Lennick M, Fletcher PW, Blume AJ, Schaffer L, Goldstein NI (2002) Peptides identify the critical hotspots involved in the biological activation of the insulin receptor. J Biol Chem 277:22590–22594

Sankaran S, Panigrahi S, Mallik S (2011) Odorant binding protein based biomimetic sensors for detection of alcohols associated with Salmonella contamination in packaged beef. Biosens Bioelectron 26:3103–3109

Scarberry KE, Dickerson EB, McDonald JF, Zhang ZJ (2008) Magnetic nanoparticle-peptide conjugates for in vitro and in vivo targeting and extraction of cancer cells. J Am Chem Soc 130:10258–10262

Schafer A, Pahnke A, Schaffert D, van Weerden WM, de Ridder CMA, Rodl W, Vetter A, Spitzweg C, Kraaij R, Wagner E, Ogris M (2011) Disconnecting the yin and yang relation of epidermal growth factor receptor (EGFR)-mediated delivery: a fully synthetic, EGFR-targeted gene transfer system avoiding receptor activation. Hum Gene Ther 22:1463–1473

Scott JK, Smith GP (1990) Searching for peptide ligands with an epitope library. Science 249:386–390

Shi H, Kwok RTK, Liu J, Xing B, Tang BZ, Liu B (2012) Real-time monitoring of cell apoptosis and drug screening using fluorescent light-up probe with aggregation-induced emission characteristics. J Am Chem Soc 134:17972–17981

Smith GP (1985) Filamentous fusion phage: novel expression vectors that display cloned antigens on the virion surface. Science 228:1315–1317

Smith GP, Scott JK (1993) Libraries of peptides and proteins displayed on filamentous phage. Methods Enzymol 217:228–257

Takakusagi Y, Kobayashi S, Sugawara F (2005) Camptothecin binds to a synthetic peptide identified by a T7 phage display screen. Bioorg Med Chem Lett 15:4850–4853

Tsampalas M, Gridelet V, Berndt S, Foidart JM, Geenen V, d'Hauterive SP (2010) Human chorionic gonadotropin: a hormone with immunological and angiogenic properties. J Reprod Immunol 85:93–98

Underwood S, Mulvaney P (1994) Effect of the solution refractive index on the color of gold colloids. Langmuir 10:3427–3430

Vassar R, Bennett BD, Babu-Khan S, Kahn S, Mendiaz EA, Denis P, Teplow DB, Ross S, Amarante P, Loeloff R, Luo Y, Fisher S, Fuller L, Edenson S, Lile J, Jarosinski MA, Biere AL, Curran E, Burgess T, Louis JC, Collins F, Treanor J, Rogers G, Citron M (1999) Beta-secretase cleavage of Alzheimer's amyloid precursor protein by the transmembrane aspartic protease BACE. Science 286:735–741

Vickers CJ, Gonzalez-Paez GE, Wolan DW (2013) Selective detection of caspase-3 versus caspase-7 using activity-based probes with key unnatural amino acids. ACS Chem Biol 8:1558–1566

Vosyka O, Vinothkumar KR, Wolf EV, Brouwer AJ, Liskamp RMJ, Verhelst SHL (2013) Activity-based probes for rhomboid proteases discovered in a mass spectrometry-based assay. Proc Natl Acad Sci U S A 110:2472–2477

Wang M, Zhang G, Zhang D, Zhu D, Tang BZ (2010) Fluorescent bio/chemosensors based on silole and tetraphenylethene luminogens with aggregation-induced emission feature. J Mater Chem 20:1858–1867

Wang W, Wei Z, Zhang D, Ma H, Wang Z, Bu X, Li M, Geng L, Lausted C, Hood L, Fang Q, Wang H, Hu Z (2014) Rapid screening of peptide probes through in situ single-bead sequencing microarray. Anal Chem 86:11854–11859

Wang X, Peng L, Liu R, Gill SS, Lam KS (2005) Partial alloc-deprotection approach for ladder synthesis of "one-bead one-compound" combinatorial libraries. J Comb Chem 7:197–209

Wang ZH, Wang WZ, XL B, Wei ZW, Geng LL, Wu Y, Dong CY, Li LQ, Zhang D, Yang S, Wang F, Lausted C, Hood L, Hu Z (2015) Microarray based screening of peptide nano probes for HER2 positive tumor. Anal Chem 87:8367–8372

Warren AD, Kwong GA, Wood DK, Lin KY, Bhatia SN (2014) Point-of-care diagnostics for non-communicable diseases using synthetic urinary biomarkers and paper microfluidics. Proc Natl Acad Sci U S A 111:3671–3676

Whaley SR, English DS, EL H, Barbara PF, Belcher AM (2000) Selection of peptides with semiconductor binding specificity for directed nanocrystal assembly. Nature 405:665–668

Wong RC, Tse HY (2009) Lateral flow immunoassay. Humana Press, New York, NY

Yu M, Stott S, Toner M, Maheswaran S, Haber DA (2011) Circulating tumor cells: approaches to isolation and characterization. J Cell Biol 192:373–382

Zauner T, Berger-Hoffmann R, Muller K, Hoffmann R, Zuchner T (2011) Highly adaptable and sensitive protease assay based on fluorescence resonance energy transfer. Anal Chem 83:7356–7363

Zhang CQ, Zheng L, Nurnberg J, Vacari BM, Zhou JZ, Wang Y (2014) Cleavage of pro-tumor necrosis factor alpha by ADAM metallopeptidase domain 17: a fluorescence-based protease assay cleaves its natural protein substrate. Anal Biochem 445:14–19

Zhang P, Cheetham AG, Lock LL, Li Y, Cui H (2015) Activatable nanoprobes for biomolecular detection. Curr Opin Biotechnol 34:171–179

Zhao WT, Yao CL, Luo XT, Lin L, Hsing IM (2012) Staining-free gel electrophoresis-based multiplex enzyme assay using DNA and peptide dual-functionalized gold nanoparticles. Electrophoresis 33:1288–1291

Zhou W, Gao X, Liu D, Chen X (2015) Gold nanoparticles for in vitro diagnostics. Chem Rev 115:10575–10636

Zou Q, Yang K-L (2016) Identification of peptide inhibitors of penicillinase using a phage display library. Anal Biochem 494:4–9

Microfluidic Immunoassay Devices as Next-Generation Cancer and Medical Diagnostics Platform

13

Toshihiro Kasama, Yoshinobu Baba, and Manabu Tokeshi

13.1 Introduction

Immunoassays are applied for medical diagnostics, food safety testing, drug discovery, biological researches, etc. (Wild 2005) and they show some of the most remarkable activities in the field of lab-on-a-chip systems and micro total analysis systems (Bange et al. 2005; Chin et al. 2007; Henares et al. 2008; Tachi et al. 2007). Miniaturization of immunoassay systems enables rapid and highly sensitive analysis with a small amount of sample and reagents.

Since the first study of chip-based immunoassay (Chiem and Harrison 1997), immobilization of antibody on the surface of microbeads has contributed to the improvement of detection sensitivity and assay time (Sato et al. 2000, 72; Moorthy et al. 2004; Haes et al. 2006; Shin et al. 2007; Thompson and Bau 2010). This is called the bead-bed format. We have published papers on the subject of the

T. Kasama (✉)
School of Engineering, The University of Tokyo, Tokyo, Japan

Graduate School of Engineering, Nagoya University, Nagoya, Japan
e-mail: kasama.toshihiro@kk.alumni.u-tokyo.ac.jp

Y. Baba
Graduate School of Engineering, Nagoya University, Nagoya, Japan

ImPACT Research Center for Advanced Nanobiodevices, Nagoya University, Nagoya, Japan

Graduate School of Medicine, Nagoya University, Nagoya, Japan

National Institute of Advanced Industrial Science and Technology, Takamatsu, Japan
e-mail: babaymtt@apchem.nagoya-u.ac.jp

M. Tokeshi
ImPACT Research Center for Advanced Nanobiodevices, Nagoya University, Nagoya, Japan

Faculty of Engineering, Hokkaido University, Sapporo, Japan
e-mail: tokeshi@eng.hokudai.ac.jp

P. Chandra et al. (eds.), *Next Generation Point-of-care Biomedical Sensors Technologies for Cancer Diagnosis*, DOI 10.1007/978-981-10-4726-8_13

bead-bed format immunoassay devices, in which the capability of this format to detect human disease markers has been demonstrated (Sato et al. 2001, 2002, 2004; Kakuta et al. 2006; Ohashi et al. 2009; Ihara et al. 2010). However, there exist some difficulties for this format in liquid handling, though highly sensitive detection and rapid assay are achieved. In order to pack the microbeads inside the microchannel, it is necessary to apply relatively high pressure. In addition, the removal of bubbles from liquid is difficult. Therefore, the development of novel immunoassay chips is desired for clinical applications such as point-of-care (POC) testing (Delamarche et al. 2005; Linder et al. 2005; Hosokawa et al. 2006; Gervais and Delamarche 2009). On the other hand, three-dimensional (3-D) hydrogel-based immunoassay chips have been reported (Zubtsov et al. 2006; Sung et al. 2009). They showed that the immobilization of antibodies within 3-D hydrogel structures offers several advantages, such as high immobilization capacity and high antibody activity, over 2-D immobilization.

In this chapter, we introduce two types of new immunoassay microdevices, both of which can overcome difficulties mentioned above. One is 3-D hydrogel structures holding antibody-immobilized microbeads. Another device is 3-D hydrogel structures with chemically bonded antibodies.

13.2 Microbead-Embedded Immunoassay Devices

13.2.1 Fabrication Procedure of Microbead-Embedded Immunoassay Devices

Antibody-immobilized beads were prepared with polystyrene beads (1 μm in diameter) and antibody solution in a 1.5 mL microtube. The antibody solution was added to the microbeads, and the suspension was rotated gently at room temperature, followed by overnight incubation at 4 °C. After the incubation, the antibodies were immobilized on the microbeads, but there was space among the antibody molecules for nonspecific adsorption of proteins such as antigen and detection antibody. In order to prevent nonspecific adsorption, the microbeads were immersed in 1% BSA for 45 min at room temperature.

Photocross-linkable prepolymer, photoinitiator, and Millipore water were mixed in another 1.5 mL microtube and the mixture was stirred using a vortex mixer. The mixture was added to the antibody-immobilized microbead solution. Then, the mixture was stirred gently at room temperature. This solution was used to fabricate microbead-embedded immunoassay devices. Fabrication steps of the devices are shown in Fig. 13.1. First, 250 nL of the solution was injected into the microchannel by using a pipette. Second, UV light (365 nm, 20 mW) was irradiated through a photomask covering the microchannel. This process took approximately 10 s. The exposed areas became hydrogel structures which included many antibody-immobilized microbeads. Third, the non-polymerized solution was sucked by using a vacuum pump, and the surface of the microchannel was flushed

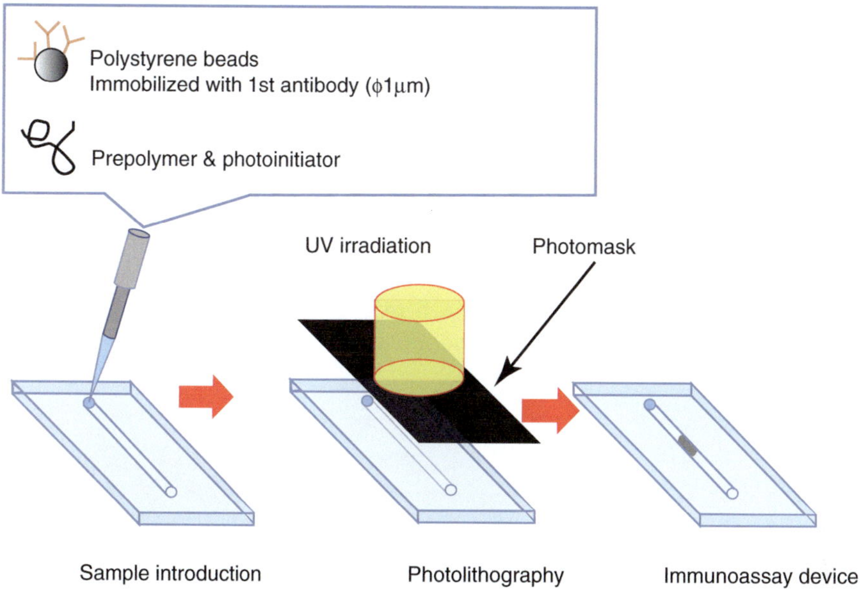

Fig. 13.1 Fabrication steps of the microbead-embedded immunoassay devices

with PBS. Finally, 1% BSA in PBS was injected and then kept in the microchannel for 1 h at room temperature. BSA prevented nonspecific binding of antigens and detection antibodies to the surface of the microchannel and hydrogel structures. After removing the BSA solution, the microchannel was flushed with PBS. Although the hydrogel structures were physically fixed between the roof and the floor of the microchannel, they did not move or break during the immunoassay.

13.2.2 Assay Procedures

First, 250 nL of the sample solution was injected into the microchannel with a pipette. After incubation, free antigen molecules in the sample solution were sucked with an aspirator, and then the microchannel was flushed three times with PBS. Second, 250 nL (1 μg mL^{-1}) of the fluorescent-labeled detection antibody solution was injected into the microchannel. After the incubation, the microchannel was flushed three times with PBS to remove the free fluorescent-labeled secondary antibody molecules. Finally, the fluorescence signal from the microbeads in hydrogel structures was detected by using a fluorescence microscope equipped with a CCD camera and three lasers (488, 532, and 632.8 nm). By using ImageJ software, the fluorescent intensity per unit area was calculated for each hydrogel structure.

13.2.3 Immuno-Pillar Device

Firstly, we developed pillar-like hydrogel structure by using a photomask shown in Fig. 13.2. We call this immunoassay chip the immuno-pillar device (Ikami et al. 2010). Each immuno-pillar has a dimension of 200 μm in diameter and 40 μm in height. The arrangement of five immuno-pillars is shown in Fig. 13.3.

13.2.3.1 Immunoassay of Disease Markers

First, we evaluated the performance of the immuno-pillar device for standard C-reactive protein (CRP) solutions (1% BSA in PBS). CRP is a well-known disease marker relating cardiac events and inflammation. The calibration curves for

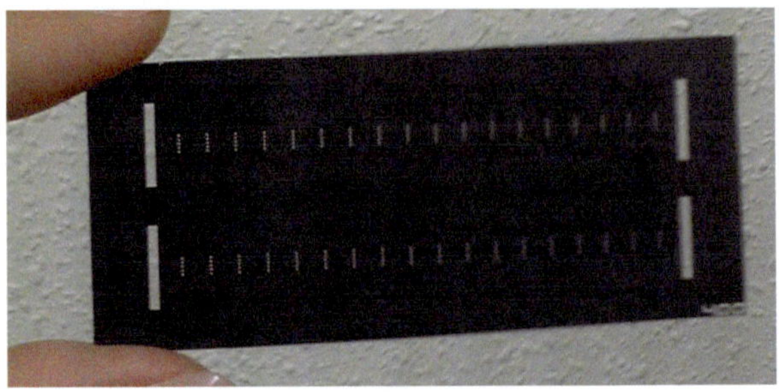

Fig. 13.2 Picture of photomask prepared for fabricating the immuno-pillar devices

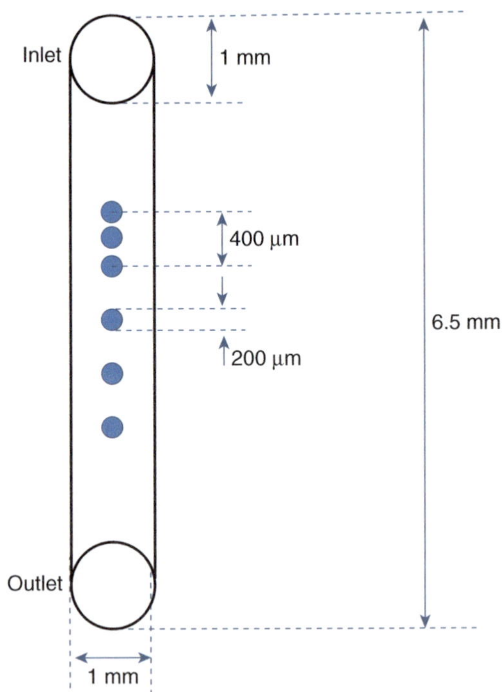

Fig. 13.3 Schematic representation of the immuno-pillar device

standard CRP solutions are shown in Fig. 13.4a–c. In our assay procedures, total assay time was calculated by adding all the times for the incubations, washings, and detection. The fluorescence intensity represents the average of the fluorescence signal intensities of 3–5 immuno-pillars. The error bar denotes their standard deviation. The background is the sum of the autofluorescence of plastic substrate, antibody-immobilized microbeads, and UV curable resin. The total assay times were 4 min, 8 min, and 12 min, respectively. Despite the very short assay times, fluorescence signal and CRP concentration have positive correlation. The calibration curve of (B) (total assay time: 8 min) was very similar to that of (A) (12 min). From (A), (B), and (C) (4 min), the limit of detection (LOD), which gave a signal at 3 SDs (standard deviations) above the background, was estimated to be 100 pg mL^{-1}; the slope of the calibration curve of (C) was gentle. The immuno-pillar devices demonstrated the ability to detect disease marker with high sensitivity and rapidity in spite of easy assay procedure. For actual diagnosis of several diseases, the cutoff values of CRP concentration are higher than 100 pg mL^{-1} (Gabay and Kushner 1999). Also, we could change the detection range of the immuno-pillar device by using a lower concentration of fluorescence-labeled secondary antibody or a lower power of the excitation laser beam (data not shown). In particular, shifting the detection range to higher sample concentrations is easier than that to lower concentrations. The features of the immuno-pillar device of rapid assay and high sensitivity were derived from the immuno-pillar itself and the 1 μm diameter polystyrene microbeads for the immobilization of capture antibodies. The pore size of the immuno-pillars was likely 100 nm or more because the fluorescence beads with diameter of 100 nm leaked from the immuno-pillars in our preliminary experiments. The diffusion kinetics of the antigen and antibody within the immuno-pillars was not slow (Fig. 13.5). Therefore, protein molecules such as the antigen and antibody could easily penetrate into the immuno-pillars and could diffuse within the immuno-pillars. According to our calculation for the present experimental conditions, the number of microbeads within the immuno-pillar was estimated to be about 32,700. By using the surface of these microbeads, the number of reaction sites for an antigen–antibody reaction was dramatically increased.

Next, we tested the performance of the immuno-pillar device for serum samples which were spiked with the known concentrations of CRP. Figure 13.4 d–f shows the calibration curves for serum samples with CRP. The immuno-pillar devices showed good performance also for serum samples. Influence of proteins in the serum may cause the scattering of the signal intensity in the high-concentration region. The LOD for the total assay time of 4, 8, and 12 was 100 pg mL^{-1}.

In addition, we also evaluated the performance of the immuno-pillar devices for the standard and serum samples of alpha-fetoprotein (tumor marker) and prostate-specific antigen (prostate cancer marker). These results are summarized in Table 13.1. In summary, we can conclude that the immuno-pillar devices had great potential for tests of serum samples and would be suitable as an immunoassay device for POC diagnostics because it was quick, had high sensitivity, was easy to use, and needed only small sample and reagent volumes.

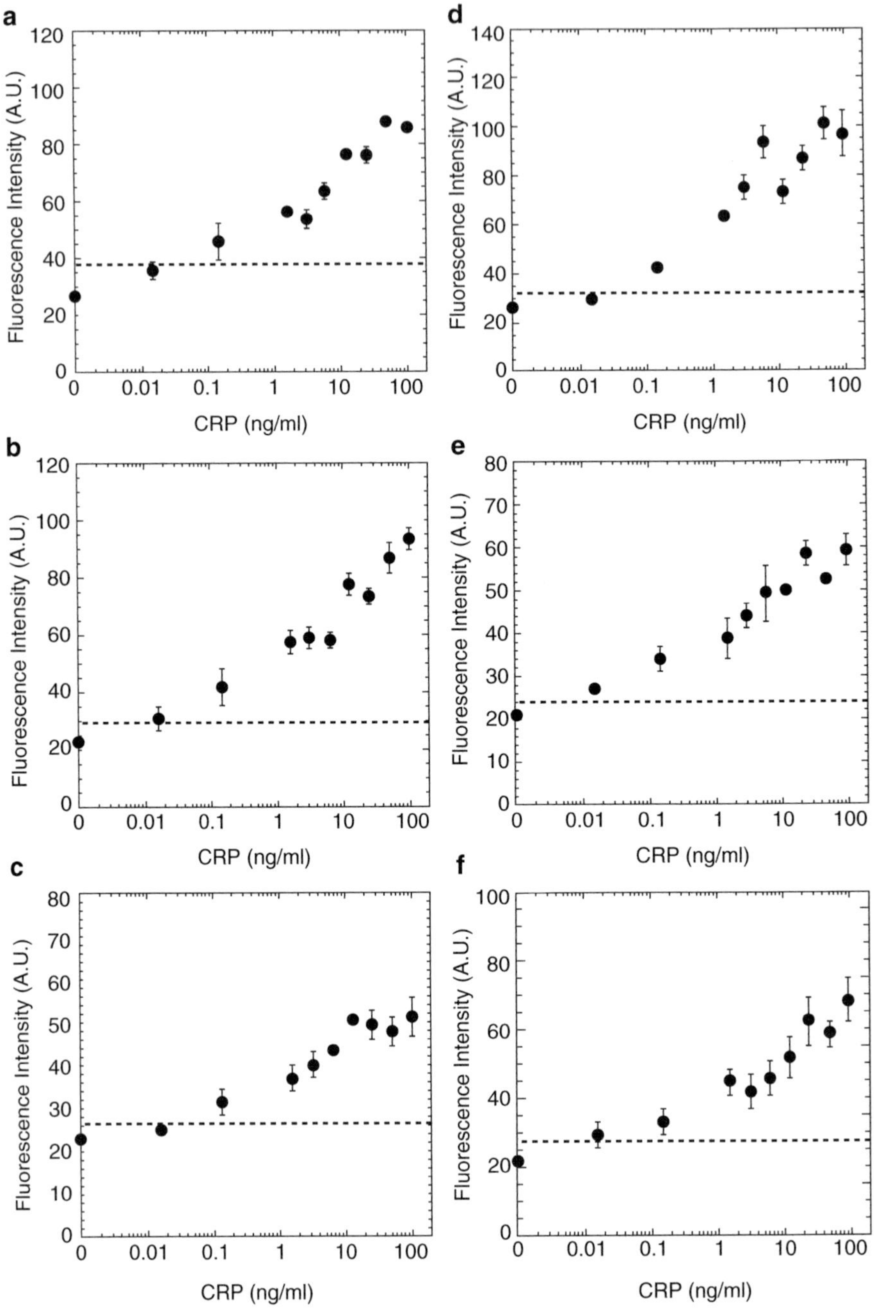

Fig. 13.4 Calibration curves obtained with CRP of standard samples (**a**–**c**) and of serum samples (**d**–**f**) (adapted from reference Ikami et al. (2010)). Total assay times were (**a**) 12 min, (**b**) 8 min, (**c**) 4 min, (**d**) 12 min, (**e**) 8 min, and (**f**) 4 min. The *dashed line* represents the signal level at 3 SDs above the background

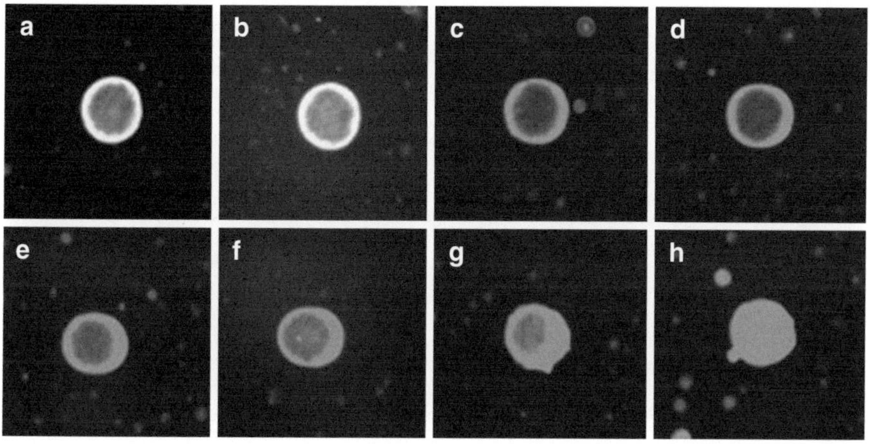

Fig. 13.5 Fluorescence images of the immuno-pillar at the fluorescence-labeled antibody immersion time of (**a**) 40, (**b**) 60, (**c**) 80, (**d**) 100, (**e**) 120, (**f**) 140, (**g**) 160, and (**h**) 180 s (reproduced with permission from reference Ikami et al. (2010))

Table 13.1 Detection sensitivity of the immuno-pillar devices (adapted from reference Ikami et al. (2010))

Sample		Tota assay time		
		4 min	8 min	12 min
CRP	In 1% BSA-PBS	~100 pg mL^{-1}	~100 pg mL^{-1}	~100 pg mL^{-1}
	In serum	~100 pg mL^{-1}	~100 pg mL^{-1}	~100 pg mL^{-1}
AFP[a]	In 1% BSA-PBS	~100 pg mL^{-1}	~100 pg mL^{-1}	~100 pg mL^{-1}
	In serum	~1 ng mL^{-1}	~1 ng mL^{-1}	~100 pg mL^{-1}
PSA[a]	In 1% BSA-PBS	~5 ng mL^{-1}	~1 ng mL^{-1}	~100 pg mL^{-1}
	In serum	~5 ng mL^{-1}	~5 ng mL^{-1}	~100 pg mL^{-1}
Triplex	In serum	—	—	~100 pg mL^{-1}

[a]In the assay of AFP and PSA, the concentration of the fluorescent-labeled detection antibody solution was 50 μg mL^{-1} and 50 μg mL^{-1}, respectively

13.2.3.2 Multiplex Immunoassay of Disease Markers

The immuno-pillar devices are also available to perform multiplex assay. For example, if the immuno-pillars hold three kinds of microbeads (three different antibodies are immobilized), a triplex assay becomes possible. Schematic illustration of the immuno-pillar device for the triplex assay is depicted in Fig. 13.6a. Fabrication process of this immuno-pillar device is the same as that of the above-mentioned devices. Therefore, the number of each kind of microbeads in the immuno-pillar is one-third, ca. 10,000. We fabricated a suitable immuno-pillar device and performed triplex simultaneous assay for CRP, AFP, and PSA. 250 nL of serum solution which was spiked with CRP, AFP, and PSA was used as the sample. 250 nL of the mixture solution of fluorescence dye-labeled antibodies was used as the detection antibody solution. The incubation time was constant at 5 min. Thus, the assay for one sample

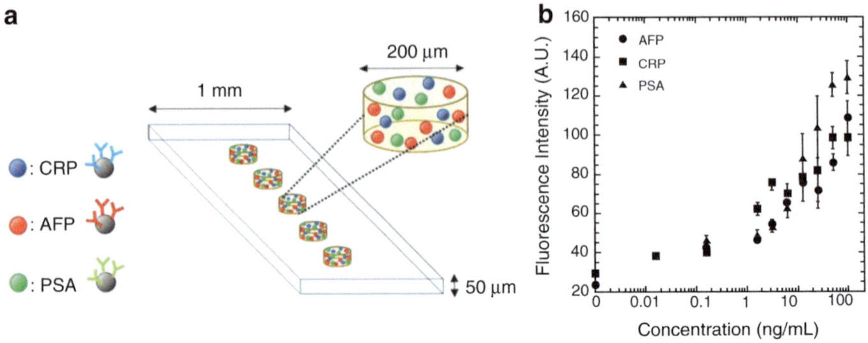

Fig. 13.6 (**a**) Simplified schematic of the immuno-pillar device for the triplex assay. (**b**) Calibration curves of CRP, AFP, and PSA (adapted from reference Ikami et al. (2010))

was finished in 15 min. The results of the multiplex assay are shown in Fig. 13.6b. All three calibration curves showed the positive correlations between the fluorescence signal and the sample concentration, and the LOD for each was 100 pg mL^{-1}. It should be noted that the LODs for three markers in the multiplex assay were almost the same as that of the single assay. From this analysis, we could conclude that the immuno-pillar devices had great potential also for multiplex assay of serum samples. Moreover, optimization of the number of microbeads and/or the concentrations of detection antibodies may lead to the shortening of assay time.

13.2.3.3 Immunoassay of Toxins in Food

Immuno-pillar devices could be applied to detect toxins in dairy products (Jin et al. 2013). Here we attempt to detect staphylococcal enterotoxins (SEs) in milk by using chicken immunoglobulin IgY anti-SE antibody as capture antibody. IgY antibodies, unlike mammalian IgG antibodies, do not combine with protein A because they possess a different structure of the Fc region, thereby avoiding nonspecific reactivity against *Staphylococcus aureus*. We performed detection tests against SEs in milk. The total assay time was approximately 12 min. The calibration curves for SEs are summarized in Fig. 13.7. These results show that fluorescence intensities increased in a dose-dependent manner for SEs (0–100 ng/mL) in milk. In all cases, each immuno-pillar device could detect the corresponding SEs with high sensitivity. The LODs for SEA, SEB, SEC, SED, and SEE are summarized in Table 13.2. High specificity of the SE immuno-pillar devices was confirmed by measuring the cross-reactivity against the comparative antigens of 100 ng mL^{-1}.

13.2.3.4 Multiplex Immunoassay of Toxins in Food

Several kinds of SEs often coexist in polluted foods. Therefore, we evaluated the ability of the immuno-pillar devices to detect SEA, SEB, and SED simultaneously (Kasama et al. 2015a). These are the three worst factors of SE poisoning. In order to simulate contaminated milk, standard, native SEs with more than 95% purity were diffused in commercially available milk. The resulting calibration plots for SEA,

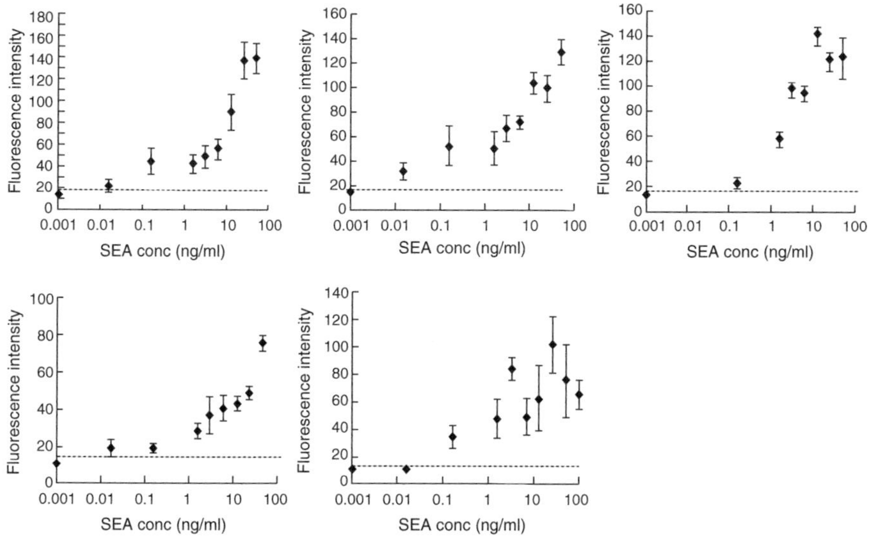

Fig. 13.7 Calibration plots of standard SEs in milk (reproduced with permission from reference Jin et al. (2013)). All tests were performed in triplicate, and *error bars* show standard deviation calculated from fluorescence intensities of 3–5 immuno-pillars. *Dashed lines* represent the signal levels at 3 SDs above the background

Table 13.2 Detection limits of immuno-pillar device (ng mL^{-1}) (adapted from reference Jin et al. (2013))

	SEA	SEB	SEC	SED	SEE
Immuno pillar device (in PBS)	0.01	0.1	0.1	0.1	0.1
Immuno pillar device (in milk)	0.1	0.01	0.1	0.1	0.1

SEB, and SED were obtained (Fig. 13.8). The tests for three replicates per sample were performed. The values and SDs of the fluorescence intensity were calculated from these results. The LOD for each SE was calculated to be 15.6 pg mL^{-1}, which is lower than not only those of common SE detection methods (Jin et al. 2013; Rose et al. 1989; Kuang et al. 2013), but also those of the immuno-pillar devices for single assays of SEA and SED (Jin et al. 2013). Relaxation of self-quenching (Chen and Knutson 1988) enhances fluorescence intensity, resulting in lower LODs. In the immuno-pillars, the antibody-immobilized microbeads formed clusters via hydrophobic interaction between the antibodies. In the case of multiplex immunoassay devices, microbead clusters were composed of three kinds of microbeads supporting each anti-SE antibody. Consequently, the distance between the same fluorescence-labeled antibodies was relatively extended in the multiplex immunoassay devices. The LOD for each SE is much lower than the lowest SE concentration in major food poisoning outbreaks (380 pg mL^{-1}) (Asao et al. 2003). Therefore, contaminated food that may potentially cause a food poisoning outbreak could be immediately identified by immunoassay by using the immuno-pillar devices.

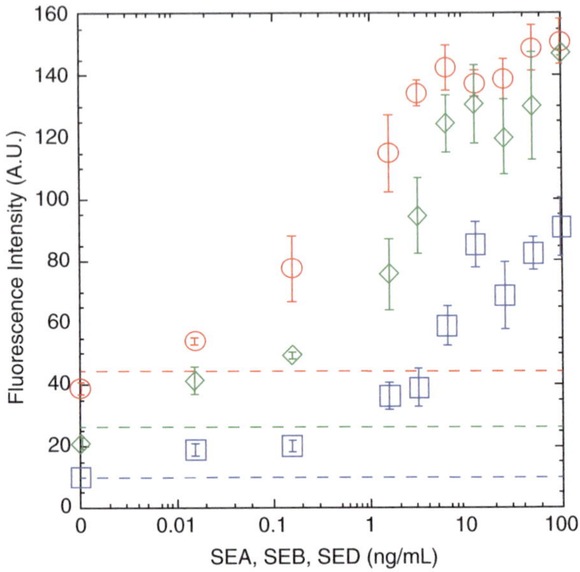

Fig. 13.8 Calibration plots for SEA (*red*), SEB (*blue*), and SED (*green*) (reproduced with permission from reference Kasama et al. (2015a)). The detection limits are represented by the *dashed lines* and are estimated as 3 SD above the backgrounds

13.2.4 Immuno-Wall Device

13.2.4.1 Immunoassay of Disease Markers

In order to improve the efficiency of bound-free (BF) separation, we modified the structure of immuno-pillars. Here, we propose immuno-wall device which has a long and thin hydrogel object inside a microchannel (Fig. 13.9) (Kasama et al. 2014). Unreacted antigens and fluorescence-labeled antibodies were completely removed by just immersing the device in a washing buffer (PBS with 0.5% Tween 20) for 1 min. In addition, the long structure also allowed us to analyze fluorescence intensity by using inexpensive desktop fluorescence scanner instead of expensive fluorescence microscopes.

First, we compared the efficiencies of BF separation of the immuno-wall and immuno-pillar devices. We prepared DyLight 650-conjugated anti-rabbit IgG antibody solution (50 μg mL^{-1}), which did not react to the antibody immobilized on the microbead surface. The devices underwent immersion in the solution (30 s) and the washing buffer (several minutes). After that, fluorescence intensity was measured by using a fluorescence microscope. The results are shown in Fig. 13.10. This figure shows that the unreacted fluorescence-labeled antibodies exited from the immuno-wall within 1 min by simply immersing the device into a washing buffer. In contrast, fluorescence-labeled antibodies could not exit from the immuno-pillar even after 9-min immersion. In order to completely remove non-reacted antibodies from immuno-pillars, it is necessary to squeeze the immuno-pillar by aspiration.

The long structure of the immuno-wall devices allowed us to determine fluorescence intensity by simple fluorescence scanner. Recently, we have developed microchip-optimized fluorescence scanner (Fig. 13.11). This scanner scanned on a line of 20 mm length and 100 μm width and obtained the profile of fluorescence

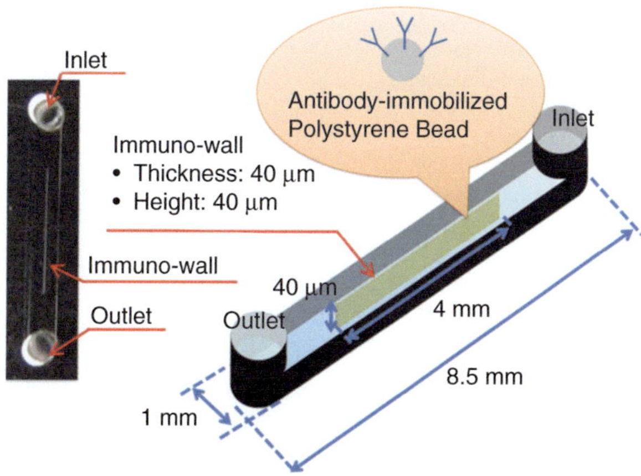

Fig. 13.9 Photograph and schematic of the immuno-wall device (reproduced with permission from reference Kasama et al. (2014)). Free antigens and fluorescence-labeled antibodies were easily removed from both sides of the immuno-wall

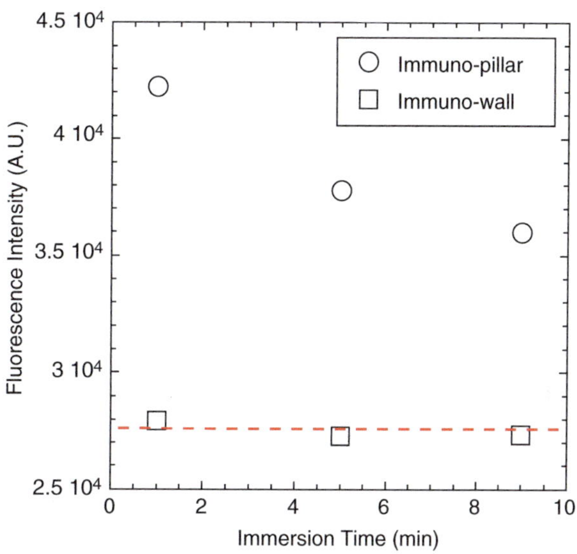

Fig. 13.10 Efficiencies of bound-free separation. The *dashed line* shows the autofluorescence intensity of the immuno-pillar and immuno-wall devices (reproduced with permission from reference Kasama et al. (2014))

intensity within 1 min. It was easy to scan across the immuno-wall devices having the dimension of 4 mm in length. In contrast, it was difficult to scan the immuno-pillar devices as small as 200 μm in diameter because this reader had no objective lens.

By using the immuno-wall devices and the fluorescence scanner, CRP assays for human sera were performed. The total assay time was 10 min. CRP in human sera were quantitatively analyzed and the calibration curve was obtained (Fig. 13.12). The fluorescence intensity was obtained by averaging the fluorescence signal intensities of 3–5 areas of the immuno-wall. We achieved the LOD of 10 ng mL^{-1}. In addition, the present immunoassay system provided good quantitative capability between 10 ng/mL and 10 μg mL^{-1}, offering the application possibility for rapid CRP test.

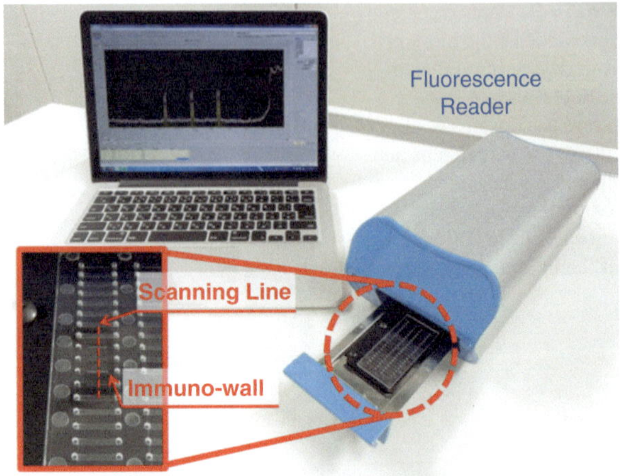

Fig. 13.11 Photograph of the fluorescence reader. Specification of the reader is as follows: weight = 2.1 kg, dimension ($W \times H \times D$) = 130 × 95 × 260 (mm), and maximal excitation wavelength = 637 nm. In the inset, the *dashed line* represents the scanning area (length = 20 mm, width = 100 μm) (reproduced with permission from reference Kasama et al. (2014))

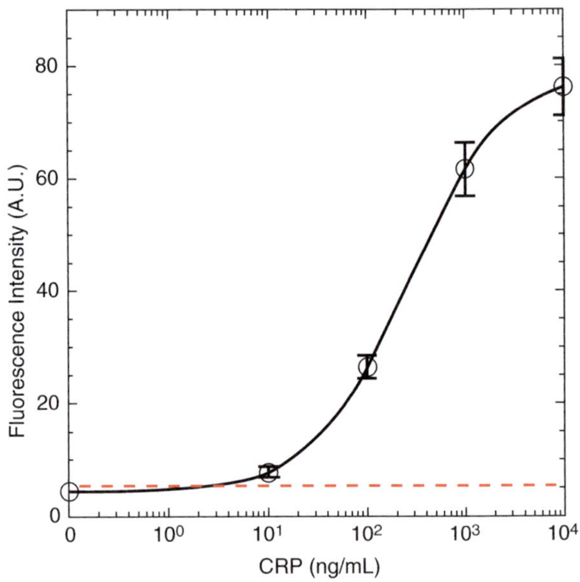

Fig. 13.12 Calibration curve for CRP in human serum. Sigmoidal curve was calculated by the four-parameter logistic equation. The immunoassay was carried out three times at each concentration (reproduced with permission from reference Kasama et al. (2014))

13.3 Immuno-Wall Devices with Chemically Bonded Antibodies

13.3.1 Fabrication Procedure of Immunoassay Devices with Chemically Bonded Antibodies

Until now, we presented the immuno-pillar devices and immuno-wall devices which held antibody-immobilized microbeads in their pores. In this section, we fabricate the immuno-wall devices with another photocross-linkable polymer, BIOSURFINE®-AWP (Toyo Gosei Co., Ltd.), which has pendant azide group. Picture and schematic of the

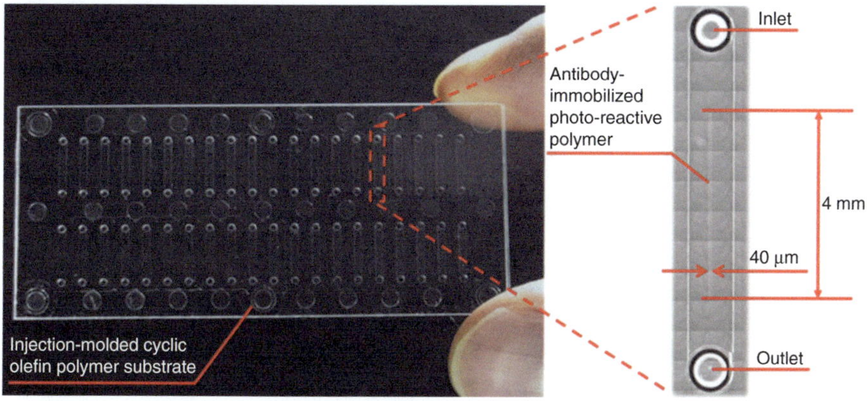

Fig. 13.13 Photograph of the immuno-wall device made with azido-unit pendant polymer (reproduced with permission from reference Kasama et al. (2015b))

immuno-wall device made with BIOSURFINE®-AWP is shown in Fig. 13.13. The device fabrication process is as follows. High-concentration (10 mg mL^{-1}) streptavidin in PBS was mixed with an equal volume of BIOSURFINE®-AWP in a microtube. The mixture was introduced into the microchannel by using the pipette. Then, the mixture was irradiated with UV light (320 nm, 20 mW) for 5 s through a photomask covering the microchannel. The irradiated BIOSURFINE®-AWP was cross-linked each other. At the same time, streptavidin molecules were photo-immobilized to the BIOSURFINE®-AWP. After the UV irradiation, the uncured BIOSURFINE®-AWP was sucked by using an aspirator. Then, immuno-wall remained at the center of the microchannel. Finally, the microchannel was washed with washing buffer. In order to immobilize capture antibody, 1 μL of biotinylated antibody solution was introduced into the microchannel and the device was settled for 60 min at room temperature. Although the diffusion of proteins including IgG antibody was observed in the immuno-wall, mostly analytes were captured at the side surface of the immuno-wall because of ultrahigh-density immobilization of capture antibody. On the other hand, large substances including cell debris could not penetrate the immuno-wall.

13.3.2 Precision Medicine of Lung Cancer

Lung cancer is the leading cause of cancer-related mortality worldwide. Approximately 85% of lung cancers are classified as non-small-cell lung cancer (NSCLC) (Ferlay et al. 2015). Somatic mutations of epidermal growth factor receptor (EGFR) are detected in approximately 10–16% of NSCLC patients in the United States and Europe (Rosell et al. 2009) and 30–50% in Asia (Sequist et al. 2007). Approximately 90% of these mutations are the substitution of leucine 858 by arginine in exon 21 (L858R point mutation) (Sequist et al. 2007) and the in-frame deletions in exon 19, especially the E746-A750 deletion (Sequist et al. 2007). Several studies revealed that these mutations have sensitivity to EGFR-tyrosine kinase inhibitors (TKIs) (Lynch et al. 2004; Paez et al. 2004). Therefore, EGFR mutation testing in the clinical setting has been important. Direct sequencing of PCR

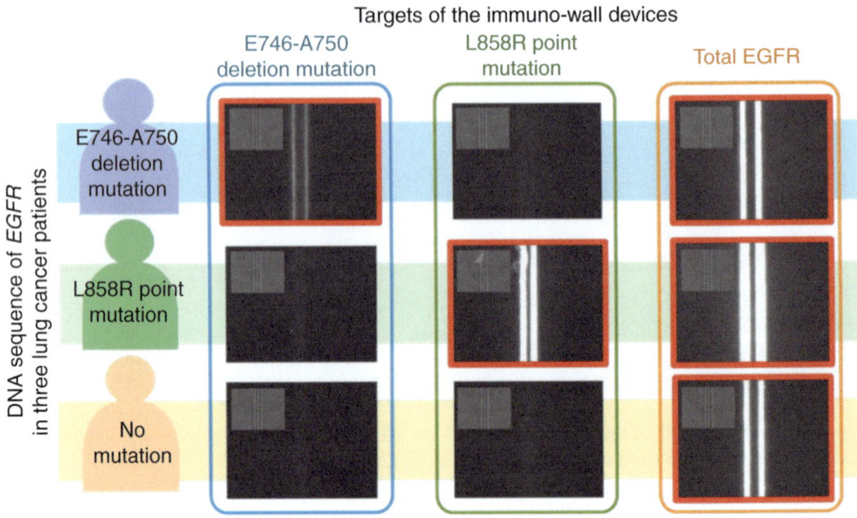

Fig. 13.14 Immunoassay results (reproduced with permission from reference Kasama et al. (2015b)). Bright-field images (*insets*) and fluorescence images of the immuno-wall devices are shown. The *red frames* of pictures show the results being positive. The side surfaces of immuno-walls exposed to the sediment lysates emitted fluorescence signals

products is one of the commonly used methods worldwide. However, its clinical application is limited due to the sensitivity depending on the proportion of tumor cells in the specimens.

By using the immuno-wall devices, the sandwich-type fluorescence immunoassay procedure was performed for the sediment lysates obtained from pleural effusion samples of three NSCLC patients. Their tumors had E746-A750 deletion-mutated EGFR, L858R point-mutated EGFR, or wild-type EGFR, respectively. The sediments in the pleural effusion were gathered by centrifugation, and then lysed with lysis buffer. Mutated EGFR-specific antibodies and total-EGFR antibody were immobilized to the immuno-walls. Another total-EGFR antibody was employed as detection antibody.

The immunoassay results are summarized in Fig. 13.14. Total assay time was less than 20 min. We can obviously recognize that the devices detected mutated EGFRs specifically. This means that the patients having responses to the EGFR-TKI are successfully distinguished.

13.3.3 Precision Surgery of Brain Tumors

Since the glioma tends to infiltrate into the normal brain tissue, it is difficult to define the edges of glioma. Therefore, the gliomas are not fully resectable, resulting in recurrence and eventual fatality. Because R132H mutation in IDH1 is observed in patients with grade II and III gliomas with approximately 65% (Suzuki et al. 2015; Gorovets et al. 2012; Hartmann et al. 2009; Parsons et al. 2008; Arita et al. 2015; Horbinski 2013), IDH1 mutation testing should help us to define the tumor boundary from the normal brain. However, the current methods for analyzing the genetic

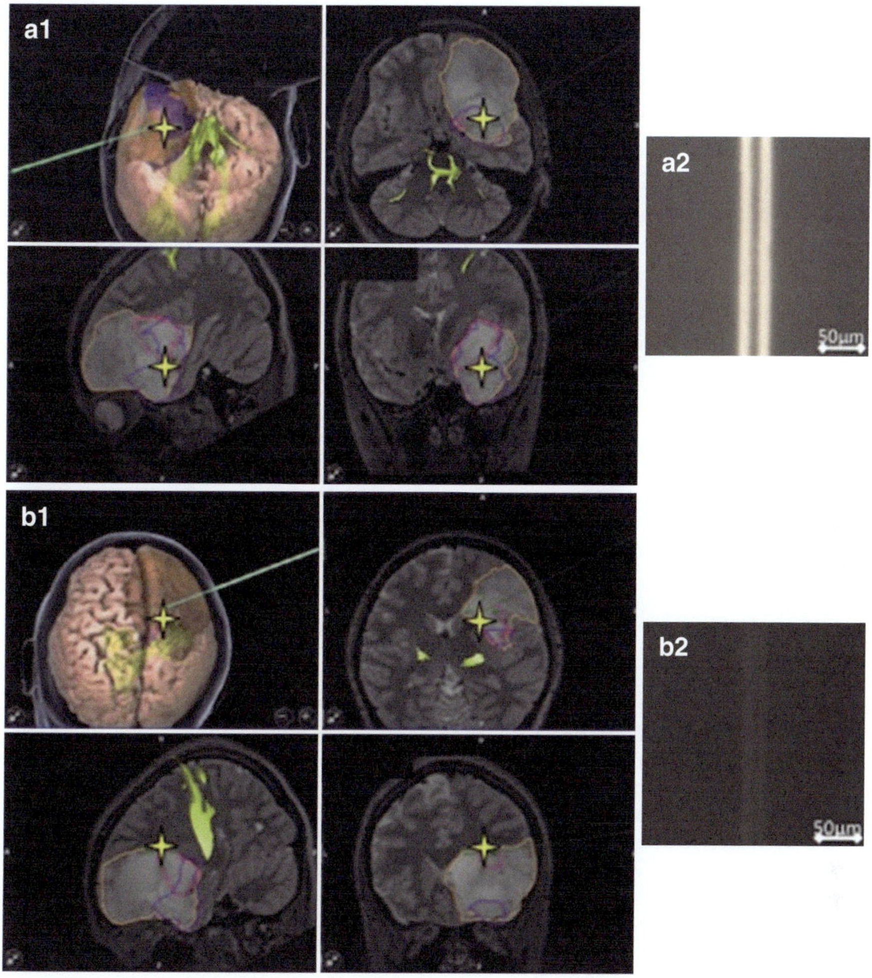

Fig. 13.15 Tumor boundary detection using the immuno-wall devices (adapted from reference Yamamichi et al. (2016)). The tumor region in a patient was roughly estimated by the magnetic resonance imaging before tumor removal surgery (A1 and B1). Two specimens were collected (*stars* in A1 and B1). A specimen obtained from the center of the tumor (*star* in A1) tested positive in the assay using the immuno-wall device (A2). On the other hand, a specimen obtained from the edge of the tumor (*star* in B1), which appeared normal, tested negative (B2)

status of glioma tissue are time consuming, 60 min at least. This makes it difficult to test IDH1 mutation during the surgery. In order to solve this problem, we fabricated the immuno-wall devices with R132H mutant IDH1-specific capture antibody and performed immunoassay for a lysate of glioma tissue obtained from brain tumor patients. The total-IDH1 antibody was adopted as the detection antibody.

The representative immunoassay results for a glioma and its edge tissue are shown in Fig. 13.15. The fluorescence was observed only on the immuno-wall device used for the center of tumor, which means that we can define the boundary between the glioma and normal brain.

13.4 Summary

Microfluidic immunoassay devices have inherent advantages such as portability and reduced sample and reagent consumption. In addition, the microchannel can restrict molecular diffusion, resulting in the rapidity and high sensitivity of the immunoassay. We believe that these unique features allow us to use microfluidic immunoassay devices in the POC cancer diagnosis.

Here we have introduced the immuno-pillar devices and the immuno-wall devices. It has been demonstrated that these microfluidic immunoassay devices have great potential for practical immunoassay and POC cancer diagnostics. Also, the precision medicine and precision surgery have been realized. This implies that the microfluidic immunoassay devices have possibility of changing the process of diagnosis.

Acknowledgement This study was supported in part by the priority research project of "The knowledge hub of AICHI," Nagoya University Hospital Funding for Clinical Research, the Translational Research Network Program from the Japan Agency for Medical Research and Development (AMED), and JSPS KAKENHI Grant Number JP16K18438.

References

Arita H, Narita Y, Yoshida A, Hashimoto N, Yoshimine T, Ichimura K (2015) IDH1/2 mutation detection in gliomas. Brain Tumor Pathol 32:79–89

Asao T, Kumeda Y, Kawai T, Shibata T, Oda H, Haruki K, Nakazawa H, Kozaki S (2003) An extensive outbreak of staphylococcal food poisoning due to low-fat milk in Japan: estimation of enterotoxin A in the incriminated milk and powdered skim milk. Epidemiol Infect 130:33–40

Bange A, Halsall HB, Heineman WR (2005) Microfluidic immunosensor systems. Biosens Bioelectron 20:2488–2503

Chen RF, Knutson JR (1988) Mechanism of fluorescence concentration quenching of carboxyfluorescein in liposomes: energy transfer to nonfluorescent dimers. Anal Biochem 172:61–77

Chiem N, Harrison DJ (1997) Microchip-based capillary electrophoresis for immunoassays: analysis of monoclonal antibodies and theophylline. Anal Chem 69:373–378

Chin CD, Linder V, Sia SK (2007) Lab-on-a-chip devices for global health: past studies and future opportunities. Lab Chip 7:41–57

Delamarche E, Juncker D, Schmid H (2005) Microfluidics for processing surfaces and miniaturizing biological assays. Adv Mater 17:2911–2933

Ferlay J, Soerjomataram I, Dikshit R, Eser S, Mathers C, Rebelo M, Parkin DM, Forman D, Bray F (2015) Cancer incidence and mortality worldwide: sources, methods and major patterns in GLOBOCAN 2012. Int J Cancer 136:E359–E386

Gabay C, Kushner I (1999) Acute-phase proteins and other systemic responses to inflammation. N Engl J Med 340:448–454

Gervais L, Delamarche E (2009) Toward one-step point-of-care immunodiagnostics using capillary-driven microfluidics and PDMS substrates. Lab Chip 9:3330–3337

Gorovets D, Kannan K, Shen R, Kastenhuber ER, Islamdoust N, Campos C, Pentsova E, Heguy A, Jhanwar SC, Mellinghoff IK, Chan TA, Huse JT (2012) IDH mutation and neuroglial developmental features define clinically distinct subclasses of lower grade diffuse astrocytic glioma. Clin Cancer Res 18:2490–2501

Haes AJ, Terray A, Collins GE (2006) Bead-assisted displacement immunoassay for staphylococcal enterotoxin B on a microchip. Anal Chem 78:8412–8420

Hartmann C, Meyer J, Balss J, Capper D, Mueller W, Christians A, Felsberg J, Wolter M, Mawrin C, Wick W, Weller M, Herold-Mende C, Unterberg A, Jeuken JWM, Wesseling P, Reifenberger G, von Deimling A (2009) Type and frequency of IDH1 and IDH2 mutations are related to astrocytic and oligodendroglial differentiation and age: a study of 1,010 diffuse gliomas. Acta Neuropathol 118:469–474

Henares TG, Mizutani F, Hisamoto H (2008) Current development in microfluidic immunosensing chip. Anal Chim Acta 611:17–30

Horbinski C (2013) What do we know about IDH1/2 mutations so far, and how do we use it? Acta Neuropathol 125:621–636

Hosokawa K, Omata M, Sato K, Maeda M (2006) Power-free sequential injection for microchip immunoassay toward point-of-care testing. Lab Chip 6:236–241

Ihara M, Yoshikawa A, Wu Y, Takahashi H, Mawatari K, Shimura K, Sato K, Kitamori T, Ueda H (2010) Micro OS-ELISA: rapid noncompetitive detection of a small biomarker peptide by open-sandwich enzyme-linked immunosorbent assay (OS-ELISA) integrated into microfluidic device. Lab Chip 10:92–100

Ikami M, Kawakami A, Kakuta M, Okamoto Y, Kaji N, Tokeshi M, Baba Y (2010) Immuno-pillar chip: a new platform for rapid and easy-to-use immunoassay. Lab Chip 10:3335–3340

Jin W, Yamada K, Ikami M, Kaji N, Tokeshi M, Atsumi Y, Mizutani M, Murai A, Okamoto A, Namikawa T, Baba Y, Ohta M (2013) Application of IgY to sandwich enzyme-linked immunosorbent assays, lateral flow devices, and immunopillar chips for detecting staphylococcal enterotoxins in milk and dairy products. J Microbiol Methods 92:323–331

Kakuta M, Takahashi H, Kazuno S, Murayama K, Ueno T, Tokeshi M (2006) Development of the microchip-based repeatable immunoassay system for clinical diagnosis. Meas Sci Technol 17:3189

Kasama T, Hasegawa Y, Kondo H, Ozawa T, Kaji N, Tokeshi M, Baba Y (2014) Development of immuno-wall devices and a mobile fluorescence reader for on-site sample-to-answer immunoassay. In: Proceedings of international conference on miniaturized systems for chemistry and life sciences, pp 935–937

Kasama T, Ikami M, Jin W, Yamada K, Kaji N, Atsumi Y, Mizutani M, Murai A, Okamoto A, Namikawa T, Ohta M, Tokeshi M, Baba Y (2015a) Rapid, highly sensitive, and simultaneous detection of staphylococcal enterotoxins in milk by using immuno-pillar devices. Anal Methods 7:5092–5095

Kasama T, Hase T, Nishiwaki N, Yogo N, Sato M, Kondo M, Kaji N, Tokeshi M, Hasegawa Y, Baba Y (2015b) Immuno-wall lab-on-chip companion diagnostic devices for rapid and low-cost detection of mutant epidermal growth factor receptors (EGFR) from cytological samples in lung cancer patients. In: Proceedings of international conference on miniaturized systems for chemistry and life sciences, pp 925–927

Kuang H, Wang W, Xu L, Ma W, Liu L, Wang L, Xu C (2013) Monoclonal antibody-based sandwich ELISA for the detection of staphylococcal enterotoxin A. Int J Environ Res Public Health 10:1598–1608

Linder V, Sia SK, Whitesides GM (2005) Reagent-loaded cartridges for valveless and automated fluid delivery in microfluidic devices. Anal Chem 77:64–71

Lynch TJ, Bell DW, Sordella R, Gurubhagavatula S, Okimoto RA, Brannigan BW, Harris PL, Haserlat SM, Supko JG, Haluska FG, Louis DN, Christiani DC, Settleman J, Haber DA (2004) Activating mutations in the epidermal growth factor receptor underlying responsiveness of non-small-cell lung cancer to gefitinib. N Engl J Med 350:2129–2139

Moorthy J, Mensing GA, Kim D, Mohanty S, Eddington DT, Tepp WH, Johnson EA, Beebe DJ (2004) Microfluidic tectonics platform: a colorimetric, disposable botulinum toxin enzyme-linked immunosorbent assay system. Electrophoresis 25:1705–1713

Ohashi T, Mawatari K, Sato K, Tokeshi M, Kitamori TA (2009) micro-ELISA system for the rapid and sensitive measurement of total and specific immunoglobulin E and clinical application to allergy diagnosis. Lab Chip 9:991–995

Paez JG, Jänne PA, Lee JC, Tracy S, Greulich H, Gabriel S, Herman P, Kaye FJ, Lindeman N, Boggon TJ, Naoki K, Sasaki H, Fujii Y, Eck MJ, Sellers WR, Johnson BE, Meyerson M (2004)

EGFR mutations in lung cancer: correlation with clinical response to gefitinib therapy. Science 304:1497–1500

Parsons DW, Jones S, Zhang X, Lin JC-H, Leary RJ, Angenendt P, Mankoo P, Carter H, Siu I-M, Gallia GL, Olivi A, McLendon R, Rasheed BA, Keir S, Nikolskaya T, Nikolsky Y, Busam DA, Tekleab H, Diaz LA, Hartigan J, Smith DR, Strausberg RL, Marie SKN, Shinjo SMO, Yan H, Riggins GJ, Bigner DD, Karchin R, Papadopoulos N, Parmigiani G, Vogelstein B, Velculescu VE, Kinzler KW (2008) An integrated genomic analysis of human glioblastoma multiforme. Science 321:1807–1812

Rose SA, Bankes P, Stringer MF (1989) Detection of staphylococcal enterotoxins in dairy products by the reversed passive latex agglutination (SET-RPLA) kit. Int J Food Microbiol 8:65–72

Rosell R, Moran T, Queralt C, Porta R, Cardenal F, Camps C, Majem M, Lopez-Vivanco G, Isla D, Provencio M, Insa A, Massuti B, Gonzalez-Larriba JL, Paz-Ares L, Bover I, Garcia-Campelo R, Moreno MA, Catot S, Rolfo C, Reguart N, Palmero R, Sánchez JM, Bastus R, Mayo C, Bertran-Alamillo J, Molina MA, Sanchez JJ, Taron M, Group SLC (2009) Screening for epidermal growth factor receptor mutations in lung cancer. N Engl J Med 361:958–967

Sato K, Tokeshi M, Odake T, Kimura H, Ooi T, Nakao M, Kitamori T (2000) Integration of an immunosorbent assay system: analysis of secretory human immunoglobulin A on polystyrene beads in a microchip. Anal Chem 72:1144–1147

Sato K, Tokeshi M, Kimura H, Kitamori T (2001) Determination of carcinoembryonic antigen in human sera by integrated bead-bed immunoasay in a microchip for cancer diagnosis. Anal Chem 73:1213–1218

Sato K, Yamanaka M, Takahashi H, Tokeshi M, Kimura H, Kitamori T (2002) Microchip-based immunoassay system with branching multichannels for simultaneous determination of interferon-γ. Electrophoresis 23:734–739

Sato K, Yamanaka M, Hagino T, Tokeshi M, Kimura H, Kitamori T (2004) Microchip-based enzyme-linked immunosorbent assay (microELISA) system with thermal lens detection. Lab Chip 4:570–575

Sequist LV, Bell DW, Lynch TJ, Haber DA (2007) Molecular predictors of response to epidermal growth factor receptor antagonists in non-small-cell lung cancer. J Clin Oncol 25:587–595

Shin K-S, Lee SW, Han K-C, Kim SK, Yang EK, Park JH, Ju B-K, Kang JY, Kim TS (2007) Amplification of fluorescence with packed beads to enhance the sensitivity of miniaturized detection in microfluidic chip. Biosens Bioelectron 22:2261–2267

Sung W-C, Chen H-H, Makamba H, Chen S-H (2009) Functionalized 3D-hydrogel plugs covalently patterned inside hydrophilic poly(dimethylsiloxane) microchannels for flow-through immunoassays. Anal Chem 81:7967–7973

Suzuki H, Aoki K, Chiba K, Sato Y, Shiozawa Y, Shiraishi Y, Shimamura T, Niida A, Motomura K, Ohka F, Yamamoto T, Tanahashi K, Ranjit M, Wakabayashi T, Yoshizato T, Kataoka K, Yoshida K, Nagata Y, Sato-Otsubo A, Tanaka H, Sanada M, Kondo Y, Nakamura H, Mizoguchi M, Abe T, Muragaki Y, Watanabe R, Ito I, Miyano S, Natsume A, Ogawa S (2015) Mutational landscape and clonal architecture in grade II and III gliomas. Nat Genet 47:458–468

Tachi T, Kaji N, Tokeshi M, Baba Y (2007) Microchip-based immunoassay. Bunseki Kagaku 56:521

Thompson JA, Bau HH (2010) Microfluidic, bead-based assay: theory and experiments. J Chromatogr B 878:228–236

Wild D (2005) The immunoassay handbook. Elsevier, Amsterdam

Yamamichi A, Kasama T, Ohka F, Suzuki H, Kato A, Motomura K, Hirano M, Ranjit M, Chalise L, Kurimoto M, Kondo G, Aoki K, Kaji N, Tokeshi M, Matsubara T, Senga T, Kaneko MK, Suzuki H, Hara M, Wakabayashi T, Baba Y, Kato Y, Natsume A (2016) An immuno-wall microdevice exhibits rapid and sensitive detection of IDH1-R132H mutation specific to grade II and III gliomas. Sci Technol Adv Mater 17:18–625

Zubtsov D, Ivanov S, Rubina A, Dementieva E, Chechetkin V, Zasedatelev A (2006) Effect of mixing on reaction-diffusion kinetics for protein hydrogel-based microchips. J Biotechnol 122:16–27

Point-of-Care Device with Plasmonic Gold Nanoarray Sensing Chip for Biomarker Detections

14

Xiaodong Zhou, Ten It Wong, Ling Ling Sun, and Jie Deng

14.1 Introduction

LSPR is generated on metal nanostructures upon the illumination of light whose energy can be absorbed by the nanostructures and cause collective electron charge oscillations on its surface. LSPR traps the light on the metal nanostructures surface within tens of nanometers, where the light is strongly enhanced by tens to hundreds of times. LSPR is particularly beneficial for chemical or biological detections. Because the sizes of the chemical or biological molecules are usually in a few nanometer range, they will be attached to the metal nanostructure surface roughly within the coverage of the strong LSPR field, thus the LSPR signal variation is relatively specific to the molecule attachment, i.e., the molecules not bound to the metal nanostructure will not much affect the detection signal, thus raw samples could possibly be tested without wash or purifications (Anker et al. 2008; Lee and El-Sayed 2006; Sherry et al. 2005).

There are two methods to use LSPR for biosensors. One is to detect the shift of the LSPR absorption due to the refractive index change upon molecule binding. This is a direct assay that requires less time and cost; however, for some applications, the sensitivity might not be enough. The other is to utilize the LSPR to excite the fluorescent labels in a sandwich assay, because the labeled sandwich assay is intrinsically more sensitive than direct assay, and LSPR field is 10–100 times stronger than the incident light, the LSPR amplifies the fluorescent labels and

X. Zhou (✉) • T.I. Wong • J. Deng
Institute of Materials Research and Engineering, A*STAR (Agency for Science, Technology and Research), 2 Fusionopolis Way, #08-03, Innovis, Singapore 138634, Singapore
e-mail: donna-zhou@imre.a-star.edu.sg

L.L. Sun
Temasek Microelectronics Centre, School of Engineering, Temasek Polytechnic, 21 Tampines Avenue 1, Singapore 529757, Singapore

P. Chandra et al. (eds.), *Next Generation Point-of-care Biomedical Sensors Technologies for Cancer Diagnosis*, DOI 10.1007/978-981-10-4726-8_14

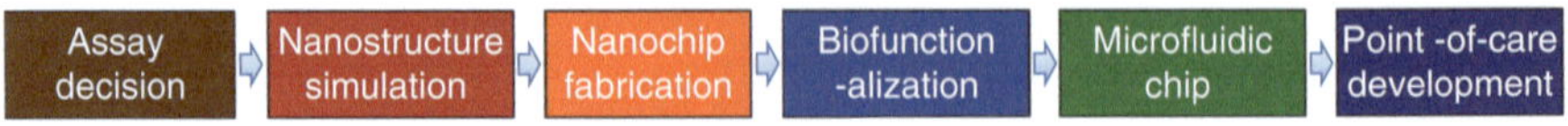

Fig. 14.1 The development process of a LSPR-based point-of-care device

achieves quite high sensitivity. Since both methods can be applied on the development of a point-of-care device, the selection of detection methods is based on sensitivity requirements for different applications. The development of a LSPR-based point-of-care device has a flow chart illustrated in Fig. 14.1. The first step is to identify the nanoparticles and the related bioassay that can meet the requirement of the specific application. After which, some plasmonic simulation is required to identify the plasmonic peak and the LSPR field distribution, thus the suitable light source can be found and the size, shape, and inter-distance of the nanoparticles can be optimized.

LSPR can be generated by metal nanoparticles in solution, or metal nanostructures (random or periodic nanoslits, nanobars, nanoholes, or nanopillars/islands) on a substrate (Kuwata et al. 2003; Mayer and Hafner 2011; Murphy et al. 2005). We established our point-of-care device with periodic gold nanoarray on glass substrate, because the substrate-based LSPR chip can be washed with incorporated microfluidic channels to reduce nonspecific binding, the periodic nanostructures have much narrower LSPR bandwidth that helps improving the detection sensitivity, and gold resists oxidation and extends the shelf-life of the fabricated sensing chips. Furthermore, gold functions with different kinds of biomolecules and eases the biofunctionalization of the nanostructures. However, periodic gold nanostructures are difficult to be fabricated, so we have developed a whole process of fabricating the gold nanopillar or nanohole array on glass.

According to the simulated plasmonic field as well as the simulated and characterized LSPR wavelength of the nanochips (Ghaemi et al. 1998; Wu et al. 2012a; Wu et al. 2012b), the bioassay is established on the top of the gold nanoarray, and a microfluidic channel with samples flow through the nanochip will be fabricated. The microfluidic chip can be actuated and detected by the automated point-of-care device, and the point-of-care device will also quantify and report the concentration of the biomarker in the specimen.

14.2 Nanochip Design, Fabrication, and Characterizations

14.2.1 Design of the Gold Nanoarray

We first designed the gold nanopillars and gold nanoholes with various pitches and sizes, and the relation of the LSPR peak location to the size, pitch of the gold nanohole arrays are shown in Fig. 14.2. The gold nanostructures can be designed by plasmonic simulations according to the Maxwell equations, which are applicable for the electromagnetic field and transmission/refraction/absorption calculations for metal nanoparticles larger than several nanometers. Commercial software available

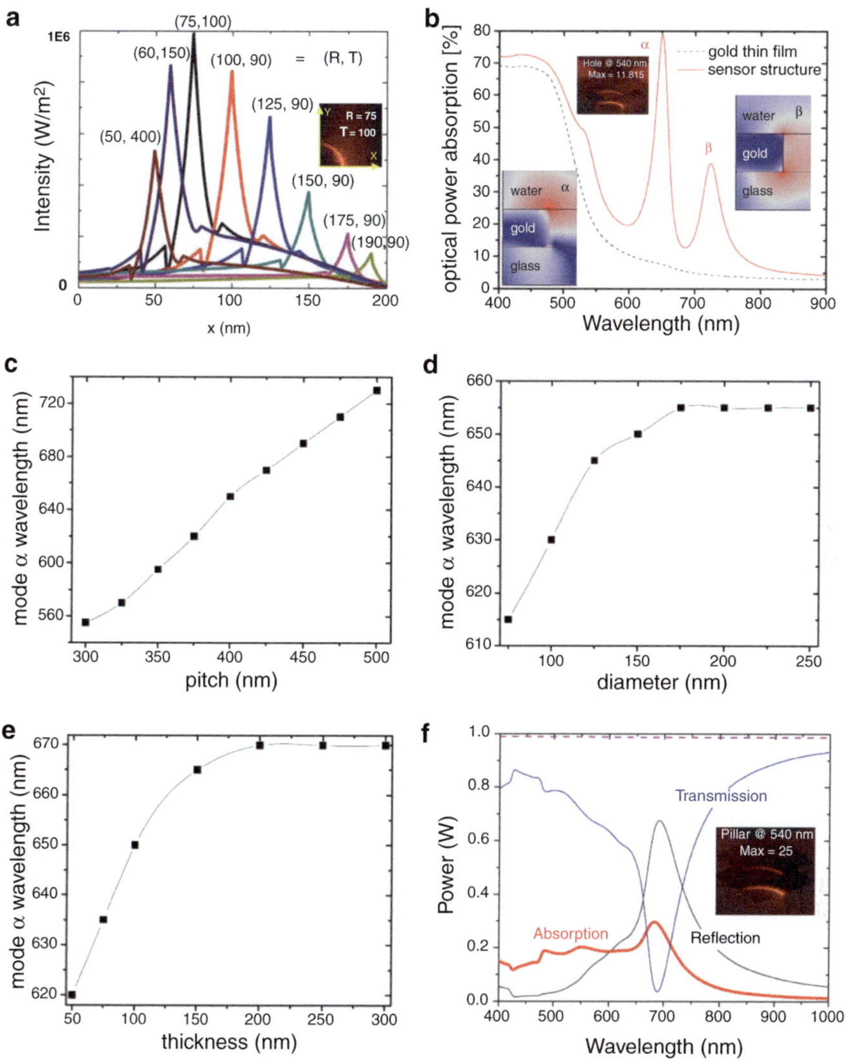

Fig. 14.2 Plasmonic peak and electromagnetic field for water side light illumination. (**a**) is at various gold nanohole dimensions with a fixed pitch of 400 nm; (**b**) gold nanohole array ($p = 400$ nm, $R = 75$ nm, $T = 100$ nm) showing the α and β modes; (**c–e**) the α peak shifts of gold nanohole array in (**b**), when only varies the pitch, diameter, or gold thickness, respectively; (**f**) The gold nanopillar array simulation with $p = 320$ nm, $R = 75$ nm, and $T = 50$ nm

for such simulations includes COMSOL, CST Microwave, etc. For the ease of mass productive nanofabrication of the sensing chips, we simulated the gold nanopillar and nanohole arrays on glass substrate, by setting the frequency-dependent dielectric constant of gold using the data from Palik handbook (Palik 1998), and the refractive indices for quartz, air, and water were taken as 1.52, 1.0, and 1.33, respectively.

For biosensing, the gold nanoarray will be immersed in buffer with its reflective index similar to water. Plasmonics are generated at the top rims (at the water–gold interface) and the bottom rims (at the gold–glass interface) of the gold nanostructures, and plasmonics on the top rims contributes more to the biosensing. The simulated top rim plasmonic field intensity is presented in Fig. 14.2a for gold nanohole arrays with a fixed pitch of 400 nm and various radii (R) and thicknesses (T). The gold nanohole array with $R = 75$ nm and $T = 100$ nm yields the strongest plasmonic field of 9.538×10^5 W/m^2 (in our simulations, the incident power was 1 nW within a quarter of a single nanohole area, i.e., 200 nm × 200 nm area). Figure 14.2a also indicates that the strongest plasmonic signal is at the rim of the gold nanoholes, and this intensity drops to a half when X-axis distance is 6 nm away from the rim. Figure 14.2b further demonstrated that two plasmonic modes α (at around 647 nm) and β (at around 730 nm) will be generated at the top and bottom rims of the gold nanoholes, respectively. In comparison, in Fig. 14.2b, 100 nm gold film on the glass substrate presents no obvious plasmonic peaks.

The influences of the pitch, diameter, and thickness to the wavelength of the α mode in Fig. 14.2c–e indicate that the pitch will influence the absorption the most, and 400 nm is the most suitable pitch for having a plasmonic peak at around 647 nm which is suitable for red fluorescent dye excitations. Figure 14.2c–e shows that 200 nm of pitch variation (from 300 to 500 nm) will cause a linear peak wavelength shift of 180 nm (from 555 to 735 nm), 50 nm of diameter variation (from 125 to 175 nm) will cause a wavelength shift of 10 nm (645–655 nm), while 50 nm of thickness variation (from 50 to 100 nm) will cause a wavelength shift of 27 nm (from 620 to 647 nm). It is also well known that engineering errors are unavoidable in the gold nanostructure fabrications. So for different nanohole parameters, the plasmonic peak absorption variation of the α mode can be roughly estimated by $\Delta\lambda = (\partial\lambda/\partial p) \times \Delta p + (\partial\lambda/\partial D) \times \Delta D + (\partial\lambda/\partial T) \times \Delta T$, with $\partial\lambda/\partial p = 0.9$, $\partial\lambda/\partial D = 0.2$ and $\partial\lambda/\partial T = 0.54$.

The gold nanopillar array with $p = 320$ nm, $R = 75$ nm, and $T = 50$ nm is also used in our work, and its transmission, refraction, and absorption peaks in Fig. 14.2f present two plasmonic peaks at 540 and 690 nm. For our work of using green light to excite the quantum dot (QD) label in the sandwich assay, this kind of gold nanopillar array provides twice stronger plasmonic field (by comparing the insets in Fig. 14.2b, f), thus outperforms the gold nanohole array of $p = 400$ nm, $R = 75$ nm, and $T = 100$ nm.

14.2.2 Gold Nanoarray Fabrication

Based on our design, the gold nanohole array (with $p = 400$ nm, $R = 75$ nm, $T = 100$ nm) and gold nanopillar array (with $p = 320$ nm, $R = 75$ nm, and $T = 50$ nm) are our targets for nanofabrication. For point-of-care device, the gold nanoarray is expected to be disposable, thus the mass fabrication of the gold nanochip is a key enabling technology for point-of-care devices. We have developed a mass fabrication method as illustrated in Fig. 14.3. It includes two parts, the first is the nickel

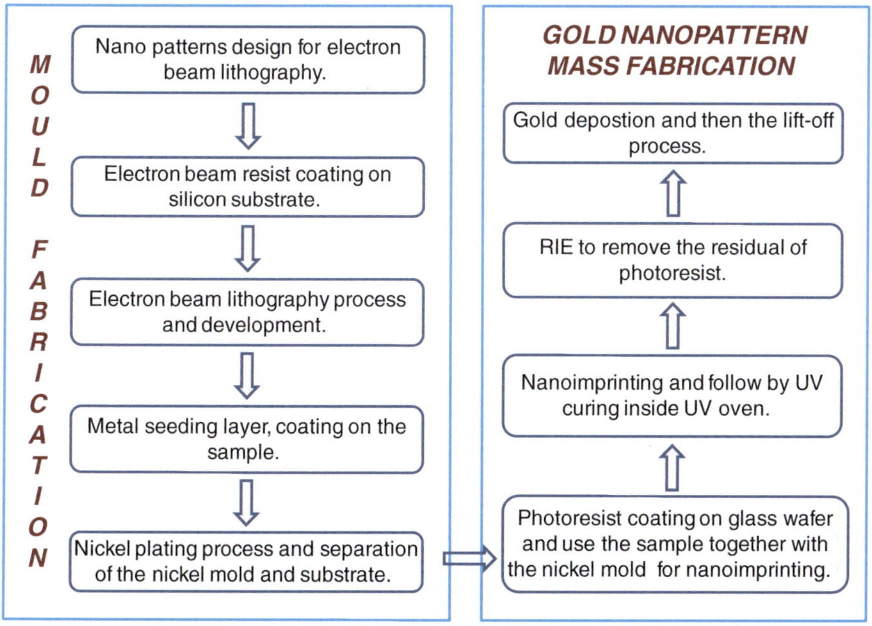

Fig. 14.3 Mass fabrication of metal nanostructures on glass wafer. The *left side* indicates the nickel mold fabrication process, while the *right* is the gold nanostructure mass fabrication process

mold fabrication, and the second part the gold nanoarray mass fabrication via nanoimprinting (Wong et al. 2013).

Electron-beam (e-beam) lithography is a technology of using focused e-beam to expose the resist which is sensitive to the e-beam (Altissimo 2010; Chen 2015). The e-beam lithography can reach a resolution of a few nanometers, and it is a revolutionary nanotechnology for obtaining precise and controllable nanostructures in any shape. E-beam lithography has been heavily applied in industry and nanotechnology research, for fabricating the accurately designed nanopatterns in plasmonics, semiconductors, surface property modified materials such as for anti-fouling or anti-fogging, and photonic waveguides. Although nanoimprinting can replace the e-beam lithography in some cases, its master mold is produced by e-beam lithography.

In our work, the e-beam lithography is used to fabricate the nanochips in two ways: the first is the fabrication of the gold nanostructure array directly by e-beam lithography (Song et al. 2015a), the other is to make a nickel mold which is used to get the gold nanostructure array through nanoimprinting (Wong et al. 2013; Song et al. 2015a; Song et al. 2015b). In either case, for reducing the cost of nanochip fabrication, e-beam lithography is conducted with 25 pieces of 1 cm × 1 cm dies arranged on one 4″ wafer. The chip arrangement is shown in Fig. 14.4a. Each die has its central 1.8 mm × 1.8 mm area filled with fabricated 140 nm × 140 nm sized

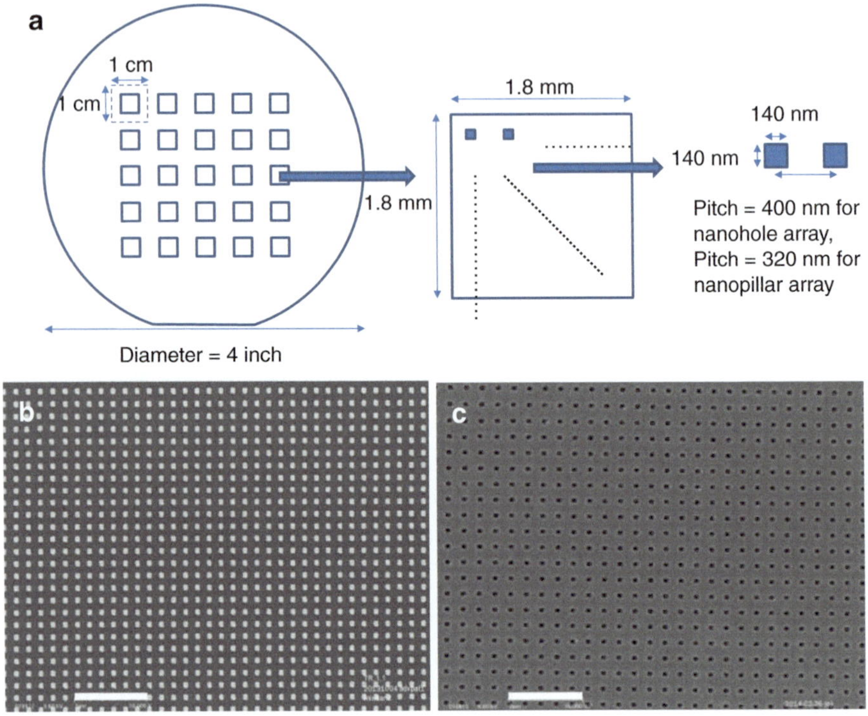

Fig. 14.4 (**a**) The arrangement of nanochips on a 4″ wafer with 25 dies, and scanning electron microscope (SEM) images of the fabricated gold (**b**) nanopillar and (**c**) nanohole arrays under 10,000× magnification, where the scale bar represents 2 μm

nanohole or nanopillar array. We fabricated the nanoarray in square shape because e-beam writes the squares slightly faster, and after nanofabrication, the gold nanopatterns will have some round corners anyway.

The main caveat of the e-beam lithography, however, is its slow writing speed, which substantially increases the cost. E-beam writing time includes the beam settling time (i.e., the alignment time and beam shift time, as well as the time for beam adjustment) and e-beam resist exposure time. According to the size of the nanopatterns, we optimized process and managed to reduce the e-beam resist exposure time by using the writing voltage of 100 keV, field size of 300 μm, lower dot map of 20k, and a comparatively high current of 800 pA on a single Gaussian beam (ELS-7000) e-beam writing system. We did a series of experiments to compare, and found that a minimum dose of 178 μC/cm^2 for positive ZEP e-beam resist is required, in order to get a reasonably high quality e-beam pattern for nanopatterns with 150 nm size. Compared with the normally used current of 200 pA and dosage of 320 μC/cm^2 for e-beam writing on the ZEP, 800 pA current and 178 μC/cm^2 dose combination will reduce the resist exposure time to 1/7 (Deng et al. 2016). This is very important, because for the 4″ wafer e-beam writing shown in Fig. 14.4a, even with our accelerated e-beam writing recipe, it took 32 h to complete. For the negative e-beam resist NEB22, it can be exposed at 800 pA current with 110 μC/cm^2 dosage. Following the e-beam exposure, the e-beam resist was developed for a few minutes to get the nanopatterns.

In order to reduce the cost while keeping the high precision of the designed gold nanohole array, nanoimprinting technology was adopted for the nanochip fabrication, because nanoimprinting can achieve sub 5 nm resolution with high throughput for nanolithography (Austin et al. 2004).

After the nanoarray pattern was written by e-beam lithography on silicon wafer, the wafer was used for electroplating to get a nickel mold. To fabricate the nickel mold, a thin metal seed layer was coated on the patterns to initiate the electroplating process. Two-step plating was adopted: at the beginning 12 μm of nickel was slowly electroplated on the silicon wafer at a low current density of 0.7 A/dm^2, followed by a high current density of 12 A/dm^2 to electroplate up to 300 μm of nickel. After which, the nickel mold was separated from the silicon wafer. Trion plasma etching system with O_2 gas was used to clean the nickel mold for 3 min, and the silicon wafer was recycled (Wong et al. 2013).

To nanoimprint, UV curable photoresist was selected, taking its advantage of using a lower pressure (10 bar) and lower temperature (as low as room temperature) for the imprinting process compared with thermal nanoimprinting (30–40 bar, >100 °C), which greatly reduces the rate of wafer breakage in nanoimprinting. The nickel mold was first coated with an anti-sticky layer of (Heptadecafluoro-1,1,2,2-tetra hydrodecyl) trichlorosilane (FDTS), then it was nanoimprinted on the UV curable photoresist coated on a 4″ glass wafer with the photoresist cured by UV light source.

After nanoimprinting, the nickel mold was easily demolded, and was cleaned for reuse. Reuse of the nickel mold is critical, because the mold is generally fabricated through e-beam lithography which is the major cost for the process to fabricate the nanochips. The cost saving for fabricating the nanochip is tremendous if a nickel mold can be reused, and nickel mode is reported to be durable for 10,000 times of imprinting without obvious damage (Rai-Choudhury 1997).

The residue of the UV curable photoresist on the glass wafer was removed by etching in an Oxford reactive ion etching system. The glass wafer was place in a Dexton Explorer e-beam deposition machine, to first deposit a few nanometer (<5 nm) of chromium to enhance the adhesion of the gold on glass, then deposit the gold film on the top of the chromium layer. Finally, the resist was lifted off at room temperature to obtain the gold nanoarray.

The fabricated gold nanopillar or nanohole arrays on the 4″ glass wafer were diced into 1 cm × 1 cm chips by diamond blade using a Disco dicing saw. The fabricated patterns were viewed under scanning electron microscope (SEM) or atomic-force microscopy (AFM) for checking their morphology and quality. The fabricated gold nanopillar and nanohole arrays are presented in Fig. 14.4b and Fig. 14.4c, respectively.

14.3 Point-of-Care Device Based on LSPR Wavelength Shift

A point-of-care device using the LSPR wavelength shift is developed for the measurement of cardiac troponin I (cTnI) in serum, as illustrated in Fig. 14.5. In Fig. 14.5a, white light transmitted through an optical fiber bundle was collimated

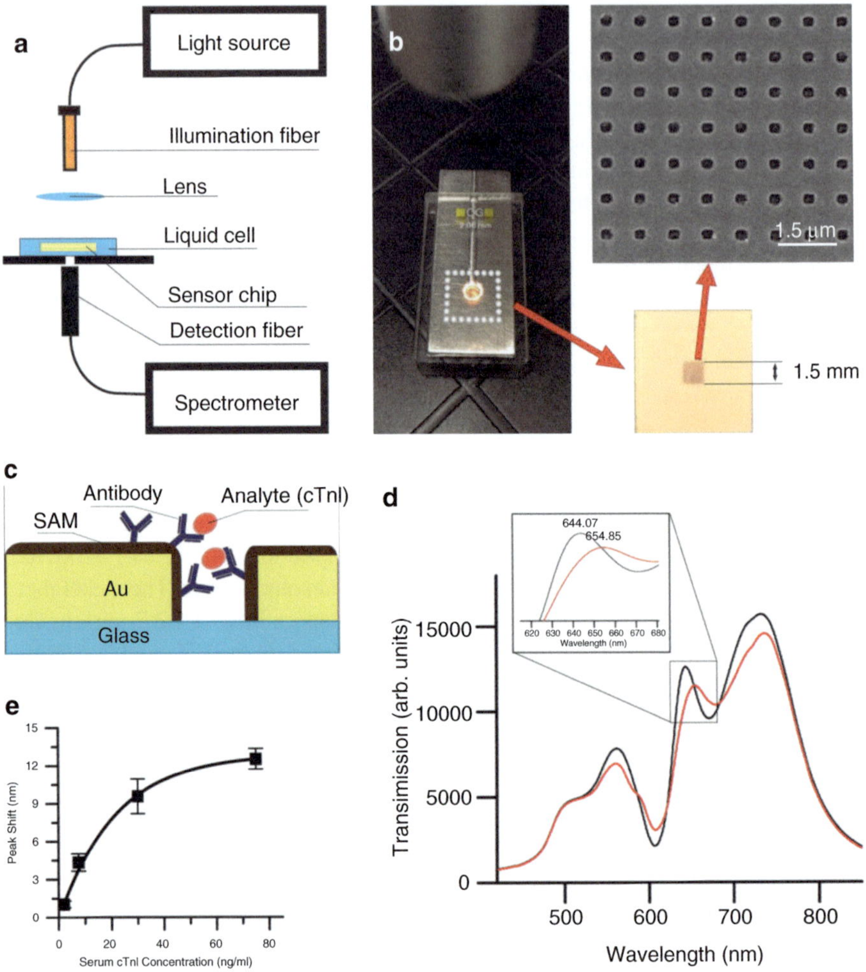

Fig. 14.5 The point-of-care device based on the LSPR shift of a nanohole array (reproduced with permission from reference Ding et al. (2015)). (**a**) Optical setup for the nanohole sensor detection. (**b**) A liquid cell seals the gold nanohole array (with $p = 400$ nm, $R = 75$ nm, and $T = 100$ nm filling a sensing area of 1.5 mm × 1.5 mm) for the optical detection. (**c**) is the bioassay on the gold surface, (**d**) shows the transmission spectrum of the nanochip upon cTnI (at 30 ng/mL concentration) binding, and (**e**) is the characterization curve of the detected cTnI concentration versus the LSPR peak shift

and illuminated on the top of the gold nanohole array ($p = 400$ nm, $R = 75$ nm, and $T = 100$ nm) placed inside a transparent flow cell made of thermoplastic elastomer (Fig. 14.5b). The transmitted light spectrum was collected by another fiber bundle and recorded by a UV-vis spectrometer HR2000+ (Ocean optic, Dunedin, FL, USA) in the range of 300–1000 nm. Each frame of transmission was collected using an acquisition time of 20 ms, and the final spectrum was produced by averaging 100 frames within 2 s. Spectra were analyzed and evaluated using OriginPro 9, and the transmission peaks were identified using a Lorentz-based method (Ding et al. 2015).

The direct assay in Fig. 14.5c was established by first forming an amine-reactive self-assembly monolayer (SAM) incubating the nanochips in ethanolic solution of 0.4 mM of 10-carboxy-1-decanethiol and 1.6 mM of 1-octanethiol for 12 h at room temperature, then washed thoroughly with pure ethanol and dried in room temperature. Then the chip was incubated in a mixture of sulfo-*N*-hydroxysuccinimide (sulfo-NHS) and 1-ethyl-3-(3-dimethylaminopropyl) carbodiimide (EDC) (Bio-Rad, Hercules, CA, USA) to activate the carboxylic group of the SAM for 15 min. Next, 50 μL of 200 μg/mL anti-troponin antibody 560 (Hytest, Finland) in acetate buffer at pH 4.5 was spotted on the chip surface and incubated for 30 min. Finally, the sensor chip was immersed in 1 M of ethanolamine-HCl solution (Bio-Rad, Hercules, CA, USA) for 15 min to deactivate the unreacted esters, followed by a rinse with deionized water and dried with nitrogen gas at room temperature. The sensing chip is blocked by 1% bovine serum albumin (BSA) solution for 15 min to prevent the nonspecific binding, then 50 μL of human cTnI standard in serum (Phoenix Pharmaceuticals, CA, USA) in the concentrations of 75, 30, 7.5, or 2.5 ng/mL was applied and incubated for 30 min. The sensor chip was rinsed with PBS solution three times after incubation, and inserted into the measurement cell containing PBS buffer to record its transmission spectrum.

The binding of the cTnI on anti-troponin antibody on the sensor chip will shift the plasmonic peak at the wavelength of 644.07–654.85 nm (Fig. 14.5d). The peak at 644.07 nm is the α peak in Fig. 14.2b which has the plasmonic field concentrated on the gold–water interface where the bioassay is anchored. Thus the LSPR biosensing for this plasmonic peak monitoring is relatively sensitive. Figure 14.5e shows the characterization curve for cTnI detection, with error bars obtained by measuring the same troponin concentrations with three individual nanochips.

The performance of this point-of-care device is found to have a full half-magnitude width (FHMW) of 32.84 nm, figure of merit (FOM) of 10.5, σ of signal at 0.017 nm, and signal-to-noise ratio equals to 256 (at 7.5 ng/mL troponin in serum), and the limit of detection (LOD) based on three times of σ is identified to be 0.55 ng/mL.

14.4 Point-of-Care Device Based on Fluorescent Label Enhancement

We have developed several sandwich bioassays based on plasmonic-enhanced fluorescent labels to respectively detect the prostate specific antigen (PSA) for prostate cancer, thrombin for blood clots, and procalcitonin (PCT) for sepsis. When establishing such a bioassay, a few factors are considered. The first is that the distance of the fluorescent label should be around 10–20 nm to the gold nanostructure surface, so that the dyes are enhanced by the plasmons instead of being quenched by the strong plasmons. The second is that the excitation wavelength of the dye must be close or slightly shorter than the LSPR generated by the gold nanostructures, so as to make sure that the light being used to excite the fluorescent dyes are amplified by the plasmon. Here we use quantum dot (QD) as labels, because compared with organic fluorescent dyes, the QDs are about 20 times brighter, and they have a broadband of excitation wavelength range (Song et al. 2015a).

These three kinds of biomarkers are all detected with a sandwich assay on the gold nanoarray surface using the QD as fluorescent label. However, they are still slightly different. The first type of assay demonstrated for PSA detection is to use the tris(2-carboxyethyl) phosphine (TCEP) solution to cleave the PSA monoclone capturing antibody (cAb) into a half and utilize the –SH link to bind the PSA capturing antibody, which is an effective way to reduce the distance between the QD label and the gold surface (Song et al. 2015a). The second is to use aptamer to capture the thrombin, because compared with capturing antibodies, aptamers are relatively cost-effective and with high repeatability, durability, reproducibility, and it is relatively easy for the surface regeneration when the nanochip is planned to be reused (Song et al. 2015b). The third is to use the whole PCT capturing antibody to capture the PCT on the gold surface (Sun et al. 2016). All these three kinds of bioassays can be detected by a microscope or a camera, and further the point-of-care device can be fabricated based on imaging of the fluorescent labels.

14.4.1 Bioassay for PSA Detection with Cleft Antibody

PSA is a US FDA approved cancer biomarker indicating the risk and existence of prostate cancer for males, with a cutoff value of 4 ng/mL for prostate cancer screening. Since the antibody is about 14 nm high, a standard sandwich assay will make the QD labels probably more than 30 nm above the gold nanostructure surface and hinder the enhancement by the plasmons. In order to further reduce the distance between the QDs and the gold surface, in our method, the whole capturing antibody was cleft into a half by TCEP solution (Song et al. 2014). This was achieved by adding 5 μL of TCEP into 50 μL of 250 μM cAb in PBS (1.26 mg/mL, 8.4 μM). After incubating at room temperature for 30 min to allow the mixtures to fully react, G25 column was used to purify the cleft cAb. The cleft cAb solution was stored at 4 °C, and diluted to 50 μg/mL in 4-(2-hydroxyethyl)-1-piperazineethanesulfonic acid (HEPES) buffer (prepared by mixing 10 mM HEPES with 150 mM NaCl) before use.

Compared with the whole capturing antibody, the cleft antibody is about half of its size, and since the –SH link will be attached to the gold surface, the cleft cAb are with their fragment antigen-binding sites facing upwards on the gold surface. The cleft cAbs in an orientation controlled state have more binding sites compared with the randomly orientated whole cAb, which will further increase the sensitivity for biosensing. With the cleft cAb, a sandwich assay of "cleft cAb – PSA – biotinylated dAb – QD with streptavidin" was immobilized on the gold surface (Fig. 14.6a), and the distance of the gold nanostructure to the QD is about 20–25 nm, which is within the distance for the QDs to be effectively excited by plasmons (Song et al. 2015a). The QD-655 has an emission peak at the wavelength of 655 nm. Practically, we used green light at the wavelength of 540 nm to illuminate the chip from the bioassay side and detect the QD emission from the same side of the chip. As the light sheds from the top of the chip, the plasmonic fields for the gold nanohole array ($p = 400$ nm, $R = 75$ nm, and $T = 100$ nm) and gold nanopillar array ($p = 320$ nm,

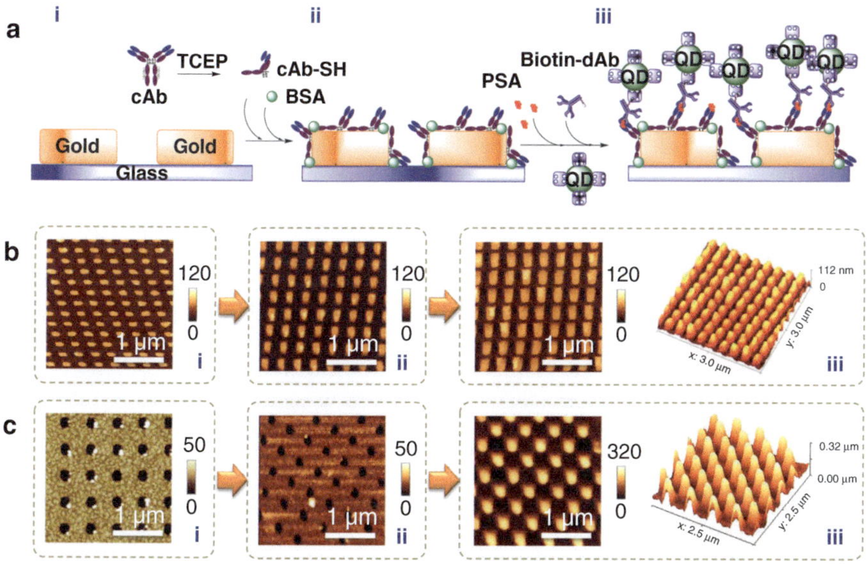

Fig. 14.6 (**a**) The sandwich assay established on gold nanoarray surface with the TCEP cleft cAb. (**b**, **c**) are the AFM images of the biofunctionalization steps (i)–(iii) in (**a**), for gold nanopillar array (p = 320 nm, R = 75 nm, and T = 50 nm) and nanohole array (p = 400 nm, R = 75 nm, and T = 100 nm), respectively. Reproduced with permission from reference Song et al. (2015a)

R = 75 nm, and T = 50 nm) are shown in Fig. 14.2b and Fig. 14.2f, respectively. The simulations show that at the wavelength of 540 nm, the gold nanopillar array can generate 25 times field enhancement. For the gold nanohole array, it is about 12 times. However, the maximum plasmonic fields are at the bottom rims of the gold nanopillar or nanohole array, and the bottom rim intensities are twice of the fields at the top rims.

Before surface modification of the gold nanochips, nanochips were cleaned by UV/O_3. After water rinse, 30 μL of cleft cAb in HEPES was dripped on the nanochip and soaked overnight in fridge at 4 °C, then it was rinsed by HEPES. BSA solution (3 mg/mL) was applied on the chip for 10 min for surface blockage, and HEPES buffer was used to rinse the chip three times to remove the unbound cleft cAb (step (ii) in Fig. 14.6a).

To attach PSA to the cleft cAb, PSA solution was prepared in HEPES buffered saline (HBS), with the concentrations of 100 ng/mL, 50 ng/mL, 10 ng/mL, 5 ng/mL, 1 ng/mL, 100 pg/mL, and 0 pg/mL, where 0 pg/mL was used as a negative control for PSA detection. 30 μL of PSA was dripped onto the cleft cAb decorated nanochip. After 1 h incubation at room temperature, the chip was rinsed three times with HBS. To detect the PSA concentration, 20 μL of biotinylated dAb at the concentration of 20 μg/mL was applied on the chip, incubated for 30 min, and rinsed by HBS. Finally, 15 μL of 20 nM streptavidin conjugated QD-655 was pipetted to the nanochip, and the chip was washed after 30 min, enabling the binding of the biotinylated dAb to QDs with streptavidin (step (iii) in Fig. 14.6a).

Each step of the nanochip immobilization was characterized by wet AFM with the quantitative imaging mode (QITM) to minimize the lateral forces applied to the sample. The AFM results in Fig. 14.6b for nanopillars and Fig. 14.6c for nanoholes show that for the assay using a PSA concentration of 100 ng/mL, the QDs are aggregating on the top of the gold surface. This is because the biotin molecules on one QD are conjugated to the streptavidin molecules on another QD, and cause the QDs to conjugate one to another. The height of the QDs piled on the gold surface is comparable to the height of QDs closely packed on the available flat gold surface. The QD aggregation can be reduced by stringent rinse of the chip after QD attachment, changing the buffer of the QDs to reduce their self-aggregation, or further diluting the concentration of the QD solution.

The surface modified nanochip is detected under a Nikon Ti Eclipse fluorescent microscope. In this system, the light from the 130 W mercury lamp passes through a band-pass filter 540/25 nm on the filter cube, and the filtered green light hits the nanochip with liquid side facing the light. The light reflected from the nanochip, including the QD emission and the scattered light, will pass through the long pass filter 605/25 nm on the filter cube for detection. The red QD emission light at 655 nm will concurrently enter an EMCCD camera to capture the fluorescent images, and a spectrograph connected to a CCD detector for QD emission spectral record. Both CCDs are cooled down to −70 °C to minimize the signal noise. The input slit for spectral detection is 500 μm wide.

In Fig. 14.7a for the PSA with the QD sandwich assay on the gold nanopillar array, the bright-field images of the nanochips look similar for all PSA concentrations, while the dark-field QD fluorescent images have bright signal in the nanopillar area while is almost totally dark on glass. This indicates that a strong QD signal is observed at 100 ng/mL of PSA, and the chip with PSA as low as 5 ng/mL can be discerned by eyes in the dark-field image. The nonspecific binding is low, as indicated by the black control chip (blank) in dark-field images. Figure 14.7b is the characterization curve for PSA concentrations for gold nanopillar array, plotted with the count intensity normalized to 100 ng/mL. Longer integration time helps to suppress the noise and enhance the weak signal, thus at 100 ms, lower LOD of 100 pg/mL is achieved according to the three-time noise level. The signal is nonlinear for the integration time of 20 ms due to the insufficient signal integration. Co-effect of the QD aggregation at a high PSA concentration (Fig. 14.6b) and tweezers effect of the plasmonic field (Juan et al. 2011) also contributes to the nonlinearity of the 20 ms characterization curve.

For the gold nanohole array, the QD bioassay existed on both the gold nanohole array and its outside gold film, and the uneven gold nanoparticles on the gold film also help enhance the QD excitation to a certain extent. The gold nanohole area in Fig. 14.7c showed a much stronger fluorescence emission than the outside gold film. In our control experiment, the totally dark image representing low nonspecific binding in the bioassay. In Fig. 14.7d, at the fluorescence integration time of 20 ms, the response of the QD emission on the gold

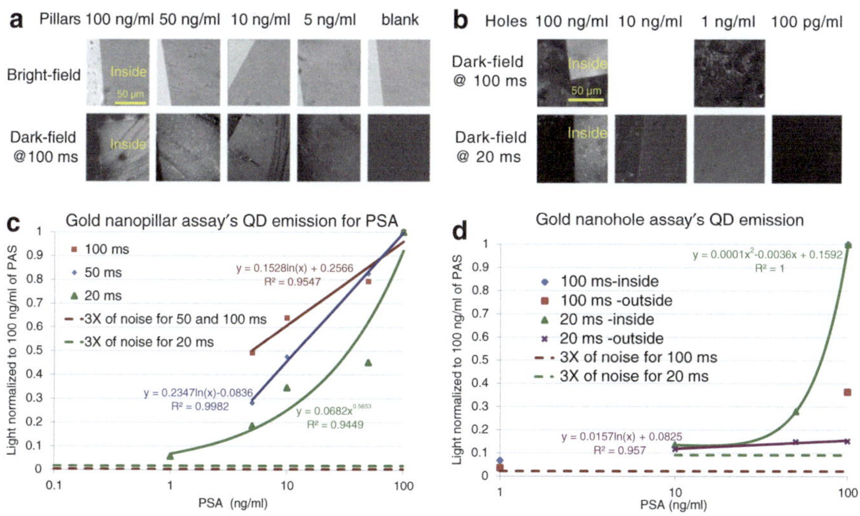

Fig. 14.7 (**a**) The bright-field and dark-field fluorescent images for gold nanopillar array (*left* shows glass area, *right* shows the gold nanopillar area) of QDs on chips for various PSA concentrations, with 100 ms integration time. (**b**) is the characterization curve of the QD assay on nanopillar array. (**c**) The dark-field images (*left* shows glass area) with integration time of 100 and 20 ms for gold nanohole array. (**d**) is the characterization curve of the QD assay on gold nanohole array. Scale bars for all images in (**a**) and (**b**) are the same. Reproduced with permission from reference Song et al. (2015a)

nanoholes is nonlinear to the PSA concentration and 1 ng/mL of PSA can be detected, while on the gold film the QD emission is very low and is relatively linear to the PSA concentration.

The gold nanopillar array outperforms the gold nanohole array with better linearity of the PSA detection curve, higher sensitivity, and lower noise level, because at the wavelength of 540 nm, the calculated plasmonic field intensity for the gold nanopillar array is in average 20–25% higher than that for the gold nanohole array. Furthermore, the gold nanopillars also have a much higher surface and volume coverage of the QDs enhanced by the plasmonic field than the gold nanoholes. According to Fig. 14.2, only the adjacent area within a distance of 10 nm to the top and bottom rims of a gold nanohole or nanopillar is significantly enhanced by plasmons. Based on the geometry of our designed gold nanoholes and nanopillars, the surface coverage of the QDs that can be excited by plasmons is about 10.21% for square gold nanoholes and 34.45% for gold nanopillars, while the volume coverage of the plasmonic field for QD excitation is about 0.515% for the square gold nanoholes, and 11.81% for the gold nanopillars. This infers that plasmonic enhancement of the QD emission is the determinant of the QD bioassay's performance. Once the gold nanostructures are designed with the strongest plasmonic peak coincident with the QD excitation wavelength, the QD bioassay's performance will be greatly improved.

14.4.2 Bioassay for Thrombin Detection with Aptamer

Thrombin is suitable for sandwich detection, because it has two aptamer binding sites: the fibrinogen recognition exosite (FRE) for a 15-mer *oligonucleotide* (APT_{15}) binding and the heparin binding site for a 29-mer *oligonucleotide* (APT_{29}) binding (Fig. 14.8). The two binding sites locating at opposite positions of thrombin enable the construction of sandwich assay for sensitive and selective thrombin detection by optical or electrochemical methods (Song et al. 2015b).

Figure 14.8 shows a schematic drawing of the sandwich bioassay on the gold nanohole array chip. 5′-Thiol 15-mer primary aptamer (APT_{15}-SH: 5′-HS-C6-TTTTTTTTTT-GGTTGGTGTGGTTGG-3′) was first established on gold surface, then the surface was blocked by thiolated poly(ethylene glycol) (PEG-750) and peptide-ethylene glycol mixed matrix (mixture of CV_3T-EG_4 and PEG-750) to reduce the nonspecific binding. Thrombin was applied, and the bound thrombin was stained by the QD-655 and biotinylated 29-mer secondary aptamer (APT_{29}: 5′-biotin-AGTCCGTGGTAGGGCAGGTTGGGGTGACT-3′) mixture (APT_{29}-QD655).

In detail, before functionalizing the nanochip, the purchased APT_{15}-SH with disulfide protection should be pretreated by cleaving the dithiol protecting group, which was performed by diluting 20 μL of aptamer (100 μM) into 100 μL DI water, adding 9 μL of TCEP (6 mM) for 30 min, and purifying with a 1 mL G-25 column. The APT_{15}-SH was diluted to 1 μM with DI water before use.

To prepare the APT_{29}-QD655, 2 μL of QD655-streptavidin conjugate (1 μM) was added into 100 μL of Tris buffer 1 (pH 7.4, 50 mM Tris·HCl, 100 mM NaCl, 5 mM KCl), and APT_{29}-biotin (100 μM) was also diluted to 100 nM by Tris buffer 1. 120 μL of the diluted aptamer was mixed with 100 μL of the above diluted QD655. The mixture was kept at least for 10 min for its conjugation.

The gold nanohole array (p = 400 nm, R = 75 nm, and T = 100 nm) chip was cleaned by UV/O_3 for 8 min, followed by dripping 70 μL of APT_{15}-SH (1 μM) for 15 h. After thoroughly cleaning with water, the chip was immersed in 10 mM PEG-750 solution (α-methoxy-ω-mercapto PEG, 750 Da, Rapp Polymere GmbH,

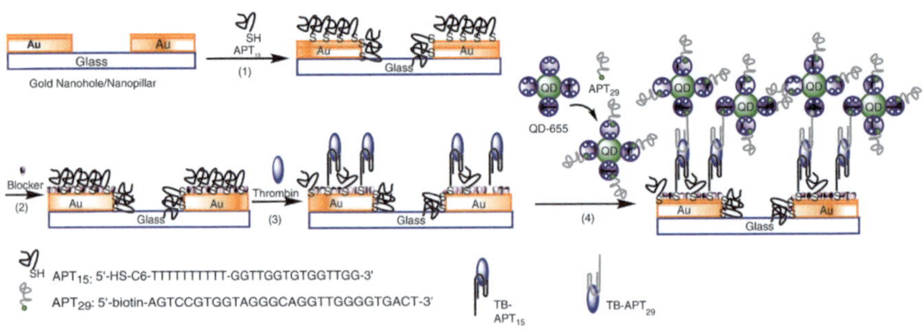

Fig. 14.8 QD-655-based sandwich bioassay for thrombin detection with aptamers on gold nanohole array (p = 400 nm, R = 75 nm, and T = 100 nm). Reproduced with permission from reference Song et al. (2015b)

Germany) for 10 min and washed with deionized water. The chip was soaked in a mixture of 20% CV_3T-EG_4 (Peptidesynthetics, UK)–80% PEG-750 in deionized water (1 μM) for 30 min and rinsed with water, after which the chip is ready for use.

Thrombin samples with various concentrations were prepared in Tris buffer 1. 30 μL of thrombin was pipetted on the above surface modified nanohole array and kept for 1 h at room temperature. Then the chip was rinsed with Tris buffer 2 (pH 7.4, 50 mM Tris·HCl, 0.1% tween 20) for three times, and stained by 20 μL of APT_{29}-QD655 for 30 min. The chip was rinsed thoroughly with Tris buffer 2, and finally the fluorescent signal was read by a fluorescent microscope. The sensor chip can be regenerated by soaking in Glycine solution (10 mM, pH 1.5) for 2 min.

The AFM characterization of the bioassay in Fig. 14.9a shows that the gold nanohole array surface after APT_{15}-SH and blocker anchoring, the aptamers aggregated at the nanohole sides with sizes of 50–150 nm and height of 20 nm, due to the end-to-end DNA stacking interactions occurring on duplex DNA, or ssDNA whose local density is close to duplex DNA (Kosaka et al. 2013). The aptamer islands occurred on the edges of the gold nanoholes for their lowest surface tension. After the QDs sandwich assay is established, the thrombin and QDs occupies and closely packed on the top of the aptamer islands, with a maximum height up to 150 nm, and a smaller aptamer island will have a lower QD aggregation height.

On the gold film surface, QDs form a uniform layer with only a few nanometers high, which implies an easier immobilization of DNA on the flat gold surface than gold nanohole array surface. The QD signal in the bioassay on the nanoholes is much stronger than gold film, due to the plasmonic on the gold nanoholes and the blocker layer which is favorable to gold nanoholes.

As presented in Fig. 14.9b, the thrombin can at least be detected by naked eyes under the Nikon Ti Eclipse fluorescent microscope (used and described for the abovementioned PSA detection) at the thrombin concentration of 10 ng/mL. The more sensitive way to quantify the bioassay is the fluorescent spectrum as plotted in Fig. 14.9c with 100 ms integrated time. The unnoticeable signals on the control of the gold nanohole array and the gold film testified the low nonspecific binding of the assay.

The characterization curves of these bioassays together with their standard deviations are plotted in Fig. 14.9d, showing the fluorescent intensity inside the gold nanohole array drops almost linearly as the thrombin concentration decreases. The prolonged integration time improves the sensitivity, and the LODs for the integration time periods of 20, 50, and 100 ms are, respectively, 5, 2, and 1 ng/mL, according to their respective three-time noise levels. Considering the sparse aptamer islands in Fig. 14.9a, the sensitivity of the bioassay can be increased with longer aptamer incubation time for achieving a ssDNA monolayer or a higher density aptamer layer. The sensitivity can also be increased by designing the plasmonic peak of the nanohole array close to the QD excitation wavelength or reducing the QD aggregation by stringent rinsing.

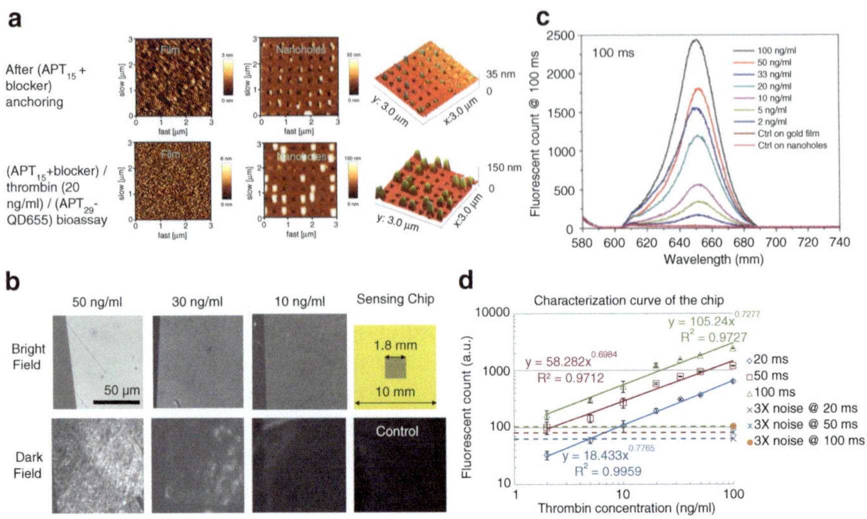

Fig. 14.9 (**a**) AFM characterization of the thrombin bioassay on gold nanohole array ($p = 400$ nm, $R = 75$ nm, and $T = 100$ nm) and gold film. (**b**) The bright-field and dark-field fluorescent images (*left* shows glass area, *right* shows the gold nanohole area) of QDs on chips for various thrombin concentrations, with 50 ms integration time. Scale bar is the same for all images. (**c**) is the fluorescent spectra with integration time of 100 ms for gold nanohole array. (**d**) Characterization curves of the QD assay on gold nanohole array for thrombin detection. (**a**, **c**, and **d**) are reproduced with permission from reference Song et al. (2015b)

14.4.3 Bioassay for Procalcitonin Detection with Whole Antibody

Procalcitonin (PCT) is a 116 amino acid residue peptide with molecular weight of about 13 kDa and a precursor of calcitonin. Because cytokines and endotoxin released during bacterial infections inhibit the final step synthesis of calcitonin, PCT level will increase in patients with bacterial infections. PCT increase is an early and highly specific indicator to clinically relevant bacterial infections and sepsis (such as C-reactive protein (CRP) and white blood cell count), and a low serum PCT level accurately rules out the diagnosis of bacteremia (Chirouze et al. 2002). PCT-guided antibiotic therapy, which the physicians follow an antibiotic treatment algorithm based on the PCT value, has demonstrated a significant reduction compared with standard therapy in antibiotic prescription at inclusion, duration of antibiotic therapy, total antibiotic exposure days, and the length of staying in the ICU without adverse outcome of increased mortality. Since PCT-guided antibiotic therapy is more efficient, it reduces medical cost, and decreases the excessive use of antibiotics, which is believed to be the main cause of spread of antibiotic-resistant bacteria (Kopterides et al. 2010). Thus a point-of-care device for sepsis detection is expected to improve the ease, speed, and cost of procalcitonin tests, increase the use of procalcitonin-based antibiotic therapy, and reduce the misuse of antibiotics.

The point-of-care device for PCT detection is developed by our team as presented in Fig. 14.10 (Sun et al. 2016). For preparation of the gold nanopillar array chip

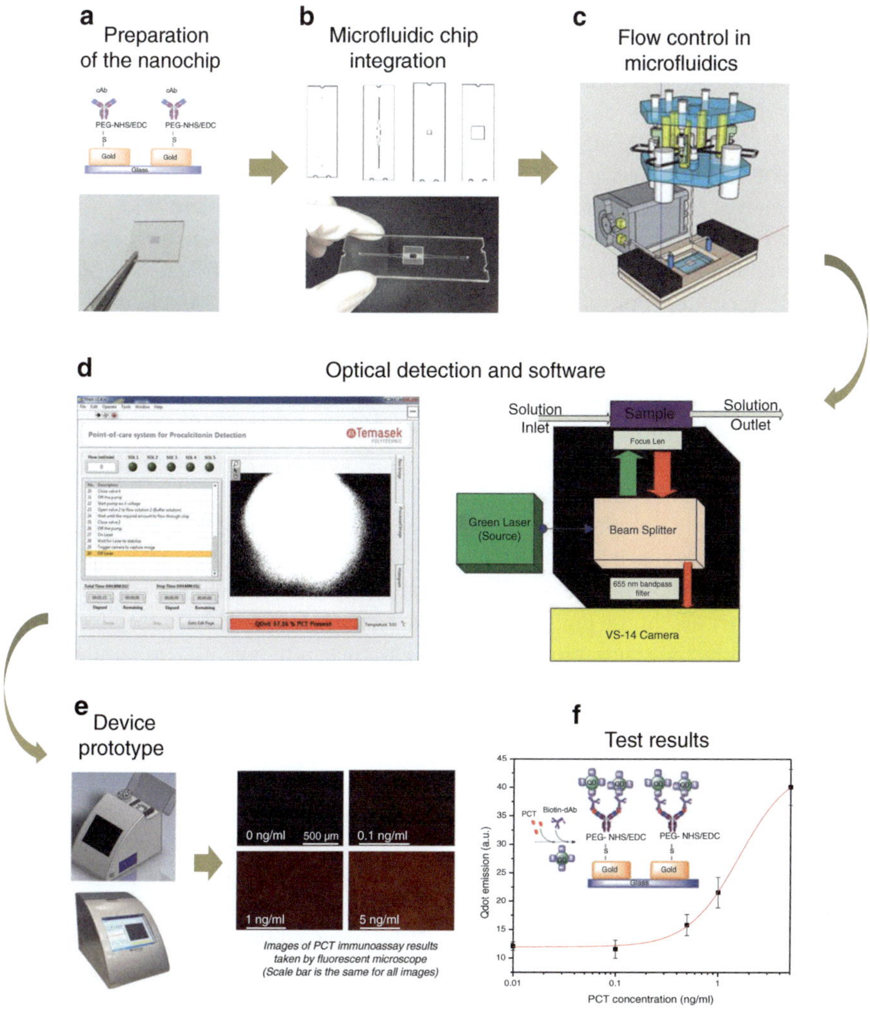

Fig. 14.10 The development of the point-of-care device for PCT detection, with plasmonic-enhanced dye emission. (**a**) The fabricated gold nanopillar array and its surface preparation; (**b**) integration of the nanochip with a microfluidic chip; (**c**) the microfluidic chip holder and its liquid delivery system; (**d**) optical setup and the Labview software; (**e**) the prototype and (**f**) the test results of the point-of-care device

($p = 320$ nm, $R = 75$ nm, and $T = 50$ nm) in Fig. 14.10a, the chip was in turn cleaned with isopropanol, acetone, and deionized water, and dried at room temperature with nitrogen gas. The chip was incubated with 1 mM (in ethanol) PEG7 thiol acid (thiol-COOH, Poly-pure AS, Norway) overnight at room temperature to form an amine-reactive self-assembly monolayer (SAM) and washed thoroughly with ethanol and dried. Than a mixture of 75 mM of sulfo-*N*-hydroxysuccinimide (sulfo-NHS) and 15 mM of 1-ethyl-3-(3-dimethylaminopropyl) carbodiimide (EDC) (Bio-Rad,

Hercules, CA, USA) was dripped on the surface to activate the carboxylic group of the SAM for 15 min. Next, 30 μL of 50 μg/mL anti-PCT (Anti-Procalcitonin mouse monoclonal antibody, Abcam) was spotted on the sensor surface and incubated for 2 h. Finally, the sensor chip was immersed in 1 M of ethanolamine-HCl solution (Bio-Rad, Hercules, CA, USA) for 15 min to deactivate the excessive carboxylic acid groups, followed by a rinse with deionized water and dried with nitrogen gas.

The above prepared nanochip was integrated with a microfluidic chip in Fig. 14.10b. The microfluidic chip was fabricated with sandwiched low-cost PMMA polymer and adhesive Mylar® sheet, each having CO_2 laser fabricated cut-through patterns. The patterns of the four layers are also shown in Fig. 14.10b, with layer 1 using PMMA to have through holes for the inlet and outlet of the microfluidic, layer 2 on a double-side adhesive Mylar® defining the microfluidic channel, layer 3 on a single-side Mylar® to expose the nanopillar array structure to the fluidic channel, and layer 4 with PMMA to accommodate the nanochip in its cavity. The four layers were laminated and bound together by the Mylar® adhesive layers, with the nanopillar array inside the fluidic channel.

The fabricated microfluidic chip will be sealed inside a microfluidic chip holder (the plate with two black bars on its sides in Fig. 14.10c, the holder was invented by SIMTech, A*STAR, and now is commercially available), which provides the interface to connect the fluidic pipes for different washing buffer and bio-reagents to the microfluidic chip. Because the microfluidic chip has only one inlet, in our design, a small 8-into-1 manifold is connected to the microfluidic chip holder. Each fluidic pipe at the multiple-channel-end of the manifold connects to a different reagent, and has an individual valve controlled by home-made Labview software (electronically activated through the communication of the Labview software to an USB6002 data acquisition hardware purchased from National Instruments) to selectively connect the reagent to the microfluidic channel. A mechanical jig (the hexagonal shape in Fig. 14.10c) is designed and fabricated for holding a vial and a valve for each bio-reagent in one column of the jig. When a bio-reagent is selected to flush the microfluidic chip, the valve for this reagent will be open while the valves for the rest reagents will be shut, and the time for injecting the selected reagent will be controlled by parameters input (or preset) in the Labview software. A commercial available peristaltic pump (grey box in Fig. 14.10c) is also controlled by the home-made Labview software and the USB6002 data acquisition hardware to determine the time and pumping speed when a reagent is drawn into the microfluidic chip. The automation system formed with the bio-reagents, microfluidic chip and its holder, the valves and the pump enables the labor free operation of the sandwich assay establishment and detection on the chip.

The optical system configuration in Fig. 14.10d is similar to the Nikon Ti Eclipse fluorescent microscope used for PSA and thrombin detection, except being far less expensive and highly compact. The 532 nm wavelength laser passes through the beam-splitter via an optical fiber cable and illuminates on the nanochip inside the microfluidic chip. Upon the establishment of the sandwich assay for PCT detection, the emission of the QD-655 at the wavelength of 655 nm will transmit through a 655 nm band-pass filter and be detected by an intensity recordable CCD camera. The optical system in Fig. 14.10d is packaged in a black box made of light absorption material to avoid light scattering and reflection to achieve high signal-to-noise ratio. The on and off of the laser source and the exposure of the camera are also controlled by the home-made Labview software. So after the bioassay establishment for PCT detection, the optical signal will be read and processed automatically in the software shown on the panel of the computer screen in the point-of-care device. The software is multifunctional for the tasks of running the sequences for establishing the bioassay on the nanochip after analyte injection, turning on the laser and camera at the right time to take the QD emission image with 4 s of exposure time, analyzing the image to correlate to the PCT concentration to the light intensity in the image, displaying the quantified PCT concentration on the screen and saving the data.

For the PCT test, the gold nanopillar array integrated microfluidic chip was place in its holder, and a drop of specimen (such as PCT in serum) was injected into a vial through a cover at the top of the point-of-care device (Fig. 14.10e). Once the start button was pressed, the microfluidic system first injected 100 μL of 1% BSA solution through the nanopillar array surface to incubate for 10 min to block any non-specific binding. After 2 min flush of phosphate-buffered saline with Tween-20 (PBST), analyte with PCT biomarker was drawn to the microfluidic channel and incubated for 10 min to react with the capture antibodies previously immobilized on the nanopillar array. PBST was flowed for 2 min to remove unreacted PCT, and 50 μL of biotin-dAb (20 μg/mL, biotin conjugated procalcitonin detection antibody, Raybiotech) was injected and incubated for 10 min. After that, the chip was flushed with PBST again for 2 min, followed by applying 50 μL of streptavidin conjugated QD-655 (20 nM, Life Technology) and incubating for 10 min. The unbound QD-655 was flushed away by 2-min PBST washing. The sandwich cAb-PCT-(biotin-dAb)-(QD-655) immunoassay was built on the nanopillar chips (Fig. 14.10f) and ready for fluorescent automatic reading by the built-in CCD camera.

Various PCT concentrations of 10, 5, 1, 0.5, and 0.1 ng/mL were tested with the point-of-care device. The images are shown in Fig. 14.10f, and the light intensity analyzed by the Labview program is used to draw the characterization curve. For each PCT concentration, three individual microfluidic chips were tested to obtain the error bar. The QD emission versus PCT concentration indicated a LOD at 0.5 ng/mL, which is enough for the indication of sepsis in clinics. This point-of-care device has relatively high sensitivity, is labor free with fully automated test process, and can obtain the results with 30 min. Because the nanochip and the microfluidic chip are mass produced, the test cost is also relatively low at around ten US dollars per test. Because the PCT detection is based on CCD image intensity, it is feasible to further develop the nanochip for multiple biomarker detections without much additional cost by spotting different cAbs at various locations of the nanochip.

Conclusion

This book chapter introduces the principle, design, fabrication, surface functionalization of the LSPR biosensing nanochips, as well as its incorporation into a microfluidic chip for point-of-care devices.

Irregular nanopatterns tend to broaden the LSPR peak and reduce the Q factor of the plasmonic field, so our work demonstrates the simulation and design of periodic gold nanohole and nanopillar arrays for biosensing, which helps to increase the sensitivity of the LSPR biosensors. To reduce the cost of the LSPR biosensors, we developed methods for mass production of the gold nanoarrays on glass substrate by fabricating the 4″ nickel mold for nanoimprinting, and the gold nanostructures fabricated are with high quality compared with the original designs, proven by SEM and AFM images, as well as the characterized optical spectra.

The gold nanoarray is demonstrated for point-of-care detections by a direct assay for cTnI detection by LSPR peak wavelength shift, and by plasmonic-enhanced fluorescence detection for PSA, thrombin, and PCT detections. The LODs of the gold nanoarray chip for cTnI, PSA, thrombin, and PCT are, respectively, at the concentrations of 0.55, 0.01, 1, and 0.5 ng/mL. The difference in LODs is related to the bioassay design, biomarker property, and detector sensitivity. These LODs all meet the clinical sensitivity of these biomarkers in blood, thus our research has shown great potential of applying the gold nanoarray-based point-of-care devices in clinics.

Acknowledgments The authors would like to express their gratitude to the A*STAR, Singapore, for funding the project 102 152 0014, and Ministry of Education, Singapore, for the project MOE2013-TIF-1-G-024.

References

Altissimo M (2010) E-beam lithography for micro−/nanofabrication. Biomicrofluidics 4:026503

Anker JN, Hall WP, Lyandres O, Shah NC, Zhao J, Van Duyne RP (2008) Biosensing with plasmonic nanosensors. Nat Mater 7:442–453

Austin MD, Ge H, Wu W, Li M, Yu Z, Wasserman D, Lyon SA, Chou SY (2004) Fabrication of 5 nm linewidth and 14 nm pitch features by nanoimprint lithography. Appl Phys Lett 84(26):5299–5301

Chen Y (2015) Nanofabrication by electron beam lithography and its applications: a review. Microelectron Eng 135:57–72

Chirouze C, Schuhmacher H, Rabaud C, Gil H, Khayat N, Estavoyer JM, May T, Hoen B (2002) Low serum procalcitonin level accurately predicts the absence of bacteremia in adult patients with acute fever. Clin Infect Dis 35:156–161

Deng J, Wong TI, Sun LL, Quan C, Zhou X (2016) Acceleration of e-beam lithography by minimized resist exposure for large scale nanofabrication. Microelectron Eng 166:31–38

Ding T, Hong M, Richards AM, Wong TI, Zhou X, Drum C (2015) Quantification of a cardiac biomarker in human serum using localized surface plasmon resonance. PLoS One 10(3):e0120974

Ghaemi HF, Thio T, Grupp DE, Ebbesen TW, Lezec HJ (1998) Surface plasmons enhance optical transmission through subwavelength holes. Phys Rev B 58(11):6779–6782

Juan ML, Righini M, Quidant R (2011) Plasmon nano-optical tweezers. Nat Photonics 5:349–356

Kopterides P, Siempos II, Tsangaris I, Tsantes A, Armaganidis A (2010) Procalcitonin-guided algorithms of antibiotic therapy in the intensive care unit: a systematic review and meta-analysis of randomized controlled trials. Crit Care Med 38(11):2229–2241

Kosaka PM, González S, Domínguez CM, Cebollada A, San Paulo A, Calleja M, Tamayo J (2013) Atomic force microscopy reveals two phases in single stranded DNA self-assembled monolayers. Nanoscale 5:7425–7432

Kuwata H, Tamaru H, Esumi K, Miyano K (2003) Resonant, light scattering from metal nanoparticles: practical analysis beyond Rayleigh approximation. Appl Phys Lett 83:4625

Lee K-S, El-Sayed MA (2006) Gold and silver nanoparticles in sensing and imaging: sensitivity of plasmon response to size, shape, and metal composition. J Phys Chem B 110(39):19220–19225

Mayer KM, Hafner JH (2011) Localized surface plasmon resonance sensors. Chem Rev 111:3828–3857

Murphy CJ, Sau TK, Gole AM, Orendorff CJ, Gao J, Gou L, Hunyadi SE, Li T (2005) Anisotropic metal nanoparticles: synthesis, assembly, and optical applications. J Phys Chem B 109:13857–13870

Palik E (1998) Handbook of optical constants of solids. Elsevier, New York

Rai-Choudhury P (ed) (1997) Handbook of microlithography, micromachining and microfabrication. IET, Hertfordshire, p 349

Sherry LJ, Chang S-H, Schatz GC, Van Duyne RP, Wiley BJ, Xia Y (2005) Localized surface plasmon resonance spectroscopy of single silver nanocubes. Nano Lett 5(10):2034–2038

Song HY, Hobley J, XD S, Zhou X (2014) End-on covalent antibody immobilization on dual polarization interferometry sensor chip for enhanced immuno-sensing. Plasmonics 9(4):851–858

Song HY, Wong TI, Sadovoy A, Wu L, Bai P, Deng J, Guo S, Wang Y, Knoll W, Zhou X (2015a) Imprinted gold 2D nanoarray for highly sensitive and convenient PSA detection via plasmon excited quantum dots. Lab Chip 15(1):253–263

Song HY, Wong TI, Guo S, Deng J, Tan C, Gorelik S, Zhou X (2015b) Nanoimprinted thrombin aptasensor with picomolar sensitivity based on plasmon excited quantum dots. Sens Actuators B 221:207–216

Sun LL, Ng W, Zhou X, Leo YS, Wong TI (2016) Plasmonic biosensing based microfluidics point-of-care system for procalcitonin detection. Singapore provisional patent application 10201606001Q

Wong TI, Han S, Wu L, Wang Y, Deng J, Tan CYL, Bai P, Loke YC, Yang XD, Tse MS, Ng SH, Zhou X (2013) High throughput and high yield nanofabrication of precisely designed gold nanohole array for fluorescence enhanced detection on biomarkers. Lab Chip 13(12):2405–2413

Wu L, Bai P, Li EP (2012a) Designing surface plasmon resonance of subwavelength hole arrays by studying absorptance. J Opt Soc Am B 29(4):521–528

Wu L, Bai P, Zhou X, Li EP (2012b) Reflection and transmission modes in nanohole-array-based plasmonic sensors. IEEE Photonics J 4(1):26–33

SERS Biosensing and Bioimaging: Design and Applications in Cancer Diagnostics

15

Kien Voon Kong

15.1 Introduction

Enhancement of cancer diagnostics based on biomarker detection and nanoparticle tags has been a focus of current scientific and medical research. Raman spectroscopy was one of the first methods used to identify different diseases based on their unique Raman signatures. The method is convenient and noninvasive, and has had many practical uses in surgery for monitoring respiratory and anesthetic gas mixtures. Unfortunately, owing to its weak efficiency and low sensitivity, Raman spectroscopy is not the optimal choice for routine use in biomedicine.

Surface-enhanced Raman spectroscopy is the advanced alternative to Raman spectroscopy. The high sensitivity, spatial resolution, and chemical sensitivity of SERS facilitate classification and analyte detection through amplification. SERS may well be at the frontline of medical diagnostics in near future.

This chapter focuses on the utilization of SERS and its role in cancer detection based on biomarkers and nanoparticle tags. Nanometer-sized particles, when conjugated with peptides, monoclonal antibodies, or other small molecules that function as biomolecular targeting ligands, can then be used to detect tissues and cells of interest. Among the many recent advances that have led to the development of different types of nanoparticles, SERS nanoparticles for in vivo biomarkers sensing is of our particular interest for this chapter. A brief history of SERS at the start of the chapter will be followed by a discussion of its unique optical properties, enhancement mechanisms, and usage, with an emphasis on the major SERS-active nanostructure materials. Usage of SERS substrates and nanostructures in tumor and biomarker detection will follow, along with its application and potential in biomedicine. A description of SERS nanoparticles for in vitro diagnostics, in vivo

K.V. Kong
Department of Chemistry, National Taiwan University, Taipei, Taiwan
e-mail: kvkong@ntu.edu.tw

P. Chandra et al. (eds.), *Next Generation Point-of-care Biomedical Sensors Technologies for Cancer Diagnosis*, DOI 10.1007/978-981-10-4726-8_15

spectroscopic detection, label-free molecular detection, and image-guided cancer surgery will follow. This chapter will end with a discussion on future challenges and potential of SERS in human study and application, with final remarks.

15.2 Advantages of SERS

In 1973, Martin Fleischmann and his colleagues first observed SERS as strong Raman spectra of pyridine adsorbed onto a rough silver electrode surface (Fleischmann et al. 1974). This discovery, however, was not fully explained until 1977, when two groups independently saw that the enhanced signals could be accounted for merely by the concentration of the scattering species. Each group later proposed a theory to explain the observed enhancement. Jeanmaire and Van Duyne proposed a mechanism based on electromagnetic field enhancements, while J. Alan Creighton suggested a chemical enhancement (Albrecht and Creighton 1977; Jeanmaire and Van Duyne 1977). Both theories are recognized today as explanations for the amplification of Raman scattering intensities. Of the two studies, George C. Schatz and Moskovits found that electromagnetic field enhancement contributed much greater to enhancement of Raman signal (King and Schatz 1979; Moskovits 1978). In later experiments, other molecules were observed to adsorb on metals such as silver, gold, and aluminum (Henglein 1993; Li et al. 2010a; Linnert et al. 1993; Mulvaney 1996). Usage of SERS has since then grown exponentially.

Compared to the multitude of other methods used in biomedical imaging and detection, SERS-based platforms have the advantages of fast speed and high sensitivity. SERS was demonstrated to have high enhancement factors and sensitivities beyond that of fluorescence. While fluorescent-imaging is more time-consuming, SERS is almost instantaneous, and its signals are much more resilient against photobleaching and photo-degradation that result from the use of other techniques (Doering and Nie 2003).

SERS particles can be designed for simultaneous excitation with a single laser beam in the near-infrared region (NIR), and their SERS signals cover approximately the same region (200–3500 cm^{-1}). The high effectiveness of SERS in the NIR region is significant because auto-fluorescence and light attenuation from tissues in this region are low. Since it is not susceptible to the interference of water, SERS is well suited for biological imaging (Aroca et al. 2004). SERS has also overcome the detection limits of many of the current imaging technologies.

15.3 Mechanisms in SERS

Both electromagnetic effect and chemical effect operate simultaneously in SERS. Exciting surface Plasmon resonances of nanoparticles by electromagnetic mechanism leads to a huge increase in the electromagnetic field strength at the particle surface. The chemical enhancement mechanism arises when molecules localize at specific surface sites, including terraces and steps, electronically couple with the surface, producing an enhancement effect to the Raman signal. We take a further look at electromagnetic as well as chemical enhancements in this section.

Metallic particles must rely on their plasmonic properties to increase the local strength of an electromagnetic field. Surface Plasmon polaritons spread along the metallic surface plane and dielectric interface. Here, the field strength exponentially decreases into the dielectric having a decay length of approximately 200 nm (Willets and Duyne 2007). Localized surface plasmons, on the other hand, concentrate fields locally around a metallic nanoparticle. The electron cloud of the particle is pushed and pulled with the oscillating polarity of the incident field, leading to the particle to emit its own dipole field, which in turn enhances the incident electromagnetic field (Kelly et al. 2003; Liz-Marzán 2006). When the incident radiation is closer to the localized surface plasmon resonance (LSPR) frequency, field enhancements become much greater. LSPR frequency is dependent on the nanoparticles' optical properties as well as on its surroundings (Henglein 1993). Electrostatic approximation of the dipole radiation can be used to assess the magnitude of the electromagnetic enhancement. The metal sphere will emit its own dipole field upon irradiation, thus enhancing the local optical field that is used to enhance a nearby molecule (Fig. 15.1) (Schlücker 2014).

As particles decrease in size, less surface area is available to collect the incident light. The highest enhancements for gold and silver nanoparticles with spherical shapes are typically observed in *h* size range from 40 to 60 nm in diameter (Hong and Li 2013; Kreibig and Fragstein 1969; Stamplecoskie et al. 2011). Theoretically, the optimal condition for maximal SERS intensities occurs when the LSPR peak has a high overlap with the laser line and Raman-scattered frequencies, though this is not always the case (Li et al. 2014; McFarland et al. 2005). Experimental observations, however, have exhibited that for single particles, Plasmon location is optimal between that of the laser line and Raman bands (Sivapalan et al. 2013).

The electromagnetic fields surrounding nanostructures are greatest where sharp edges or vertices are located, allowing enhancement factors as high as 10^{10} (Fig. 15.2) (McMahon et al. 2010; Xu et al. 1999, 2000). The strength of EM field, however, drops with increasing distance from the particle's surface. According to theoretical calculations, triangles and cubes offer maximal local field enhancements

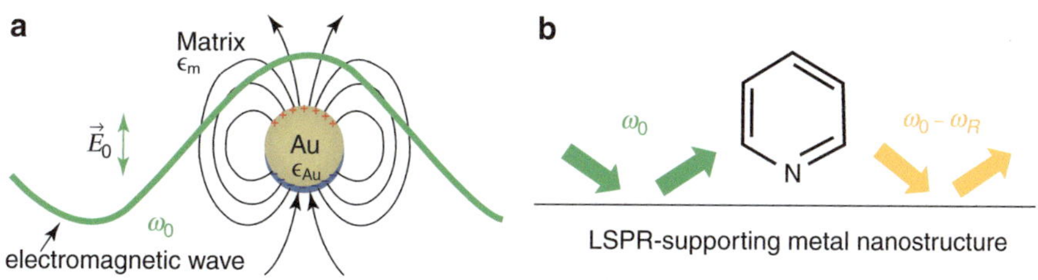

Fig. 15.1 (**a**) Electromagnetic enhancement in SERS by LSPR on a gold nanoparticle. (**b**) The enhancement of the "incoming" field (ω_{inc}, *green*) with the elastic light scattering on a LSPR-supporting metal nanostructure. Reprinted with permission by Wiley-VCH Verlag GmbH & Co

Fig. 15.2 Simulation of the distribution of $|E|^2$ intensity of local (*top*) and nonlocal (*bottom*) for various shapes of nanowire (diameters = 50 nm). The nanowire is outlined in *white*. The direction and polarization of incident light are indicated (*k*: wave vector, *E*: electric field, *B*: magnetic field density). All intensity profiles are normalized by their maximal intensity. Although the structures with sharper tips show greater maximal enhancements, these fields are highly localized and drop off instantaneously with distance. Reprinted with permission from McMahon et al. (2010). Copyright 2010 American Physical Society

7–16 times greater than circles. These enhancements become similar at distances just 0.5 nm away from the surface (McMahon et al. 2010). For these particles to provide greater SERS signals, it is crucial that Raman reporter molecules reach and adsorb to these specific locations along the edges and vertices.

The electronic structural changes when a free molecule is attached onto a metallic surface gives rise to the mechanism known as chemical enhancement (CE). Albrecht and Chreighton first proposed this mechanism, which was later confirmed by experimental data (Albrecht and Creighton 1977; Furtak and Macomber 1983; Lombardi et al. 1984). The CE mechanism has a much shorter ranged effect compared with the EM mechanism, as the CE mechanism is on the Angstrom length scale and requires the reporter to be either chemically bonded or directly adsorbed on to the metallic surface (Jensen et al. 2008; Morton and Jensen 2009). CE mechanism enhancements are also weaker than those of the EM mechanism. A theoretical description that incorporates the molecular polarizability, charge transfer, and the resonances of the particle Plasmon in a single expression was developed by Lombardi and Birke (Jensen et al. 2008; Lombardi and Birke 2008, 2009; Valley et al. 2013), but because the occurrences of each component are not independent of each other, the magnitude of each component with respect to the overall CE has proved difficult. As shown in Fig. 15.3, CE is attributed to four mechanisms: (a) interaction of molecule and nanoparticle in ground state, (b) excitation wavelength resonant with HOMO (Highest Occupied Molecular Orbital) → LUMO (Lowest Unoccupied Molecular Orbital) transition, (c) excitation wavelength resonant with charge-transfer transitions (charge-transfer (CT) resonance) between nanoparticle and molecule, and (d) excitation wavelength resonant with a strong field of Plasmon excitations in the nanoparticle.

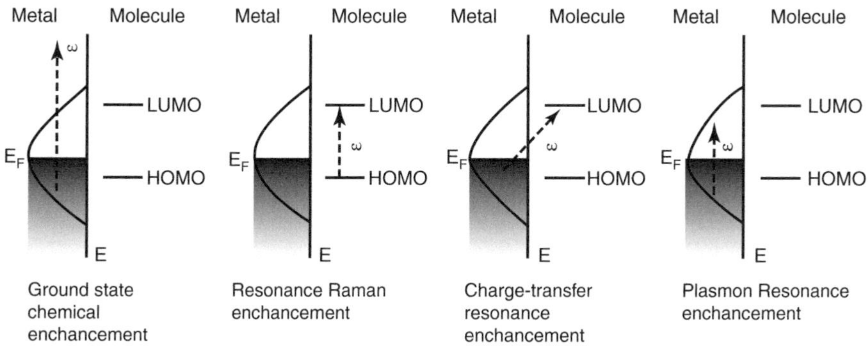

Fig. 15.3 Illustration of interrelated processes for SERS. The enhancement factor for charge-transfer (CT) resonance is typically dependent on the energy gap of HOMO-LUMO. Reprinted with permission from Royal Society of Chemistry

15.4 SERS Nanoparticle Tags

SERS tags possess high sensitivity and enhanced multiplexing and quantitative abilities (Doering and Nie 2003; Doering et al. 2007; Keren et al. 2008; Qian et al. 2008; Qian and Nie 2008; Zavaleta et al. 2009). SERS tags are thus the preferred choice over fluorescent probes in molecular imaging applications as well as in spectroscopic biological detection.

Controlled aggregation to form dimers or higher order multimeric structures creating hot spots produces enhancement factors which may have several orders of magnitudes greater than the single particle (Lee et al. 2006; Wang et al. 2005). An efficient method was developed by the Xia group to create dimers of silver spheres. This required use of iron (III) nitrate to etch and dimerize polymer-stabilized silver nanocubes in an ethanol solvent (Fig. 15.4a) (Li et al. 2010b). The nanocubes are transformed into spherical shapes by the iron salt, with 66% of the product yield being dimers and the rest, monomers. Nanocubes of varying size, from 40 to 100 nm, could be used in this process to yield dimers. The Xia group later demonstrated this method to produce SERS tags of the dimers for cancer cell imaging (Xia et al. 2013). The dimers with the adsorbed Raman reporter were modified using the Stober method with a silica layer, then coated with targeting ligands and dextran. The Schlucker group also presented an efficient method for producing colloidal dimers. In this process, PVP-stabilized gold nanospheres were aggregated using sodium chloride (Steinigeweg et al. 2011). The aggregates were then fractionated into monomeric, dimeric, and trimeric solutions using density gradient centrifugation. To assemble the metal nanoparticles in a controlled fashion, DNA was also used as a scaffold (Fig. 15.4b, c) (Mirkin et al. 1996; Xu et al. 2015). DNA was also used to produce quadramers in pyramidal shapes. Controlling the amount of DNA molecules on each of the molecule surfaces also helped narrow the gap between the metal particles. Placing the reporter molecules within high-field regions of the dimers has yielded SERS enhancements as high as 10^{12} (Lim et al. 2010, 2011).

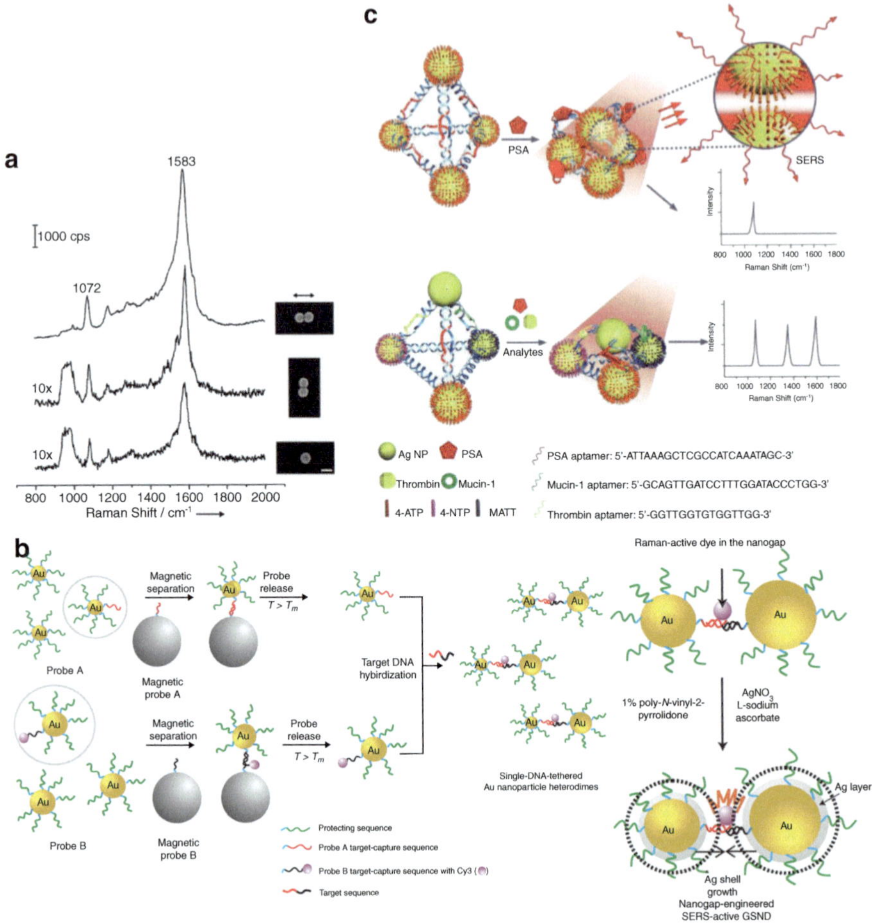

Fig. 15.4 (**a**) SERS spectra for a silver nanoparticle (*bottom*) and a dimer of silver nanoparticles where its axis is parallel (*top*) and perpendicular (*middle*) to the polarization of laser. (**b**) A schematic represents the use of DNA modification and magnetic purification of gold nanoparticles for the gold heterodimers and subsequently used for the preparation of gold–silver core–shell nanodumbbells (GSND). (**c**) Scheme of self-assembled silver pyramids mediated singleplex and multiplex SERS assay for PSA, PSA, thrombin, and mucin-1. Reprinted with permission from Wiley-VCH Verlag GmbH & Co and 2010 Nature publisher Group

The cores of SERS tags most often contain gold particles as they are considered to be nontoxic and chemically inert. Large concentrations of gold nanoparticles, greater than the amount needed for SERS detection, have previously been administered in patients during cancer clinical trials for CT imaging (Popovtzer et al. 2008). Silver nanoparticles are plasmonically more active in the visible spectrum than gold; however, their advantages are lost in the NIR region, and silver nanoparticles are not only toxic to mammalian cells, but their surfaces are also prone to oxidation

(Bondarenko et al. 2013; Greulich et al. 2012). Thus, gold is one of the safer and more preferred choices in clinical applications in the NIR region.

SERS tags are best prepared using reporters with inherent large Raman cross sections. A reporter with an adsorption spectrum that overlaps with the laser line of the wavelength being used can invoke surface enhance resonance Raman scattering (SERRS), thus furthering the total enhancement 10–100 fold (McNay et al. 2011). The use of resonant reporters also helps ensure that the reporter signal outweighs Raman bands of environmental contaminants that may potentially adsorb to the surface of the particle. To counteract against such contaminants as well as prevent desorption from the metal surface, reporter molecules should have moieties containing nitrogen and/or sulfur atoms that form strong surface bonds. Increased surface binding affinity also invokes greater overall enhancements. Chemical enhancements (CE) arise from a ligand-metal bond formation and electromagnetic (EM) fields are greatest when the molecule is anchored to the surface of the metal. Surface adsorption can be facilitated by electrostatic interactions between reporters and their respective ligands. A rational design for SERS reporters that exhibit signal intensity as well as enhanced stability has yet to be reported. Reporters based on triphenylmethine parent structures with a lipoic acid anchor group were developed by Olivo and colleagues (Samanta et al. 2011). These reports produced strong SERS signals and exhibited an improved stability, which was afforded by the bidentate thiol binding of the lipoic acid group to the surface of the metal. Olivo and colleagues also developed NIR resonant dyes that exhibited a sensitivity 12-fold greater than the average, owing to the stability of a dithiol anchoring group (Samanta et al. 2011). This method could be further used to functionalize other dyes that also contain a lipoic acid anchoring group while causing little minimal perturbation to the original electronic structure of the dyes. Further research pertaining to NIR resonant dyes of different electronic structures would significantly benefit in multiplex SERS tagging applications.

15.5 Label-Free Detection and Identification

SERS enables both identification and detection without the hassle of labeling analytes, providing thorough spectroscopic information, along with information about the orientation and conformation of the molecules that are adsorbed (Das et al. 2009; Yu and Golden 2007). This method has been used to identify pathogens by comparing the relative intensities of multiple bands of molecular components of pathogens.

Genotypic analysis using polymerase chain reaction (PCR) is employed for identifying bacteria within an infection. PCR first amplifies the nucleic acid sequences of the organism, and these genetic strands are then run through a microarray for classification (Nadkarni et al. 2002). The primer, which determines the nucleic acid sequence that needs to be amplified, is required by PCR to be readily available for the DNA or RNA of the bacteria of interest. Some bacteria samples may also require a long period of time to culture. This time frame proves disadvantageous. In some cases, the duration of this period may exceed that of the therapeutic window in which clinicians must intervene to prevent the infection from worsening.

Clear diagnoses and earlier intervention may be aided by a SERS method that helps detect and identify the bacterial strains. This SERS platform would greatly aid clinicians when facing cases of bacterial infection. For simple sampling of the bacterial cells on a SERS active-substrate, clinicians can use the Raman fingerprint to determine the relative bacteria concentrations within the cellular membrane (Jarvis and Goodacre 2004; Zeiri et al. 2004). This, however, may not be applicable to all bacterial cells. Cell lysates, for one, owing to the array of components that compete for surface adsorption and as thus denatures the proteins and alters the spectra over a period of time, are much more difficult to pinpoint let alone study. Intact membrane components, because of fewer exposed components and the fact that proteins within the cellular membrane are less likely to denature upon adsorption, exhibit more consistency in their SERS spectra. Overall, bacterial strain identification will rely on the reproducibility and uniqueness of the SERS spectra among the various species of pathogens.

In SERS detection and identification of bacterial strains, it is crucial to note the unique spectral features of each sample that allows specificity and high certainty in their classification. In one research, a multivariate analysis technique was developed by Ziegler and colleagues, which compared SERS data obtained from a vast record of spectra of different pathogenic species (Patel et al. 2008). Each bacterial strain that was studied by the group is considered to possess its own particular spectral character. This characteristic is assigned a "barcode" for identification and classification. Upon filtering for noise, the normalized SERS spectrum is analyzed as a function of frequency through its second derivative. Black or white assigned blocks with the Raman spectra curvature create the barcode that marks the bacterial strain. The barcode is then run through a record of known barcodes, and using the principal component analysis (PCA), the bacterial species can be identified. A low-cost, battery-powered, portable Raman system has been developed, and it can be used for the detection of bacterial species used in biological warfare, such as anthrax spores (Fig. 15.5) (Zhang et al. 2005).

Cancer is one of the top three leading causes of death around the world, and as such, a rapid, efficient, and simple diagnosis is crucial and highly desirable for detecting it in its early stages. This would not only facilitate follow-up treatment procedures, but also provide cost-relief for management and care of cancer patients. Traditional diagnostic methods have drawbacks and limitations, ranging from inefficiency to the lengthy periods of time they require for detection; thus SERS has gradually risen as the alternative approach for cancer cell detection (Lin et al. 2011, 2012; Yan and Reinhard 2010). Since the SERS spectra allows for differentiation of bacterial cell types as previously stated, there is no question that the same methodology can be applied in detection and identification of cancer cells. SERS has the potential for differentiating cell types of mammals, and has previously been used to differentiate normal cells from cancerous ones. Due to the enhanced metabolic rates of cancer cells compared to that of normal cells, a significant difference in their SERS spectra has been repeatedly examined. This also owes to the fact that cancer

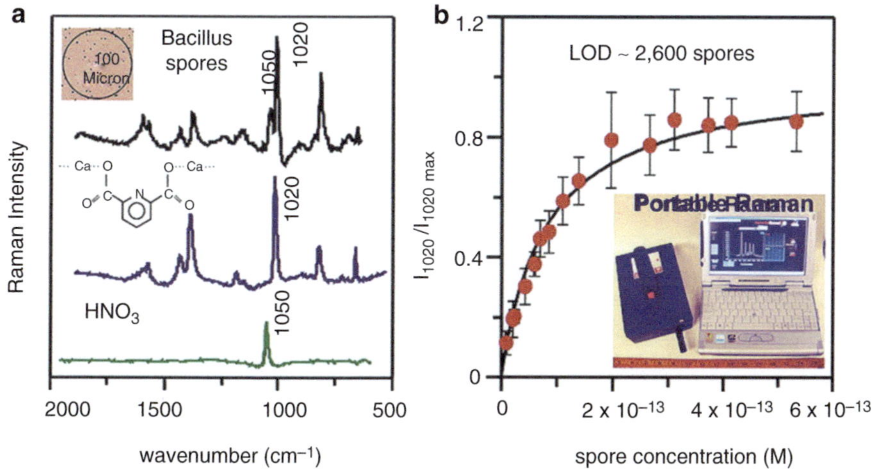

Fig. 15.5 (**a**) SERS spectra of spore suspended on a AgFON substrate, the calcium dipicolinate (CaDPA) (0.5 mM), and HNO_3 (0.02 M). (**b**) A low-cost, battery-powered, portable Raman system for rapid detection. Reprinted with permission from American Chemical Society

cells have a distinct preference in metabolic pathways and surface receptor regulation (Lin et al. 2011). While mass spectrometry is commonly used in biomarker identification and classification, SERS also offers a similar profiling method (DeBerardinis et al. 2008; Urayama et al. 2010). SERS has the advantage of rapid detection and facile implementation; however, SERS spectra of cells can evolve over time, especially cancer cells that metastasize. Renal cells, for example, begin secreting molecules in response to environmental conditions that may cause complications in the cells' reproducibility.

15.6 Multiplexed Tagging and Diagnostics

SERS particles can be excited with a single laser beam in the NIR region (around 785 nm) and their emitted signals cover about the same spectral region (200–1800 cm^{-1}). In this section, we discuss the different methods of SERS tagging from molecular to cellular detection, as well as their use and the benefits they provide for biomedicine.

Fluorescently labeled antibodies are commonly used in immunophenotyping to identify various cell types for disease diagnosis and treatment. As previously stated, SERS nanotags have narrow spectral signals independent of the external environment, and are thus suitable for detection of multiple markers in one setting. A study of multiplexed phenotyping was conducted by Maiti et al., where the tags were functionalized with lipoic acid derivatives of triphenylmethine and cyanine dyes then PEGylated, in which the distal ends were functionalized with EGFR and HER2 antibodies (Maiti et al. 2011). The two tags with different antibodies were used to identify two distinct cell lines, OSCC and SKBR-3. This strategy enabled the authors to correctly identify the cells, and is a method that can be expanded further to identify cancer types and subtypes, as well as cancer stem cell markers. Flow cytometry measures the probe emission from thousands of individual cells, and thus provides more accurate details of a whole ensemble instead of the fewer number cells in vitro. For these reasons, flow cytometry is more relevant and adaptable for clinical settings (Chattopadhyay et al. 2008; Nolan and Sebba 2011). Coupling this with SERS nanotags produces a multi-parameter flow cytometry. The instrumentation for SERS flow cytometry was pioneered by Nolan and colleagues for analyzing SERS nanotags adhered to micrometer-sized beads or single cells (Nolan and Sebba 2011). MacLaughlin et al. later reported the use of SERS flow cytometry to identify leukemia and lymphoma cells (MacLaughlin et al. 2013). SERS flow cytometry offers a greater number of probes to enable better identification accuracy, presenting a more efficient and cost-effective platform for diagnoses.

Methods for biomarker detection aim to identify molecular indicators of disease in clinical samples such as urine or blood (Anderson and Anderson 2003; Stern et al. 2010). A number of biological analytes, such as those associated with cancer, have been identified in the concentration rage of a few ng/mL. Methods ranging from radioactive immunoassays, enzyme-linked immunoassay (ELISA), mass spectrometry (MS), Western blot, or a combination of these have been performed for detection of biomarkers (Ambrosi et al. 2010; Fortin et al. 2009). Mass spectrometry provides great sensitivity, but requires that protein samples are purified before detection analysis (Chan et al. 2009). While immunoassay methods are less time-consuming, they typically lack the sensitivity and quantitation abilities of mass spectrometry, and can only operate within a small dynamic range (Jia et al. 2009). SERS methods have ultrahigh sensitivity and can be multiplexed over many samples. Its wide working range enables biomarker detection in whole blood, making SERS the preferred method for detection.

Grubisha and group previously performed a study that demonstrated the benefits of using SERS tags in detection of biomarkers over a large dynamic range (Grubisha et al. 2003). Derivatives of the Raman reporter dithiobis (benzoic acid) were used. The derivatives formed a monolayer on 30 nm gold particles where the distal ends of the molecule were biofunctionalized with antibodies, allowing a larger concentration of the reporter on the surface of the particle, and thus leading to brighter probes. These probes were then used to detect prostate-specific antigen (PSA). Normal PSA levels fall between 4 and 10 ng/mL (Polascik et al. 1999). The SERS-based method has a working range from 1 ng/mL to 10 μg/mL, capable of monitoring both healthy and diseased PSA levels (Ward et al. 2001). Li et al. used gold nanostar SERS nanotags and gold nanotriangle arrays to demonstrate a sensitivity increase of the SERS-ELISAs (Li et al. 2013). Detection antibodies were attached to the gold nanostar SERS tags while captured antibodies were anchored to the gold nanotriangles on the substrate. Increased sensitivity was thought to have arisen from the tips of the nanostar coupling with the nanotriangles. These assays were used to determine the biomarker VGEF in clinical blood samples and the limits of detection was found to be 7 fg/mL (Zhou et al. 2012).

Despite recent advances in fluorescence-based immunoassays, the SERS approach offers a much more rapid detection time, less washing steps, and easier operation that allows multiple detections in a single setting. Studies of multiplexed detection of cardiovascular protein biomarkers in the blood using SERS nanoparticle tags have also been reported. Considering coronary artery disease (CAD), one major problem is to predict sudden cardiac events such as plaque rupture and myocardial infarction. The great sensitivity, multiplexing, and dynamic range of SERS facilitate assessment of multiple biomarkers within a single tube. Assays that can distinguish populations that have high risk for plaque rupture are crucial in facilitating immediate diagnosis, intervention, and treatment. While scoring methods such as the Framington Risk Score determine the risk of such cardiac events, they are not precise predictors for patients diagnosed with CAD (Shlipak et al. 2008). Thus, it is of great importance that a diagnostic procedure to detect biomarkers in the blood is developed and applied to better provide for CAD patients (Eapen et al. 2013).

Detection of DNA sequences is important to molecular and cellular identification. Known polymorphisms within the DNA sequence can be used to assess susceptibility or diagnose patients of diseases such as diabetes, cancer (Zacho et al. 2011), and more. Microassays are often employed in DNA biomarker detection. In microassays, fluorescence signals the hybridization of the target to surface-anchored probe sequences. But as stated before, immunoassays, though less time-consuming, lack the sensitivity and quantitation abilities of spectrometry. SERS tagging is the better method as the target strands can easily be detected using a colloidal substrate. As opposed to planar substrates, colloidal substrates can increase the kinetic rate of hybridization. The first to demonstrate multiplexed DNA detection using SERS molecular beacons was the Vo-Dinh group (Hsin-Neng and Tuan 2009). Complementary strands to that of the intended biomarker formed hairpin loops, where one end is a silver nanoparticle, while the other end carries a Raman reporter. The gap of the reporter to metal nanoparticle can be controlled by the assembly of DNA with a hybridization process in which the hairpin structure will open and distance the reporter from metal nanoparticle. Thus, it creates a decrease in the Raman signal. Fluorescence-based probes based off similar constructions have also been developed, but in these, a fluorescent dye is on one end and a fluorescence quencher is on the other (Tan et al. 2004). Hybridization of the fluorescence-based probe turns on the fluorescence, whereas SERS is turned off because of the separation of the metal particle and the reporter. Multiplexed detection of two genes characteristic of breast cancer using the SERS-based approach had been previously reported. Probes that had two different Raman reporters were used. Their structures consisted of relevant biomarkers for cancer, the genes ERBB2-MS and KI-67-MS, and reporter dyes, Cy3 and TAMRA. It was noted that the Raman spectra of the dyes were very distinct and easily resolved in mixture, and that the probes were specific to their target, only diminishing the SERS signal when the target was present. In different cases, it may be preferable to have SERS signals that are turned on (positive) rather than SERS signals that are turned off (negative). Upon target hybridization causing aggregation, positive signal enhancement can be obtained through electromagnetic enhancements and plasmonic coupling. In one study, Graham and colleagues used

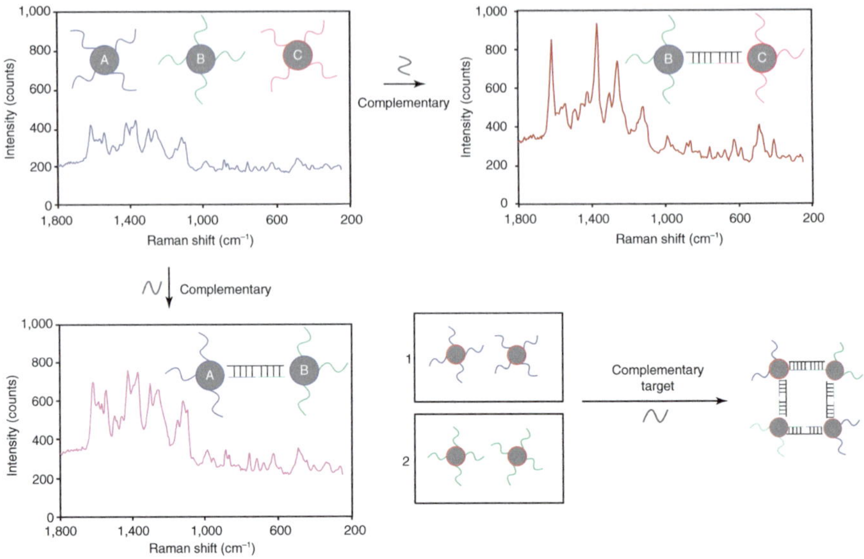

Fig. 15.6 The Raman signal could be selectively enhanced via assembly of DNA with hybridization process. By using three different DNA (oligonucleotide) sequences, dye 1 and dye 2 can be selectively enhanced in the presence of their complementary DNA sequences; the presence of complementary DNA sequence for A and B could enhance the signal of dye 1; the presence of complementary DNA sequence for D and C could enhance the signal dye 2. Reprinted with permission from Nature Publishing Group

silver nanoparticles coated with a single layer of resonant Raman reporters that functionalized with the DNA sequences (5′-end thiolated) complementary to the intended target strands. Hybridization caused the solution to change from yellow to green-blue (Fig. 15.6) (Graham et al. 2008). The SERS signal also increased multifold owing to the induced aggregation.

15.7 In Vivo Sensing of Biological Markers

SERS nanoparticles that are spectrally encoded are very suitable for spectroscopic detection and in vivo tumor targeting. In this section, we explore how SERS functions in tumor targeting as well its application in image-guided surgery.

Solid tumors have been discovered to have highly permeable vasculatures with a number of fenestrations, and owing to increased intra-tumoral pressures, functional lymphatic vessels are often absent in solid tumors, which decrease the outflow of particles that may have entered their permeable vasculature (Boucher and Jain 1992). Thus, enhanced permeation and retention (EPR) of nanoparticles are often seen in solid tumors (Iyer et al. 2006; Maeda 2001). Passive targeting is relying on the EPR effect to ensure that nanoparticles delivered will stay within the tumor, whereas active targeting is using antibodies conjugated to the nanoparticle surface

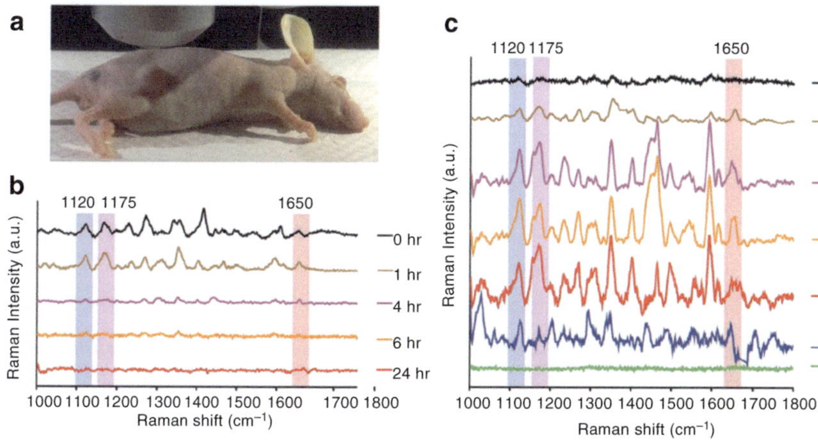

Fig. 15.7 SERS nanotags were injected into mouse model. (**a**) Photograph of tumor-bearing mouse model. (**b**) Collection of SERS spectra from the tumor-bearing mouse. Due to the lack of targeting property, the signal is not detectable after 6 h injection. (**c**) SERS spectra for targeted SERS nanotags bound to TGFbRII, CD44, and EGFR biomarkers. The signal is detectable up to 48 h. Reprinted with permission from Nature Publishing Group

to aid in increasing accumulation. In a research by Nie and coworkers, SERS tags were used in vivo tumor targeting, and differences between the active and passive targeting were examined (Qian et al. 2008). Although the passively targeted particles were discovered to accumulate at similar levels as the actively targeted ones, remarkably decreased signals of passive particles were observed subsequently, which was due to the passively targeted probes washing out of the tumor. Dinish et al. reported similar results in a recent study, in which actively targeted particles led to longer retention times within tumors (Fig. 15.7) (Dinish et al. 2014). They demonstrated in vivo multiplexed detection of EGFR, CD44, and TGFbRII through the use of three biocompatible SERS nanotags that had Raman reporters (Cy5, MGITC, Rh6G), and observed that the antibody-coated nanotags exhibited SERS signal for as long as 2 days, while the antibody-free nanotags had detectable signals only for 4–6 h.

For numerous cancer patients, surgery is the first line of treatment that is followed by chemotherapy and radiation. Surgical resection of the tumor leads to a 45% chance of survival for the patient, and as thus provides a strong advantage among many cancers today (De Grand and Frangioni 2003). Complete resection, in which the surgeon must remove the entire tumor without any traces, is one of the most important factors in increasing survival of the patient (Evans 2003; Karakiewicz et al. 2005). Complete resection of tumors of the prostate, colon, lung, and pancreas, as compared to an incomplete resection, has been shown to lead to a three- to five-fold improvement in patient survival. While intraoperative MRI assists in surgical resection of cancer tumors, the method prolongs surgery time and increases associated costs of the resection (Ramina et al. 2010). On the other hand, using optical methods in resection provides real-time imaging with high resolution at lower costs

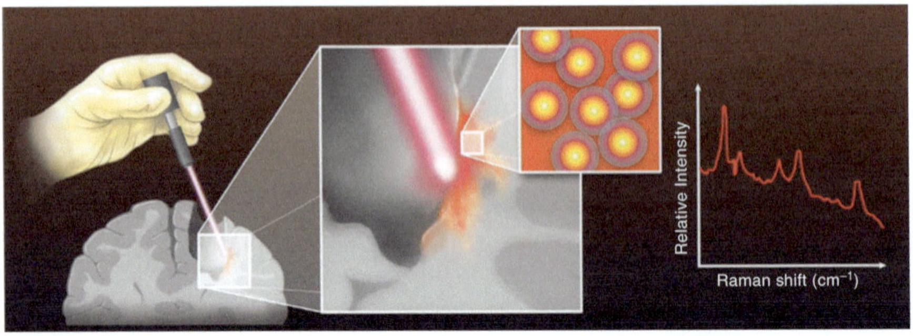

Fig. 15.8 Schematic represents the use of a handheld Raman scanner and SERS nanoparticles for guiding surgical resection of brain tumors. Reprinted with permission from American Chemical Society

of surgery. With its high sensitivity of detection and spatial resolution, intraoperative SERS is optimal when compared with other imaging modalities. Mohs and coworkers reported the development of a handheld spectroscopic device that operates in the NIR region to detect both fluorescent and Raman signals (Mohs et al. 2010). The spectra obtained with the device were compared to those obtained by a standard Raman spectrophotometer, and were found to have less than 1.0% of variance between them.

The ultrahigh sensitivity of SERS tags while using the spectroscopic pen enabled detection of cancer cell invasions to areas outside of the main tumor even after a complete resection. SERS-guided surgery provides the added advantage in discovering microscopic metastases and residual tumor cells that may have resulted from an incomplete resection. Kircher's group recently developed a handheld Raman scanner to guide brain tumor resections (Karabeber et al. 2014). Mice with implanted glioblastomas were used in their demonstration (Fig. 15.8). This particular cancer was used as it has an 80–90% chance of recurrence in the same area, which can incur major problems in the event of an incomplete resection (Petrecca et al. 2013). Kircher's group tested three surgical scenarios: resection without Raman imaging, resection using a Raman microscope, and resection using the handheld Raman device during surgery. Raman imaging by microscope and by handheld device generated significant results, providing better determination of tumor margins while also identifying the positive margins left behind during the surgery. The handheld Raman device was used to illuminate the brain tissue at short distances, and enabled detection of small tumor foci, or clusters of tumor cells, that had went unnoticed. Since injected particles are unlikely to cross the blood-brain barrier, nanoparticles are more promising, as they can locate the brain with the help of monocytes. Since very few particles successfully reach the intended target site, optical-guided surgery requires the high sensitivity of SERS nanotags.

Surface-enhanced resonance Raman scattering (SERRS) "nanostars," was reported by Harmsen et al. Nanostars are significantly more sensitive than previous SERS nanotags (Harmsen et al. 2015b). Other than human sarcoma xenograft

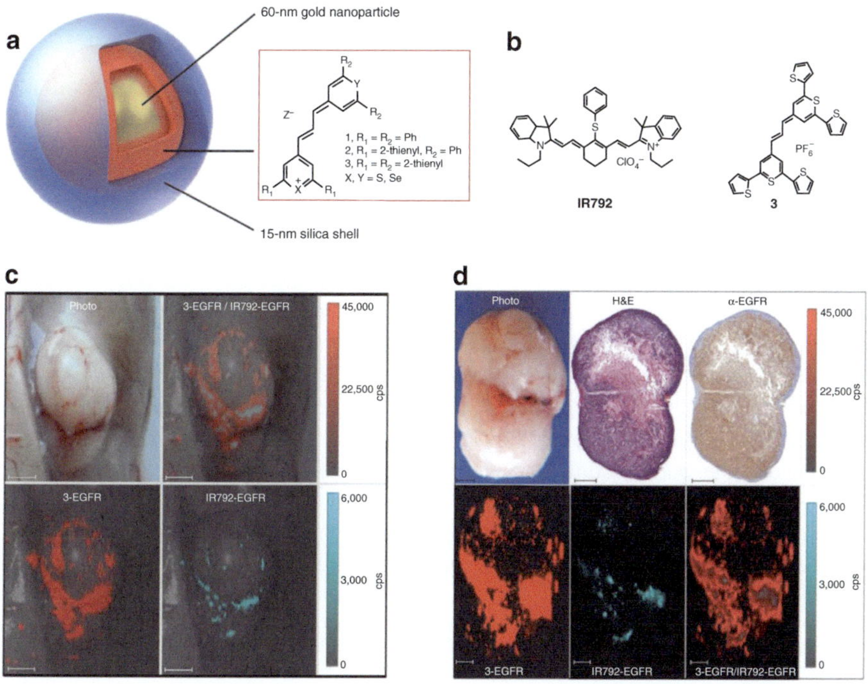

Fig. 15.9 (**a**) A 60-nm gold core is enclosed in a chalcogenopyrylium dye-silica shell with 15 nm thickness. (**b**) Thiopyrylium dye 3 and commercial dye IR792 are optimized for high-sensitive in vivo detection. (**c**) An A431 tumor xenograft in nude mice was used for the comparison between epidermal growth factor receptor (EGFR)-targeted SERS nanoprobe of IR792- or 3-. The imaging of tumors was performed by Raman after 18 h. Higher signal was observed for nanoprobe with chalcogenopyrylium dye 3 (*red*) as compared to IR792-based SERS nanoprobe (*cyan*). All scale bars represent 2.0 mm. (**d**) Raman imaging was performed on the excised tumor. The sample was fixed in 4% paraformaldehyde, and stained with hematoxylin and eosin staining (H&E) and anti-EGFR. High concentration of EGFR-targeted nanoprobe was detected throughout the tumor, indicating the expression of EGFR is homogenous. Necrotic region within the tumor showed low Raman signal, which is due to the low accumulation of nanoprobe (scale bars = 1.0 mm). Reprinted with permission from Nature Publishing Group

model, Harmsen et al. demonstrated the efficacy of SERRS using mouse models of breast cancer, sarcoma, prostate cancer, and pancreatic cancer. Intravenously injected SERRS nanostars offer high precision detection of both microscopic lesions and macroscopic tumors (Fig. 15.9). Harmsen et al. also demonstrated the use of near-infrared 2-thienyl-substituted chalcogenopyrylium dyes adsorbed onto gold nanoparticles to produce SERS nanoprobes for in vivo biological marker detection could reach attomolar sensitivity level (Harmsen et al. 2015a).

Development of minimally invasive surgical techniques is at the frontline of biomedical research. Such techniques would not only decrease patient down time and discomfort, but also help lower the overall costs of treatment. Image-guided surgery can be adapted into a clinical setting, and holds great potential for future cancer diagnosis.

15.8 Concluding Remarks

SERS is capable of providing rich spectroscopic information with high specificity and enhanced sensitivity. Recent advancements have led to the development of SERS substrates and nanoparticles that are now used for label-free detection and multiplexed tagging, which have proven to be beneficial in biomedical applications.

Compared with traditional imaging probes, SERS nanoparticles offer ultrahigh detection sensitivity, label-free detection, spectroscopic multiplexing, and multivalent targeting, among its numerous other advantages. These features would not only aid in image-guided surgery and highly sensitive tracking and detection of tumor cells, but also enable better identification and classification. SERS has experienced numerous milestones since its discovery, and continues to be the central focus of activities ranging from cellular tagging, single-molecule and single-particle spectroscopy, as well as image-guided surgery. The many advantages SERS offers facilitate its adoption into the medical field, but this clinical transition still may take time to be fully implemented. Thus, systematic nano-toxicological studies, such as nanoparticle distribution, metabolism, excretion, pharmacokinetics, and pharmacodynamics, must be carried out to expand its use not only in biomedicine, but also for potential human applications.

Acknowledgments This work was supported by the Ministry of Science and Technology (MOST), Taiwan, under Grant No. 104C3562-1 and the Food and Drug Administration (FDA), Ministry of Health and Welfare, Taipei, Taiwan, under Grant No. 105TFDA-A-105.

References

Albrecht MG, Creighton JA (1977) Anomalously intense Raman spectra of pyridine at a silver electrode. J Am Chem Soc 99:5215–5217

Ambrosi A, Airò F, Merkoçi A (2010) Enhanced gold nanoparticle based ELISA for a breast cancer biomarker. Anal Chem 82:1151–1156

Anderson NL, Anderson NG (2003) The human plasma proteome: history, character, and diagnostic prospects. Mol Cell Proteomics 2:50

Aroca RF, Ross DJ, Domingo C (2004) Surface-enhanced infrared spectroscopy. Appl Spectrosc 58:324A–338A

Bondarenko O, Juganson K, Ivask A, Kasemets K, Mortimer M, Kahru A (2013) Toxicity of Ag, CuO and ZnO nanoparticles to selected environmentally relevant test organisms and mammalian cells in vitro: a critical review. Arch Toxicol 87:1181–1200

Boucher Y, Jain RK (1992) Microvascular pressure is the principal driving force for interstitial hypertension in solid tumors: implications for vascular collapse. Cancer Res 52:5110–5114

Chan ECY, Koh PK, Mal M, Cheah PY, Eu KW, Backshall A, Cavill R, Nicholson JK, Keun HC (2009) Metabolic profiling of human colorectal cancer using high-resolution magic angle spinning nuclear magnetic resonance (HR-MAS NMR) spectroscopy and gas chromatography mass spectrometry (GC/MS). J Proteome Res 8:352–361

Chattopadhyay PK, Hogerkorp C-M, Roederer M (2008) A chromatic explosion: the development and future of multiparameter flow cytometry. Immunology 125:441–449

Das G, Mecarini F, Gentile F, De Angelis F, Mohan Kumar HG, Candeloro P, Liberale C, Cuda G, Di Fabrizio E (2009) Nano-patterned SERS substrate: application for protein analysis vs. temperature. Biosens Bioelectron 24:1693–1699

De Grand AM, Frangioni JV (2003) An operational near-infrared fluorescence imaging system prototype for large animal surgery. Technol Cancer Res Treat 2:553–562
DeBerardinis RJ, Lum JJ, Hatzivassiliou G, Thompson CB (2008) The biology of cancer: metabolic reprogramming fuels cell growth and proliferation. Cell Metab 7:11–20
Dinish US, Balasundaram G, Chang Y-T, Olivo M (2014) Actively targeted in vivo multiplex detection of intrinsic cancer biomarkers using biocompatible SERS nanotags. Sci Rep 4:4075
Doering WE, Nie S (2003) Spectroscopic tags using dye-embedded nanoparticles and surface-enhanced Raman scattering. Anal Chem 75:6171–6176
Doering WE, Piotti ME, Natan MJ, Freeman RG (2007) SERS as a foundation for nanoscale optically detected biological labels. Adv Mater 19:3100–3108
Eapen DJ, Manocha P, Patel RS, Hammadah M, Veledar E, Wassel C, Nanjundappa RA, Sikora S, Malayter D, Wilson PWF, Sperling L, Quyyumi AA, Epstein SE (2013) Aggregate risk score based on markers of inflammation, cell stress, and coagulation is an independent predictor of adverse cardiovascular outcomes. J Am Coll Cardiol 62:329–337
Evans RA (2003) Positive surgical margins and ipsilateral breast tumor recurrence predict disease-specific survival after breast-conserving therapy. Cancer 98:2522–2523. author reply 2523–2524
Fleischmann M, Hendra PJ, McQuillan AJ (1974) Raman spectra of pyridine adsorbed at a silver electrode. Chem Phys Lett 26:163–166
Fortin T, Salvador A, Charrier JP, Lenz C, Lacoux X, Morla A, Choquet-Kastylevsky G, Lemoine J (2009) Clinical quantitation of prostate-specific antigen biomarker in the low nanogram/milliliter range by conventional bore liquid chromatography-tandem mass spectrometry (multiple reaction monitoring) coupling and correlation with ELISA tests. Mol Cell Proteomics 8:1006–1015
Furtak TE, Macomber SH (1983) Voltage-induced shifting of charge-transfer excitations and their role in surface-enhanced Raman scattering. Chem Phys Lett 95:328–332
Graham D, Thompson DG, Smith WE, Faulds K (2008) Control of enhanced Raman scattering using a DNA-based assembly process of dye-coded nanoparticles. Nat Nanotechnol 3:548–551
Greulich C, Braun D, Peetsch A, Diendorf J, Siebers B, Epple M, Koller M (2012) The toxic effect of silver ions and silver nanoparticles towards bacteria and human cells occurs in the same concentration range. RSC Adv 2:6981–6987
Grubisha DS, Lipert RJ, Park H-Y, Driskell J, Porter MD (2003) Femtomolar detection of prostate-specific antigen: an immunoassay based on surface-enhanced Raman scattering and immunogold labels. Anal Chem 75:5936–5943
Harmsen S, Bedics MA, Wall MA, Huang R, Detty MR, Kircher MF (2015a) Rational design of a chalcogenopyrylium-based surface-enhanced resonance Raman scattering nanoprobe with attomolar sensitivity. Nat Commun 6:6570
Harmsen S, Huang R, Wall MA, Karabeber H, Samii JM, Spaliviero M, White JR, Monette S, O'Connor R, Pitter KL, Sastra SA, Saborowski M, Holland EC, Singer S, Olive KP, Lowe SW, Blasberg RG, Kircher MF (2015b) Surface-enhanced resonance Raman scattering nanostars for high-precision cancer imaging. Sci Transl Med 7:271ra277
Henglein A (1993) Physicochemical properties of small metal particles in solution: "microelectrode" reactions, chemisorption, composite metal particles, and the atom-to-metal transition. J Phys Chem 97:5457–5471
Hong S, Li X (2013) Optimal size of gold nanoparticles for surface-enhanced Raman spectroscopy under different conditions. J Nanomater 2013:9
Hsin-Neng W, Tuan V-D (2009) Multiplex detection of breast cancer biomarkers using plasmonic molecular sentinel nanoprobes. Nanotechnology 20:065101
Iyer AK, Khaled G, Fang J, Maeda H (2006) Exploiting the enhanced permeability and retention effect for tumor targeting. Drug Discov Today 11:812–818
Jarvis RM, Goodacre R (2004) Discrimination of bacteria using surface-enhanced Raman spectroscopy. Anal Chem 76:40–47
Jeanmaire DL, Van Duyne RP (1977) Surface Raman spectroelectrochemistry. J Electroanal Chem Interfacial Electrochem 84:1–20

Jensen L, Aikens CM, Schatz GC (2008) Electronic structure methods for studying surface-enhanced Raman scattering. Chem Soc Rev 37:1061–1073
Jia C-P, Zhong X-Q, Hua B, Liu M-Y, Jing F-X, Lou X-H, Yao S-H, Xiang J-Q, Jin Q-H, Zhao J-L (2009) Nano-ELISA for highly sensitive protein detection. Biosens Bioelectron 24:2836–2841
Karabeber H, Huang R, Iacono P, Samii JM, Pitter K, Holland EC, Kircher MF (2014) Guiding brain tumor resection using surface-enhanced Raman scattering nanoparticles and a hand-held raman scanner. ACS Nano 8:9755–9766
Karakiewicz PI, Eastham JA, Graefen M, Cagiannos I, Stricker PD, Klein E, Cangiano T, Schröder FH, Scardino PT, Kattan MW (2005) Prognostic impact of positive surgical margins in surgically treated prostate cancer: multi-institutional assessment of 5831 patients. Urology 66:1245–1250
Kelly KL, Coronado E, Zhao LL, Schatz GC (2003) The optical properties of metal nanoparticles: the influence of size, shape, and dielectric environment. J Phys Chem B 107:668–677
Keren S, Zavaleta C, Cheng Z, de la Zerda A, Gheysens O, Gambhir SS (2008) Noninvasive molecular imaging of small living subjects using Raman spectroscopy. Proc Natl Acad Sci 105:5844–5849
King FW, Schatz GC (1979) Theory of Raman scattering by molecules adsorbed at electrode surfaces. Model calculations for resonance Raman scattering by an adsorbed diatomic. Chem Phys 38:245–256
Kreibig U, Fragstein CV (1969) The limitation of electron mean free path in small silver particles. Z Phys 224:307–323
Lee SJ, Morrill AR, Moskovits M (2006) Hot spots in silver nanowire bundles for surface-enhanced Raman spectroscopy. J Am Chem Soc 128:2200–2201
Li JF, Huang YF, Ding Y, Yang ZL, Li SB, Zhou XS, Fan FR, Zhang W, Zhou ZY, WuDe Y, Ren B, Wang ZL, Tian ZQ (2010a) Shell-isolated nanoparticle-enhanced Raman spectroscopy. Nature 464:392–395
Li W, Camargo PHC, Au L, Zhang Q, Rycenga M, Xia Y (2010b) Etching and dimerization: a simple and versatile route to dimers of silver nanospheres with a range of sizes. Angew Chem Int Ed 49:164–168
Li M, Cushing SK, Zhang J, Suri S, Evans R, Petros WP, Gibson LF, Ma D, Liu Y, Wu N (2013) Three-dimensional hierarchical plasmonic nano-architecture enhanced surface-enhanced Raman scattering immunosensor for cancer biomarker detection in blood plasma. ACS Nano 7:4967–4976
Li M, Kang JW, Dasari RR, Barman I (2014) Shedding light on the extinction-enhancement duality in gold nanostar-enhanced Raman spectroscopy. Angew Chem 126:14339–14343
Lim D-K, Jeon K-S, Kim HM, Nam J-M, Suh YD (2010) Nanogap-engineerable Raman-active nanodumbbells for single-molecule detection. Nat Mater 9:60–67
Lim D-K, Jeon K-S, Hwang J-H, Kim H, Kwon S, Suh YD, Nam J-M (2011) Highly uniform and reproducible surface-enhanced Raman scattering from DNA-tailorable nanoparticles with 1-nm interior gap. Nat Nanotechnol 6:452–460
Lin D, Feng S, Pan J, Chen Y, Lin J, Chen G, Xie S, Zeng H, Chen R (2011) Colorectal cancer detection by gold nanoparticle based surface-enhanced Raman spectroscopy of blood serum and statistical analysis. Opt Express 19:13565–13577
Lin J, Chen R, Feng S, Pan J, Li B, Chen G, Lin S, Li C, Sun L-Q, Huang Z, Zeng H (2012) Surface-enhanced Raman scattering spectroscopy for potential noninvasive nasopharyngeal cancer detection. J Raman Spectrosc 43:497–502
Linnert T, Mulvaney P, Henglein A (1993) Surface chemistry of colloidal silver: surface plasmon damping by chemisorbed iodide, hydrosulfide (SH-), and phenylthiolate. J Phys Chem 97:679–682
Liz-Marzán LM (2006) Tailoring surface plasmons through the morphology and assembly of metal nanoparticles. Langmuir 22:32–41
Lombardi JR, Birke RL (2008) A unified approach to surface-enhanced Raman spectroscopy. J Phys Chem C 112:5605–5617
Lombardi JR, Birke RL (2009) A unified view of surface-enhanced Raman scattering. Acc Chem Res 42:734–742

Lombardi JR, Birke RL, Sanchez LA, Bernard I, Sun SC (1984) The effect of molecular structure on voltage induced shifts of charge transfer excitation in surface enhanced Raman scattering. Chem Phys Lett 104:240–247

MacLaughlin CM, Mullaithilaga N, Yang G, Ip SY, Wang C, Walker GC (2013) Surface-enhanced Raman scattering dye-labeled Au nanoparticles for triplexed detection of leukemia and lymphoma cells and SERS flow cytometry. Langmuir 29:1908–1919

Maeda H (2001) The enhanced permeability and retention (EPR) effect in tumor vasculature: the key role of tumor-selective macromolecular drug targeting. Adv Enzyme Regul 41:189–207

Maiti KK, Samanta A, Vendrell M, Soh K-S, Olivo M, Chang Y-T (2011) Multiplex cancer cell detection by SERS nanotags with cyanine and triphenylmethine Raman reporters. Chem Commun 47:3514–3516

McFarland AD, Young MA, Dieringer JA, Van Duyne RP (2005) Wavelength-scanned surface-enhanced Raman excitation spectroscopy. J Phys Chem B 109:11279–11285

McMahon JM, Gray SK, Schatz GC (2010) Calculating nonlocal optical properties of structures with arbitrary shape. Phys Rev B 82:035423

McNay G, Eustace D, Smith WE, Faulds K, Graham D (2011) Surface-enhanced Raman scattering (SERS) and surface-enhanced resonance Raman scattering (SERRS): a review of applications. Appl Spectrosc 65:825–837

Mirkin CA, Letsinger RL, Mucic RC, Storhoff JJ (1996) A DNA-based method for rationally assembling nanoparticles into macroscopic materials. Nature 382:607–609

Mohs AM, Mancini MC, Singhal S, Provenzale JM, Leyland-Jones B, Wang MD, Nie S (2010) Hand-held spectroscopic device for in vivo and intraoperative tumor detection: contrast enhancement, detection sensitivity, and tissue penetration. Anal Chem 82:9058–9065

Morton SM, Jensen L (2009) Understanding the molecule–surface chemical coupling in SERS. J Am Chem Soc 131:4090–4098

Moskovits M (1978) Surface roughness and the enhanced intensity of Raman scattering by molecules adsorbed on metals. J Chem Phys 69:4159–4161

Mulvaney P (1996) Surface plasmon spectroscopy of nanosized metal particles. Langmuir 12:788–800

Nadkarni MA, Martin FE, Jacques NA, Hunter N (2002) Determination of bacterial load by real-time PCR using a broad-range (universal) probe and primers set. Microbiology 148:257–266

Nolan JP, Sebba DS (2011) Chapter 20—Surface-enhanced Raman scattering (SERS) cytometry. In: Zbigniew Darzynkiewicz EHAOWT, Donald W (eds) Methods in cell biology. Academic Press, Cambridge, MA, pp 515–532

Patel IS, Premasiri WR, Moir DT, Ziegler LD (2008) Barcoding bacterial cells: a SERS-based methodology for pathogen identification. J Raman Spectrosc 39:1660–1672

Petrecca K, Guiot M-C, Panet-Raymond V, Souhami L (2013) Failure pattern following complete resection plus radiotherapy and temozolomide is at the resection margin in patients with glioblastoma. J Neurooncol 111:19–23

Polascik TJ, Oesterling JE, Partin AW (1999) Prostate specific antigen: a decade of discovery-what we have learned and where we are going. J Urol 162:293–306

Popovtzer R, Agrawal A, Kotov NA, Popovtzer A, Balter J, Carey TE, Kopelman R (2008) Targeted gold nanoparticles enable molecular CT imaging of cancer. Nano Lett 8:4593–4596

Qian XM, Nie SM (2008) Single-molecule and single-nanoparticle SERS: from fundamental mechanisms to biomedical applications. Chem Soc Rev 37:912–920

Qian X, Peng X-H, Ansari DO, Yin-Goen Q, Chen GZ, Shin DM, Yang L, Young AN, Wang MD, Nie S (2008) In vivo tumor targeting and spectroscopic detection with surface-enhanced Raman nanoparticle tags. Nat Biotechnol 26:83–90

Ramina R, Coelho Neto M, Giacomelli A, Barros E, Vosgerau R, Nascimento A, Coelho G (2010) Optimizing costs of intraoperative magnetic resonance imaging. A series of 29 glioma cases. Acta Neurochir 152:27–33

Samanta A, Maiti KK, Soh K-S, Liao X, Vendrell M, Dinish US, Yun S-W, Bhuvaneswari R, Kim H, Rautela S, Chung J, Olivo M, Chang Y-T (2011) Ultrasensitive near-infrared raman reporters for SERS-based in vivo cancer detection. Angew Chem Int Ed 50:6089–6092

Schlücker S (2014) Surface-enhanced Raman spectroscopy: concepts and chemical applications. Angew Chem Int Ed 53:4756–4795

Shlipak MG, Ix JH, Bibbins-Domingo K, Lin F, Whooley MA (2008) Biomarkers to predict recurrent cardiovascular disease: the heart and soul study. Am J Med 121:50–57

Sivapalan ST, DeVetter BM, Yang TK, van Dijk T, Schulmerich MV, Carney PS, Bhargava R, Murphy CJ (2013) Off-resonance surface-enhanced Raman spectroscopy from gold nanorod suspensions as a function of aspect ratio: not what we thought. ACS Nano 7:2099–2105

Stamplecoskie KG, Scaiano JC, Tiwari VS, Anis H (2011) Optimal size of silver nanoparticles for surface-enhanced Raman spectroscopy. J Phys Chem C 115:1403–1409

Steinigeweg D, Schütz M, Salehi M, Schlücker S (2011) Fast and cost-effective purification of gold nanoparticles in the 20–250 nm size range by continuous density gradient centrifugation. Small 7:2443–2448

Stern E, Vacic A, Rajan NK, Criscione JM, Park J, Ilic BR, Mooney DJ, Reed MA, Fahmy TM (2010) Label-free biomarker detection from whole blood. Nat Nanotechnol 5:138–142

Tan W, Wang K, Drake TJ (2004) Molecular beacons. Curr Opin Chem Biol 8:547–553

Urayama S, Zou W, Brooks K, Tolstikov V (2010) Comprehensive mass spectrometry based metabolic profiling of blood plasma reveals potent discriminatory classifiers of pancreatic cancer. Rapid Commun Mass Spectrom 24:613–620

Valley N, Greeneltch N, Van Duyne RP, Schatz GC (2013) A look at the origin and magnitude of the chemical contribution to the enhancement mechanism of surface-enhanced Raman spectroscopy (SERS): theory and experiment. J Phys Chem Lett 4:2599–2604

Wang H, Levin CS, Halas NJ (2005) Nanosphere arrays with controlled sub-10-nm gaps as surface-enhanced Raman spectroscopy substrates. J Am Chem Soc 127:14992–14993

Ward AM, Catto JWF, Hamdy FC (2001) Prostate specific antigen: biology, biochemistry and available commercial assays. Ann Clin Biochem 38:633–651

Willets KA, Duyne RPV (2007) Localized surface plasmon resonance spectroscopy and sensing. Annu Rev Phys Chem 58:267–297

Xia X, Li W, Zhang Y, Xia Y (2013) Silica-coated dimers of silver nanospheres as surface-enhanced Raman scattering tags for imaging cancer cells. Interface Focus 3(3):20120092

Xu H, Bjerneld EJ, Käll M, Börjesson L (1999) Spectroscopy of single hemoglobin molecules by surface enhanced Raman scattering. Phys Rev Lett 83:4357–4360

Xu H, Aizpurua J, Käll M, Apell P (2000) Electromagnetic contributions to single-molecule sensitivity in surface-enhanced Raman scattering. Phys Rev E 62:4318–4324

Xu L, Yan W, Ma W, Kuang H, Wu X, Liu L, Zhao Y, Wang L, Xu C (2015) SERS encoded silver pyramids for attomolar detection of multiplexed disease biomarkers. Adv Mater 27:1706–1711

Yan B, Reinhard BM (2010) Identification of tumor cells through spectroscopic profiling of the cellular surface chemistry. J Phys Chem Lett 1:1595–1598

Yu Q, Golden G (2007) Probing the protein orientation on charged self-assembled monolayers on gold nanohole arrays by SERS. Langmuir 23:8659–8662

Zacho J, Yazdanyar S, Bojesen SE, Tybjærg-Hansen A, Nordestgaard BG (2011) Hyperhomocysteinemia, methylenetetrahydrofolate reductase c.677C>T polymorphism and risk of cancer: cross-sectional and prospective studies and meta-analyses of 75,000 cases and 93,000 controls. Int J Cancer 128:644–652

Zavaleta CL, Smith BR, Walton I, Doering W, Davis G, Shojaei B, Natan MJ, Gambhir SS (2009) Multiplexed imaging of surface enhanced Raman scattering nanotags in living mice using non-invasive Raman spectroscopy. Proc Natl Acad Sci 106:13511–13516

Zeiri L, Bronk BV, Shabtai Y, Eichler J, Efrima S (2004) Surface-enhanced Raman spectroscopy as a tool for probing specific biochemical components in bacteria. Appl Spectrosc 58:33–40

Zhang X, Young MA, Lyandres O, Van Duyne RP (2005) Rapid detection of an anthrax biomarker by surface-enhanced Raman spectroscopy. J Am Chem Soc 127:4484–4489

Zhou L, Ding F, Chen H, Ding W, Zhang W, Chou SY (2012) Enhancement of immunoassay's fluorescence and detection sensitivity using three-dimensional plasmonic nano-antenna-dots array. Anal Chem 84:4489–4495

Microfluidic Paper-Based Analytical Devices for Point-of-Care Diagnosis

16

Zhuan Zhuan Shi, Yao Lu, and Ling Yu

16.1 Introduction and General Overview

Point-of-care testing (POCT) shows its significant importance in academic and social affairs (Jansen et al. 1998; Wu et al. 1999). From commercialized products to laboratory prototypes, the creative atmosphere in this research field has maintained its vitality and drawn increasing attention from researchers in related areas. There have been many inventions that have changed the practice of medicine at the point of care in either rural or developed areas, and one prominent candidate for POCT is paper-based microfluidic analytical devices, also called Chip-on-a-Paper, or paper-based POC.

The development of paper chips dates back to the seventeenth century, when British chemist Robert Boyle fabricated the pH test strip, aka litmus paper. In 1949, Müller and his colleagues invented paper-thin layer chromatography for eluting dye by impregnating paraffin barriers on paper to form fluid paths (Müller and Clegg 1949). These pioneering works demonstrated that the confined hydrophilic region of a cellulose paper sheet can speed up the solution-diffusion process and reduce sample consumption. The application of paper as an analytic tool has been revitalized since the scientific term "microfluidic paper-based analytic devices (μPADs)" was defined by Martinez et al. in 2007 (Martinez et al. 2007).

Compared to traditional substrate materials, such as quartz, glass, silicon, and polymers, in the fabrication of POCT devices, the highly appreciated characteristics of paper are, but not limited to, the following (Martinez et al. 2007; Tobjörk and Österbacka 2011; Yetisen et al. 2013; Nguyen et al. 2014):

Z.Z. Shi • Y. Lu • L. Yu (✉)
Faculty of Materials and Energy, Institute for Clean Energy and Advanced Materials, Southwest University, Chongqing 400715, China
e-mail: lingyu12@swu.edu.cn

P. Chandra et al. (eds.), *Next Generation Point-of-care Biomedical Sensors Technologies for Cancer Diagnosis*, DOI 10.1007/978-981-10-4726-8_16

1. Cheap and abundant sources: paper is ubiquitous and can be mass produced.
2. Chemical and physical processability: paper has good biocompatibility and can be easily modified by chemical and physical treatments.
3. Disposability and recyclability.
4. Energy-saving assay: paper is cellulose and it can transport liquids by capillary forces without any external pumping forces.

Apparently, simplicity, a long-lasting trend in the development of all microfluidic devices (Sun et al. 2014; Petryayeva and Algar 2015), is no exception in the realm of paper microfluidics, which itself represents a cheap and fast way of doing experiments. From pH testing to filtrate solutions, the beneficial features of cellulose paper, such as its absorptivity, storage, and ability to diffuse liquid, have made it indispensable in the construction of POCTs (Mabey et al. 2004; Ali et al. 2009; Martinez et al. 2009). Socioeconomic concerns comprise one of the driving forces promoting the research and application of μPADs. First, μPADs can provide a wide range of low-cost portable devices for clinical diagnostics (Parolo and Merkoçi 2013; Petryayeva and Algar 2015), environmental monitoring (Alkasir et al. 2012; Jayawardane et al. 2014; Nath et al. 2015), and food safety analysis (Chen et al. 2015; Gomes and Sales 2015; Zhang et al. 2015). More appreciably, the shortage of medical personnel and equipment in undeveloped regions could be offset by promoting the application of μPADs. Meanwhile, μPADs also can be conveniently deployed in laboratories to reduce the cost of assays. Owing to its importance in facilitating all such advances, the past several decades have witnessed explosive development and application in the field of μPAD development. The aim of this chapter will be to illustrate the state-of-the-art of μPAD technology, discuss assay formats, and describe the commercial successes enjoyed by and challenges faced by μPADs.

16.2 Fabrication of μPAD

Paper is a permeable porous material composed of a solid fiber networks. The porous geometry and the hydrophilic nature of cellulose determine the fluidic properties of paper (Bruzewicz et al. 2008; Carrilho et al. 2009; Lu et al. 2009; Cate et al. 2014). This is the fundamental chemical nature of paper that allows making use of it to build fluid paths. The ultimate goal of μPAD fabrication is to form a specific hydrophilic area on a paper sheet. The fabrication methods should be in line with the format and application of the as-prepared μPADs. Surveying the reported state-of-art fabrication methods, two fabrication principles are generally followed.

16.2.1 Paper Shaping

Shaping paper by cutter, knife, and/or scissors is a straightforward approach to preparing paper artware (Fenton et al. 2008; Fu et al. 2010; Renault et al. 2013). The strength of physically shaped paper is obtained by eliminating wet

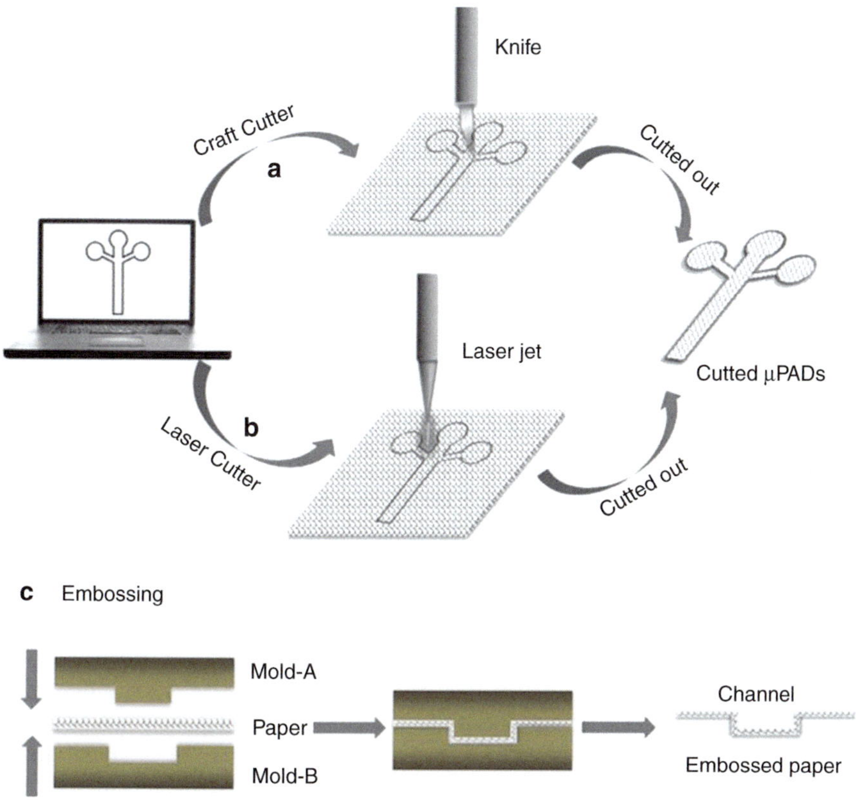

Fig. 16.1 Building fluid paths on cellulose paper by physical treatments. (**a**) Computer-controlled craft-knife cutter machine directly draws lines and creates holes in a sheet of cellulose paper. The unwanted parts are then removed to form a multichannel structure. (**b**) The laser nozzle of a laser cutter draws lines on a paper sheet. The high energy of the laser can directly shape the paper with high precision. (**c**) Building of an open-channel microfluidic device by embossing channels on omniphobic paper. The cross-sectional schematic of the embossing process shows a sheet of paper placed between two plastic molds that are pressed

chemicals, which can potentially damage the cellulose matrix. To improve the speed, precision, and production efficiency of paper shaping, computer-controlled cutting machines have been used instead of hole punches and handheld blades to fabricate μPADs (Nie et al. 2013; Evans et al. 2014a, b). In principle, a cutter machine faithfully transfers computer-designed structures onto a paper substrate. Two kinds of cutter machines can be applied for paper-device fabrication:

1. *Craft-knife cutter*: In this type of machine, the blade of the mechanical cutter can cut the paper substrate on demand. The flexibility and complexity demanded by paper cutting can be satisfied through use of a computer-controlled knife (see Fig. 16.1a). The major challenge of craft-knife cutting is the mitigation of the

warping and tearing that usually occur during the cutting process (Fenton et al. 2008). Multiple passes with an optimized cutting force can be applied to reduce this damage.

2. *Laser-cutting device*: In laser cutting, the focused laser beam strikes the material, which then either burns or vaporizes away, leaving an edge with a high-quality surface (see Fig. 16.1b). Advantages of laser cutting over mechanical knife cutting include easier work-holding and reduced interference caused by a contaminated knife. The precision of laser cutting may be better than that of mechanical cutting, since the possibility of paper warping and wear can be minimized by a small heat-affected zone of laser system. Normally, a laser cutter is more expensive than a craft-knife cutter, which would restrain the application of laser cutting in a resource-limited setting (Cate et al. 2014).

Direct cutting or shaping by a craft-knife or laser cutter can be used to fabricate μPADs for multiplexed sample analysis (Nie et al. 2013). However, these paper-based devices suffer from low mechanical stability and need solid supports as unwanted paper material is removed. To solve this problem, a cutting-technique-prepared paper strip can be laminated with a plastic backing, in a similar way to identification-card production, to provide mechanical strength to the paper device (Spicar-Mihalic et al. 2013). Pressure molding techniques, such as embossing, hold promise for generating fluid paths from omniphobic paper, whose surface has been modified by treatment with a hydrophoic fluorinated alkylsilane. An open-channel microfluidic paper device can be fabricated by sandwiching a sheet of paper between molds with complementary structures (Thuo et al. 2014) (see Fig. 16.1c). Optimizing the pressure imposed on the mold-paper-mold sandwich structure can build an identical paper-channel depth, which is clearly different from paper devices built by direct cutting. Moreover, fluid-flow behavior in these so-fabricated paper channels is similar to that observed in polydimethylsiloxane- (PDMS-) based open-channel devices, on which laminar flow, in the form of droplet generation through shearing in a T-device, can be observed. This kind of embossing technique, coupled with functionalized paper sheet, has potential for building complicated fluid networks for biosensing applications.

16.2.2 Defining Hydrophobic/Hydrophilic Regions on Paper

Apart from shaping paper into structures, such as channels and holes, by direct cutting, defining/patterning hydrophobic/hydrophilic regions on a paper sheet is one important strategy to use in developing μPADs. Based on the available methods of manipulating hydrophobic materials onto a paper sheet, the patterning approaches can be grouped into two categories: (1) *Bottom-up patterning*, in which stepwise addition/deposition/grafting of hydrophobic materials onto a specified area of a paper sheet is carried out; and (2) *top-down patterning*, in which hydrophobic materials are grafted to fully cover the paper sheet, followed by selective removal of part of the hydrophobic materials to reveal the hydrophilic area of the paper.

16.2.2.1 Bottom-Up Patterning

The principle of the bottom-up patterning approach is that hydrophobic materials are selectively implanted into a cellulose matrix to confine the hydrophilic region on a paper sheet (see Fig. 16.2a). Depending on the materials' properties, several techniques, such as screen printing (Dungchai et al. 2011), injection printing (Abe et al. 2008), contact printing, and dip coating, can be applied to deliver hydrophobic materials onto a paper sheet. The pioneering and most often used hydrophobic material is wax, which can be easily melted and can penetrate into a paper substrate through heating to form a hydrophobic barrier to confine fluid paths. Wax drawing not only can be used as a prototyping method to replace printing, but can also be used to add hand-drawn features to wax-printed μPADs (Lu et al. 2009). The wax-dipping method, which uses a magnetic fixed metal mask to cover the hydrophilic zones and dips the fixed mask into molten wax, is a low-cost alternative to the aforementioned techniques (Songjaroen et al. 2011). However, such wax-patterning methods usually suffer from low resolution, although their simple and economical characteristics make them attractive. The wax printing achieved by commercial wax printers is a highly appreciated approach (Carrilho et al. 2009; Lu et al. 2009). In this method, solid wax lines are printed on a paper sheet, which are then heat-melted to form a wax barrier on the paper. Using a wax-printing method, channels of different thicknesses and size channels can be created (Li and Liu 2014). Collectively, the merits of wax-patterning methods are (1) simple printing and baking processes, (2) rapid fabrication requiring only 5–10 min, and (3) elimination of organic solvents.

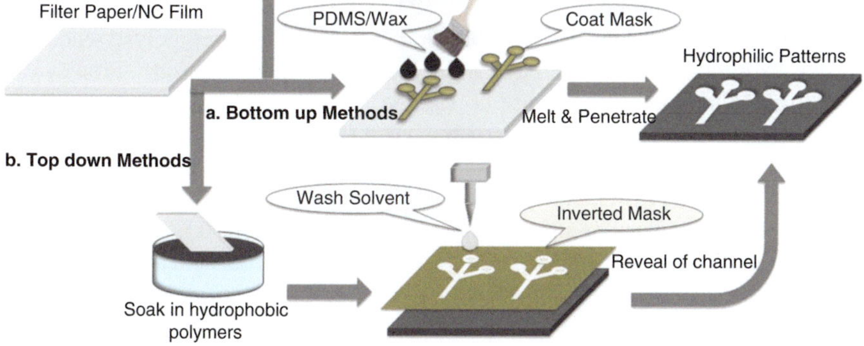

Fig. 16.2 Patterning hydrophobic barriers on paper sheets by (**a**) selectively implanting bottom-up hydrophobic materials onto paper sheets by screen printing, injection printing, wax printing, dip coating, and flexographic printing; and (**b**) utilizing a top-down method in which a paper sheet is fully covered by hydrophobic ink, followed by removal of the ink in specific regions to reveal the hydrophilic cellulose paper. *PDMS* polydimethylsiloxane, *NC* nitrocellulose

PDMS is a popular material for microfluidic device fabrication (Fujii 2002). Unlike solid wax, the PDMS ink can completely penetrate the paper before heating. Thus, printing PDMS resin onto paper and curing is another way of building a hydrophobic barrier. Screen printing (Ma et al. 2014), injection print, and stamping methods are used to selectively deliver PDMS resin onto a cellulose matrix. Progressively, other polymers have been explored to fabricate μPADs. Alkyl ketene dimer (AKD) (Li et al. 2010), which can be polymerized with heat, is deposited onto paper by injection printing to define hydrophobic regions. In a similar way, UV curable inks can be directly printed onto a paper sheet to fabricate a paper device (Maejima et al. 2013; Yamada et al. 2014). Innovative methods also have been tried, e.g., lacquer spraying (Nurak et al. 2013), a classic long-lasting wood finish, has been explored for patterning paper and has demonstrated itself to be the basis of a consistent and reproducible method for creating a hydrophobic barrier on paper.

The main concern with bottom-up strategies to build hydrophobic barriers on paper is the lateral resolution of the fluid path (channel width). The diffusion of the hydrophobic ink along the horizontal direction of the paper challenges the size limitation (width) of the microchannel that can be built. The impact of lateral diffusion should be specifically considered in wax printing and PDMS printing. Next, inkjet printing usually requires several printing runs to generate devices, and such multistep printing can potentially cause low print resolution (Cate et al. 2014). The positive aspect of this situation is that printing techniques, such as wax printing, screen printing, and inkjet printing, can be scaled up and adapted for high throughput and low-cost fabrication of paper devices. One candidate for high-throughput paper-device production is flexographic printing (Olkkonen et al. 2010), a well-established method for printing letters or images on almost any type of substrate, including plastic, metallic films, and paper. Fast and high-throughput production of μPADs may be realized using this method because commercially available flexographic printers used in industrial settings can print at speeds higher than 300 m/min on different substrates such as paper and plastic (Cate et al. 2014). A successful trial of using flexographic printing to fabricate μPADs is printing with polystyrene ink, which is dissolved in toluene, on cellulose paper. The solidified polystyrene then creates hydrophobic barriers that partially or completely penetrate through the paper substrate. However, before it can be widely used in scaled-up μPAD fabrication, several technical requirements need to be satisfied: (1) improving the flexibility and availability of the individual printing plates to lower the cost of operating a specialized flexographic printer, and (2) exploring the possibility of printing different kind of reagent at a time with acceptable resolution. It is anticipated that progress in flexographic printing can facilitate the commercialization of paper-based-device fabrication.

16.2.2.2 Top-Down Patterning

The principle of top-down patterning is that a paper sheet is fully impregnated with hydrophobic material. The hydrophobic material is then selectively removed to reveal the hydrophilic region (see Fig. 16.2b). The representative top-down approach is photolithography, which uses light to transfer a geometric pattern from a photomask to a light-sensitive chemical "photoresist" on the substrate. It is a standard microfabrication process that is used to pattern parts of a thin film or the bulk of a substrate. The UV-cured photoresist is a hydrophobic layer that can reform the hydrophilic region of the paper (Martinez et al. 2007). In principle, the photoresist, normally an epoxy-based negative photoresist designated SU-8, is coated on a paper substrate. A photomask with a desired pattern is used to control the photoresist-saturated paper that is exposed to UV light. The UV-light-treated paper sheet is then placed in a developer, such as 1-methoxy-2-propanol acetate, and the uncured-photoresist can subsequently be removed by the solvent to reveal the hydrophilic region, while the cross-linked photoresist forms hydrophobic barriers within the paper. A similar photolithography patterning technique has been established to fabricate μPADs by selectively curing photo-cross-linkable methylacrylic anhydride on paper. The other top-down strategy is a "wet-etch" process. In this approach, a solvent is delivered onto a specific region of a hydrophobic ink-impregnated paper sheet. The solvent can specifically wash off the precoated hydrophobic ink to reveal the native paper matrix. For example, polystyrene is precoated on the paper, and toluene as a solvent is then printed onto the precoated paper to dissolve and remove the polystyrene.

The fabrication resolution of top-down approaches relies heavily on expensive lithography equipment and reagents. Utilization of an organic solvent, which can potentially damage the integrity of the paper matrix, is a major issue. In addition, the cured photoresist usually suffers from low mechanical elasticity and can break upon bending, dooming its potential for building flexible paper-based devices. Compared to photolithography, the "wet-etching" strategy involves less wet chemical processing and lithographic machinery. It can be carried out, combined with digital inkjet printing and even flexographic printing techniques, to deliver a specific solvent onto the hydrophobic ink-impregnated paper to scale up to a high-throughput capability in the fabrication process.

16.2.3 Loading of Biological and Chemical Components

Paper is a good material for building a suitable place, container, or strip to conduct functional assays. Except for fabrication methods for building paper structures, effective loading of biological or chemical reagents to the desired region of the patterned-paper sheet is crucial. Inkjet printing can deposit biochemical reagents on paper (Maejima et al. 2013; Yamada et al. 2014), and is both practical and cost-effective. The principles for deposition of biochemical reagents anchored on a paper matrix are in line with the immobilized methods established in immunoassay (Gerbers et al. 2014), DNA microarray (Rosa et al. 2014), etc. Strategies such as physical absorption and entrapment are routinely used to fabricate μPADs. Moreover, due to the well-characterized chemical properties of cellulose paper, it is possible to modify the paper fiber, making it a better substrate with which to covalently capture biochemical molecules. For example, the available functional groups on cellulose paper include a backbone of hydroxyl groups and the reducing end of the cellulose ring. These can be chemically modified to form carbonyl or carboxyl groups, which can then covalently bind with amine-terminated molecules such as proteins or DNA. Alternatively, cellulose paper can be treated with polyamide-epichlorohydrin (PAE) (Saito and Isogai 2007) and polyvinyl amine (PVA) (Feng et al. 2006) to produce hydrophobic esters for biomolecule anchoring.

16.3 μPAD Formats

The format of a μPAD is determined by the chosen application and detection scheme. Progress in μPAD format has progressed from paper-only devices to hydride-based devices through the integrating of paper strips and electronic components.

16.3.1 Paper Only

A two-dimensional (2D) strip is a simple format for paper-based μPADs in which the fluid paths are built on the same layer (depth) of the paper (Fig. 16.3a). The sample solution can diffuse in the hydrophilic region because of capillary action. In general, 2D μPADs can be used to carry out sequential steps such as washing, sample pretreatment, and signal enhancement. To boost the multiplexing capability of a 2D μPAD, structures with multiple channels can be designed and fabricated. In addition, multiple reagents can be spotted on a 2D shaped porous substrate to achieve multiplexed lateral flow assays. Because of their straightforward structural design, 2D μPADs can be fabricated using either top-down or bottom-up strategies, such as paper cutting and shaping, wax printing, photolithography, inkjet printing, and screen printing.

Three-dimensional (3D) μPADs are assembled from multiple layers of patterned/structured paper (Martinez et al. 2008). The fluid networks are embedded at different layers of the paper. The flow transports not only laterally but also vertically, thus enabling the sample flow from one inlet to different outlets or detection zones. The stereo structure has more potential for implementing several assays without interfering with each other. To realize a 3D structure, three strategies have been established to assemble 3D μPADs.

1. Stacking multiple layers of patterned paper sheets using double-sided adhesive tape (Martinez et al. 2008): Vertically stacking several layers of paper with the assistance of double-sided adhesive tape is a straightforward way of building 3D μPADs. Aligning the hydrophilic spots on different layers allows fluid flow vertically to link the independent channels on the multiple layers, thus guiding the flow transport between the layers of paper (Fig. 16.3b).
2. Refining the thickness/depth of the hydrophobic layer within the paper matrix (Renault et al. 2014): This strategy makes full use of the thickness of the paper sheet. By precisely controlling the hydrophobic layer formed within the paper matrix, multiple fluid paths can be sophisticatedly formed at different layers of a single paper sheet (Fig. 16.3c). Normally, the hydrophobic ink, such as wax or PDMS, can be delivered on the front and back sides of the paper sheet. By tuning

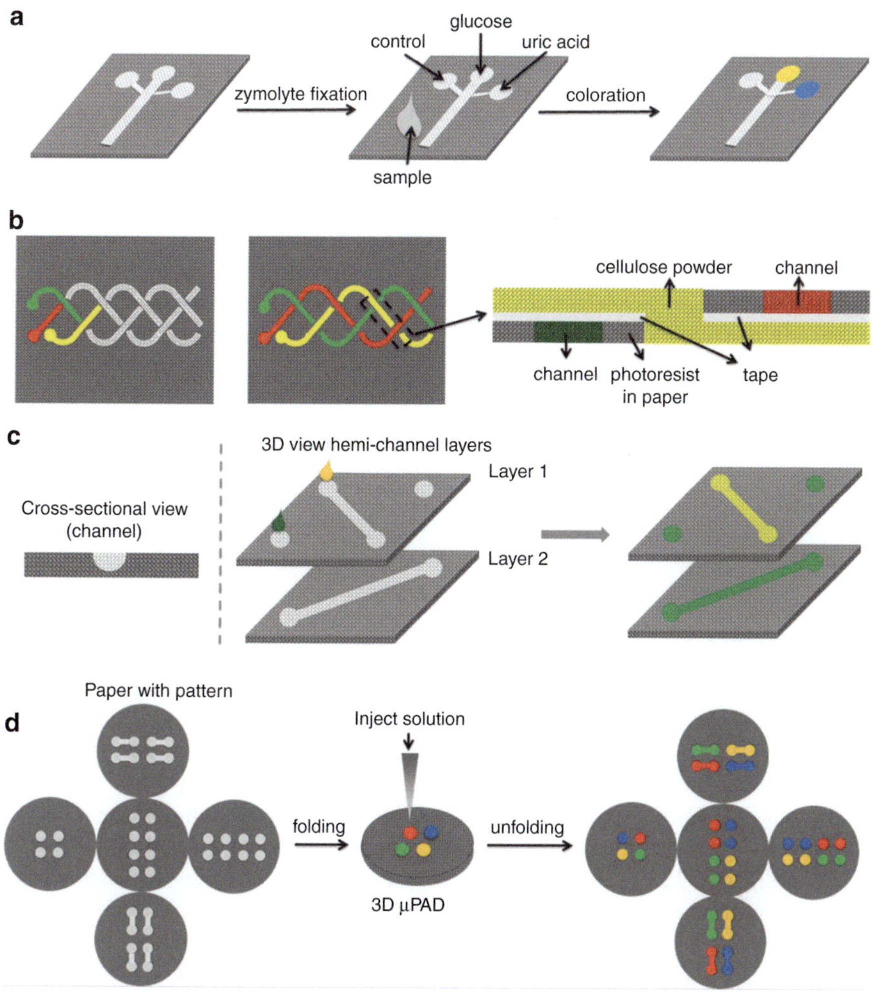

Fig. 16.3 Format of paper-based devices: (**a**) 2D paper-based device in which microstructures are formed on the same plane of a single, flat paper sheet; 3D paper-based device. (**b**) Stacking of patterned paper sheets—paper sheets containing hydrophobic/hydrophilic patterns are vertically stacked with the assistance of double-sided adhesive tape (adapted from reference Martinez et al. (2008)); (**c**) Controlling the depth of hydrophobic ink penetrated into the paper matrix—hydrophobic ink is delivered on both the front and back sides of the paper sheet, and amount and speed of ink penetration into the paper matrix to form the hydrophobic barrier at different depths (planes) of a single paper sheet is precisely controlled (adapted from reference Renault et al. (2014)); (**d**) Origami-paper-based device—a single sheet of flat paper with patterned channels is assembled by simple paper folding (adapted from reference Liu and Crooks (2011))

the amount of ink placed on the sheet, the ink saturation gradients, and the temperature for ink curing or melting, the depth of the ink that penetrates into the paper matrix to form the hydrophobic barrier can be controlled. The device is outwardly a single, flat paper sheet, but the fluid paths lie at different depths of the paper matrix to form a 3D fluid network.

3. Origami-paper device (Liu and Crooks 2011): The key advantage of origami paper is the ability to fabricate multiple structures on a flat sheet of paper and vertically assemble them by folding. In this method, all of the μPAD functional units are fabricated on a single flat sheet of paper in one photolithographic step. First, the flat sheet of paper is folded to align the functional units vertically (Fig. 16.3d). The fluid, then, is not only diffused laterally on every layer, but is also transported vertically at the alignment region. If all the hydrophobic barriers could be cured at the same time, no matter how complex the structures are, the fabrication process could be completed within a few minutes. The origami method of fabricating 3D μPADs is simple and inexpensive, yet effective.

16.3.2 Integrating Paper and Electronic Components

The paper and electronic units in hydride-based μPADs are designed to perform electrochemical, electrochemiluminescence, and photoelectrochemical-based detection (Dossi et al. 2013; Santhiago et al. 2014; Shi et al. 2015). It is anticipated that their successful integration can boost the sensitivity, portability, and digital analysis capability of μPADs. Thus, conducting circles and electrodes are fabricated on paper substrates to build electronics units on μPADs. The screen-printing technique is most often used to load conductive materials, such as carbon ink or silver ink, onto a paper matrix to form the desired screen-printed electrodes (SPEs) and wires (Yang et al. 2014). Surprisingly, pencil lead can also be used as an economical source for drawing conductive lines, dots, and pads on paper (Santhiago and Kubota 2013; Dossi et al. 2014). Conductive nanomaterials, such as gold nanoparticle (AuNP) ink, can be printed on paper by a calligraphic pen to build electrodes (Liana et al. 2013). However, many efforts have been directed to modifying the screen-printing carbon electrode with functional conductive materials. AuNPs, gold nanorods (Ma et al. 2015), gold and manganese oxide nanoparticles (Li et al. 2014b), gold–palladium alloy nanoparticles (Li et al. 2014a), and platinum nanospheres have been grown on SPEs to improve their performance as detectors.

The fluid paths and electronics components can be arranged either on the same paper sheet (2D device) or on different layers (3D device). For a 2D device, the electrodes are separated from each other by a paper channel on the same plane (Dungchai et al. 2009) (see Fig. 16.4a). For a 3D integrated device, electrodes can be fabricated on a separated paper sheet and vertically assembled with the fluid networks (Lu et al. 2012; Zhang et al. 2013) (see Fig. 16.4b). Otherwise, a single sheet of paper folded into a 3D device that changes shape, and fluidic and electrical connectivity, by simply folding and unfolding the structure (Wang et al. 2016) (see Fig. 16.4c). Because of the hydrophilic nature of paper, the sample and electrolytes can diffuse between layers to effect the analysis.

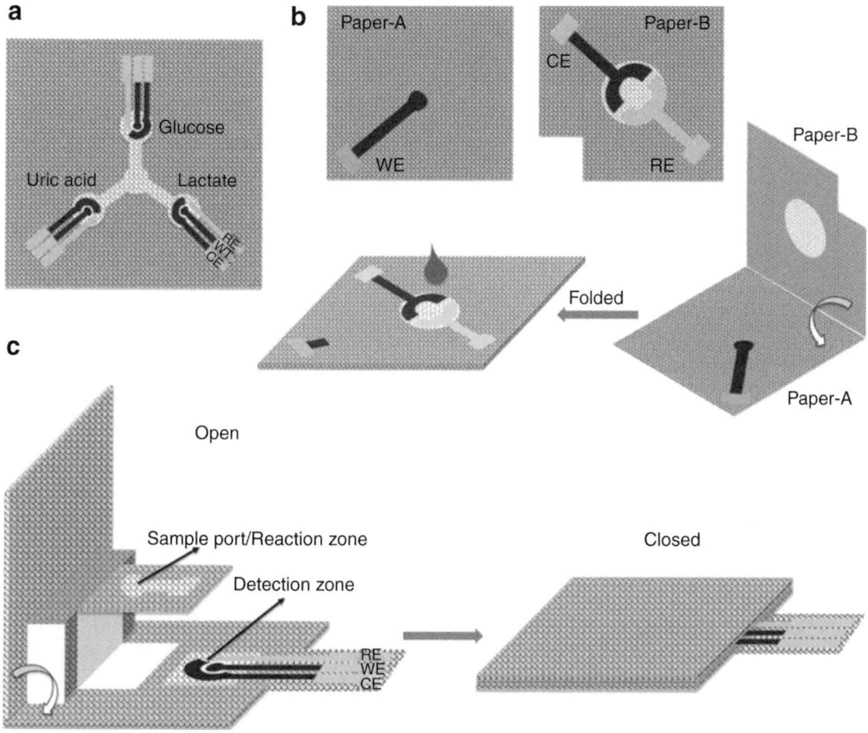

Fig. 16.4 Format of paper-electronic components integrated devices: (**a**) 2D integrated device in which electrodes and paper-channel are formed on the same plane of a single, flat paper sheet (adapted from reference Dungchai et al. (2009); 3D integrated device: (**b**) vertically stacking of electrodes and reaction zone, which are patterned on separated paper sheets (adapted from reference Zhang et al. (2013)); (**c**) origami-integrated device—a single sheet of paper with patterned channels and electrodes is assembled by simple paper folding (adapted from reference Wang et al. (2016)). *WE* working electrode, *CE* counter electrode, *RE* reference electrode

16.4 Detection Schemes Conducted Using μPADs

The signal-reporting system utilized in μPADs is in line with progress made in developing other POCT tools. Several detection methods have been realized in μPADs, such as colorimetric, chemiluminescence (CL), electrochemiluminescence (ECL), and electrochemical (EC) detection. Based on the signal generated during a biochemical assay, the detection methods can be categorized as either (1) optical signal sensing or (2) electrical signal sensing. Basically, to satisfy the low-cost requirement of point-of-care applications, signal detection should not be conducted by large, expensive pieces of equipment. Colorimetric and EC detection are favored because of their simplicity and/or high sensitivity and selectivity.

16.4.1 Optical Sensing

16.4.1.1 Colorimetric Detection

Colorimetric detection is the most commonly used detection scheme in paper-based μPADs due to its simple operation and straightforward signal readout. The other important factor to be considered is that paper strips are favored for glucose (Zhu et al. 2014a), uric acid (Demirel and Babur 2014), and lactic acid detection. The core of those enzyme-based reactions is the production of hydrogen peroxide, which participates in horseradish peroxidase- (HRP-) led oxidation. The substrates are oxidized by HRP using hydrogen peroxide as the oxidizing agent, yielding characteristic color, fluorescence, and luminescence signals (see Fig. 16.5). As they are well characterized in biological assay techniques, such as enzyme-linked immunosorbent assay (ELISA), the HRP-conjugated antibody, ligands, and oligonucleotide provide the specific recognition and also catalyze the versatile substrate to generate characteristic color changes, which are proportional to the concentrations of analyte. The visual color changes can be either discriminated by the naked eye, or can be recorded by scanners or cameras and converted to digital results for quantitative analysis.

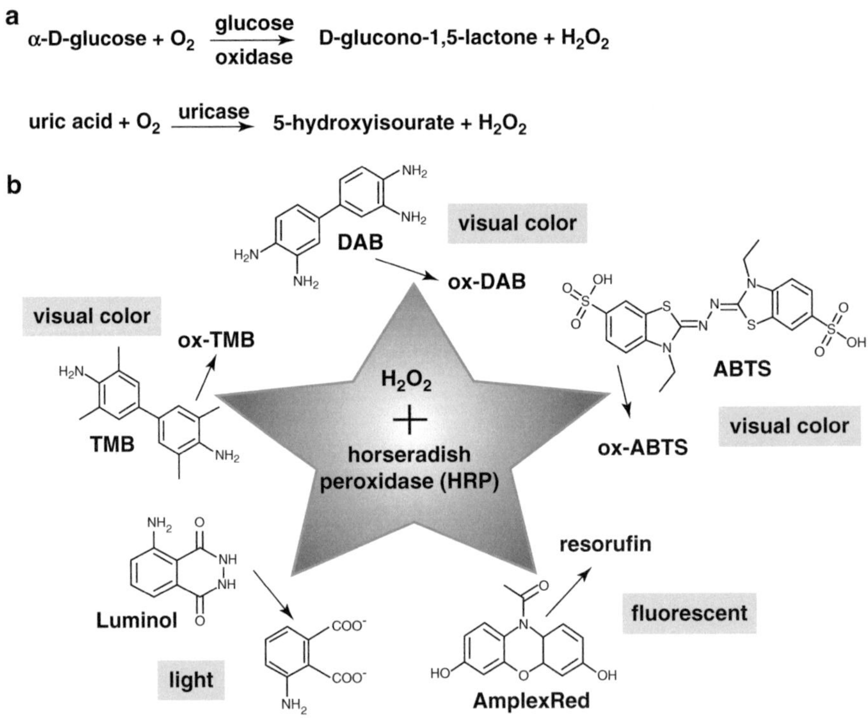

Fig. 16.5 Enzyme-based detection: (**a**) hydrogen peroxides are produced during glucose-oxidase- or uricase-mediated oxidation; (**b**) hydrogen peroxides participate in HRP catalyzing of the conversion of substrates to generate visual color changes [e.g., 3,3′,5,5′-tetramethyl benzidine—TMB, 3,3′-diaminobenzidine—DAB, 2,2′-azino-bis(3-ethylbenzothiazoline-6-sulfonic acid)—ABTS], producing light (e.g., luminol) or a highly fluorescent signal (e.g., Amplex® Red)

Next, chemical dyes, which have been used as indicators to determine the existence of a chemical reaction, are also widely used to construct a colorimetric-based μPADs. The classical example is phenol red for pH sensing. For instance, multiplexing of pH and nitrite sensing by colorimetric signal can be read and analyzed by a smartphone (Lopez-Ruiz et al. 2014). In this application, the pH-sensing zones of a μPAD have been immobilized with the pH indicators phenol red and chlorophenol red. While in nitrite-sensitive zones, Griess-reaction-related reagents are preloaded. The analyte dropped into the sample zone can flow along the channels toward the pH- and nitrite-sensing zones. After reaction, the visible color can be captured and analyzed by a smartphone, demonstrating POCT capability.

The next group of colorimetric change is given by function nanoparticles, which have higher extinction coefficients than ordinary dyes. Gold colloid labeled antibody has long been used to build lateral flow tests (Yu et al. 2015). The aggregation of the gold nanoparticles at the detection zone brings about a pinkish-brown color that can be either discriminated by the naked eye or quantified by a scanner.

Apart from improving the performance of the signal-generation system achieved, for example, by increasing the labeling efficiency of an enzyme or gold colloid, modification of a paper substrate to improve the colorimetric signal-to-noise ratio is another important development trend. Biomolecules can be adsorbed on the paper surface by van der Waals forces and electrostatic forces. Because of the hydrophilic characteristic of paper, physical adsorption is not stable. Moreover, the weak negative charge surface makes only adsorbing positive ions or cationic molecules possible. Thus, chitosan as a natural biopolymer has been used to modify paper-based analytical devices to improve the analytical performance of colorimetric measurements (Gabriel et al. 2016). Next, chemical modification of paper matrix, such as hydroxyl on the cellulose, to enhance the sensing material loading is favored because of the well-characterized chemical property of paper and covalent conjugation proved stable immobilization. For example, the oxidized cellulose can produce acid groups to form Schiff base between the cellulose and biomolecules (Su et al. 2007). Apart from chemical modifying paper fiber, introducing of chemical or biochemical compounds onto the paper to enhance the effective sensing material immobilization is another strategy. For instance, a paper surface can be functionalized with zwitterionic poly(carboxybetaine) through surface-initiated atomic-transfer radical polymerization to decrease surface fouling and speed up flow transportation in paper channels (Zhu et al. 2014b). Otherwise, cellulose binding modules (CBMs) can be genetic engineered to capture enzymes, antibodies (Cao et al. 2007), cells (Craig et al. 2007), and bacteriophages (Tolba et al. 2008). Due to the fast development in nanoparticle synthesis and functionalization, the concept of using carrier particles to build sensitive colorimetric sensing strip is getting popular. In principle, the sensing material is covalently coupled to the colloidal particles and then loaded on cellulose. There are several advantages of using carrier particles for improving the signal-to-noise ratio: firstly, the carrier colloid particles could be delivered to the paper matrix by printing, which is one important technique to fabricate paper device. Secondly, colloidal particles are more easy to trap on the pore surface of the paper compared to biological molecules in solution. Thirdly, the blocking and detection functions can also be carried out on carrier particles.

Common carrier particles include gold nanoparticles (Zhao et al. 2008), SiO_2 nanoparticles (Evans et al. 2014a), ceria nanoparticles (Ornatska et al. 2011), porous sol-gel particles (Bang et al. 2008), and so on.

Owing to the revolutionary technological advancement in mobile communication, smartphone-based analysis shows tremendous potential for POCT. The colorimetric signal can be easily captured in a timely manner using a smartphone at the assay point. Moreover, smartphone apps are becoming capable of sophisticated image analysis, including the ability to quantify colorimetric change, which can thus lead to the ability to perform quantitative or semiquantitative analysis (see Fig. 16.6a).

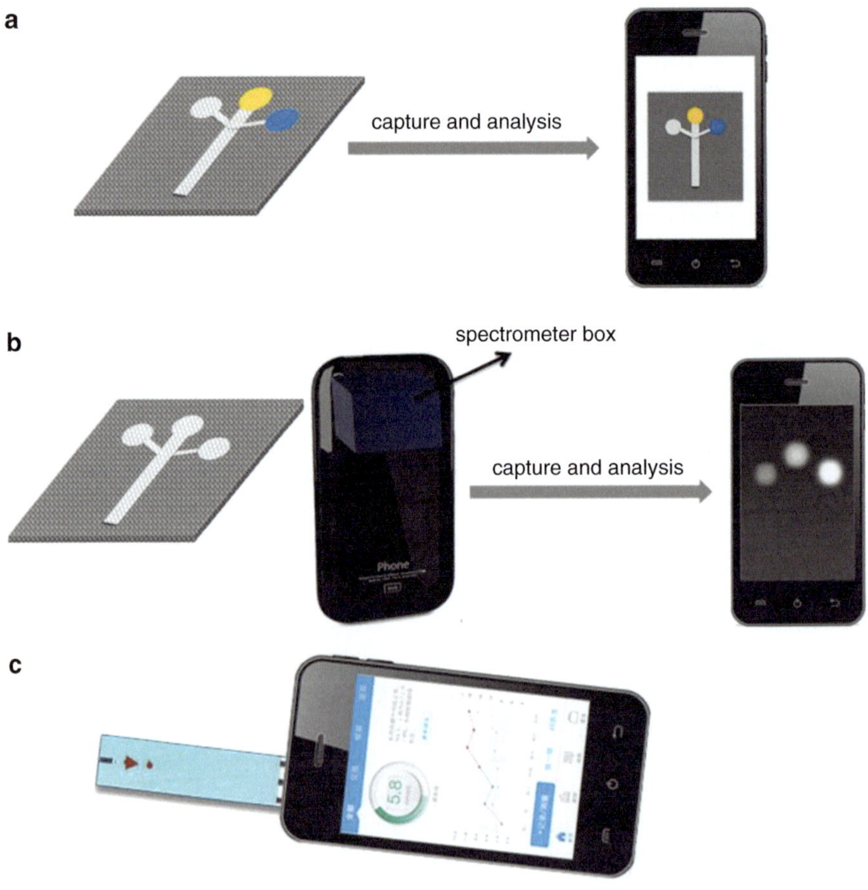

Fig. 16.6 Potential of smartphone-based analysis to promote the affordability and availability of sensitive POCT. (**a**) Smartphone camera coupled with an image-process app can capture and analyze colorimetric changes; (**b**) smartphone-docked miniaturized spectrometer for conducting absorbance and fluorescence measurements; (**c**) smartphone-docked miniaturized interface to read the electrochemical signals

16.4.1.2 Fluorescence Detection

Fluorescent dyes are one of the well-characterized probes used in cell, protein (Yamada et al. 2014), and DNA analysis (Scida et al. 2013). Using fluorophores with different excitation/emission wavelengths, multiplexing can be achieved. Inherited from procedures established in cell biology, fluorescence-based μPADs have been demonstrated in the analysis of biological samples, such as cancer biomarkers and bacteria (Rosa et al. 2014). For example, a competitive hybridization assay is conducted on a 3D paper-based platform where quencher-labeled ssDNA and capture ssDNA are competing to hydrolyze with fluorophore-labeled ssDNA (Scida et al. 2013). A target analyte can replace the quencher-labeled ssDNA from fluorophore-labeled ssDNA due to it having a higher matching degree than the quencher, producing a fluorescent signal that is linearly proportional to the concentration of target DNA. Except for natural fluorophore conjugates, nanoparticles with a characteristic photoluminescence property have been explored for application in μPADs. Quantum dots (QDs) (Noor and Krull 2014) or semiconductor nanocrystals are particularly promising for optical applications due to their high extinction coefficient. The protein oligonucleotides are linked with QDs and function as a signal reporter. Tuning the size of the QDs can produce different emission wavelengths, visualizable as distinct colors, proving a multiplexing capability. Moreover, the surface of QDs can be functionalized to ensure an effective labeling with biological probes.

However, the whitening additives in paper can increase the background fluorescence and thus compromise the sensitivity and specificity of the assay. In addition, the experimental setting of fluorescence detection is more complicated than that of colorimetric detection. In the near future, a smartphone-docked, miniaturized fluorospectrometer is highly anticipated to take full advantage of sensitive fluorescence-based sensing for POCT applications (see Fig. 16.6b).

16.4.1.3 Luminescence Detection

Chemiluminescence (CL) is the emission of light as the result of a chemical reaction. The applications of chemiluminescence in analytical biochemistry have been demonstrated in western-blot and cell-ELISA. Although there is light emitted in the assay, no excitation light and optical filter are involved compared to fluorescent detection. In addition, CL reagents are usually inexpensive and the detection is highly sensitive (Yu et al. 2011a, b), making it a promising candidate for constructing sensitivity assays on μPADs. Two sets of CL are favored in μPAD applications. First, enhanced chemiluminescence is a commonly used technique in biological assays. In this setting, the HRP-conjugated-antibody, oligonucleotide probes participate in specifically recognizing the molecule of interest and catalysis of the conversion of the CL substrate to produce a light signal (Zhou et al. 2014). Next, luminol chemistry is applied for blood, hemoglobin, and metal-ion detection based on the principle that luminol with hydrogen peroxide in the presence of metal ions, such as iron or copper, produces a luminescence signal. For example, a CL-based μPAD is designed to simultaneously and quantitatively test for glucose and uric acid

(Yu et al. 2011a, b). The μPAD is fixed in a cassette, and the cassette could be closed using a black metallic cover that has a sample-injection hole. The sample-injection area is aligned with the photomultiplier of the analyzer. Once the sample migrates toward the CL detection area and generates a CL signal, the signal is recorded by a computer.

Electrochemiluminescence (ECL) is a detection method that uses electrochemical reactions to generate luminescence (Delaney et al. 2011; Doeven et al. 2014; Ge et al. 2014). The high sensitivity and specificity provided by an ECL signal reporter is due to the low background optical signal, which can be minimized by controlling the electrode potential (Richter 2004; Deng et al. 2009; Forster et al. 2009; Feng et al. 2014). Owing to rapid developments in printing technology and the creation of functional conductive nanomaterials, a μPAD coupled with electronic components can be fabricated cost-effectively. Growing interest has been focused on the integration of ECL with μPADs. The electrochemically triggered light emission also can be recorded by a mobile phone camera. To reinforce the point-of-care applicability of ECL-based detection, expensive potentiostats are not favored for controlling the potential on the electrodes of a paper-based device (Delaney et al. 2013). An innovation is to trigger the electrochemical reaction by operating the electrode potential from the audio socket of a phone. Audio functional properties, such as frequency, amplitude, and duration of square-wave pulses, can be transduced to the working electrode. This demonstrates the powerful aspects of applying smartphones in POCT; such application not only fulfills the requirement for signal detection, but also that to control the testing.

16.4.2 Electrochemical Sensing

Progress in electrochemical (EC) sensing has been both rapid and prodigious in recent decades. A glucose strip coupled with a handhold electronic reader has achieved huge commercial success, demonstrating the strengths of μPADs coupled with electrochemical detection schemes. The use of wax printing to build the fluidic paths and screening print to create electrodes can be used to fabricate paper-based electrochemical sensors for the detection and measurement of glucose, uric acid, DNA, and cancer biomarkers. Leveraging the synergy between new sensing materials, device architectures, and fabrication techniques has led to the invention of paper-based electrochemical microfluidic devices with high sensitivity and selectivity. For instance, a 3D origami-based multiplex EC μPAD has been fabricated using a nanoporous silver-paper electrode for disease detection and early diagnosis (Li et al. 2013). Tumor-associated biomarkers are selectively tagged with nanoporous gold-chitosan hybrids, which contain absorbed metal ions. The metal ions can be detected by square-wave voltammetry (SWV), and the biomarker concentrations are linearly correlated with the measured peak currents of the metal ions. Because of the good biocompatibility of cellulose paper, culturing cells in μPADs is a promising platform for investigating cellular activity and screening potential therapeutic drugs. The integrated electrochemical sensor can monitor cellular apoptosis, hydrogen peroxide release, and glycan production under drug challenges (Liu et al. 2014; Shi et al. 2014; Su et al. 2015).

The bottleneck to wider application of EC-based μPADs is the availability of cost-effective, miniaturized, and portable electrochemical analyzers. Currently, most of the EC-based μPADs that have been demonstrated work in conjunction with a desktop electrochemical workstation. While the reported EC-based μPADs read verifiable signals, such as amperometric signals, impedance, conductivity, and signals from both cyclic and square-wave voltammetry, the question of whether it is possible to build a miniaturized portable device to read multiple signals at a reasonable price still remains. It is anticipated that in the near future progress in the electronics sector will facilitate the development of a handheld electrochemical reader like a glucose meter that can take full advantage of the sensitivity and quantitative analysis capability of an electrochemical biosensor.

16.5 Classification of μPADs

New sensing concepts have been put forward by rapid developments in materials science and mechanical engineering. Exciting μPAD designs for biomedical analysis, environmental supervision, food sanitation monitoring, etc. are being innovated in the practical application pipeline of paper-based analytical device research. An appropriate classification of these devices will help detail the importance, challenges, and future trends of μPADs development.

First, the devices can be classified based on their purpose or fields to which they are to be applied, e.g., a paper strip for urine testing or a μPAD for detecting illegal additives in raw meat. This is a satisfactory way for nonprofessionals to understand this area of endeavor, since fewer details are able to be discussed without sacrificing the required depth of understanding. Otherwise, paper-based devices can be categorized based on the biological or chemical nature of the target of analysis, as in the case of a paper strip for enzyme detection or a paper-based device for DNA testing.

As simplicity is highly related to manufacturing cost and functionality is determined by analytical accuracy, both aspects are important for eventual commercialization. Considering the fabrication/preparation process, paper-based devices can be divided into two categories as follows (see Fig. 16.7): (a) On-demand devices, which are blank-paper platforms without predeposited biological and chemical reagents. Depending on the samples to be tested, the detection reagents are chosen and introduced into the devices by users prior to the test, either before or after the addition of test samples. (b) Ready-to-use devices, which are designed as complete analytical sensors by integrating indication reagents into the detection zones of the devices. Based on the particular detection chemistry incorporated, this type of device is used to detect specific analytes in test samples (Li et al. 2012). Although this classification does not reveal the nature of the analyzers or the application field

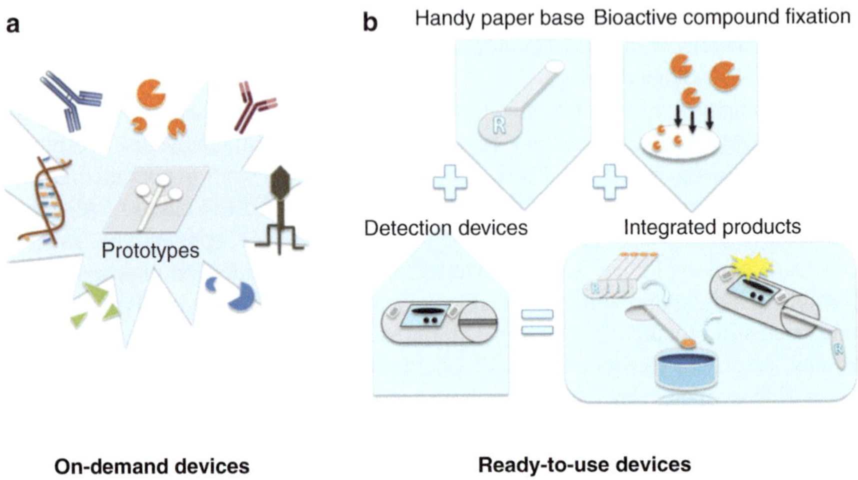

Fig. 16.7 Classification of μPADs according to specifics of their preparation and application: (**a**) On-demand devices, which are blank-paper-based platforms without predeposited biological and chemical reagents. (b) Ready-to-use devices, which have indication reagents integrated into their detection zones

for the device, it helps in constructively understanding a critical challenge; that is, how to make the trade-off between simplicity and functionality of newly designed paper-based analytical devices commercially viable?

Simplicity and low cost are two predominate factors influencing the development of paper-based microfluidics. New fabrication methods include ink printing, paper folding and cutting, pressure-based fiber-structure shaping, and other potentially feasible techniques that are continually being created and improved. It is reasonable to predict that fabrication techniques can sustain the rapid development of on-demand devices. In the near future, efforts should be focused on simplifying and standardizing the assay procedures and reagents.

From the POCT point of view, the second group of devices, namely ready-to-use devices, should be usable by end-users without complex procedures. However, in addition to paper platform manufacturing, the biggest issue with ready-to-use devices is how to effectively preload and preserve the functional indicator reagents, which are normally expensive compared to a paper-based platform. The biochemical molecules comprising such reagents are fundamental to the specific reorganization, binding, catalyzing, and conversion required to realize target detection and signal transmission, such as color development, fluorescence, and luminescence generation. Because such molecules are either extracts from living biomaterials or are synthesized from sophisticated chemical reactions, their isolation and purification are time consuming and labor intensive, increasing the cost of mass producing the devices utilizing them. In addition, for the application of a ready-to-use paper-based device, one key function is simple and/or cost-effective signal recognition and transfer. Most of the existing desktop ultraviolet spectroscopy or fluorescent microscopy devices and scanners would not be a perfect match for ready-to-use μPADs for the mass population of end-users. To this end, simple and portable handheld devices are highly anticipated in order to translate biosignals into visual information for interpretation by the naked eye or comprehensive data for analysis by the handheld devices.

Collectively, fixation of functional biomolecules and signal recognition and transfer components are the key requirements that increase the complexity and cost of paper-based devices. Paper itself is indeed inexpensive and easily accessible, but functionalizing it is costly, a factor that would be further increased with the desire to achieve higher accuracy and reproducibility. Although it is an unrealistic goal to set a universal standard that combines simplicity and functionality in paper-based devices, tremendous efforts have been directed to finding techniques that fulfill the goals of simplicity and functionality. For example, ink printing and pressure molding are paper-manufacturing techniques. They are economically viable candidates for reducing the cost of designing laboratory techniques and products that would be widely accepted by industry. In the future, synthetic paper may become the very first μPADs-making step rather than starting with ready-made paper. This can be achieved by advancements in fabric and/or organic polymer weaving techniques. The second trend is characterized by integration; that is, combining different functional parts to increase application efficiency and/or accuracy. Revolutions in telecommunication, especially the progressively more powerful smartphone, are affordable for the majority of the end-user population, significantly opening up greater possibilities in POCT. Owing to the widespread use of smartphones with reduced signal recognition and transference costs, POCT is truly within reach of masses of end-users of practical daily testing applications (see Fig. 16.6).

16.6 Market Impact of μPADs

Progress in fabrication, cellulose-paper functionalization, and biomolecule preloading are all aimed at improving target-detection sensitivity in low-cost analytical devices. A growing number of miniaturized analytical devices have been developed to meet increasing demands from the medical, food safety supervision, environmental monitoring, and homeland security industries.

Currently, the majority of POCT devices are designed for biomedical applications. The first paper-based bioassay was introduced in 1957 for the identification of glucose in urine (Free et al. 1957). This assay was developed into a commercial product in the mid-1960s to diagnose and assist in the management of diabetes. The same concept has been used to develop a "dipstick" to detect urinary albumin and pH by examining distinct color changes. Colorimetric-based urinary dipsticks can multiplex ten compounds, including nitrites, ketones, urobilinogen, and bilirubin (Free et al. 1957). The underpinning of the commercial success of these dipsticks is the color-coded chart provided with the sticks that allowed the user to determine the presence of an analyte at a defined location along the length of the strip. Now, their use is widely accepted by the medical community. Although dipsticks and lateral flow strips dominate the rapid diagnostics market, some μPADs have emerged as versatile POCT platforms for the detection of glucose, uric acid, hemoglobin, and nitrites (Table 16.1).

Table 16.1 Commercial products of paper-based analytic devices for biomedical applications

Diseases or applications	Company	Product name	Sample types	Target chemicals	Assay time	Other features
Blood Glucose test strips	Abbott	Freestyle Lite test strips	Whole blood	Glucose		Need freestyle lite blood glucose monitoring system
	Accu Chek	Aviva Plus strips	Whole blood	Glucose	5-s	Finger, palm or forearm testing Need Accu Chek Aviva Plus meter
	Bayer & Contour	Bayer Contour NEXT test strip	Whole blood	Glucose		Evaluates a single blood sample seven times for exceptionally accurate results. No coding required Need Contour Next meters
	MediSense® Optium™	MediSense Optium Blood Glucose test strips	Whole blood	Glucose	5-s	Need MediSense Optium Blood Glucose Meter
Urinalysis Test Strips	Bayer	Keto-Diastix Ketone and Glucose Reagent Strips for Urinalysis 50ct	Urine	Ketone (acetoacetic acid) and glucose		Show presence and concentration of two chemicals
	Teco	URS-1G Glucose Test Strips	Urine	Glucose	30 s–2 min	Clear and accurate results for glucose levels
	Siemens	Multistix 10 SG reagent strips for urinalysis	Urine	Glucose, bilirubin, acetoacetic acid, specific gravity, blood, pH, protein, urobilinogen, nitrite, leukocytes	Not mentioned	

Pregnancy Test	OSOM®	OSOM®hCG Card Pregnancy Test	Urine	Human chorionic gonadotropin (hCG)	In 3 min	Easy-to-read black on white results, 20 mIU/mL
	NOVAtest	LH131	Urine	LH (luteinizing hormone)	In less than 5 min	Sensitivity: 15 mIU/mL
Viral infections	Alere Determine™	HIV-1/2 Ag/Ab Combo	Whole blood/ serum/plasma	HIV-1/2 antibodies	20 min	
	Chembio Diagnostic Systems Inc.	HIV 1/2 STAT-PAK® Assay	Whole blood/ serum/plasma	HIV-1/2 antibodies	15 min	Minimal sample size required—5 μL
	Quidel Corporation	Solana Influenza A+B Assay	Nasal and nasopharyngeal swabs	Influenza A and influenza B viral RNA	Approximately 45 min	Qualitative in vitro diagnostic
	OSOM®	OSOM® RSV/ AdenoTest	Nasal swab, nasal suction specimens	Respiratory syncytial virus (RSV) and adenovirus antigens	1 min hands-on time; 10 min or less	Sensitivity versus PCR > 90% (RSV); 85% (adenovirus)
	Firstvue	Malaria Rapid Test	Blood	P. falciparum	In 10 min	Colloidal Gold Technology, monoclonal antibodies coated strips
	Firstvue	HBsAg Rapid Test	Whole blood/ serum/plasma	Hepatitis B surface antigen	Not mentioned	

(continued)

Table 16.1 (continued)

Diseases or applications	Company	Product name	Sample types	Target chemicals	Assay time	Other features
Bacterium and Parasites	OSOM®	OSOM® H. pylori Test	Whole blood/ serum/plasma	H. pylori antibodies	10 min or less	95.9% sensitivity versus biopsy/histology
	Meridian Bioscience Inc.	NEW ImmunoCard STAT!® HpSA®	Human stool	H. pylori antigens	In 5 min	Monoclonal antibody-based test
	OSOM®	OSOM® Trichomonas Test	Urine	Trichomonas antigen	In 10 min or less	Detects the antigen; does not require live organism
	Chembio Diagnostic Systems, Inc.	Chagas STAT-PAK® Assay	10 μL whole blood, 5 μL serum or plasma	Antibodies to T. cruzi	15-min	Requires no cold chain storage, uses a minimal sample size
Cancer	Firstvue	Prostate specific antigen (PSA)	Whole blood/ serum/plasma	Prostate specific antigen	In 10 min	Semi-quantitative detection

The next important market for μPADs is food safety supervision. Food presents a complex matrix and the detection of minor components such as vitamins, allergens, and herbicide or pesticide residues often requires ultrasensitive analytical devices with low detection limits. In the past, food safety supervision relied heavily on specialists and food safety bureaus because conventional analytical techniques, such as gas chromatography and spectrometer, which are often expensive, complicated, and slow. The emergence of miniaturized analytical systems enables users to accurately perform analyses with small liquid volumes and at unprecedented speeds. This development should partially placate consumers and stakeholders who are desperately looking for user-friendly, ready-to-use detection strips to safeguard food quality and uphold safety. Since pathogens such as Campylobacter, Salmonella, and Escherichia coli O157:H7 (Atalay et al. 2011) are thought to be responsible for the majority of food-borne diseases, paper-based analytical devices can be applied to specifically examine suspicious food on a daily basis and in a timely manner to prevent food-borne diseases. Such devices and methods are desirable because conventional bacterial detection methods, such as colony-, immunology-, and polymerase-chain reaction-based methods may take up to several hours or even a few days to yield results. For direct probing, the toxins produced by bacteria, e.g., Staphylococcal Enterotoxin B (SEB) produced by Staphylococcus aureus, are important targets in food safety analysis. Most desired by the food safety market are strips that can sensitively detect harmful additives, pesticide residues, and antibiotics. The food analysis μPADs developed in laboratories and launched to market to date are limited. Considering the customers' basic needs in daily life, it is easy to determine the direction for research and development. Solid foods do not have enough water and surface area to activate wet chemical reactions on paper, while liquid foods usually have their own pigments/coloration, which can potentially interfere with test accuracy. High accuracy is generally needed for testing food, since the targeted chemicals/pathogens may exist at very low concentrations in the samples. Such high accuracy and sensitivity may not be satisfied by a μPAD if its selling point is simpler fabrication and operation, low cost, and enhanced portability.

Similarly, for environmental monitoring that aims to provide quality water, soil, and air to mass populations, sensitive and affordable analytical tools are sought by customers. For example, microfluidic paper-based electrochemical devices (μPEDs) have been fabricated to selectively analyze Pb(II) in an aqueous solution containing a mixture of Pb(II) and Zn(II) (Nie et al. 2010). In addition, both electrochemical and colorimetric detection have been realized in an integrated paper device for the rapid screening for Au(III) in the presence of a common source of interference, Fe(III), in industrial waste solutions (Apilux et al. 2010). Even though μPAD devices have advantages, the number of successful examples of environmental analytical devices is relatively less than those that have found application in biomedical fields (Table 16.2). Owing to the crucial importance of a safe environment to human life, it is believed that increasingly miniaturized environmental analytical devices will find their way to market, since sufficient technological advances and accurate market positioning are deemed achievable. Similar to the two-type classification method mentioned above, the future growth of μPADs for

Table 16.2 Commercial products of paper-based analytic devices for food safety and environmental supervision applications

Applications	Company	Product name	Sample types	Target chemical	Accuracy level	Other features
Food and drink	Hydrion	Sanitizer test strips: QT-40	Non-alkaline sanitizer	Quaternary sanitizers	Comparison chart measures 0, 150, 200, 400 and 500 ppm	10 s/test
	3M	Fryer test strips	Cooking oils	Free fatty acid (FFA)	FFA concentration ratings from 2 to 7%	15 s/test
	Raburn	Chlorine test tape refill	Solution	Chlorine	Measures 10, 50, 100, and 200 ppm	
	Precision Labs Inc.	Sanitizer test strips: iodine	Solution	Iodine	Measures 12.5, 25, and 50 ppm	Reads in 60 s
	Insta-TEST®	Peroxide test strips	Solution	Peroxide	0–90 ppm	Get results in 10 s
	Micro Essential Laboratory	Hydrion (QT-10) Quat Test Paper 0–400 PPM	Solution	Quaternary sanitizers	Measures 0, 150, 200, 400 and 500 ppm	
	San Jamar	SFC1250QT Saf-Check thermometer holder and quaternary sanitizer test strip dispenser	Solution	Quaternary sanitizers	Not mentioned	All-in-one system that keeps thermometer and quaternary sanitizer test strips together, GREEN = safe; RED = change solution

Environmental	Environmental Test Systems (ETS) AQUACHEK®	AQUACHEK® SHOCKCHEK™POOL and SPA test strips	Pool water	Two harmful chemistries, total and free chlorine	Not mentioned	
	(ETS) AQUACHEK®	AQUACHEK® RED	Pool water	Total hardness, chlorine, bromine, free chlorine, pH, total alkalinity, stabilizer (cyanuric acid)	Not mentioned	Test for seven important chemistries in less than 1 min
	(ETS) Accugrow®	AccuGrow soil test strips	Soil	pH, nitrogen, phosphorus, and potassium	Not mentioned	One dip into the soil sample, no capsules or powders to dissolve

environmental and food analysis will likely to be focused on two key elements: (1) Development of instrumentation-free and elegant paper-based devices, which can be correctly operated by minimally trained personnel and the results of which can be clearly understood, and (2) portability and online data analysis to enable use in the field and interpretation of testing data for comprehensive understanding.

16.7 Future Challenges

Apart from the development of techniques to build a simple and easy-to-use test strip, the challenges that current μPADs encounter in commercialization include how to balance the cost and functionality of such a testing device. More specifically, although more and more designs are claimed to be cheaper and faster, what exactly would be a "better" test device in the interest of the public? The criterion highlighted by the World Health Organization is not which test is the cheapest, but that the cost of the test needs to provide the desired clinical benefit. This can be interpreted by the concept of the cost-to-benefit ratio. For example, averting the prescription of months of expensive, ineffective, and potentially toxic therapy by a drug-sensitivity-screening POCT justifies a higher price than a POCT that only marginally improves public health. Additionally, the cost of "discounting" a concept well known in economics and finance states that the availability of a good or service today has greater value than the same good or service in the future (Chin et al. 2012). In short, the true value of newly created POCT designs can never be defined by a simple, one-dimensional standard. Target consumers are another major factor that cannot be avoided in a discussion of the application and commercialization of μPADs. As functionalization and simplicity are difficult to achieve together, precise customer positioning is needed (see Fig. 16.8). A useful paper-based analytical device may not be the

Fig. 16.8 Balance between the functionality and simplicity of a μPAD moving toward commercialization

cheapest, but could be the one that best suits the need of the target consumers. The commercialization of new designs can be divided into multiple product lines, and the production cost, use conditions, and detection accuracy of each product line can differ based on the needs of the target consumers. For example, a POCT device can be combined with typical clinic analytical equipment to achieve high-accuracy detection that suits the needs of hospitals, while another POCT device can be combined with smartphone readers to suit the needs of nonprofessionals at the same time. Researchers exploring new POCT device ideas in laboratories also need to consider commercialization possibilities from the start of their experiments.

Finally, local cultures, ethics, and political trends are also factors that influence the market for μPADs, not just in the medical space, but in the food safety and environmental monitoring spaces as well. For example, a large volume of literature emphasizes simplified manufacturing in an effort to facilitate acquisition of affordable analytical devices in rural and otherwise resource-limited areas. However, the number of comprehensive studies of such markets are lacking and are generally needed. Projecting the actual uptake and performance of such "simplified" devices in developing countries require answers to the following questions:

- To what extent are civilians able to access medical services? How many of them are willing to go and/or can afford to go to hospitals?
- How to maintain detection reliability, data/data transmission security, and thus protect security of the patients?
  For projecting the uptake and performance of food safety and environmental test devices, the following questions need to be answered:
- Does the target area have environmental and/or food safety regulations or laws?
- To what extent are any regulations and/or laws followed and to what degree have they been implemented?

Analytical devices are needed in quality and quantity, especially when healthcare, food quality, and environmental emergencies or crises occur on large scales. Such occurrences provide a good opportunity for popularizing inexpensive, portable analytical devices.

Conclusion

As a good candidate for POCT applications, paper microfluidics or μPADs achieve tremendous progress in sensing platform design and fabrication. Because low cost and easy to use are two key selling points of μPADs, socioeconomic concerns comprise one of the driving forces promoting the research and application of μPADs. Currently, the research focus is to build miniaturized handheld device to direct reading the signal on μPADs. Coupling of low cost, disposable paper strip or paper device with portable economical signal reading tool can provide a truly POCT service. The state-of-art technology discussed in this chapter is promoting the scale-up fabrication of paper device. However, apart from the technology development, the commercialization of the μPADs heavily relies on the choice of analyte/target and the potential users. In addition, an important thing to do now is to seek standardization of global regulations for POC devices, especially in the biomedical area. Unifying regulations are not needed only for measuring product quality, but they would also help deter any possible monopolization and stimulate more business startups, which would encourage minor enterprises to launch new products.

Acknowledgments Financial support from the National Natural Science Foundation of China (No. 31200700 and 21375108), Science Foundation of Chongqing (cstc2014jcyjA10070), Fundamental Research Funds for the Central Universities (XDJK2015B020, XDJK2016A010 and XDJK2016D001).

References

Abe K, Suzuki K, Citterio D (2008) Inkjet-printed microfluidic multianalyte chemical sensing paper. Anal Chem 80(18):6928–6934

Ali MM, Aguirre SD, Xu Y, Filipe CD, Pelton R, Li Y (2009) Detection of DNA using bioactive paper strips. Chem Commun 43:6640–6642

Alkasir RS, Ornatska M, Andreescu S (2012) Colorimetric paper bioassay for the detection of phenolic compounds. Anal Chem 84(22):9729–9737

Apilux A, Dungchai W, Siangproh W, Praphairaksit N, Henry CS, Chailapakul O (2010) Lab-on-paper with dual electrochemical/colorimetric detection for simultaneous determination of gold and iron. Anal Chem 82(5):1727–1732

Atalay YT, Vermeir S, Witters D, Vergauwe N, Verbruggen B, Verboven P, Nicolaï BM, Lammertyn J (2011) Microfluidic analytical systems for food analysis. Trends Food Sci Technol 22(7):386–404

Bang JH, Lim SH, Park E, Suslick KS (2008) Chemically responsive nanoporous pigments: colorimetric sensor arrays and the identification of aliphatic amines. Langmuir 24(22):13168–13172

Bruzewicz DA, Reches M, Whitesides GM (2008) Low-cost printing of poly (dimethylsiloxane) barriers to define microchannels in paper. Anal Chem 80(9):3387–3392

Cao Y, Zhang Q, Wang C, Zhu Y, Bai G (2007) Preparation of novel immunomagnetic cellulose microspheres via cellulose binding domain-protein A linkage and its use for the isolation of interferon α-2b. J Chromatogr A 1149(2):228–235

Carrilho E, Martinez AW, Whitesides GM (2009) Understanding wax printing: a simple micropatterning process for paper-based microfluidics. Anal Chem 81(16):7091–7095

Cate DM, Adkins JA, Mettakoonpitak J, Henry CS (2014) Recent developments in paper-based microfluidic devices. Anal Chem 87(1):19–41

Chen Y, Wang Y, Liu L, Wu X, Xu L, Kuang H, Li A, Xu C (2015) A gold immunochromatographic assay for the rapid and simultaneous detection of fifteen β-lactams. Nanoscale 7(39):16381–16388

Chin CD, Linder V, Sia SK (2012) Commercialization of microfluidic point-of-care diagnostic devices. Lab Chip 12(12):2118–2134

Craig SJ, Shu A, Xu Y, Foong FC, Nordon R (2007) Chimeric protein for selective cell attachment onto cellulosic substrates. Protein Eng Des Sel 20(5):235–241

Delaney JL, Hogan CF, Tian J, Shen W (2011) Electrogenerated chemiluminescence detection in paper-based microfluidic sensors. Anal Chem 83(4):1300–1306

Delaney JL, Doeven EH, Harsant AJ, Hogan CF (2013) Use of a mobile phone for potentiostatic control with low cost paper-based microfluidic sensors. Anal Chim Acta 790:56–60

Demirel G, Babur E (2014) Vapor-phase deposition of polymers as a simple and versatile technique to generate paper-based microfluidic platforms for bioassay applications. Analyst 139(10):2326–2331

Deng L, Zhang L, Shang L, Guo S, Wen D, Wang F, Dong S (2009) Electrochemiluminescence detection of NADH and ethanol based on partial sulfonation of sol–gel network with gold nanoparticles. Biosens Bioelectron 24(7):2273–2276

Doeven EH, Barbante GJ, Kerr E, Hogan CF, Endler JA, Francis PS (2014) Red–green–blue electrogenerated chemiluminescence utilizing a digital camera as detector. Anal Chem 86(5):2727–2732

Dossi N, Toniolo R, Piccin E, Susmel S, Pizzariello A, Bontempelli G (2013) Pencil-drawn dual electrode detectors to discriminate between analytes comigrating on paper-based fluidic devices but undergoing electrochemical processes with different reversibility. Electroanalysis 25(11):2515–2522

Dossi N, Toniolo R, Terzi F, Impellizzieri F, Bontempelli G (2014) Pencil leads doped with electrochemically deposited Ag and AgCl for drawing reference electrodes on paper-based electrochemical devices. Electrochim Acta 146:518–524

Dungchai W, Chailapakul O, Henry CS (2009) Electrochemical detection for paper-based microfluidics. Anal Chem 81(14):5821–5826

Dungchai W, Chailapakul O, Henry CS (2011) A low-cost, simple, and rapid fabrication method for paper-based microfluidics using wax screen-printing. Analyst 136(1):77–82

Evans E, Gabriel EFM, Benavidez TE, Coltro WKT, Garcia CD (2014a) Modification of microfluidic paper-based devices with silica nanoparticles. Analyst 139(21):5560–5567

Evans E, Gabriel EFM, Coltro WKT, Garcia CD (2014b) Rational selection of substrates to improve color intensity and uniformity on microfluidic paper-based analytical devices. Analyst 139(9):2127–2132

Feng X, Pelton R, Leduc M (2006) Mechanical properties of polyelectrolyte complex films based on polyvinylamine and carboxymethyl cellulose. Ind Eng Chem Res 45(20):6665–6671

Feng QM, Pan JB, Zhang HR, Xu JJ, Chen HY (2014) Disposable paper-based bipolar electrode for sensitive electrochemiluminescence detection of a cancer biomarker. Chem Commun 50(75):10949–10951

Fenton EM, Mascarenas MR, López GP, Sibbett SS (2008) Multiplex lateral-flow test strips fabricated by two-dimensional shaping. ACS Appl Mater Interfaces 1(1):124–129

Forster RJ, Bertoncello P, Keyes TE (2009) Electrogenerated chemiluminescence. Annu Rev Anal Chem 2:359–385

Free AH, Adams EC, Kercher ML, Free HM, Coo MH (1957) Simple specific test for urine glucose. Clin Chem 3(3):163–168

Fu E, Kauffman P, Lutz B, Yager P (2010) Chemical signal amplification in two-dimensional paper networks. Sens Actuators B 149(1):325–328

Fujii T (2002) PDMS-based microfluidic devices for biomedical applications. Microelectron Eng 61:907–914

Gabriel EF, Garcia PT, Cardoso TM, Lopes FM, Martins FT, Coltro WK (2016) Highly sensitive colorimetric detection of glucose and uric acid in biological fluids using chitosan-modified paper microfluidic devices. Analyst 141:4749–4756

Ge L, Yu J, Ge S, Yan M (2014) Lab-on-paper-based devices using chemiluminescence and electrogenerated chemiluminescence detection. Anal Bioanal Chem 406(23):5613–5630

Gerbers R, Foellscher W, Chen H, Anagnostopoulos C, Faghri M (2014) A new paper-based platform technology for point-of-care diagnostics. Lab Chip 14(20):4042–4049

Gomes HI, Sales MGF (2015) Development of paper-based color test-strip for drug detection in aquatic environment: application to oxytetracycline. Biosens Bioelectron 65:54–61

Jansen RT, Blaton V, Burnett D, Huisman W, Queraltó JM, Zérah S, Allman B (1998) Additional essential criteria for quality systems of medical laboratories. Clin Chem Lab Med 36(4):249–252

Jayawardane BM, Wei S, McKelvie ID, Kolev SD (2014) Microfluidic paper-based analytical device for the determination of nitrite and nitrate. Anal Chem 86(15):7274–7279

Li X, Liu X (2014) Fabrication of three-dimensional microfluidic channels in a single layer of cellulose paper. Microfluid Nanofluid 16(5):819–827

Li X, Tian J, Garnier G, Shen W (2010) Fabrication of paper-based microfluidic sensors by printing. Colloids Surf B Biointerfaces 76(2):564–570

Li X, Ballerini DR, Shen W (2012) A perspective on paper-based microfluidics: current status and future trends. Biomicrofluidics 6(1):011301

Li W, Li L, Li M, Yu J, Ge S, Yan M, Song X (2013) Development of a 3D origami multiplex electrochemical immunodevice using a nanoporous silver-paper electrode and metal ion functionalized nanoporous gold–chitosan. Chem Commun 49(83):9540–9542

Li L, Ma C, Kong Q, Li W, Zhang Y, Ge S, Yan M, Yu J (2014a) A 3D origami electrochemical immunodevice based on a Au@ Pd alloy nanoparticle-paper electrode for the detection of carcinoembryonic antigen. J Mater Chem B 2(38):6669–6674

Li L, Xu J, Zheng X, Ma C, Song X, Ge S, Yu J, Yan M (2014b) Growth of gold-manganese oxide nanostructures on a 3D origami device for glucose-oxidase label based electrochemical immunosensor. Biosens Bioelectron 61:76–82

Liana DD, Raguse B, Wieczorek L, Baxter GR, Chuah K, Gooding JJ, Chow E (2013) Sintered gold nanoparticles as an electrode material for paper-based electrochemical sensors. RSC Adv 3(23):8683–8691

Liu H, Crooks RM (2011) Three-dimensional paper microfluidic devices assembled using the principles of origami. J Am Chem Soc 133(44):17564–17566

Liu F, Ge S, Yu J, Yan M, Song X (2014) Electrochemical device based on a Pt nanosphere-paper working electrode for in situ and real-time determination of the flux of H_2O_2 releasing from SK-BR-3 cancer cells. Chem Commun 50(71):10315–10318

Lopez-Ruiz N, Curto VF, Erenas MM, Benito-Lopez F, Diamond D, Palma AJ, Capitan-Vallvey LF (2014) Smartphone-based simultaneous pH and nitrite colorimetric determination for paper microfluidic devices. Anal Chem 86(19):9554–9562

Lu Y, Shi W, Jiang L, Qin J, Lin B (2009) Rapid prototyping of paper-based microfluidics with wax for low-cost, portable bioassay. Electrophoresis 30(9):1497–1500

Lu J, Ge S, Ge L, Yan M, Yu J (2012) Electrochemical DNA sensor based on three-dimensional folding paper device for specific and sensitive point-of-care testing. Electrochim Acta 80:334–341

Ma S, Tang Y, Liu J, Wu J (2014) Visible paper chip immunoassay for rapid determination of bacteria in water distribution system. Talanta 120:135–140

Ma C, Li W, Kong Q, Yang H, Bian Z, Song X, Yu J, Yan M (2015) 3D origami electrochemical immunodevice for sensitive point-of-care testing based on dual-signal amplification strategy. Biosens Bioelectron 63:7–13

Mabey D, Peeling RW, Ustianowski A, Perkins MD (2004) Tropical infectious diseases: diagnostics for the developing world. Nat Rev Microbiol 2(3):231–240

Maejima K, Tomikawa S, Suzuki K, Citterio D (2013) Inkjet printing: an integrated and green chemical approach to microfluidic paper-based analytical devices. RSC Adv 3(24):9258–9263

Martinez AW, Phillips ST, Butte MJ, Whitesides GM (2007) Patterned paper as a platform for inexpensive, low-volume, portable bioassays. Angew Chem Int Ed 46(8):1318–1320

Martinez AW, Phillips ST, Whitesides GM (2008) Three-dimensional microfluidic devices fabricated in layered paper and tape. Proc Natl Acad Sci U S A 105(50):19606–19611

Martinez AW, Phillips ST, Whitesides GM, Carrilho E (2009) Diagnostics for the developing world: microfluidic paper-based analytical devices. Anal Chem 82(1):3–10

Müller RH, Clegg DL (1949) Automatic paper chromatography. Anal Chem 21(9):1123–1125

Nath P, Arun RK, Chanda N (2015) Smart gold nanosensor for easy sensing of lead and copper ions in solution and using paper strips. RSC Adv 5(84):69024–69031

Nguyen TH, Fraiwan A, Choi S (2014) Paper-based batteries: a review. Biosens Bioelectron 54:640–649

Nie Z, Nijhuis CA, Gong J, Chen X, Kumachev A, Martinez AW, Narovlyansky M, Whitesides GM (2010) Electrochemical sensing in paper-based microfluidic devices. Lab Chip 10(4):477–483

Nie J, Liang Y, Zhang Y, Le S, Li D, Zhang S (2013) One-step patterning of hollow microstructures in paper by laser cutting to create microfluidic analytical devices. Analyst 138(2):671–676

Noor MO, Krull UJ (2014) Camera-based ratiometric fluorescence transduction of nucleic acid hybridization with reagentless signal amplification on a paper-based platform using immobilized quantum dots as donors. Anal Chem 86(20):10331–10339

Nurak T, Praphairaksit N, Chailapakul O (2013) Fabrication of paper-based devices by lacquer spraying method for the determination of nickel (II) ion in waste water. Talanta 114:291–296

Olkkonen J, Lehtinen K, Erho T (2010) Flexographically printed fluidic structures in paper. Anal Chem 82(24):10246–10250
Ornatska M, Sharpe E, Andreescu D, Andreescu S (2011) Paper bioassay based on ceria nanoparticles as colorimetric probes. Anal Chem 83(11):4273–4280
Parolo C, Merkoçi A (2013) Paper-based nanobiosensors for diagnostics. Chem Soc Rev 42(2):450–457
Petryayeva E, Algar WR (2015) Toward point-of-care diagnostics with consumer electronic devices: the expanding role of nanoparticles. RSC Adv 5(28):22256–22282
Renault C, Li X, Fosdick SE, Crooks RM (2013) Hollow-channel paper analytical devices. Anal Chem 85(16):7976–7979
Renault C, Koehne J, Ricco AJ, Crooks RM (2014) Three-dimensional wax patterning of paper fluidic devices. Langmuir 30(23):7030–7036
Richter MM (2004) Electrochemiluminescence (ecl). Chem Rev 104(6):3003–3036
Rosa AM, Louro AF, Martins SA, Inácio J, Azevedo AM, Prazeres DMF (2014) Capture and detection of DNA hybrids on paper via the anchoring of antibodies with fusions of carbohydrate binding modules and ZZ-domains. Anal Chem 86(9):4340–4347
Saito T, Isogai A (2007) Wet strength improvement of TEMPO-oxidized cellulose sheets prepared with cationic polymers. Ind Eng Chem Res 46(3):773–780
Santhiago M, Kubota LT (2013) A new approach for paper-based analytical devices with electrochemical detection based on graphite pencil electrodes. Sens Actuators B 177:224–230
Santhiago M, Henry CS, Kubota LT (2014) Low cost, simple three dimensional electrochemical paper-based analytical device for determination of p-nitrophenol. Electrochim Acta 130:771–777
Scida K, Li B, Ellington AD, Crooks RM (2013) DNA detection using origami paper analytical devices. Anal Chem 85(20):9713–9720
Shi Z, Wu X, Gao L, Tian Y, Yu L (2014) Electrodes/paper sandwich devices for in situ sensing of hydrogen peroxide secretion from cells growing in gels-in-paper 3-dimensional matrix. Anal Methods 6(12):4446–4454
Shi Z, Tian Y, Wu X, Li C, Yu L (2015) A one-piece lateral flow impedimetric test strip for label-free clenbuterol detection. Anal Methods 7(12):4957–4964
Songjaroen T, Dungchai W, Chailapakul O, Laiwattanapaisal W (2011) Novel, simple and low-cost alternative method for fabrication of paper-based microfluidics by wax dipping. Talanta 85(5):2587–2593
Spicar-Mihalic P, Toley B, Houghtaling J, Liang T, Yager P, Fu E (2013) CO_2 laser cutting and ablative etching for the fabrication of paper-based devices. J Micromech Microeng 23(6):067003
Su S, Nutiu R, Filipe CD, Li Y, Pelton R (2007) Adsorption and covalent coupling of ATP-binding DNA aptamers onto cellulose. Langmuir 23(3):1300–1302
Su M, Ge L, Kong Q, Zheng X, Ge S, Li N, Yu J, Yan M (2015) Cyto-sensing in electrochemical lab-on-paper cyto-device for in-situ evaluation of multi-glycan expressions on cancer cells. Biosens Bioelectron 63:232–239
Sun J, Xianyu Y, Jiang X (2014) Point-of-care biochemical assays using gold nanoparticle-implemented microfluidics. Chem Soc Rev 43(17):6239–6253
Thuo MM, Martinez RV, Lan WJ, Liu X, Barber J, Atkinson MB, Bandarage D, Bloch JF, Whitesides GM (2014) Fabrication of low-cost paper-based microfluidic devices by embossing or cut-and-stack methods. Chem Mater 26(14):4230–4237
Tobjörk D, Österbacka R (2011) Paper electronics. Adv Mater 23(17):1935–1961
Tolba M, Brovko LY, Minikh O, Griffiths MW (2008) Engineering of bacteriophages displaying affinity tags on its head for biosensor applications. NSTI Nanotechnol 2:449–452
Wang CC, Hennek JW, Ainla A, Kumar AA, Lan WJ, Im J, Smith BS, Zhao M, Whitesides GM (2016) A paper-based "pop-up" electrochemical device for analysis of beta-hydroxybutyrate. Anal Chem 88(12):6326–6333
Wu AH, Apple FS, Gibler WB, Jesse RL, Warshaw MM, Valdes R (1999) National Academy of Clinical Biochemistry Standards of Laboratory Practice: recommendations for the use of cardiac markers in coronary artery diseases. Clin Chem 45(7):1104–1121
Yamada K, Takaki S, Komuro N, Suzuki K, Citterio D (2014) An antibody-free microfluidic paper-based analytical device for the determination of tear fluid lactoferrin by fluorescence sensitization of Tb^{3+}. Analyst 139(7):1637–1643

Yang J, Nam YG, Lee SK, Kim CS, Koo YM, Chang WJ, Gunasekaran S (2014) Paper-fluidic electrochemical biosensing platform with enzyme paper and enzymeless electrodes. Sens Actuators B 203:44–53

Yetisen AK, Akram MS, Lowe CR (2013) Paper-based microfluidic point-of-care diagnostic devices. Lab Chip 13(12):2210–2251

Yu J, Ge L, Huang J, Wang S, Ge S (2011a) Microfluidic paper-based chemiluminescence biosensor for simultaneous determination of glucose and uric acid. Lab Chip 11(7):1286–1291

Yu J, Wang S, Ge L, Ge S (2011b) A novel chemiluminescence paper microfluidic biosensor based on enzymatic reaction for uric acid determination. Biosens Bioelectron 26(7):3284–3289

Yu L, Shi Z, Fang C, Zhang Y, Liu Y, Li C (2015) Disposable lateral flow-through strip for smartphone-camera to quantitatively detect alkaline phosphatase activity in milk. Biosens Bioelectron 69:307–315

Zhang M, Ge L, Ge S, Yan M, Yu J, Huang J, Liu S (2013) Three-dimensional paper-based electrochemiluminescence device for simultaneous detection of Pb2+ and Hg2+ based on potential-control technique. Biosens Bioelectron 41:544–550

Zhang Y, Zuo P, Ye BC (2015) A low-cost and simple paper-based microfluidic device for simultaneous multiplex determination of different types of chemical contaminants in food. Biosens Bioelectron 68:14–19

Zhao W, Ali MM, Aguirre SD, Brook MA, Li Y (2008) Paper-based bioassays using gold nanoparticle colorimetric probes. Anal Chem 80(22):8431–8437

Zhou F, Noor MO, Krull UJ (2014) Luminescence resonance energy transfer-based nucleic acid hybridization assay on cellulose paper with upconverting phosphor as donors. Anal Chem 86(5):2719–2726

Zhu Y, Xu X, Brault ND, Keefe AJ, Han X, Deng Y, Xu J, Yu Q, Jiang S (2014a) Cellulose paper sensors modified with zwitterionic poly (carboxybetaine) for sensing and detection in complex media. Anal Chem 86(6):2871–2875

Zhu WJ, Feng DQ, Chen M, Chen ZD, Zhu R, Fang HL, Wang W (2014b) Bienzyme colorimetric detection of glucose with self-calibration based on tree-shaped paper strip. Sens Actuators B 190:414–418